COMPARATIVE PHYSIOLOGY

LOCOMOTION, RESPIRATION, TRANSPORT AND BLOOD

COMPARATIVE PHYSIOLOGY

LOCOMOTION, RESPIRATION, TRANSPORT AND BLOOD

Proceedings of the International Congress on Comparative Physiology, Acquasparta, September 1972

Editors:

Liana Bolis
University of Rome, Italy

K. Schmidt-Nielsen
Duke University, Durham, U.S.A.

S.H.P. Maddrell
Agricultural Research Council, Cambridge, G. B.

1973

NORTH-HOLLAND PUBLISHING COMPANY — AMSTERDAM • LONDON

AMERICAN ELSEVIER PUBLISHING COMPANY, INC. — NEW YORK

Library of Congress Catalog Card Number 73 83687

ISBN North-Holland 0 7204 4138 2
ISBN American Elsevier 0 444 10556 5

Publishers:

NORTH-HOLLAND PUBLISHING COMPANY - AMSTERDAM
NORTH-HOLLAND PUBLISHING COMPANY, LTD. - LONDON

Sole distributors for the U.S.A. and Canada:

AMERICAN ELSEVIER PUBLISHING COMPANY, INC.
52 VANDERBILT AVENUE
NEW YORK, N.Y. 10017

PRINTED IN THE NETHERLANDS

IN MEMORY OF
PROFESSOR TREVOR SHAW, F.R.S.

1928 – 1972

PREFACE

Comparative physiology has a reputation for being concerned with the "unusual" animal. The number of known animal species exceeds one million; which ones are unusual?

Almost every conceivable habitat on this planet has been invaded by living organisms that seem uniquely adapted to their particular niche, and yet, their physiological mechanisms are based on the same fundamental principles. In a sense, then, none is unique. We find, however, that a special way of life often requires a particular mechanism to be more highly developed or more clearly expressed in one animal than in another. Therefore, by comparing different animals we can often understand physiological processes that may remain obscure if we study only the traditional "experimental" animal. This comparison of different animals in order to gain insight is the essence of comparative physiology.

Comparative physiology thus studies a wide variety of organisms. At times the result is, in appearance as well as in reality, a fragmentation which leaves us baffled and confused. We should try to progress beyond this stage.

Meetings of outstanding workers in various fields should result in stimulation and exchange of ideas, which in turn should clarify the physiological characteristics of animals and lead to increased understanding of general principles. The present volume expresses the results of such a meeting of comparative physiologists from a variety of fields, organized under the untiring and insightful efforts of Professor Liana Bolis.

Knut Schmidt-Nielsen

LIST OF CONTRIBUTORS

Alexander, R. McN., 1
Alvarado, F., 451
Antonini, E., 477, 523
Barra, D., 477
Bolis, L., 583
Bonaventura, J., 477, 493
Brunori, M., 477, 523
Capraro, V., 399
Cavagna, G. A., 43
Chiancone, E., 523
Citterio, G., 57
Colombo, G., 387
Dejours, P., 117
Di Prampero, P. E., 99
Esposito, G., 399
Fänge, R., 135
Ghiretti, F., 509
Ghiretti-Magaldi, A., 509
Giardina, B., 477, 523
Giebisch, G., 241
Honegger, P., 441
Huggel, H. J., 591
Hughes, G. M., 187
Kitto, G. B., 493
Kümmel, G., 221
Larsson, A., 619
Luisier, A. L., 465
Maetz, J., 293
Mayer-Gostan, N., 293
Nachtigall, W., 77
Nilsson, S., 323
Noble-Nesbitt, J., 333
Parsons, D. S., 417
Passow, H., 543
Peristiany, J. G., 591
Petris, A., 387
Pic, P., 293
Piiper, J., 161
Pinamonti, S., 387
Rennie, D. W., 99
Riordan, J. R., 543
Robinson, W. L., 465
Salvato, B., 509
Scheid, P., 161
Schoffeniels, E., 353
Taylor, C. R., 23
Tucker, V. A., 63
Weis-Fogh, T., 113

CONTENTS

Comparative Physiology, eds. L. Bolis, K. Schmidt-Nielsen and S. H. P. Maddrell

MUSCLE PERFORMANCE IN LOCOMOTION AND OTHER STRENUOUS ACTIVITIES

R.Mcneill ALEXANDER
Department of Pure and Applied Zoology
University of Leeds, England

INTRODUCTION

A muscle cannot exert as great a force when it is shortening rapidly, as when it is contracting isometrically or shortening slowly. Experiments with a variety of muscles have shown that the relationship between the force P and the rate of shortening V is closely represented by Hill's equation

$$V = (P_o - P)b/(P + a) \qquad (1$$

P_o is the force exerted in isometric contraction (i.e. it is the value of P when V = 0), while a and b are constants characteristic of the particular muscle (1,20). The power output of the muscle is given by PV, and is greatest at a particular value of P which can be shown by differentiation to be $[(a^2 + aP_o)^{1/2} - a]$.

It will be apparent from later sections of this paper that for a great many of the muscles which have been investigated, a/P_o lies between 0.15 and 0.25. The value of P for maximum power output is 0.27 P_o when a/P_o = 0.15 and 0.31 P_o when a/P_o = 0.25.

For many muscles, high power output is the property which seems likely to be selected for most strongly. It can be expected that the moment arms of such muscles will have become adjusted in the course of evolution so that in strenuous activity P/P_o is close to the value of about 0.3 at which maximum power can be developed. Other muscles work against mainly inertial loads, accelerating the whole body or some part of it. They may develop P_o at the beginning of contraction while the mass is still at rest, but the force must fall as the mass accelerates. It can be shown that if the mass is to be accelerated over a given distance to the highest possible final velocity, by a muscle for which $a/P_o = 0.2$, the optimum moment arm is that which makes the final value of P equal 0.2 P_o (14). The mean value of P, over the whole period of acceleration, is then about 0.35 P_o.

DEFINITIONS

The terms stress and strain are used repeatedly in this paper and require definition. The tensile stress in a material is the tensile force acting across unit cross-sectional area. If the cross-sectional area is liable to change substantially (as in a muscle fibre which swells as it shortens) the area used in calculating the stress must be specified carefully. When elastomers are being considered, stress is sometimes expressed in terms of the cross-sectional area of the stressed material, and sometimes of the unstressed material. In this paper, stress is expressed in terms of initial cross-sectional area (i.e.of the cross-sectional area while the muscle fibres are extended at the beginning of contraction). Stress is defined as the force exerted lengthwise along the muscle fibres, per unit initial cross-sectional area of fibre. Similarly strain is defined as the change in length of a muscle fibre, as a fraction of its initial length. The strain which occurs when length decreases is normally regarded as negative, but will be treated as positive throughout this paper.

Note that these definitions refer to stresses and strains along the length of the muscle fibres, rather than necessarily along the length of the muscle. This will facilitate comparisons between muscles composed of similar fibres, differently arranged. Some of the muscles considered in this paper are parallel-fibred, and calculation of stress and strain from force and amount of shortening are straightforward. Others are pinnate. Consider the symmetrical pinnate muscle illustrated in Fig. 1.

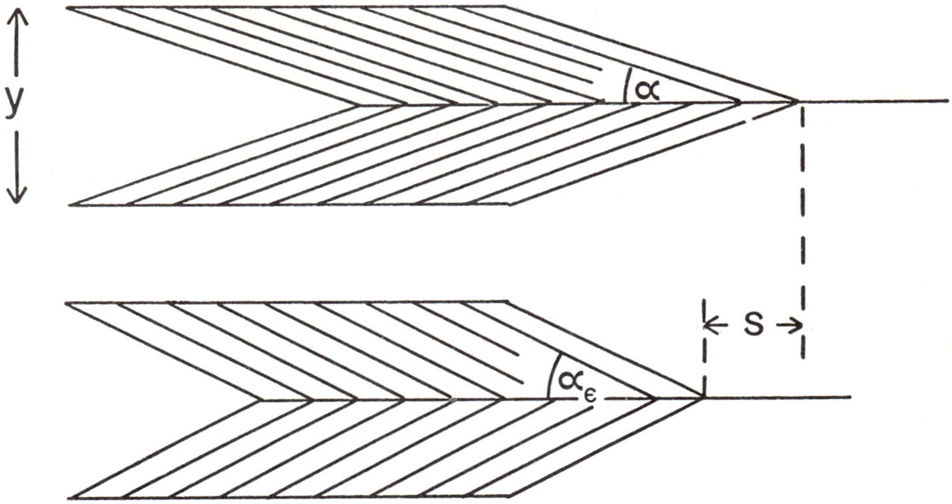

Figure 1 - These diagrams of a pinnate muscle are explained in the text.

It has breadth y, and there is an area T on each side of its central tendon on which muscle fibres insert. Its fibres shorten, undergoing strain ε and moving the tendon of insertion through distance s. This involves an increase in the angle of pinnation from the initial value α to α_ε. At strain ε a stress σ acts in the muscle fibres and the whole muscle exerts a force P. Then it can be shown that

$$P = 2T\sigma \sin\alpha \; \cos\alpha_\varepsilon$$

(5) and if α and α_ε are small and not too different so that $\cos\alpha_\varepsilon \simeq \cos\alpha$,

$$P \simeq T\,\sigma \sin 2\alpha$$

Also

$$s = \left\{ \cos\alpha - \left[(1-\varepsilon)^2 - \sin^2\alpha\right]^{1/2}\right\} y/2 \sin\alpha \quad (2$$

(ref. 3 where the quantity $n = 1-\epsilon$ is used). By rearrangement of this equation

$$(1-\epsilon)^2 = 1-(2s/y)\sin 2\alpha = (2s/y)^2 \sin^2\alpha \qquad (3$$

The product $\sigma\epsilon$ gives the work done by unit volume of muscle, in a contraction at constant stress σ through strain ϵ . The power exerted during the contraction by unit volume of muscle can similarly be obtained by multiplying the stress by the strain rate, $d\epsilon/dt$. If the density of the muscle is ρ the work per unit weight is $\sigma\epsilon/\rho$ and the power per unit weight $(\sigma/\rho)\, d\epsilon/dt$.

This power, exerted during contraction, must be distinguished from the mean power exerted in repetitive activity, over a period in which contractions are separated by intervals when the muscle is being extended and exerting no power.

Vertebrate striated muscles

The aim of this paper is to relate the performance of muscles in normal activities of intact animals, to the mechanical properties of the muscles. A difficulty frequently encountered is that the properties of the particular muscle in question have not been investigated. This difficulty is not too serious in the case of vertebrate striated muscles, which are strikingly uniform in many structural features, and in certain properties.

Such varied muscles as frog leg muscles and rabbit psoas (25), teleost myomeres and pigeon pectoralis (34) are strikingly similar in the dimensions, spacing and arrangement of the actin and myosin filaments. There is no appreciable difference in these respects between fast and slow fibres from birds (30). Maximum isometric stresses between 190 and 400 kN m^{-2} have been measured in experiments with fast and slow striated muscle from frogs (17), tortoise muscle (44) and both fast-twitch and slow-twitch muscles from rats (43). Some higher values which have been published refer to pinnate muscles but have been calculated as if for parallel-fibred muscles. Lower values have been obtained with chicken muscles, possible due to deterioration after excision (15,30). Most of the vertebrate striated muscles investigated have been found to have a/P_o between 0.15 and 0.25. Rates of contraction are much more variable: for instance, the maximum strain rates of certain fast and slow muscle fibres in frog legs have been found to be 6 and 0.1 s^{-1}, respectively, at 20°C (17).

When a muscle fibre shortens, its component sarcomeres necessarily shorten too. The relationship between sarcomere length and isometric tetanic force has been investigated very carefully in experiments with frog semitendinosus fibres (16,18). Contractile force (as distinct from passive elastic force) can only be exerted at sarcomere lengths between about 3.7 and 1.3 μm. Contraction between these limits would involve a strain of 0.65. The contractile force is 50% of the maximum value at sarcomere lengths of about 2.9 and 1.5 μm. Contraction between these limits would involve a strain of 0.5. The relationship between sarcomere length and isometric force, determined for frog muscle, has been explained in considerable detail in terms of the dimensions of the filaments. Since these dimensions seem to be so uniform among vertebrate striated muscles, it seems likely that the relationship is similarly uniform.

The similarity in properties between the vertebrate striated muscles which have been investigated physiologically, and the similarity in structure between them and other vertebrate striated muscles, enable one to assume with reasonable confidence that the component fibres of any particular vertebrate striated muscle have certain properties. They can be assumed to be able to develop a maximum isometric stress in the neighbourhood of 300 kN m^{-2}, and to have a value of a/P_o which is near 0.2. These properties imply that the stress at which maximum power can be developed will be around 90 kN m^{-2}. It can also be assumed that if any movement involves strains much greater than 0.5, the stress which can be exerted in at least one of the extreme positions will be low.

Jumping by frogs

Mrs. L.J. Calow and I have used a force platform to record the three components of the force exerted on the ground by frogs (*Rana temporaria*) taking off at the beginning of a jump (14). We have used the records, together with cinematograph film taken simultaneously and anatomical data, to estimate the stresses, strains and strain rates which occur in the principal muscles involved. We found strains from 0.1 to 0.5 and maximum strain rates (attained in the later stages of take-off) from 0.6 to 5 s^{-1}. Both strains and strain rates were lower for parallel-fibred muscles than for pinnate ones. The estimated stress averaged 110 kN m^{-2} in the plantaris longus, 100 kN m^{-2} in the extensors of the knee and 60 kN m^{-2} in the extensors of the hip, in the later stages of take-off. We confirmed in separate experiments with excised muscles that at least most of the muscles involved can develop maximum isometric stresses of around 300 kN m^{-2}. The stresses observed in the later stages of take-off are therefore close to those predicted for optimum performance in inertial loading. However, there was no indication that substantially larger stresses acted earlier in take-off, when strain rates were lower.

Running by quail

Mrs.J.Clark and I are engaged in an investigation of the mechanics of running by quail (*Coturnix coturnix*). Force platform records, cinematograph film and anatomical data are being used to determine the stresses and strains occurring in major leg muscles, in fast running. Preliminary results indicate strains of 0.1 to 0.4 in various muscles, strain rates of 0.5 to 3 s^{-1} and stresses (in extensor muscles while the foot is on the ground) of 50-100 kN m^{-2}.

Flight of pigeons

Pennycuick (32) estimated the power requirements for flight at various speeds by pigeons (*Columba livia*). He based his estimates on films of pigeons flying in a wind-tunnel but also made use of data derived from experiments on gliding in the tunnel (31). The highest speed at which his birds flew seems to have been 18 m s^{-1} (this is the highest speed for which he gives data on wing movements). He estimated that the power required for flight at this speed was 13 W. The main flight muscles of birds are the pectoralis and supracoracoideus which typically weigh 38 g and 7 g respectively, or a total of 90 g for the two sides of the body taken together, in Columba livia (33). The estimated power output at 18 m s^{-1} is thus 13/0.09 = 145 W kg $muscle^{-1}$. The frequency of the wing beat is 5.6 s^{-1} so the work per stroke is 145/5.6 = 26 J kg^{-1}. Each wing moves up and down through an angle of 100°. Movement through the maximum possible angle of 142° involves 28.5 mm movement of the pectoralis tendon (33) so movement through 100° probably involves about 20 mm movement of the tendon. I find that the mean width y of the pectoralis muscle is about 50 mm and that the mean angle of pinnation of its fibres, with the wing depressed, is about 35°. From these data and equation 3, it can be estimated that 20 mm movement of the tendon involves a strain of 0.29. As the up and down strokes are similar in duration, the strain rate must be about 3 s^{-1}. If the muscle does 26 J kg^{-1} (2.6×10^4 J m^{-3}) per stroke the mean stress must be $2.6 \times 10^4/0.29 = 9 \times 10^4$ N m^{-2} (90 kN m^{-2}).

Pennycuick (32) estimated the power required for fast vertical climbing, as well as for forward flight. He estimated that 20 W was required for the maximum rate of climb which he observed. The frequency of the wing beat averaged 9.4 s^{-1} in his films of climbing, and the wings seem to have been moving through the maximum possible angle of 142°. The power per unit weight of muscle works out at 220 W kg^{-1} and the work per stroke 24 J kg^{-1}. The strain and stress can be calculated as before; the strain is 0.37, the strain rate 7 s^{-1} and the stress 65 kN m^{-2}.

Swimming of fish

Most fishes have a superficial band of red muscle along each flank, but only white fibres in the deeper parts of the myomeres. The red fibres run longitudinally but the white fibres run at various angles, in complicated patterns (4). One pattern is found all along the swimming muscle of selachians and various of the more primitive bony fishes, but only in the caudal peduncle in most teleosts. Another (helical) pattern is found in most teleosts, anterior to the caudal peduncle. When a fish swims its median plane undulates. At any point along the body the radius of curvature of the median plane fluctuates. Its minimum value (i.e.its value when the body is most bent) seems generally to be around five times the width of the body, in films of selachians and teleosts swimming. (This generalization does not apply to the slender caudal peduncles of teleosts.) It can be calculated that bending the initially straight body to this radius involves a strain of about 0.10 in the superficial red fibres and 0.08 in the white fibres if they are arranged in the selachian pattern but 0.02-0.03 if they are arranged in the helical pattern. The strain involved in changing from a maximum bend to the left, to a maximum bend to the right, is 0.20 for the red fibres, about 0.15 for white fibres arranged in the selachian pattern and about 0.05 for white fibres arranged in the helical pattern. These are the strains which apparently occur in normal swimming.

The red muscle is aerobic and is apparently responsible for swimming at speeds up to the maximum at which oxygen balance can be maintained. For many small teleosts this maximum is 3-4 body lengths s^{-1} (see for instance ref. 11) and requires about 3 cycles of tail movements per second (9).The strain rate in the red muscle at this speed is 1.2 s^{-1}. The white muscle is anaerobic and makes short bursts of higher speed possible. The maximum burst speed is typically 12 lengths s^{-1} in teleosts and requires 8 tail cycles s^{-1} (9). At this speed the strain rate is about 0.8 s^{-1} in helically arranged white fibres and 3.2 s^{-1} in longitudinal red fibres. The red fibres may well not develop much stress at such high speeds.

The oxygen consumption of a 30 g _Onchorhynchus_ swimming at 4 lengths s^{-1} exceeded the resting consumption by about 700 mg kg^{-1} hr^{-1} (11). By assuming that the swimming muscles use all of this and are 20% efficient, it can be estimated that the power required to propel the fish is about 0.018 W. The power required at 12 lengths s^{-1} can only be estimated less directly. In the range of Reynolds number with which we are concerned (around 10^5),the drag coefficients of rigid streamlined bodies are roughly proportional to (Reynolds number)$^{-0.25}$ (see ref.3). Hence the drag on a particular body is proportional to (velocity)$^{1.75}$ and the power required to propel it to (velocity)$^{2.75}$. The power required to propel the 30 g fish at 12 lengths s^{-1} can be estimated at $3^{2.75}$ times as much as at 4 lengths s^{-1}, or 0.37 W. A typical 30 g fish would contain about 18 g swimming muscle so this is about 20 Wkg^{-1}. Since there are about 8 tail cycles s^{-1} the work per cycle is about 2.5 J kg^{-1} or 2,500 J m^{-2}. Helically-arranged muscle fibres doing the work while undergoing a strain of 0.05 would have to develop a stress of 2,500/0.05 $=5 \times 10^4$ N m^{-2} = 50 kN m^{-2}.

Feeding by teleosts

Most teleosts generally feed by sucking food into their mouths. The pressure changes in the mouth cavity which occur as the food is sucked in have been recorded from a number of species (5,6). It was calculated that, to generate the pressures recorded in the strongest sucks, the sternohyoideus and rectus abdominis muscles of the orfe *Idus* must exert stresses of 250 kN m^{-2}, those of *Tilapia* 300 kN m^{-2} and those of the angelfish *Pterophyllum* 370 kN m^{-2}. The maximum isometric stresses which the muscles could develop seem unlikely to be much above these values.

As well as these muscles, the dorsal trunk muscles play a major part in feeding. They tilt the snout up. The hyoid bars are so arranged that this movement will cause expansion of the buccal and opercular cavities if the urohyal bone (on which the sternohyoideus and rectus abdominis insert) is prevented from moving anteriorly. The sternohyoideus and rectus do not have to shorten and I believe (though I have not previously expressed this view) that they work essentially isometrically in feeding. Superimposed outlines of successive frames from films, such as ref.5 Fig.4b, seem to confirm this belief. A bulge in the ventral profile indicates that the tongue is no nearer the pectoral girdle when the buccal cavity is expanded than it was before the feeding movement started. If the muscles are indeed working isometrically, it is not surprising that they can exert such high stresses.

The dorsal trunk muscles which tilt the head certainly shorten. No attempt has been made to estimate the strain but the stress, at least in *Idus*, seems to be rather less than in the sternohyoideus and rectus; it can be estimated from data given in ref.5 as 190 kN m^{-2}.

INVERTEBRATE MUSCLES

Invertebrate muscles are far less uniform than the vertebrate muscles considered so far. Some are cross-striated and some obliquely striated. They differ considerably among themselves and from vertebrate striated muscle in the lengths of the filaments and in the numbers of each type of filament in unit cross-sectional area. Even the thickness and composition of the thick (myosin or paramyosin) filaments are variable. It is therefore necessary to consider what is known of the properties of each muscle separately.

Jet propulsion by cephalopods

Cephalopods swim by jet propulsion, ejecting water from the mantle cavity through the funnel. Trueman and Packard (41) recorded pressure changes in the mantle cavities of various cephalopods swimming (tethered) in an aquarium. They recorded pressure pulses of up to 30 kN m^{-2} in the squid *Loligo* and estimated that these implied stresses of up to 150 kN m^{-2} in the circular muscles of the mantle wall. They also estimated that the volume of water expelled at each swimming stroke from the mantle cavity of a 350 g *Loligo* is 200 cm^3. The pressure difference between the mantle cavity and the outside water is not constant while the water is being expelled but the shape of the pressure record (ref.41 Fig.3b) indicates that most of the water can be expected to leave the cavity while the pressure difference is near its maximum. Expulsion of 2×10^{-4} m^3 water across a pressure difference of 3×10^4 N m^{-2} involves work amounting to 6 J. The muscles responsible weigh about 0.12 kg (35% of the body weight) so the work done by a single contraction is 50 J kg^{-1} (5×10^4 J m^{-3}). If the stress was 1.5×10^5 N m^{-2} the strain must have been $5 \times 10^4/1.5 \times 10^5 = 0.33$. This is in good agreement with Packard's (29) observation that the diameter of the body diminishes by 30%. The pressure pulse lasts about 0.18 s so the strain rate is about $0.33/0.18 = 1.8$ s^{-1}.

Trueman & Packard (41) give data on *Sepia* as well as on *Loligo*. These data have been used as a basis for calculations like those just presented, and the results are displayed in Table 1.

The circular muscles of the mantles of cephalopods are obliquely striated (23). Their mechanical properties do not seem to have been investigated but probably resemble those of the funnel retractor muscles studied by Lowy and Millman (26). These parallel-fibred muscles developed maximum isometric stresses up to 510 kN m^{-2} in experiments with *Octopus* and 220 kN m^{-2} with *Sepia*. (The experiments gave very variable results, presumably due to damage or deterioration of some specimens. It seems likely that the highest stresses recorded are nearer to the stresses which can be developed in the intact animal, than are the mean stresses recorded.) a/P_o was found to be 0.13 and the relationship between strain and isometric stress was found to be very much as for frog striated muscle.

If *Sepia* mantle muscle has properties similar to those of *Sepia* funnel retractor, the stress it exerts in jet propulsion may be close to the optimum for power output. It is possible that if *Loligo* muscle is capable of developing isometric stresses similar to those recorded from *Octopus* funnel retractor, the high stress it develops in jet propulsion may also be close to the optimum. However, this is only a possibility: the maximum isometric stress for *Loligo* muscle is not known.

Burrowing by worms

Trueman (40) and Seymour (39) recorded pressures from the coelom of the lugworm *Arenicola*, burrowing in sand. Peaks of pressure up to 15 kN m^{-2} above ambient sometimes occurred, but peaks of 7 kN m^{-2} were more usual. These peaks seem to have occurred when part of the worm was being dilated, by contraction of longitudinal muscles, to enlarge the burrow and compact its walls. Since the trunk of *Arenicola* has no septa the pressure is the same all along it, and a limited part of it can be dilated only if the diameter of other parts is restricted by tension in the circular muscles. Trueman estimated that these muscles would have to exert a stress of 300 kN m^{-2} to withstand a pressure of 10 kN m^{-2} so they would have to exert 450 kN m^{-2} to withstand the maximum pressure of 15 kN m^{-2} recorded by Seymour. Their contraction was not far from isometric; indeed they generally lengthened slightly during the pressure peak. A pressure of 18 kN m^{-2} which may perhaps represent the maximum attainable by isometric contraction of the circular muscles was recorded when *Arenicola* was put in a potassium chloride solution.

Pressures in the earthworm *Lumbricus* were recorded by Seymour (38). This worm has septa all along the body so nearby segments may differ in pressure. Since cannulae can only be placed in segments which are above ground the pressures generated when underground segments are dilated are unknown. The highest pressure recorded in a worm above ground (squirming violently) was 7.5 kN m^{-2}. Seymour calculated that this implied stresses of 130 kN m^{-2} and 27 kN m^{-2} in the circular and longitudinal muscles, respectively. If as seems likely the longitudinal muscles can exert as much stress as the circular ones, much larger pressures should be possible in segments which are being dilated in the burrow. Pressures up to 2.5 kN m^{-2} were recorded in worms crawling in the surface.They imply stresses up to 44 kN m^{-2} in the circular muscles. A film of an earthworm crawling (19) shows segments shortening to 0.4 of their greatest lengths. The longitudinal muscles must undergo strains of about 0.6 and (if the volume of the segment remains constant) the circular ones must undergo strains of about 0.4. Complete cycles of contraction and extension occupy 5 s or more.

Earthworm body wall muscle is obliquely striated with quite densely packed actin filaments and extremely long thick ones (28). Its structure suggests the ability to exert much higher stresses than the highest pressures recorded from earthworms indicate (see ref. 27). It should also allow very large strains without substantial diminution of isometric force.

Jumping by locusts

Brown (12) used a force platform to record the forces which locusts (*Schistocerca*) exert on the ground when they take off for a jump. He recorded peak forces, in strong jumps, of about 17 times the weight of the body. This finding corresponds closely with an estimated that a constant force of 15 times the weight of the body, exerted throughout leg extension, would be required to accelerate the insect to the observed take-off velocity (3).

Brown found that the knee extends through about 2 radians during take-off, and that the moment arm of the knee extensor muscle about the joint varies with joint position between 0.3 and 0.6 mm. The tendon of the muscle must therefore move about 1 mm as the knee extends. The muscle is pinnate with initial angle of pinnation 15° and breadth (y) in a 3 g locust about 2 mm. From these data and equation 3 it can be calculated that the strain is about 0.25. Extension takes about 20 ms (13) so the strain rate is about 13 s^{-1}.

It can be calculated from the observed forces and the geometry of the leg that at the instant when the knee is half extended the knee extensor muscle must exert a force of about 5 N (3). The area T of its central tendon is about 35 mm^2. Application of equation 2 indicates a stress of about 250 kN m^{-2}.

The mechanical properties of locust knee extensor do not seem to have been investigated,but they are probably similar to those of locust flight muscle. Both are cross-striated muscles with sarcomeres about 4 µm long (24,42).The flight muscle has properties very like frog fast muscle. Weis-Fogh (42) found that it could exert a maximum isometric stress of 160 kN m^{-2} at 11°C and estimated that it could probably exert 300 kN m^{-2} at 25°C. He found that the relationship between strain and isometric stress was very like the relationship for frog semitendinosus fibres.

Brown (13) presented evidence that the locust jump makes use of the principle of the catapult. The muscle develops high tension before the leg starts extending, and shortening of the muscle depends largely on elastic recoil of the stressed materials. The velocity attained depends on the initial stress in these materials and on their visco-elastic properties, more than on the contractile properties of the muscle. A theoretical discussion of muscles subject to inertial loading which has already been referred to, led to the conclusion that in the optimum conditions for maximum final velocity, the mean force should be about 0.35 P_o and the final force about 0.2 P_o. This theory is inapplicable to the case of the locust jump because it takes no account of catapult effects. If the mechanism were a simple catapult initially extended by a force P_o, one might expect the force at the halfway point of elastic recoil to be about 0.5 P_o. The estimated value of 250 kN m^{-2} is therefore rather higher than one would expect.

A blood-sucking bug

The bug *Rhodnius* feeds on mammal blood by sucking it through a stylet which it inserts through the skin of its host. The stylet is so fine that penetration is likely to pass unnoticed by the host. Bennet-Clark (10) calculated from its dimensions and from the maximum observed rate of feeding that the feeding pump must set up a pressure difference along the stylet of at least 200 kN m^{-2} (2 atm). This estimate is based on the supposition that the blood has the same viscosity as water, and it would probably be more realistic to assume an apparent viscosity twice that of water in a tube of the diameter of the stylet (37). The pressure difference would then be 400 kN m^{-2}. Bennet-Clark estimated from the dimensions of the pump that the strain undergone by its muscle in each filling stroke was 0.27 and that to produce a pressure difference of 400 kN m^{-2} the muscle must exert a stress 160 kN m^{-2}. The frequency of the pumping action is about 3 s^{-1}. If as Bennet-Clark suggests the filling stroke is much slower than the recovery stroke (an elastic recoil) it must take about 0.3 s and the strain rate must be about 0.9 s^{-1}.

I have no information about the structure or properties of this muscle.

The song of a cicada

Pringle (35) studied the tymbals of cicadas of the genus *Platypleura*. He measured the force required to operate them and the distance moved by the insertion of the tymbal muscles. His results indicate that the tymbal muscle must exert a stress of 60 kN m^{-2} and undergo a strain of 0.015 each time it shortens. The pulse frequencies of the songs of the two species he studied were 390 and 230 s^{-1}, so around 1/600 s would be available for each shortening, and the strain rate would be about 10 s^{-1}.

The tymbal muscle is fibrillar, with sarcomeres about 2.6 µm long. A report of experiments on cicada tymbal muscle gives no information on the stress it can exert (2), but this is probably similar to the stress exerted by *Lethocerus* flight muscles which are also fibrillar, with about the same sarcomere length (8). Pringle and Tregear (36) carried out experiments in which glycerol-extracted *Lethocerus* muscles were made to oscillate in the manner characteristic of fibrillar muscle. They found mid-length stress differences between contraction and relaxation of up to 200 kN m^{-2}, depending on the conditions of oscillation. The maximum isometric stress is probably higher.

CONCLUSIONS

The data presented or derived in previous sections are summarized in Table 1. It concerns what seems to be the most strenuous activities of the various muscles. Many of the stresses are between 60 and 100 kN m^{-2}, and most of the strains are between 0.2 and 0.4. The vertebrate muscles and (less certainly) the insect ones can be expected to develop maximum isometric stresses in the range 200-400 kN m^{-2}, so the estimated stresses

Table 1
A summary of the data presented in the text

Animal	Activity	Muscle	Stress kN m^{-2}	Strain	Strain rate, s^{-1}
Rana	jumping	leg extensors	60-110	0.1-0.5	0.6-5
Coturnix	running	leg muscles	50-100	0.1-0.4	0.5-3
Columba	fast flight	pectoralis	90	0.29	3
"	climb	pectoralis	65	0.37	7
teleost	swimming	myomere (white)	50	0.05	0.8
teleosts	feeding	sternohyoideus etc.	250-370	~0	~0
Loligo	jetting	mantle	150	0.33	1.8
Sepia	jetting	mantle	80	0.19	1.3
Arenicola	burrowing	circular	450	~0	~0
Lumbricus	crawling	circular	44	0.4	0.1
Schistocerca	jumping	knee extensor	250	0.25	13
Rhodnius	feeding	pharyngeal pump	160	0.27	0.9
Platypleura	singing	tymbal	60	0.015	10

are not far from the ideal stress for maximum power output. Vertebrate striated muscle, cephalopod funnel retractor and locust flight muscle have similar relationships between stress and isometric force so it is not surprising that most of the vertebrate, cephalopod and non-fibrillar insect muscles undergo similar strains. The estimated strains are small enough for substantial stresses to be maintained throughout.

The teleost sternohyoideus, etc. and *Arenicola* circular muscles work more or less isometrically, and the stresses recorded are correspondingly high. The high stress in locust knee extensor has been discussed in relation to the catapult mechanism of jumping. *Platypleura* tymbal muscles undergo very small strains, but this is probably a characteristic of fibrillar muscle.

The strain rates which are recorded are very varied. Attempts to account for the differences should take account of differences in temperature and also in body size (on which see, for instance, refs.7 and 21). The high apparent strain rate in the locust depends largely on elastic recoil rather than contractile properties of the muscle.

REFERENCES

1. Abbot,B.C. & Wilkie,D.H. (1953) The relation between velocity of shortening and the tension-length curve of skeletal muscle. J.Physiol.*120*, 214-223.

2. Aidley,D.J. & White,D.C.S. (1969) Mechanical properties of glycerinated fibres from the tymbal muscle of a Brazilian cicada. J.Physiol. *205*, 179-192.

3. Alexander,R.McN. (1968) *Animal Mechanics.* London: Sidgwick & Jackson.

4. Alexander,R.McN. (1969) The orientation of muscle fibres in the myomeres of fishes. J.mar. biol.Assoc.U.K. *49*, 263-290.

5. Alexander,R.McN. (1969) Mechanics of the feeding action of a cyprinid fish. J.Zool.Lond.*159*,1-15.

6. Alexander,R.McN. (1970) Mechanics of the feeding action of various teleost fishes. J. Zool.Lond. 162, 145-156.

7. Alexander,R.McN. (1971) Size and Shape.London, Edward Arnold.

8. Ashhurst,D.E. (1967) The fibrillar flight muscles of giant water-bugs: an electron-microscope study. J.Cell Sci. 2, 435-444.

9. Bainbridge,R. (1958) The speed of swimming of fish as related to size and to the frequency and amplitude of the tail beat. J.exp.Biol. 35, 109-133.

10. Bennet-Clark,H.C. (1963) Negative pressures produced in the pharyngeal pump of the blood-sucking bug, Rhodnius prolixus. J.exp.Biol.40, 223-229.

11. Brett,J.R. (1964) The respiratory metabolism and swimming performance of young sockeye salmon. J.Fish.Res.Bd.Can. 21, 1183-1226.

12. Brown,R.H.J. (1963) Jumping arthropods.Times Sci.Rev.,(Summer),6-7.

13. Brown,R.H.J. (1967) Mechanism of locust jumping. Nature (Lond.) 214, 939.

14. Calow,L.J. & Alexander,R.McN. A mechanical analysis of the hind leg of a frog. In preparation.

15. Canfield,S.P. (1971) The mechanical properties and heat production of chicken latissimus dorsi muscles during tetanic contractions. J.Physiol. 219, 281-302.

16. Edman,K.A.P. (1966) The relation between sarcomere length and active tension in isolated semitendinosus fibres of the frog. J.Physiol.183, 407-417.

17. Floyd,K. & Smith,I.C.H. (1971) The mechanical and thermal properties of frog slow muscle fibres. J.Physiol. 213, 617-631.

18. Gordon,A.M., Huxley,A.F. & Julian,F.J. (1966) The variation in isometric tension with sarcomere length in vertebrate muscle fibres. J.Physiol. 184, 170-192.

19. Gray,J. & Lissmann,H.W. (1938) Studies in animal locomotion. VII. Locomotory reflexes in the earthworm. J.exp.Biol. 15, 506-516.

20. Hill,A.V. (1938) The heat of shortening and dynamic constants of muscle. Proc.R.Soc.B. 126, 136-195.

21. Hill,A.V. (1950) The dimensions of animals and their muscular dynamics. Science Progr. 38,209-230.

22. Hoyle,G. (1955) Neuromuscular mechanisms of a locust skeletal muscle. Proc.R.Soc.B. 143, 343-367.

23. Hoyle,G. (1964) Muscle and neuromuscular physiology. In Physiology of Mollusca, vol.I, pp. 313-352, ed. Wilbur,K.M. & Yonge,C.M. New York, Academic Press.

24. Huddart,H. & Oates,K. (1970) Ultrastructure of stick insect and locust skeletal muscle in relation to excitation-contraction coupling. J. Insect Physiol. 16, 1467-1483.

25. Huxley,H.E. (1960) Muscle cells. In The Cell, vol.IV, pp.365-481, ed. Brachet,J. & Mirsky, A.E., New York, Academic Press.

26. Lowy,J. & Millman,B.M. (1962) Mechanical properties of smooth muscles of cephalopod molluscs. J.Physiol. 160, 353-363.

27. Lowy,J., Millman,B.M. & Hanson,J. (1964) Structure and function in smooth tonic muscles of lamellibranch muscles. Proc.R.Soc.B. 160, 525-536.

28. Mill,P.J. & Knapp,M.F. (1970) The fine structure of obliquely striated body wall muscles in the earthworm, Lumbricus terrestris Linn. J. Cell Sci. 7, 233-261.

29. Packard,A. (1969) Jet propulsion and the giant fibre response of Loligo. Nature Lond. 221, 875-877.

30. Page,S.G. (1969) Structure and some contractile properties of fast and slow muscles of the chicken. J.Physiol. 205, 131-145.

31. Pennycuick,C.J. (1968) A wind-tunnel study of gliding flight in the pigeon. J.exp.Biol. 49, 509-526.

32. Pennycuick,C.J. (1968) Power requirements for horizontal flight in the pigeon. J.exp.Biol.49, 527-555.

33. Pennycuick,C.J. & Parker,G.A. (1966) Structural limitations on the power output of the pigeon's flight muscles. J.exp.Biol. 45, 489-498.

34. Pepe,F.A. (1971) Structure of the myosin filament of striated muscle. Progr.Biophys. 22,77-96.

35. Pringle,J.W.S. (1954) The mechanism of the myogenic rhythm of certain insect striated muscles. J.Physiol. 124, 269-291.

36. Pringle,J.W.S. & Tregear,R.T. (1969) Mechanical properties of insect fibrillar muscle at large amplitudes of oscillation. Proc.R.Soc.B. 174, 33-50.

37. Prothero,J.W. & Burton,A.C. (1962) The physics of blood flow in capillaries. II. The capillary resistance to flow. Biophys.J. 2, 199-212.

38. Seymour,M.K. (1969) Locomotion and coelomic pressure in *Lumbricus terrestris* L. J.exp.Biol. 51, 47-58.

39. Seymour,M.K. (1971) Burrowing behaviour in the European lugworm *Arenicola marina* (Polychaeta: Arenicolidae). J.Zool.Lond. 164, 93-132.

40. Trueman,E.R. (1966) Observations on the burrowing of *Arenicola marina* (L.). J.exp.Biol. 44, 93-118.

41. Trueman,E.R. & Packard,A. (1968) Motor performances of some cephalopods. J.exp.Biol. 49, 495-507.

42. Weis-Fogh,T. (1956) Tetanic force and shortening in locust flight muscle. J.exp.Biol. 33, 668-684.

43. Wells,J.B. (1965) Comparison of mechanical properties between slow and fast mammalian muscles.

J.Physiol. 178, 252-269.

44. Woledge,R.C. (1968) The energetics of tortoise muscle. J.Physiol. 197, 685-707.

Comparative Physiology, eds. L. Bolis, K. Schmidt-Nielsen and S. H. P. Maddrell

ENERGY COST OF ANIMAL LOCOMOTION [(a)]

C.R. TAYLOR
Museum of Comparative Zoology, &
Biological Laboratories,
Harvard University,
Cambridge, Massachusetts, U.S.A.

INTRODUCTION

Recently Taylor, Schmidt-Nielsen, and Raab (18) have developed a useful empirical relationship between the energetic cost of four-legged running in mammals and body weight. In this paper I would like to consider the applicability of this relationship to non mammalian vertebrates, to animals with specialized limb configurations, and to bipedal as well as quadrupedal locomotion. I should like to conclude with a discussion of some important biological consequences of this relationship.

Predicting Energetic Cost of Running from Body Weight

Steady state oxygen consumption increases linearly with increasing running speed in a wide variety of running animals (18) (Fig. 1A).

As a result, the slopes of these relationships are constants whose units are ml O_2 $(g{\cdot}km)^{-1}$. Oxygen consumption increases much more rapidly with increasing speed in smaller animals than in larger ones and the slopes are therefore steeper (Fig.1A). Net cost of running, defined as oxygen consumption during running [ml O_2 $(g{\cdot}h)^{-1}$] minus oxygen consumption during rest [ml O_2 $(g{\cdot}h)^{-1}$] divided by running velocity $(km{\cdot}h^{-1})$ has frequently been used

---

(a) This paper was written with support from NSF Grant GB 27539.

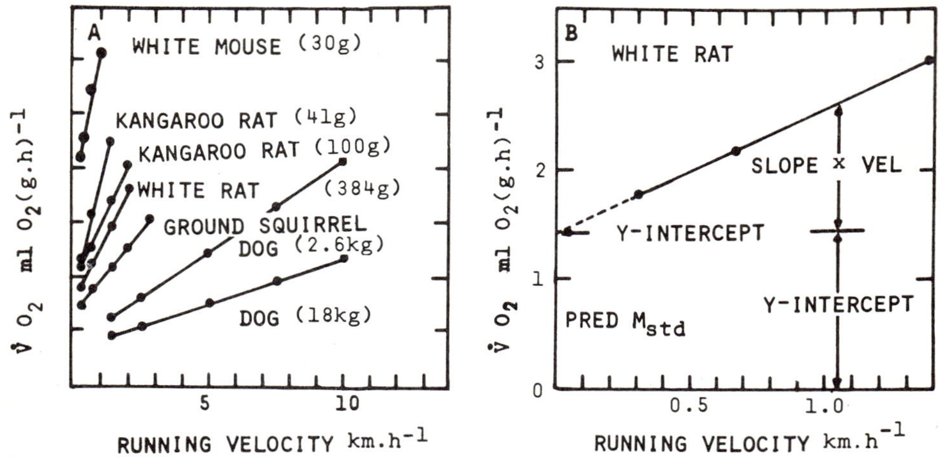

<u>Figure 1</u> -(A) Steady state oxygen consumption of quadrupedal mammals as a function of running velocity (redrawn from Taylor <u>et al.</u> (18).

(B) Steady state oxygen consumption of the white rat as a function of running velocity (redrawn from Taylor <u>et al.</u> (18). Oxygen consumption at any velocity can be calculated by adding (slope x velocity) to the y-intercept (oxygen consumption at the extrapolated zero running velocity). The y-intercept is greater than the observed resting metabolism or the standard metabolism predicted by Kleiber's equation (10).

as a measure of energy cost of running. Like the slope, its units are also ml O_2 $(g \cdot km)^{-1}$. The net cost of running and the slope of the relationship between oxygen consumption and running velocity would be identical if the y-intercept (extrapolated

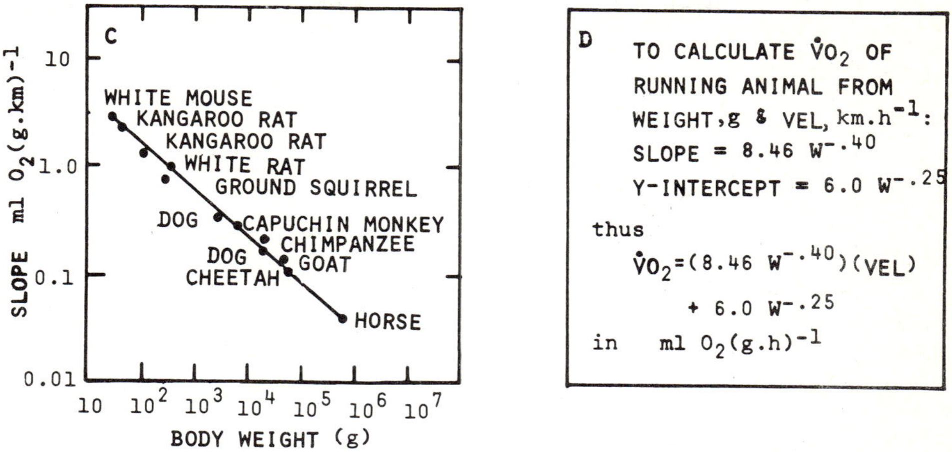

Figure 1 - (C) Slope of the relationship between steady state oxygen consumption and running velocity plotted as a function of body weight on logarithmic co-ordinates. (Data from 3,17,18. The values for the cheetah and goat are unpublished observations from my laboratory).

(D) Calculation of oxygen consumption of running mammals from body weight and velocity.

zero running velocity) were equal to resting metabolism. Unfortunately, this is not the case, as Figure 1B illustrates. The y-intercept is greater than the observed resting metabolism. It has been suggested that this reflects the energetic cost of maintaining a running posture (14). The deviation between the value for oxygen consumption at extrapolated zero running velocity and that predicted for Standard Metabolism (M_{std}) by Kleiber's equation (10) is remarkably constant (y-intercept = = 1.7 M_{std}). This constancy proves useful in predicting the oxygen consumption of running mammals, for it enables us to calculate the y-intercept from an animal's weight.

The slopes of the relationship between oxygen consumption and running velocity yield a straight line when plotted against body weight on logarithmic co-ordinates (Fig. 1C). This enables us to formulate a simple exponential equation for predicting the slope from body weight [$M_r = 8.46\ W^{-0.40}$; M_r, ml O_2 $(g \cdot km)^{-1}$; W, g] . The oxygen consumption of a running animal equals slope times velocity plus the y-intercept. Since both the slope and the y-intercept can be predicted from body weight, it is now possible to calculate oxygen consumption of running animals from body weight and running velocity (Fig.1D).

The physical parameters underlying this regular relationship remain elusive. However, as Vance Tucker discusses in another article in this volume, a similar relationship has been found between body weight and minimum cost of flying in birds. The relationship for birds can be explained in physical terms using conventional aerodynamic theory for propeller-driven craft. I am hopeful that before too long the running relationship will also be explained in physical terms.

Lizards:Is it energetically costly for them to run?

There are at least two reasons why one might expect the energetic cost of running in lizards to be different from that in mammals: 1) lizards have a much lower resting metabolism than mammals of comparable weight even at the same body temperature (6); and 2) many lizards have a sprawling wide-track gait, which would seem to require a constant expenditure of energy simply to maintain the lizard's body off the ground (mammals with an erect gait should not have to expend energy for this purpose) (Fig.2).

Robert T. Bakker, a graduate student working in my laboratory, took it upon himself to get lizards to run on a treadmill (1). He was able to

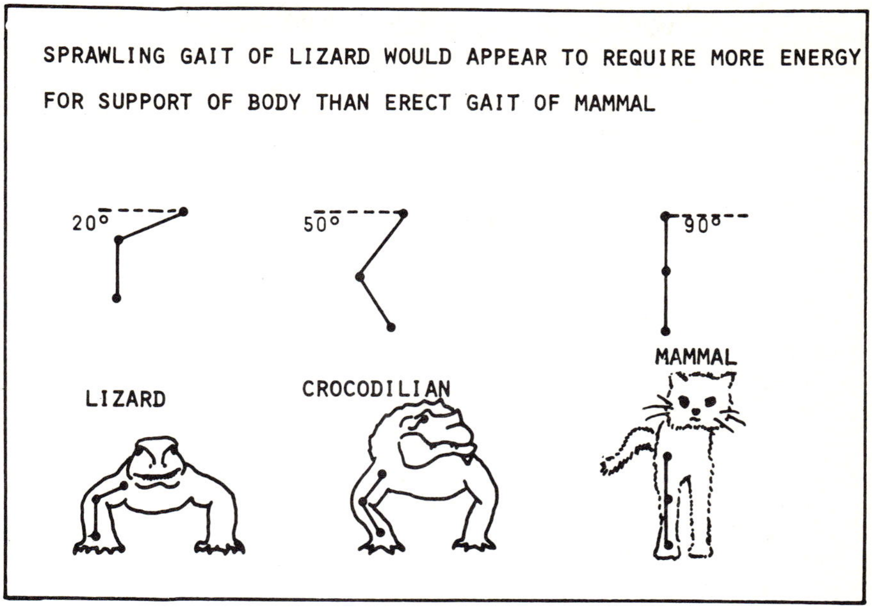

Figure 2 - The sprawling wide-track gait of a lizard would seem to require the constant expenditure of energy to keep the body off the ground. The angle between the upper limb and the ground is small, and the weight of the body acts to collapse the joint between the upper and lower limbs. In many mammals the angle approaches 90°, and the limbs act as pillars. The crocodilians are intermediate between the two extremes.

find five species (out of a very large number) which ran well on the treadmill. These animals ranged in weight from 14 to 1,200 grams. He found that steady state oxygen consumption increased linearly with speed in all his lizards (Fig.3), as was the case with mammals. Also the rate of increase was greater in smaller lizards than in larger ones. The slope of the relationship between oxygen consumption and running velocity was similar to

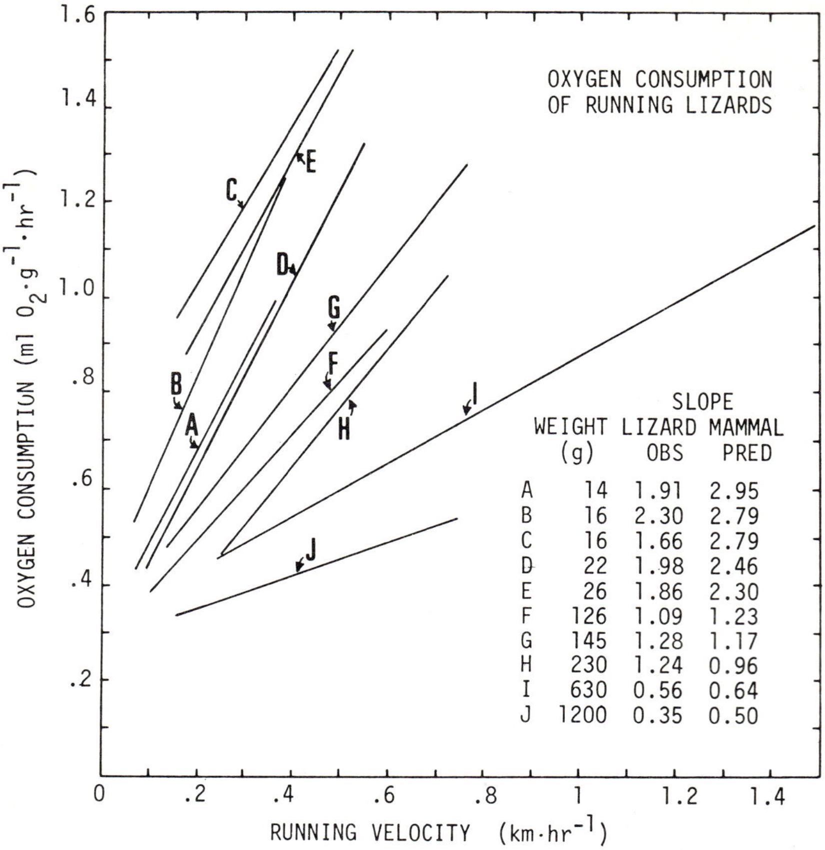

Figure 3 - Steady state oxygen consumption of ten lizards plotted as a function of running speed (from 1). Observed slopes of the relationships are given at the bottom of the graph and their units are ml O_2 $(g{\cdot}km)^{-1}$. The slopes for mammals of similar weight, predicted from the relationship of Taylor *et al.* (18) are also given. The generic names of the lizards are: (A) *Gerrhonotus*; (B) *Gerrhonotus*; (C) *Tupinambis*; (D) *Varanus*, sp.1; (E) *Varanus*, sp. 1; (F) *Ctenosaura*; (G) *Varanus*, sp.2; (H) *Varanus*, sp.1; (I) *Varanus*, sp.2; (J) *Tupinambis*.

that found in mammals of the same weight (Fig.4). However, the y-intercept was much smaller, about 1/5 that observed in mammals (Fig.5). Thus a li-

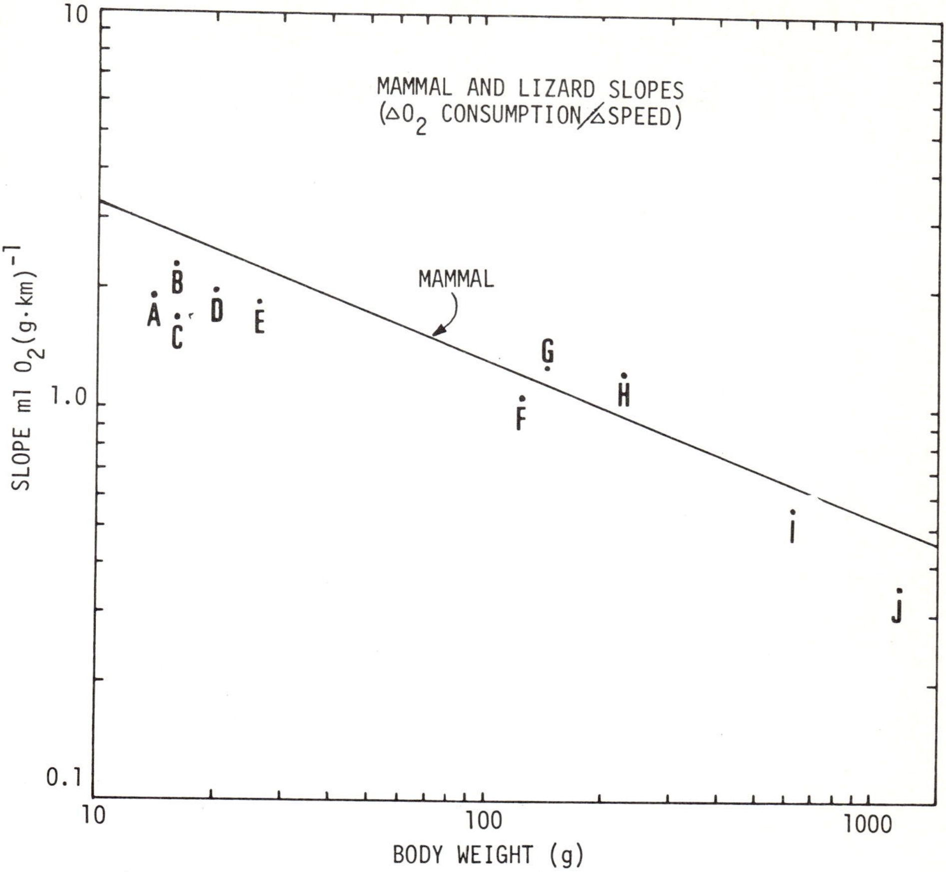

Figure 4 - The slopes of the relationships between steady state oxygen consumption of running lizards and running speed are plotted as a function of body weight on logarithmic co-ordinates (data from 1). These points fall around the line found for mammals, which is drawn on the graph.

zard expends much less energy than a mammal of the same size while running at the same velocity. Both the predicted resting metabolism and the difference between y-intercept and predicted resting metabolism in the lizard are about 1/5 that of a mammal of similar size. Despite the sprawling, wide-track locomotion, lizards do not expend large amounts of energy in keeping their body off the ground while running.

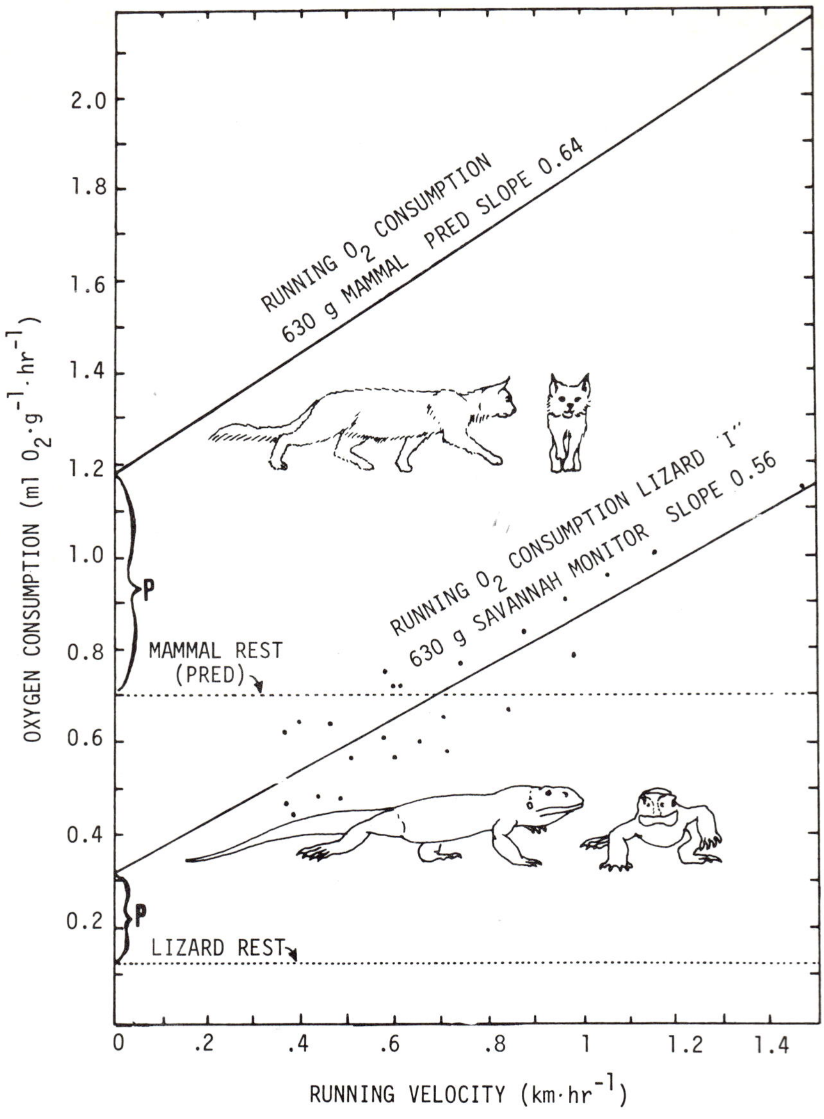

Figure 5 - A lizard expends less energy to run than a mammal of the same weight. Steady state oxygen consumption of a monitor lizard (1) is plotted as a function of running velocity and is compared with a predicted energy expenditure of a mammal of the same weight (18). The difference is due to the lower y-intercept and not to the slope.

Limb configuration, energy cost of running

The regular relationship between body weight and energy cost of running makes one curious about animals that weigh the same but have different limb configurations. One would expect, for example,that a gazelle (Gazella sp.) or a goat (Capra hircus) would require much less energy to run at a given speed than a lion (Felis leo) or a cheetah (Acinonyx jubatus). Functional anatomists have argued convincingly that two ways to decrease the kinetic component of energy expended in locomotion are: 1) to move the muscle mass close to the pivot point; and 2) to decrease the mass of the distal segments of the limb (7). This is seen clearly in gazelles and goats. In lions and cheetahs, on the other hand, the distal segments of the limbs are massive, for they must be strong enough to knock down and to withstand the tremendous forces involved in rapid acceleration. Also, the muscle mass is much further from the pivot point than in the gazelles and goats.

The average moment of inertia of the four limbs of a cheetah was found to be approximately 14 times that of goat limbs (in dissections of two animals of about the same weight and with limbs of approximately the same length). One would expect, therefore, that the cheetah would increase oxygen consumption much more rapidly with increasing speed than a goat of the same weight, particularly at slower speeds where the gaits are similar.

We spent nearly a year and suffered a few minor physical disabilities in training cheetahs to run on treadmills while we measured their steady state oxygen consumption. The results of these experiments were quite unexpected. Both the cheetah and the goat (animals of similar weight) consumed oxygen at the same rate when they ran at the same speed (Fig.6). The slopes of the relationships between oxygen consumption and running speed were approximately what would have been predicted from the body weight of the animals. Thus this relationship between energetic cost of running and body weight appears more general than one would have expected.

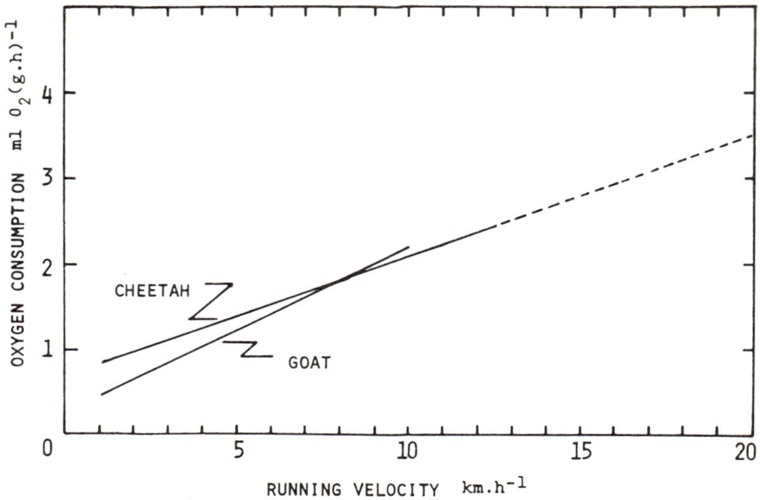

Figure 6 - Steady state oxygen consumption as a function of running speed in the cheetah (Acinonyx jubatus) and the goat (Capra hircus). The slope of the relationship between $\dot{V}O_2$ and running velocity for the cheetahs was 0.14 ml O_2 $(g{\cdot}km)^{-1}$ compared to a predicted value of 0.13 ml O_2 $(g{\cdot}km)^{-1}$. The slope for the goat was 0.20 ml O_2 $(g{\cdot}km)^{-1}$ compared to a predicted of 0.13. Data from unpublished observations in my laboratory.

Bipedal versus quadrupedal locomotion

Two bipedal animals in which the energy cost of running has been investigated, man (13) and the rhea (an ostrich-like bird) (16), expend about twice as much energy in running as quadrupedal animals of the same weight. Much of the mechanical work involved in man's bipedal running goes into alternately accelerating and decelerating the body with each stride (5). One might expect that four legs would enable quadrupedal animals to achieve a more even running speed and, thereby, minimize this expenditure of energy.

We decided to explore the difference between bipedal and quadrupedal locomotion by selecting two animals which normally run either bipedally or quadrupedally, the chimpanzee (Pan troglodytes) and the organ grinder monkey (Cebus capuchin) (17). Again to our surprise we found no difference between the energy expended in bipedal or quadrupedal locomotion in either of these animals over a range of running speeds (Figs. 7 and 8). The slope of the

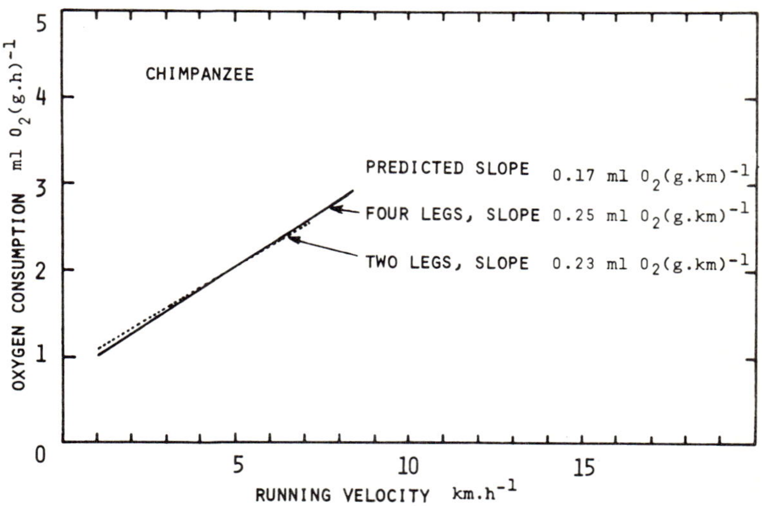

Figure 7 - Steady state oxygen consumption of chimpanzees (Pan troglodytes) running on four legs and on two legs as a function of running velocity. The slope of the relationship between $\dot{V}O_2$ and running velocity for four-legged running was 0.25 ml O_2 $(g \cdot km)^{-1}$ and the y-intercept was 0.79 ml O_2 $(g \cdot h)^{-1}$ compared with a slope of 0.23 ml $O_2 (g \cdot km)^{-1}$ and a y-intercept of 0.90 ml O_2 $(g \cdot h)^{-1}$ for two-legged running. The slope predicted from body weight was 0.17 ml O_2 $(g \cdot km)^{-1}$. Data plotted from (17).

relationship between oxygen consumption and running velocity was almost exactly that predicted on the basis of body weight for the organ grinder monkey and about 50% greater than predicted for the chimpanzee.

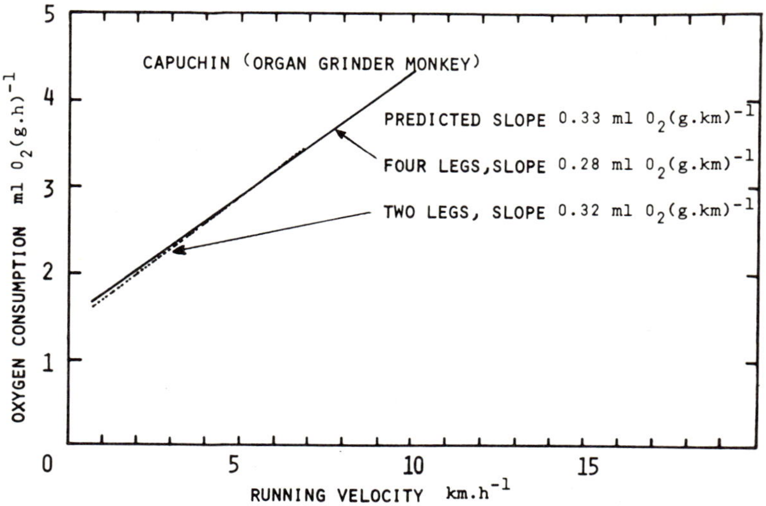

Figure 8 - Steady state oxygen consumption of organ grinder monkeys (*Cebus capuchin*) running on four legs and on two legs as a function of running velocity. The slope of the relationship between $\dot{V}O_2$ and running velocity for four-legged running was 0.28 ml O_2 $(g{\cdot}km)^{-1}$ and the y-intercept was 1.50 ml O_2 $(g{\cdot}h)^{-1}$ compared with a slope of 0.32 ml O_2 $(g{\cdot}km)^{-1}$ and a y-intercept of 1.42 ml O_2 $(g{\cdot}h)^{-1}$ for two-legged running. The slope predicted from body weight was 0.33 ml O_2 $(g{\cdot}km)^{-1}$. Data replotted from (17).

Some consequences of the relationship between cost of running and body weight

The relationship between cost of running and body weight has many interesting implications. For example, one can: 1) explain why small animals can run up steep inclines with ease while large animals cannot; 2) estimate how fast animals of different size can run, or 3) approximate the energy expenditure of animals running at top speed.

I was very impressed to observe small animals

running vertically at approximately the same maximum speed that they ran horizontally. I suppose the reason that this seems surprising is that you or I must increase our metabolism markedly upon encountering even a slight incline. How do small animals accomplish this remarkable feat? It can be explained quite simply using the relationship between energy cost of running and body weight.

As we have observed, small animals expend much more energy to move a gram a horizontal kilometer than large ones. For example, energetic cost of level running in a 30 g mouse is about eight times as great as in a 17,500 g chimpanzee. The gravitational work involved in lifting one gram one kilometer vertically, however, is the same for the two animals. If the efficiency with which the muscles perform this vertical work is similar in the two animals, then the cost of lifting one gram one kilometer would also be similar. Thus the relative increase in metabolism for the vertical component of uphill running in the mouse would be 1/8th that of the chimpanzee.

This theoretical argument turned out to be valid (15) and Figure 9 illustrates that at a given speed a mouse consumes oxygen at approximately the same rate while running: up a 15° incline, on the level, or down a 15° incline. The energy cost for the chimpanzee running up the 15° incline, on the other hand, was about twice that for level running while the cost for running down the incline was about half.

The additional energy expended in lifting one gram one kilometer above that for level running was about 3.8 cal $(g \cdot km)^{-1}$ hours at all speeds for both the mice and chimpanzees (Table I). It is extremely interesting to note that both chimpanzees and mice recovered about 95% of the potential energy which was stored running uphill on the way down.

The relationship between energetic cost of running and body weight also throws some light on a question which has interested biologists and laymen alike: How fast can animals of different size run? A.V. Hill (8) has argued that the top speed of animals should be independent of size. This seems intuitively unreasonable, for we know that a horse runs faster than a mouse. Now we can

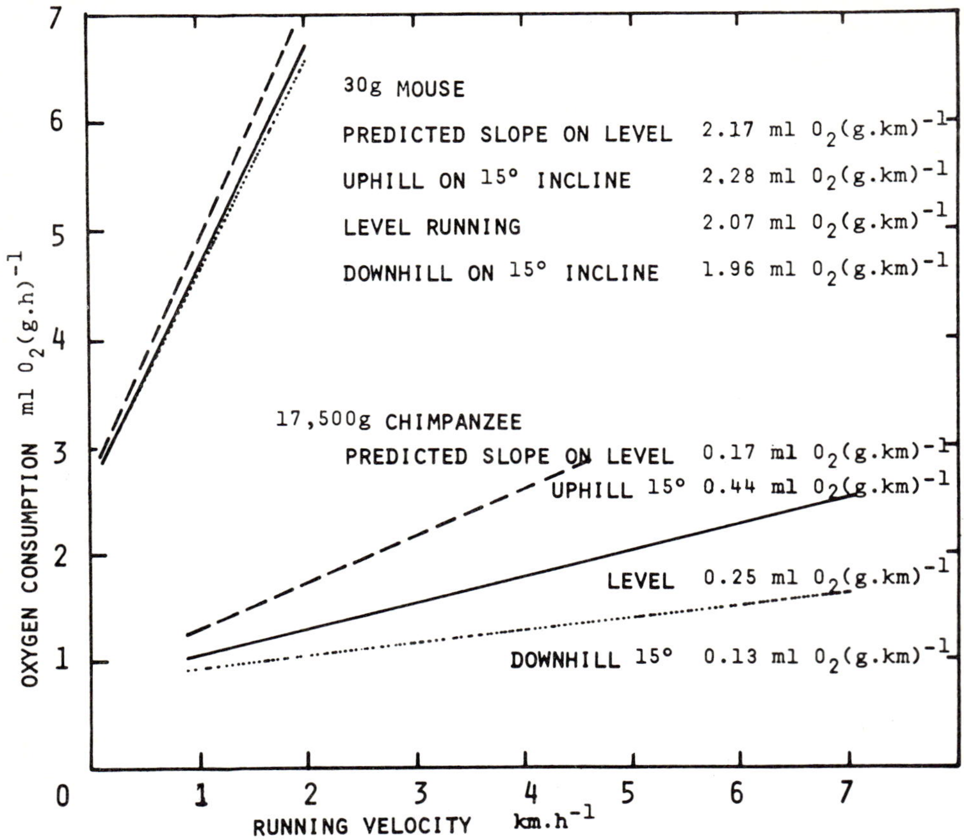

Figure 9 - Steady state oxygen consumption at various running velocities for 30 g mice and 17,500g chimpanzees: running on the level (solid lines); running uphill at a +15° incline (dashed lines); and running downhill at a -15° incline (dotted lines). Slopes and y-intercepts for the relationship between $\dot{V}O_2$ and running velocity for chimpanzees: slope for level running 0.25 ml O_2 $(g{\cdot}km)^{-1}$, y-intercept 0.79 ml O_2 $(g{\cdot}h)^{-1}$; slope for running uphill on +15° incline 0.44 ml O_2 $(g{\cdot}km)^{-1}$, y-intercept 0.85 ml O_2 $(g{\cdot}h)^{-1}$; slope for running downhill on -15° incline 0.13 ml O_2 $(g{\cdot}km)^{-1}$, y-intercept 0.79 ml O_2 $(g{\cdot}h)^{-1}$. For the mice:slope for level running 2.07 ml $(g{\cdot}km)^{-1}$, y-intercept 2.57 ml O_2 $(g{\cdot}h)^{-1}$; slope for running uphill on +15° incline 2.28 ml O_2 $(g{\cdot}km)^{-1}$, y-intercept 2.68 ml O_2 $(g{\cdot}h)^{-1}$; slope predicted for level running on the basis of body weight (18) are also given.

TABLE I - Energy required and mechanical efficiency for the vertical component of running up a 15° incline (top of table); and energy recovered and mechanical efficiency of recovery of stored energy running down hill (bottom of table). Values for man and sheep are from the literature.

Animal	Wt. (g)	watt·h $(g \cdot km)^{-1}$	cal $(g \cdot km)^{-1}$	
		RUNNING UPHILL ON +15° INCLINE		
		Energy to raise 1 g 1 km vertically		Mechanical work / Energy Expended x 100
Mice	30	4.54×10^{-3} (b)	3.90	60%
Chimpanzee	17,500	4.11×10^{-3}	3.61	66%
Sheep (2)	29,000	7.48×10^{-3}	6.45	36%
Man (12)	70,000	7.98×10^{-3}	6.88	34%
		RUNNING DOWNHILL ON -15° INCLINE		
		Energy recovered lowering 1 g 1 km vertically		Energy recovered / Mech.Energy Stored x 100
Mice	30	2.37×10^{-3} (b)	2.04	87%
Chimpanzee	17,500	2.59×10^{-3}	2.18	95%
Sheep (2)	29,000	-----	----	---
Man (12) (c)	70,000	0.93×10^{-3}	0.80	34%

(b) These values were obtained by using slopes for uphill, downhill, and level running. They are only approximations, since slopes were not significantly different at 90% level of confidence.

(c) Efficiency of recovering energy for man running downhill varies with the steepness of the incline, from near 100% on a shallow incline of -3° to 0° on an incline of about -18°. On inclines steeper than -18° it requires more energy to run downhill than to run on the level.

calculate that the mouse expends a great deal more energy per unit weight running at any speed than the horse (both in absolute terms and relative to its resting metabolism). Small animals would not be able to run at the same speed as large ones unless they could increase their metabolism by relatively greater amounts. Figure 10 illustrates how fast animals of different size could run if

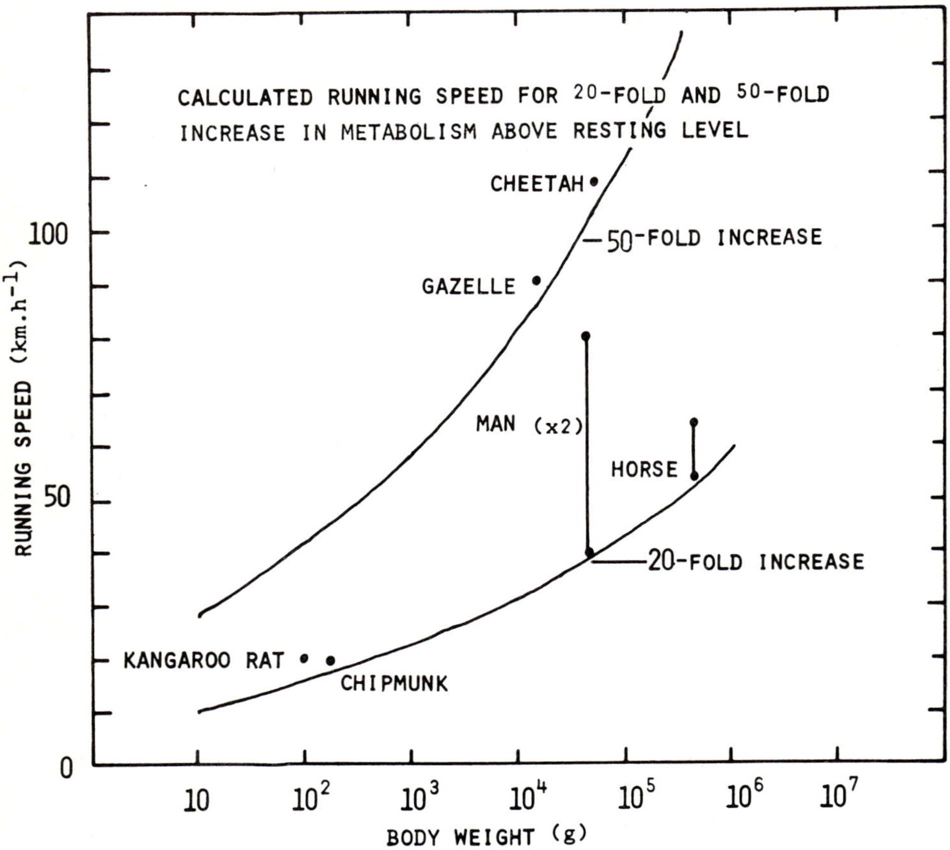

Figure 10 - Maximum speed which animals could run if they increased their metabolism by 20- and 50- fold over resting metabolism (predicted by Kleiber's equation (10)). Oxygen consumption during running was calculated using the relationship of Taylor *et al.* (18). Points for maximum observed running speed (9) are plotted for a number of animals. Values for both maximum sustained and maximum speed are plotted for man and horse. Values for man are multiplied by two, since man consumes twice as much energy as predicted (18).

they could increase their sustained metabolism by 20 fold over predicted standard metabolism or their peak metabolism by 50 fold using the predicted oxygen consumption while running. It is clear that large animals should be able to run more rapidly than small ones.

Another question which has interested biologists is the extent to which animals can increase metabolism over resting levels, a concept often referred to as "metabolic scope". The 20 and 50-fold figures which we used to calculate maximum speed are based on data for maximum aerobic and aerobic plus anaerobic metabolic rates in man, and little information exists on other species. It is possible to gain some insight into this question by calculating the metabolism of an animal running at top speed. Many reliable records of maximum speed of running animals are reported in the literature and I have used these to prepare Figure 11, where predicted metabolism at top running speed divided by predicted standard metabolism is plotted as a function of log body weight. These calculations show a bell-shaped curve, with the fastest animals being intermediate in size (5-100 kg). These intermediate sized animals appear able to increase their peak metabolism by about 50-60 fold, as is the case in man. Small animals (30-1,000 g) and very large animals (>300 kg) appear able to increase their peak metabolism by only about 20-fold.

The reason small animals are unable to achieve the same relative increase in metabolism as large animals is probably because of the higher metabolic rate of their tissues. In man the amount of stored glycogen, the maximum tolerable concentration of lactic acid, and the maximum rate at which glycogen can be broken down to lactic acid appear to limit the relative increase in metabolism available from anaerobic sources. One would expect that all of these parameters would scale linearly with weight rather than with $W^{-0.25}$ as metabolism does. If this is the case, then as animals get smaller and their metabolic rate per gram gets larger, anaerobic sources of energy will become relatively less important. The argument here is identical to the one made earlier in the paper about relative cost of running uphill for large and small animals.

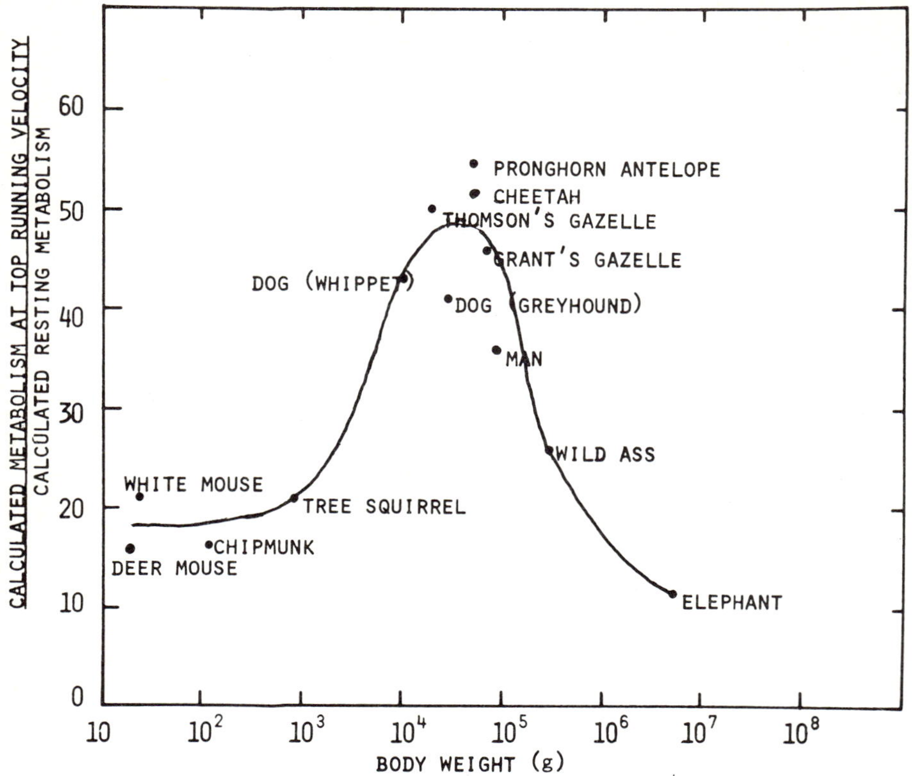

Figure 11- Relative increase in metabolism at top running speed over predicted resting metabolism as a function of log body weight. Top running speeds are those reported by Howell (9) and by Layne and Benton (11). Oxygen consumption during running was calculated using the relationship of Taylor *et al.* (18) and resting oxygen consumption was calculated using Kleiber's equation (10).

The fact that very large animals do not run as fast as intermediate sized animals cannot be explained in energetic terms. This suggests one looks at structural limitations.

A general relationship for cost of running

The results from lizards, cheetahs, and goats

and on bipedal and quadrupedal locomotion, indicate that the relationship between body weight and energetic cost of running is far more general than might have been anticipated. It also becomes obvious that an understanding of this relationship will not be possible until one is able to quantify all component parameters in the energy cost of locomotion.

REFERENCES

1. Bakker, R.T. (1972) Locomotor energetics of lizards and mammals compared. The Physiologist 15.
2. Blaxter, K.L. (1962) The Energy Metabolism of Ruminants. London: Hutchinson.
3. Brody, S. (1945) Bioenergetics and Growth. New York: Reinhold.
4. Cavagna, G.A. & Margaria, R. (1955) Mechanics of walking. J.Appl.Physiol. 21, 271-278.
5. Cavagna, G.A., Saibene, F.P. & Margaria, R. (1964) Mechanical work in running. J.Appl.Physiol. 19, 249-256.
6. Dawson, T.J. (1972) Primitive mammals and patterns in the evolution of thermoregulation. In Essays on Temperature Regulation, ed. Bligh, J. & Moore, R., Amsterdam: North Holland Publishing Company.
7. Hildebrand, M. (1960) How animals run. Sci.Am. 202, 148-157.
8. Hill, A.V. (1950) The dimensions of animals and their muscular dynamics. Sci.Prog. 38, 209-230.
9. Howell, A.B. (1944) Speed in Animals. Chicago, Illinois: Chicago University Press.
10. Kleiber, M. (1961) The Fire of Life. An introduction to Animal Energetics. New York: Wiley & Sons.
11. Layne, J.N. & Benton, A.H. (1954) Some speeds of small animals. J.Mammal. 35, 103-104.

12. Margaria,R. (1938) Sulla fisiologia e specialmente sul consumo energetico della marcia e della corsa a varie velocità ed inclinazioni del terreno. Atti della R.Accademia dei Lincei 7, 297-368.

13. Margaria,R., Cerretelli,P., Aghemo,P. & Sassi, G. (1963) Energy cost of running. J.Appl. Physiol. 18, 367

14. Schmidt-Nielsen,K. (1972) Locomotion: energy cost of swimming, flying and running. Science 177, 222-228.

15. Taylor,C.R., Caldwell,S.L. & Rowntree,V.J. (1972) Running up and down hills: some consequence of size. Science 178, 1096-1097.

16. Taylor,C.R., Dmi'el,R., Fedak,M. & Schmidt-Nielsen,K. (1971) Energetic cost of running and heat balance in a large bird, the rhea. Am.J. Physiol. 221, 597-601.

17. Taylor,C.R. & Rowntree,V.J. (1973) Running on two or on four legs: which consumes more energy? Science 179, 186-187.

18. Taylor,C.R., Schmidt-Nielsen,K. & Raab,J.L. (1970) Scaling of energetic cost of running to body size in mammals. Am.J.Physiol. 219, 1104-1107.

Comparative Physiology, eds. L. Bolis, K. Schmidt-Nielsen and S. H. P. Maddrell

HUMAN LOCOMOTION

G.A. CAVAGNA
Istituto di Fisiologia Umana,
Università di Milano & Centro
di Studio per la Fisiologia del
Lavoro Muscolare C.N.R.,
Milano, Italy.

SUMMARY

1.- In walking and running most of the external mechanical work done by the muscles at each step is required to sustain the vertical displacements and the forward speed changes of the center of gravity of the body.

2.- Internal work is mainly done to sustain the kinetic energy changes of the limbs relative to the center of gravity.

3.- In walking at low and moderate speed (up to 7 km/hr) the potential energy stored during the rise of the center of gravity is utilized to accelerate the body forwards and vice-versa,during the deceleration, kinetic energy is used to increase the potential energy (as in a pendulum): muscular activity is required only to maintain the shift between the two forms of energy.

4.- In running on the contrary the center of gravity is simultaneously lifted and accelerated forwards by the muscular push, and potential energy cannot be transformed into kinetic energy. The external mechanical work done by the muscles per km is correspondingly increased from 0.1 kcal/kg.km (as in walking up to 6-7 km/hr) to 0.25 kcal/kg.km (running from 10 to 20 km/hr).

5.- As the velocity of progression increases, the deceleration of the body forwards, which follows the impact with the ground, becomes progressively greater and absorbs more and more of the forward push of the muscles: eventually it limits the speed of the run. At the maximal speed attained (34 km/hr) the average power developed during the push amounts to 3-4 HP.

6.- The ratio between the positive work done in running and the energy expenditure is of the order of 0.4, i.e. greater than the maximal efficiency of the transformation of chemical energy into mechanical work by the contractile component of muscle. This result can best be interpreted by assuming that part of the positive work is sustained by the elastic energy stored in the series elastic elements of muscles when these are stretched during the deceleration of the body forwards.

7.- Experiments done on isolated muscle have in fact shown that the positive work done by a muscle immediately after stretching can be much greater than that done by the same muscle shortening in the same conditions, but from a state of isometric contraction.

8.- This finding however is not only due to the greater values of force attained during stretching and implying a greater storage of elastic energy; in fact both the series elastic and the contractile component of the muscle appear to be transitorily modified by the previous stretching: as a result, the first releases more mechanical energy for the same change in tension, and the second allows the muscle to generate a tension equal to or even greater than the isometric force at the same length.

One way of studying locomotion is to measure the mechanical work that the muscles must perform to sustain the movement of the body. This approach allows one to understand the causes of metabolic energy expenditure, the factors eventually limiting the speed and shows how the mechanical characteristics of muscle, mainly studied in isolated muscles, are actually operating in the whole organism.

Main sources of energy expenditure in human locomotion

Whereas in the locomotion of fishes and birds the drag through the medium represents a major cause of energy expenditure and eventually limits their speed of progression, the work due to the friction against the air and the ground represents only a small fraction of the total work done in walking and in running. Practically no mechanical work is done by the muscles against the force of friction between the foot and the ground because the foot does not move appreciably relative to the ground (except when skidding takes place or when the ground yields as when walking on sand). Friction and pressure drag do exist between the moving body and the air, but, as it will be shown below, the speed of progression, at least in human locomotion, hardly attains values high enough to involve a relevant energy expenditure due to air resistance: even at the highest speed attained in running this amounts to only about 13% of the total energy cost (26). The main factors affecting human locomotion are: 1) gravity and - particularly at high speed - 2) the kinetic energy changes of the body and of the limbs. In order to understand how this happens we will simplify the complex mechanics of locomotion by taking into consideration only: a) the displacement of the center of gravity of the body in the sagittal plane (vertical and horizontal) and b) the movements of the limbs relative to the center of gravity of the body. We will neglect: c) the lateral displacement of the center of gravity, d) friction within the body (between the muscles and the surrounding tissues and in the articulations), e) the rotation of the body and f) the energy required to sustain the isometric contractions. Even though a) and b) represent the main sources of energy expenditure in human locomotion, it must be kept in mind that the value of mechanical work which we will arrive at will represent a minimum figure.

Since the force responsible for the displacement of the center of gravity of the body is the resultant of all the external forces acting between the body and its surrounding, item a) will require "external work". The work necessary to sustain the kinetic energy changes of the limbs relative to the center of gravity, b), is done by internal forces not contributing directly to the displacement of the center of gravity and it will be called "internal work".

Both external and internal work are done by the muscular force. Work is positive when the active muscles shorten (so that their force and the displacement of their extremities have the same direction): in this case chemical energy is transformed into mechanical work and heat. The work done by the muscular force is negative when the muscle, in spite of actively generating tension and utilizing chemical energy,is forcibly stretched by some other force; in this case the muscular force and the displacement of the muscle's extremities have opposite directions and work is done on the muscle: this 1) can be degraded into heat, 2) can be stored as elastic energy and 3) may possibly be utilized to drive some chemical reactions backwards,though this possibility has not been demonstrated yet with any certitude (19).

Work against gravity and work due to velocity changes in walking and running

These were first determined by Fenn (15, 16), on several students running at their maximal speed. A similar procedure was then followed for walking and running at different speeds by Cavagna,Saibene and Margaria (5,6).

Only the work necessary to sustain the displacements of the centre of gravity (external work) will be described here since it gives the most relevant information for the mechanics of locomotion. The work necessary to sustain the kinetic energy changes of the limbs relative to the centre of gravity will be taken into consideration only when comparing the mechanical work with the energy expenditure (Fig.4).

External work

The essential feature of human locomotion (both walking and running) is that at each step cycle the centre of gravity of the body moves up and down and the speed in the forward direction also oscillates, reaching a maximum and a minimum value at each step. This happens even when locomotion takes place on level ground and when the speed of progression (averaged over the step cycle) is maintained constant. The vertical oscillations of the centre of gravity correspond to changes of the positional or potential energy of the body and imply work against gravity. The variations of the speed of progression between a maximum and a minimum, on the other hand, imply work to sustain the kinetic energy changes of the body. From this point of view therefore human locomotion would seem rather uneconomical as compared for example with the motion of a wheel where, in order to maintain the movement, work is required only to overcome friction.In walking at low and average speeds,however, the external work that the muscles must perform to sustain progression is conveniently reduced to a minimum by the fact that the increases in the potential energy of the body are met by simultaneous decreases of its kinetic energy and vice-versa (5,7). In other words the potential and the kinetic energy levels of the center of mass of the body change exactly out of phase as shown in Fig.1 by the tracings W_v and E_k. It follows that, in walking, the work against gravity, (given by the amplitude of the curve W_v) and the work due to the velocity changes (amplitude of the curve E_k) do not have to be sustained completely by muscular action: in fact the transformation of kinetic energy into potential energy and vice-versa is achieved mainly through the supporting action of the rigid skeletal structure. Muscular activity on the contrary is completely responsible for the changes of the total mechanical energy level of the body (potential plus kinetic, curve E_{tot} in Fig.1). (Elastic energy is not included in the total mechanical energy since it is difficult to measure directly. As it will be seen below it can be evaluated indirectly.) These are two phases of the step cycle in which the muscles perform external positive work (rises $\underline{a}$ and $\underline{b}$ of the curve E_{tot} in fig.1); they are separated by two interposed phases in which negative work is performed. At low to average speeds of walking the positive external work that the muscles must do to maintain progression $\underline{a}$ + $\underline{b}$ is smaller not only than the sum of the positive work done against gravity and of that necessary to sustain the kinetic energy changes, but also smaller than each one of the two taken separately (7).

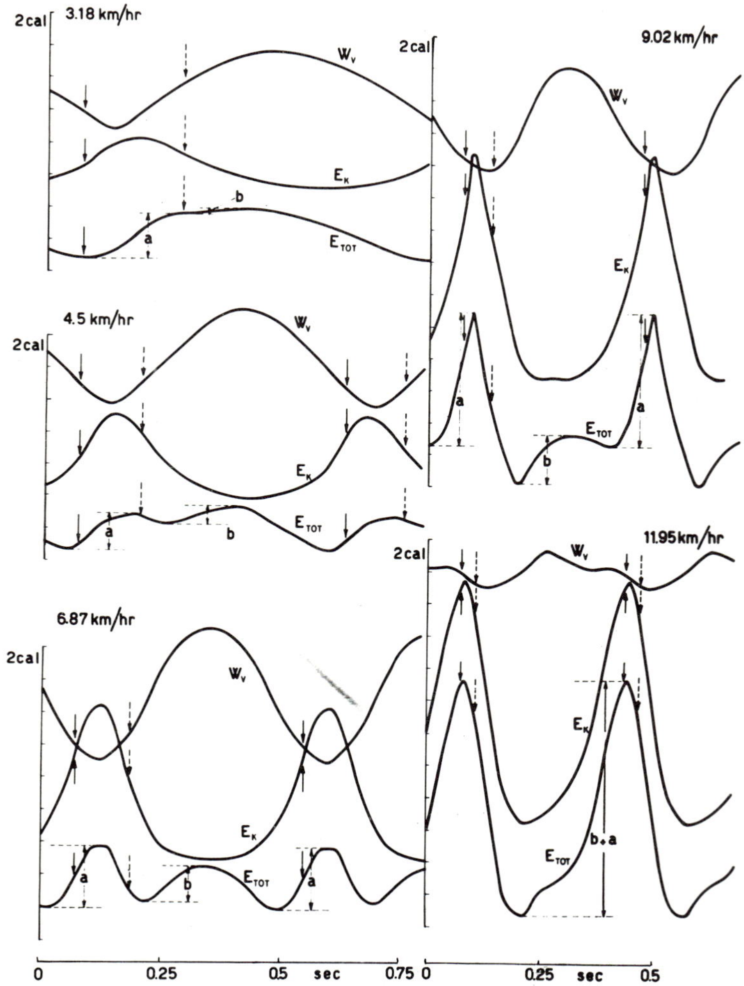

Figure 1 - The curve $W_V = \int F_V \, ds_V$ indicates the work done at each instant during walking, by the vertical component of the external force, F_V = weight + m . a_V, to sustain the vertical displacement of the centre of gravity, s_V. Since the vertical acceleration a_V is small the curve W_V has about the same trend of the potential energy level E_P =

weight. s_V, but it allows one to calculate more correctly the instantaneous change of the total mechanical energy level (7). E_k indicates the kinetic energy as calculated from the forward component of the speed of the center of gravity. $E_{tot} = W_V + E_k$ shows how the total mechanical energy level changes under the action of the muscular force. The increments a and b of E_{tot} indicate the positive external work done by the muscles at each step to sustain the displacement of the centre of gravity in a sagittal plane. The different groups of curves refer to different walking speeds as indicated. The arrows indicate the instant in which the heel touches the ground (continuous) and the other foot leaves the ground (interrupted). W_V and E_k are worked out from the records obtained by means of a platform sensitive to the vertical and the forward components of the force impressed by the feet on the ground. In the calculation air resistance is neglected and it is assumed that the body is rigid: these assumptions imply an error which is tolerable in sprint running (11) and probably negligible in walking.
(From Cavagna,G.A. & Margaria,R., J.Appl. Physiol.,19, 249, 1964).

As the velocity of progression increases the deceleration forward due to the impact with the ground (evidenced by the deflection of curve E_k in fig.1) increases continuously whereas the potential energy change available is limited by the length of the lower limbs. It follows that at high speeds of walking (Fig.1) and in running, the acceleration forwards necessary to maintain the average speed constant, can no longer be met by the decrease in potential energy, but must come directly from the muscular push. However, because this is necessarily directed upwards rather than forwards, it leads to a rise of the body simultaneous with the forward acceleration. In contrast to walking at low and average speeds, the potential and kinetic energy changes in running take place in phase. As a result, the external work per km is about double (about 0.1 kcal/kg km in walking up to 6-7 km/hr and 0.25 kcal/kg. km in running, fig.4) (6).

Factors limiting the maximal speed attained in running

Fenn measured on his students running at their maximal speed (27 km/hr) a power output of 0.5 HP to accelerate the body forward at each step (external work) and of 1.69 HP to accelerate the limbs relative to the centre of gravity (internal work); he found only 0.1 HP spent against gravity and 0.16 HP against wind resistance (15,16).

The work, W_F, done by an athlete to sustain the forward velocity changes at the beginning of a 100 meters race is given in figure 2 for the first five seconds from the start. The external work done against gravity is not represented: it amounts to about 20-30% of W_f during the first second from the start, when the positional change of the body from the crouching position is maximal, and it decreases further as the speed of the run rises (11). The continuous line indicates, within a 10% error (11): a) the positive work that the muscles must perform to accelerate the body forwards, to overcome air resistance and to deform the body (rises of the curve) and b) the negative work, done after the impact with the ground when the body decelerates forwards (lowerings of the curve). The horizontal tracts indicate the periods in which the runner is not in contact with the ground. This curve was determined from the records obtained by means of a force platform. The lower curve (crosses) represents the kinetic energy of the subject as measured from the increase of the speed of run determined by means of sights: this was the procedure hitherto followed in the literature to study sprint running (27,17,18,4,20,24). The curve at the middle was obtained by adding to the lower curve the work done against air resistance as calculated from the equation given by Hill (21) and consistent with Fenn's (16) and Du Bois-Reymond's (13) results

$\dot{W}_{ar} = 0.242\ v_f^{\ 3}$ (where the power $\dot{W}_{ar}$ is given in watts and the speed forwards v_f in m/sec.). The difference left between the circles and the continuous line is accounted for in part by the work done in the deformation of the body and in part is due to the error involved in the experimental procedure (11).

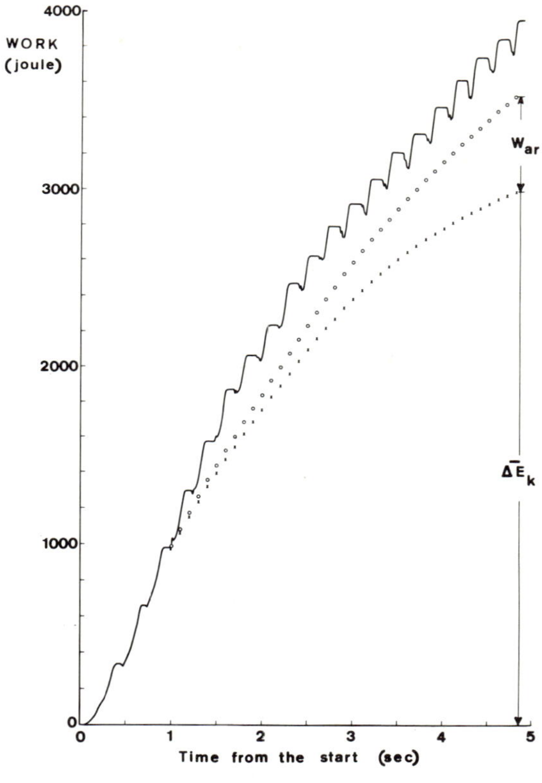

Figure 2 - The continuous line indicates the total work necessary to move the centre of gravity of the body in the forward direction as a function of the time from the start in sprint running (66 Kg athlete). The crosses indicate the fraction of the work attributable to an actual increase of the kinetic energy of the body. The curve indicated by the circles was obtained by adding to E_k the work done against air resistance, $\dot{W}_{ar}$ (modified from Cavagna, G.A., Komarek, L. & Mazzoleni S., J.Physiol.London, 217, 709, 1971).

From the curves in fig.2 it is possible to understand why in sprint running the speed increases less and less from the start to attain a constant value,as fatigue becomes a limiting factor. As is well known this finding was thought to be due to the muscle's power output being progressively affected by the increase of the speed of running and so by the speed of contraction (17). The power of the muscles propelling the body forwards is given by the slope of the rises of the curve in fig.2 and is plotted in fig.3 as a function of the speed of running. It appears that the average power developed at each push increases with speed and reaches 3-4 HP at the highest speeds: muscular function therefore does not account for the levelling off of the speed curve.

The factor responsible emerges from a closer inspection of the curves in fig.2: it turns out that about 95% of the positive work done by the muscles in propelling the body forward (given by the sum of the rises of the continuous curve in fig.2) is found as + ΔE_k at the end of the first second from the start, whereas after 3.5 to 4.5 sec 35% of it is lost during the deceleration of the body taking place at each step (lowerings of the curve), 23% is dissipated against air resistance (this subject attained a speed of about 34km/hr) and only 40% is found as + ΔE_k. The fraction of work lost during the impact with the ground is even greater in less skilful subjects. It is thus possible to conclude that the deceleration due to the impact against the ground and to a lesser extent air resistance are the main external factors limiting the speed attained by an athlete in running. Air resistance becomes even less important in untrained subjects: for them an additional important factor may be the work necessary to sustain the deformation of the body during the push (11).

Comparison between the mechanical work done and the energy expenditure

As mentioned above, at each step of walking and running the muscles perform both positive and negative work. In sprint running and in uphill locomotion the amount of positive work is greater

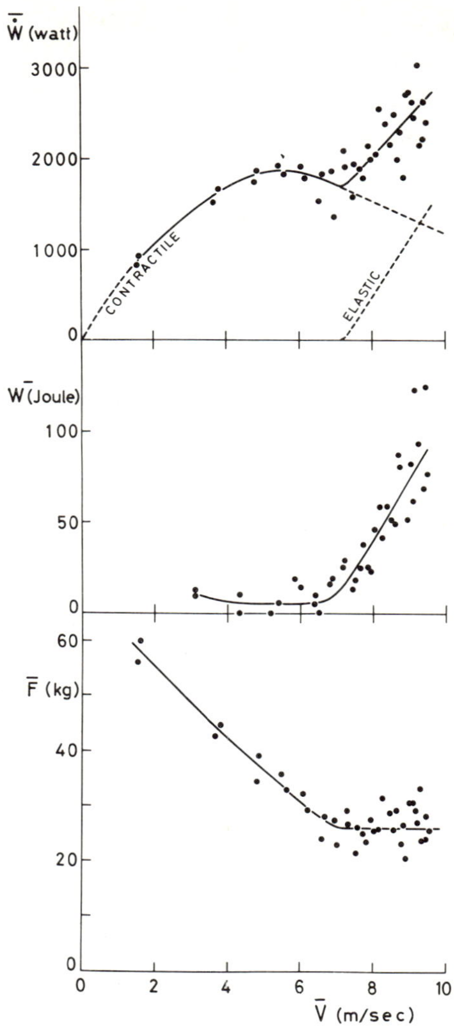

Figure 3 - The average power $\bar{\dot{W}}$ (top) and the average force F (bottom) developed by the muscles during the push, and the negative work W^- done at each step in the decelerating phase are given as a function of the speed of run for the same subject as in fig.2 (Modified from Cavagna, G.A., Komarek, L. & Mazzoleni, S., J. Physiol. London 217, 709, 1971).

than the negative work;as a result the kinetic energy, and the potential energy respectively of the body increase. When the average speed of progression is constant and locomotion takes place on level ground, positive work exceeds negative work only by the amount necessary to overcome friction: disregarding friction the positive and the negative work done by the muscles at each step are equal (as in fig.1, curve E_{tot}). In order to evaluate the energy expenditure from the mechanical work data it is therefore necessary to know the cost of both positive and negative work. As is well known, the mechanical efficiency with which chemical energy is transformed into positive work by the muscle attains a maximum of 25%; the minimal cost of positive work will therefore be 4 W^+. According to the force-velocity relationship of a muscle's contractile component, the cost of negative work varies with the speed, ranging between 20 and 40% of the cost of positive work (1,2); a minimal figure for the cost of negative work will therefore be $\frac{4}{5}$ W^+. In walking and running at a constant speed on the level, the cost of the mechanical work, positive plus negative, will therefore be

$$4\ W^+ + \frac{4}{5}\ W^+ \simeq 5\ W^+$$

In figure 4 the mechanical work done per km is given on the ordinate at the left as a function of the average speed of walking and running a steady rate on level ground. The symbols indicate the external mechanical work necessary to move the centre of gravity in a sagittal plane in walking (open symbols) and in running (closed symbols). The broken line was obtained by summing the work necessary to sustain the kinetic energy changes of the limbs relative to the center of gravity of the body (internal work) for an athlete running at 4 different speeds. The two crosses indicate: a) a value of total work (external + internal) as calculated from Elftman (14) at a walking speed of 5.6 km/hr and b) the external work necessary to displace the center of gravity in a sagittal plane as calculated by Fenn (16). According to Fenn, the internal work for the kinetic energy changes of the limbs is 0.56 kcal/kg and it is not shown in the figure. The ordinate at the right indicates the minimal energy expenditure per km and it was obtained, according to the above considerations, by multiplying by 5 the values on the ordinate at the left: in this way the symbols indicate directly on the right hand ordinate the minimum energy requirement for the mechanical work value which they indicate on the left hand ordinate. The continuous lines refer to the

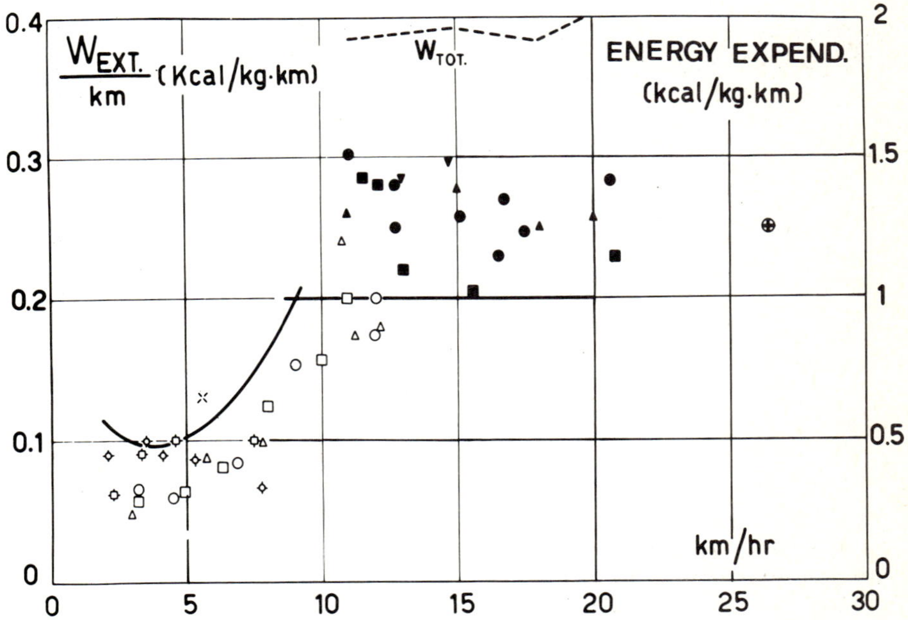

Figure 4 - Comparison between the mechanical work done per km (ordinate at the left), open symbols: external work in walking, closed: external work in running, broken line: total (external + internal work in running) and the energy expenditure (ordinate at the right using the same symbols and lines). See text. (From Cavagna, G.A., Travail mecanique dans la marche et la course. J.Physiol.Paris 61, 3, 1969).

right hand ordinate and indicate the actual energy expenditure, as measured from the oxygen consumption at steady state and in anaerobic conditions by many authors (3,25 and others afterwards).

In walking the actual energy expenditure is,as might be expected, usually greater than the minimum value predicted from the external work data; the difference seems to be greatest at very high and very low speeds where, probably, the internal work done is at a maximum.

In running, on the contrary, the actual cost of progression is paradoxically smaller than the minimum value predicted from the external work measurements. The difference becomes even greater if the internal work necessary to sustain the kinetic energy changes of the limbs is taken into account (broken line); in addition it should be pointed out that a maximal efficiency of muscular contraction was assumed in the construction of the right hand ordinate and that part of the mechanical work done (rotation of the body, lateral displacements of the center of gravity, etc.) was neglected. According to these data therefore the efficiency of running is 0.4 or greater, i.e. higher than the maximal efficiency with which the contractile component of the muscles can transform chemical energy into mechanical work. The evaluation of energy expenditure is difficult at high velocities of running such as those investigated by Fenn because of the formation of an oxygen debt which is hard to measure correctly: however little doubt exists that a trained athlete (upward triangles and broken line in fig.4) and most normal subjects can run between 10 and 20 km/hr in aerobic conditions, so that their energy expenditure can be completely evaluated from the oxygen consumption alone.

On the basis of the evidence given in fig.4, Cavagna, Saibene and Margaria (6) put forward the hypothesis that part of the positive work done by the muscles in running was not provided by the transformation of chemical energy into mechanical work, but by the potential elastic energy stored in the series elastic component of muscles when they are stretched during the preceding phase of the step in which they accomplish negative work.

This hypothesis is also in agreement with the trend of the power-velocity curve in fig.3: up to 7 m/sec the power changes with the speed of the run in a way which may depend on the well known power-velocity relationship of the contractile component of muscle; at greater speed, when the negative work done at each step begins to increase, indicating that the muscles are forcibly stretched (W^-, middle tracing), the average force developed by the muscles during the push ($\bar{F}_f$, lower tracing) stops decreasing according to the force-velocity relationship, becomes constant, and the power increases markedly. These findings suggest that at high speeds of running the elastic energy stored during stretching the contracted muscle can be utilized to keep the power output of the muscles well above the values attained by the contractile component only (11).

Effect of previous stretching on muscle's performance

G.A.Cavagna & G. Citterio

Evidence now exists that when a contracted muscle is forcibly stretched immediately before active shortening it performs a greater amount of work and/or it develops a greater mechanical power than when shortening begins from a state of isometric contraction. This has been shown both on the isolated muscle and on the muscles working *in situ* (8,10).

The simplest explanation for this effect is the following: according to the force-velocity relationship, the force with which the muscle resists lengthening attains values higher than the isometric one, the series elastic elements can thus be stretched more than during the isometric contraction and a greater amount of potential elastic energy can be stored. This elastic energy when released during the following shortening increases the amount of work done or, if the work is kept constant, the mechanical power developed.

Things however are not so simple. Consider the tracings in figure 5. These were obtained on an isolated gastrocnemius and sartorius on a Levin & Wyman ergometer. In one case the muscles were stimulated isometrically and then allowed to shorten at high speed. In the other case the muscles were stimulated isometrically at a shorter length, they were then stretched and immediately after they were allowed to shorten at the same high speed as before. As expected the work done by the previously stretched muscle, which is given by the area below the shortening curves, is greater than the work done when shortening begins from a state of isometric contraction (8). However by shifting to the left the curve obtained during shortening from a state of isometric contraction until the isometric value of the force, P_o, matches an equal value of the force developed during shortening after stretching (dotted line), it appears that the two curves diverge: a greater amount of mechanical energy is set free in the after stretch condition in spite of the fact that the initial and the final (=0) force values are now equal in both conditions. This result can be explained only by supposing that the previous stretch of the contracted muscle: 1) modifies the mechanical characteristics of the series elastic elements and/or 2) increases or makes more efficient the mobilisation of chemical energy by the contractile component.

In the experiments described in fig.5 most of the positive work done by the muscles during shortening is due to the elastic recoil of the series elastic elements: in fact the velocity of shortening imposed on the muscle is relatively very high, particularly for the gastrocnemius, and the contractile component has time to shorten only a negligible amount during the period in which the force falls from P_o to zero. This means that the shortening-distance curves in fig.5 approach the tension-length diagram of the series elastic elements: it appears then that the amount of "elastic recoil" necessary to bring the force from P_o to zero is appreciably greater for the previously stretched muscle. In other words fig.5 indicates that the previous stretching of the contracted muscle changes the mechanical characteristics of the series elastic elements making them capable of

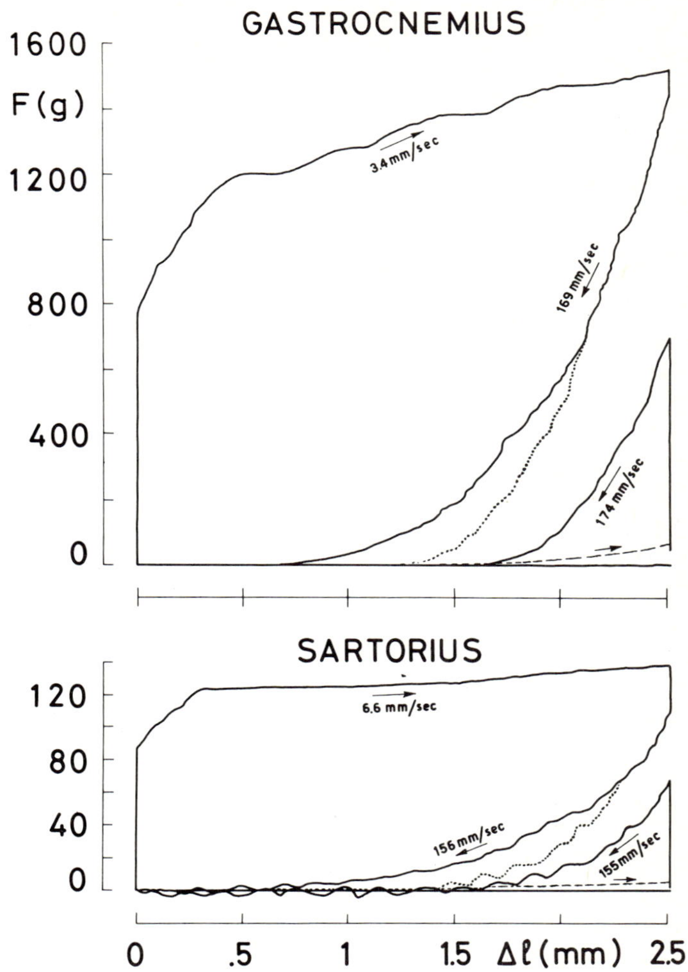

Figure 5 - Dynamic tension-length diagrams of a frog gastrocnemius (0.29 g, l_o = 2 cm) and of a toad sartorius (0.175 g, l_o = 3.7 cm) as obtained at 0°C during stretching and shortening at the speed indicated on a Levin & Wyman ergometer. The interrupted line was obtained during lengthening of the relaxed muscle and represents the tension-length diagram of the parallel elastic elements. Dotted line: see text.

a greater output of mechanical energy for the same fall of force. This is a transient effect: if an interval of time is left between stretching and shortening,the tension-length diagram of the series elastic elements obtained after stretching approaches that obtained during shortening from a state of isometric contraction.

The modification of the mechanical characteristics of the series elastic elements due to stretching implies that it is not correct to extrapolate the tension-length curve of the series elastic elements above P_o in order to calculate the elastic energy stored during stretching as hitherto assumed in the literature (22,23,9).

Possibility 2) above was tested by making the muscle shorten immediately after stretching in isotonic conditions according to the technique described by Cavagna (9). It was found that, after stretching, a frog gastrocnemius is able to lift a weight equal to or even greater than the isometric force at the same length (12). In conclusion: the experiments described seem to indicate that the previous stretching of the contracted muscle modifies transitorily both the series elastic and the contractile components of muscle making them capable of a greater output of mechanical energy.

REFERENCES

1. Abbot,B.C., Bigland,B. & Ritchie,J.M. (1952) The physiological cost of negative work. J.Physiol. (London) 117, 380-390.

2. Abbot,B.C. & Bigland,B. (1953) The effects of force and speed changes on the rate of oxygen consumption during negative work. J.Physiol. (London) 120, 319-325.

3. Atzler,E. & Herbst,R. (1927) Arbeitsphysiologische Studien. Pflügers Arch.ges.Physiol. 215, 291-328.

4. Best,C.H. & Partridge,Ruth C. (1929) The equation of motion of a runner, exerting a maximal effort. Proc.R.Soc.B 103, 218-225.

5. Cavagna,G.A., Saibene,F.P. & Margaria,R. (1963) External work in walking. J.Appl.Physiol. 18, 1-9.

6. Cavagna,G.A., Saibene,F.P. & Margaria,R.(1964) Mechanical work in running. J.Appl.Physiol.19, 249-256.

7. Cavagna,G.A. & Margaria,R. (1966) Mechanics of walking. J.Appl.Physiol. 21, 271-278.

8. Cavagna,G.A., Dusman,B. & Margaria,R.(1968) Positive work done by a previously stretched muscle. J.Appl.Physiol. 24, 21-32.

9. Cavagna,G.A. (1970) The series elastic component of frog gastrocnemius. J.Physiol.(London) 206, 257-262.

10. Cavagna,G.A., Komarek,L., Citterio,G. & Margaria,R. (1971) Power output of the previously stretched muscle. In Medicine and Sport n.6, pp. 159-167, Basel/New York: Karger.

11. Cavagna,G.A., Komarek,L. & Mazzoleni,S. (1971) The mechanics of sprint running. J.Physiol. (London) 217, 709-721.

12. Cavagna,G.A. & Citterio,G.(1971) Force-velocity relationship of frog gastrocnemius shortening in isotonic conditions immediately after stretching and from an isometric contraction. Archivio di Fisiologia 68, 316-318.

13. Du Bois-Reymond,R. (1925) Der Luftwiederstand desmenschlichen Körpers. Pflügers Arch.ges. Physiol. 208, 445-453.

14. Elftman,H. (1944) Skeletal and muscular systems: Structure and Function. In Medical Physics, pp. 1420-1430, Chicago, Year Book Publishers.

15. Fenn,W.O. (1930a) Frictional and kinetic factors in the work of sprint running. Am.J. Physiol. 92, 582-611.

16. Fenn,W.O. (1930b) Work against gravity and work due to velocity changes in running. Am.J. Physiol. 93, 433-462.

17. Furusawa,K., Hill,A.V. & Parkinson,J.L. (1927)

The dynamics of "sprint" running. Proc.R.Soc. B 102, 29-42.

18. Gertz,H. (1928) The working power in the 100 meters' race. Arch.Physiol.Skand. 55, 131-143.

19. Gillis,J.M., Maréchal,B. (1971) Influence of the tension on the incorporation of inorganic phosphate into ATP in glycerinated muscle fibres. J.Physiol.(London) 214, 41-43P.

20. Henry,F.M. & Trafton,I.R. (1951) The velocity curve of sprint running. Res.Quart.22, 409-422.

21. Hill,A.V. (1927) The air resistance to a runner Proc.R.Soc. B 102, 380-385.

22. Hill,A.V. (1950) The series elastic component of muscle. Proc.R.Soc.(Lond.) B 137, 273-280.

23. Hill,A.V. (1970) First and last Experiments in Muscle Mechanics, pp. 80-81, Cambridge, Cambridge University Press, 1st edn.

24. Ikai,M. (1967) Biomechanics of sprint running with respect to the speed curve. In Medicine and Sport, N.2, pp.282-290, Basel/New York, Karger.

25. Margaria,R. (1938) Sulla fisiologia e specialmente sul consumo energetico della marcia e della corsa a varie velocità ed inclinazioni del terreno. Atti dei Lincei 7, 299-368.

26. Pugh,L.G.C.E.(1971) The influence of wind resistance in running and walking and the mechanichal efficiency of work against horizontal and vertical forces. J.Physiol.(London) 213, 255-276.

27. Schilf,E. & Sauer,W. (1923) Über die Geschwindigkeit von Hundertmeterlaüfern und die dabei entwieckelten Kraft-und Arbeitsverhältnisse. Pflügers Arch.ges.Physiol. 200, 240-250.

Comparative Physiology, eds. L. Bolis, K. Schmidt-Nielsen and S. H. P. Maddrell

AERIAL AND TERRESTRIAL LOCOMOTION: A COMPARISON OF ENERGETICS

V.A. TUCKER
Department of Zoology,
Duke University,
Durham, North Carolina, U.S.A.

Most animals have some means of moving themselves from one place to another at the expense of metabolic energy. It is interesting to compare the effectiveness of the various devices that have evolved to accomplish locomotion. One measure of effectiveness is the amount of energy required to move a given distance. This is the measure that is usually applied in one form or another to the fuel economy of vehicles. For example, the fuel economy of European automobiles is often measured in liters of fuel consumed per 100 km traveled, and in the U.S.A., it is measured by a reciprocal quantity, miles traveled per gallon. For animals, the energy required to move a given distance is simply the ratio of the total metabolic rate (power input, P_i) during locomotion to the speed (V) of locomotion. As in automobiles, this ratio has a minimum value at a particular speed (Fig.1).

Frequently the ratio of power input to speed is divided by body weight (W) for both animals (see 16 for review) and machines (5). This quantity, $P_i/(WV)$, is a dimensionless number when weight is expressed as a force, and one system of units is used. The minimum value of $P_i/(WV)$ (the cost of transport) is functionally related to body weight for swimming, flying and running animals (16,19). Thus, for swimming fishes

$$\log (P_i/(WV)) = -0.606 - 0.295 \log W \qquad (1$$

(Data from 16, standard error of estimate = 0.068, N = 12),
for flying insects and birds

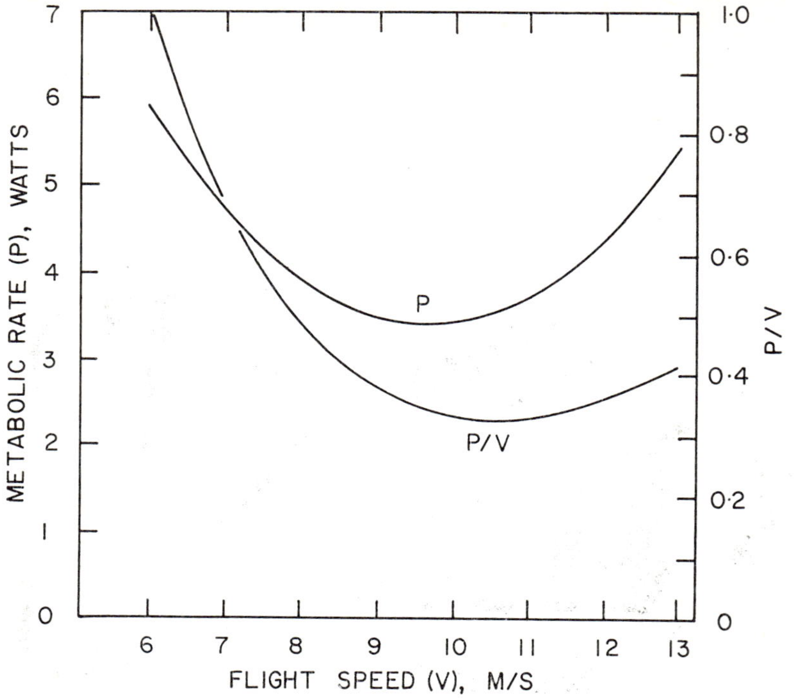

Figure 1 - Metabolic rates and ratios of metabolic rate to flight speed for a Budgerigar (mass = 0.035) flying level at various speeds in a wind tunnel (18).

$$\log (P_i/(WV)) = -0.048 - 0.227 \log W \qquad (2$$

(19), and for running mammals and a lizard,

$$\log(P_i/(WV)) = 0.879-0.680 \log W + 0.0775 (\log W)^2 \qquad (3$$

(data from 19, standard error of estimate = 0.157, N=20) (Fig.2). The constants in the above equations are for SI units - i.e. P_i in watts, W in newtons and V in meters/second. These units are used throughout this paper.

One immediately notes a striking difference between the cost of transport for swimming,flying and running animals of the same weight. For example, the cost of transport for a bird weighing 0.1 N (10 grams) is 1/27th that for a running animal of the same weight. The differences in these values are consistent with the observations that some

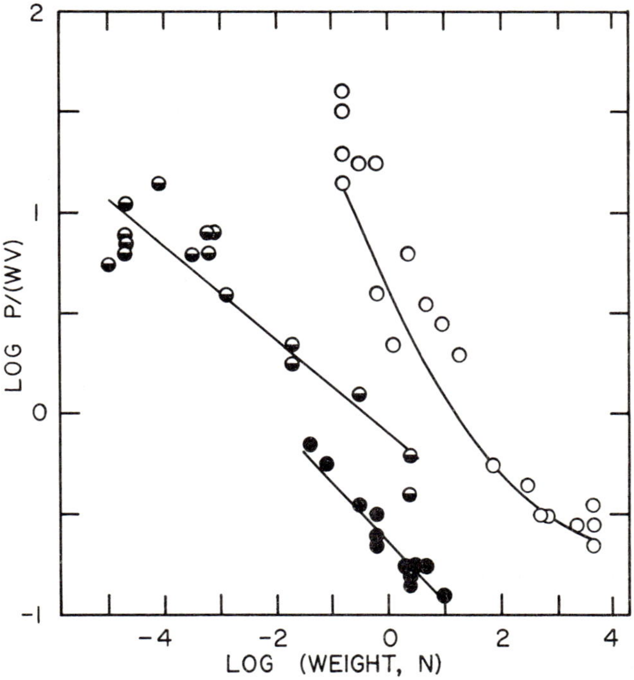

Figure 2 - Cost of transport for various swimming fishes (filled circles), flying insects and birds (half-filled circles) and walking or running mammals and a lizard (open circles).

birds migrate thousands of kilometers each year, while except for bats, small mammals do not. Why does such a striking difference exist between cost of transport for flying and running?

This question may be answered by investigating the fate of free energy released by metabolism during flying and running. Pennycuick (15) divides the power output (the rate at which kinetic energy is added to the air) of birds in level flight into the three categories: 1) induced power (P_{in}), which is the power expended to provide lift by accelerating air downwards, 2) parasite power (P_{par}), which is the power expended to overcome the drag of the body exclusive of the wings, and 3) profile power (P_{pr}), which is the power expended to overcome the profile drag of the wings. Power input, or metabolic rate, is the sum of these terms divided by an appropriate efficiency value (E) plus the basal metabolic rate (P_B). Thus

$$P_i = (P_{in} + P_{par} + P_{pr})/E + P_B \qquad (4$$

I have modified Pennycuick's theory slightly so that it fits more closely the measured values for metabolic rates and costs of transport of animals flying in wind tunnels (21). These modifications are four-fold: 1) including a span efficiency factor in the equation for P_{in}, 2) making P_{par} and P_{pr} functions of Reynolds number, 3) using a different calculation for P_{pr} and 4) replacing the term for basal metabolic rate in equation 4 with one for maintenance power (P_m). Maintenance power comprises the power input for basal metabolism plus circulation of the blood and movement of air in and out of the lungs. Thus,

$$P_i = (P_{in} + P_{par} + P_{pr})/E + P_m \qquad (5$$

When E is assigned a value of 0.2, this equation fits metabolic rates of flying birds quite well (Fig.3) as well as the empirical curve of cost of transport of flying animals (equation 2, figure 4) (21).

The higher costs of transport for running animals relative to those for birds are due to the fact that the efficiencies (as used in equation 5) are much lower for running than for flying. Using human locomotion as an example, this can be seen by rearranging equation 5 so that

$$E = \frac{P_{in} + P_{par} + P_{pr}}{P_i - P_m} \qquad (6$$

Since a runner on the ground supports his weight on a substrate rather than by accelerating air downwards, P_{in} is zero. Likewise, P_{pr} is zero, since the runner does not use wings. P_{par}, the power

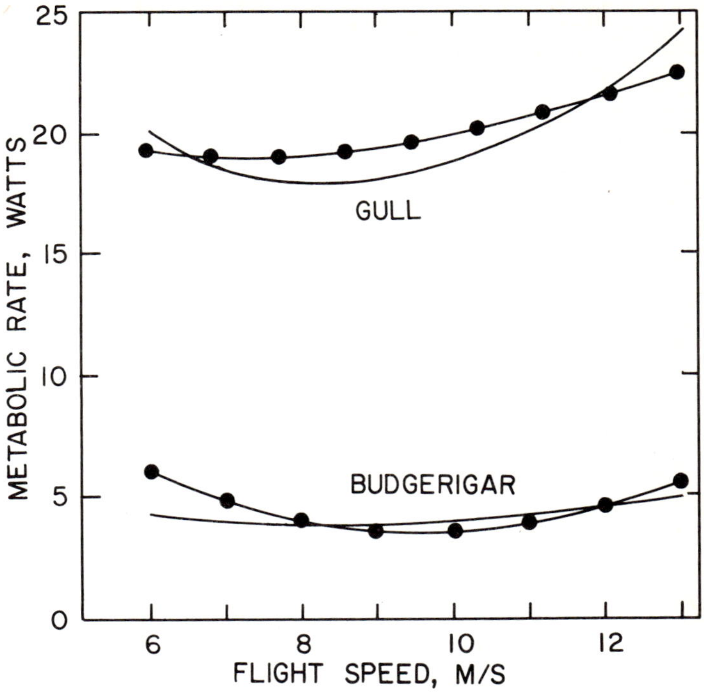

Figure 3 - Metabolic rates of Budgerigars (mass = 0.035 kg) and Laughing Gulls (mass= = 0.322 kg) flying level in a wind tunnel from measurements (beaded lines) (20) and from theory (21).

expended to overcome aerodynamic drag of the body has been estimated from wind tunnel tests. It is given by

$$P_{par} = 1/2 \rho \, C_D \, AV^3 \qquad (7$$

where ρ is the density of air, C_D is the drag coefficient (1.02) and A is the projected surface area of the body:

$$A = 0.17 \, H^2 \qquad (8$$

(H is the body height). These equations and the drag coefficient value are calculated from (3). The power input of running men (mass = 70 kg) is a function of running speed (6):

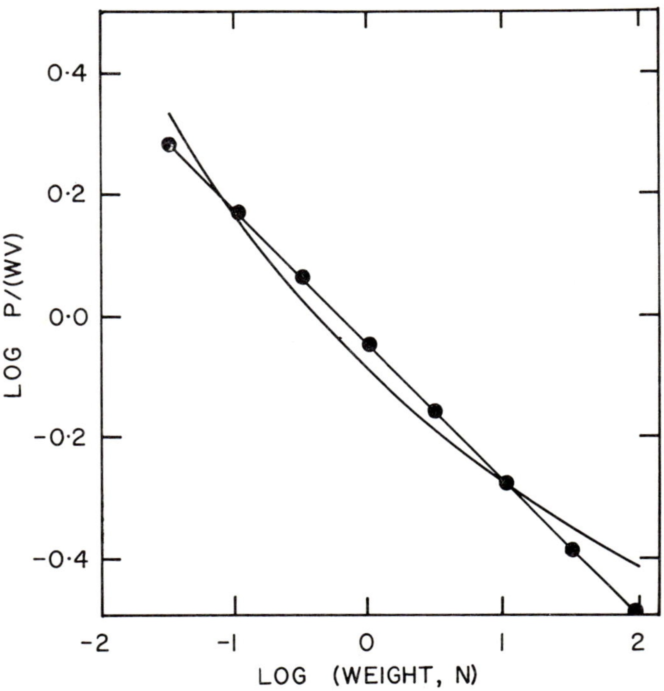

Figure 4 - Cost of transport for birds of various weights as predicted by equation 2 (beaded line) and by theory (21).

$$P_i = 83 + 2.49e^{0.917V} + 467e^{0.211V}$$

where e = 2.72.

These equations can now be substituted into equation (6) to calculate the efficiency of running at a constant speed. Taking ρ as 1.18 kg/m^3, H as 1.78 meters, P_m as the basal metabolic rate (83 watts) + 10% of P_i, and V as 5.5 m/sec, which is about the maximum speed that a good athlete can maintain for 10^3 seconds, yields an efficiency of 3.2%. This is nearly the highest efficiency that can be achieved in running level at constant velocity (Fig.5). The average person jogging for recreation for a few km at 4 m/s might achieve an efficiency of 2%.

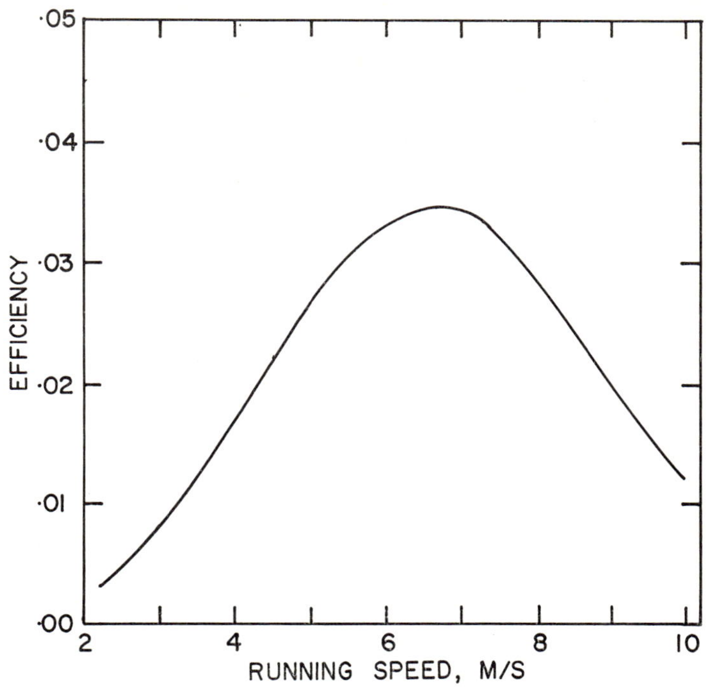

Figure 5 - The efficiency of level running for humans at various speeds.

Why should the efficiency of level running be less than 1/5 that achieved by flying birds? The efficiency of human exercise can equal or exceed that of flying birds (0.2) if the muscles are loaded appropriately. Thus, in pedaling a frictionless bicycle ergometer at constant speed against zero load, the numerator (power output) of equation 6 is zero and efficiency is zero, but increases rapidly to a maximum as load on the bicycle wheel is increased (Fig.6). Similarly, in walking up an inclined treadmill grade at constant speed, efficiency increases as the grade increases (Fig.6). In both examples, the power output is the rate at which work is done on the environment and is measured by the ergometer or by the rate of work done on the treadmill. The influence of load on efficiency is particularly clear in the case of the bicycle ergometer, where the motions of the feet are unchanged as the load varies.

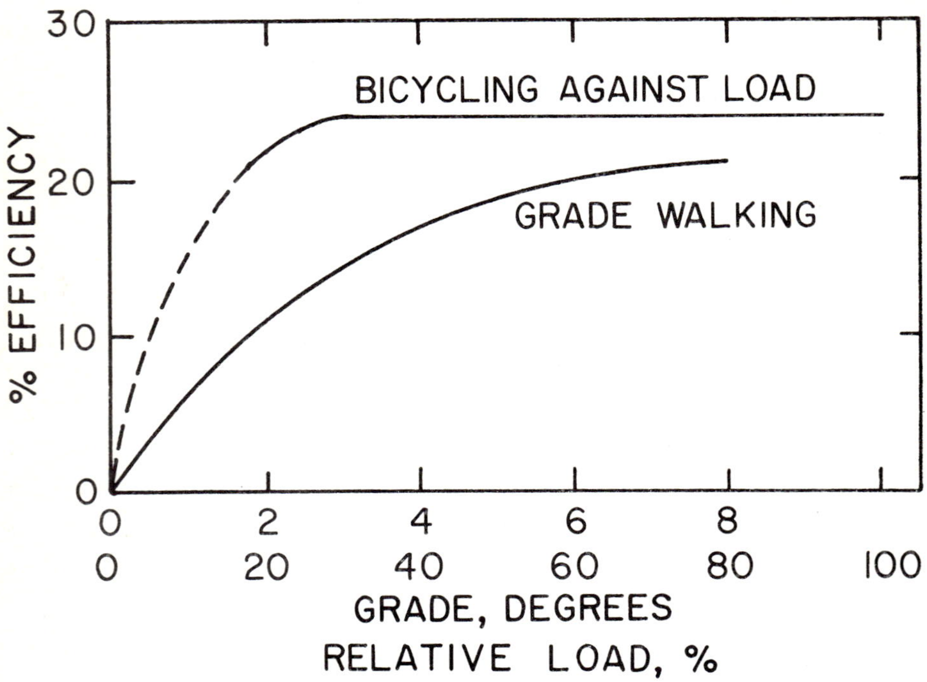

Figure 6 - Efficiency of pedaling on a bicycle ergometer at constant speed at various loads (4), and of walking up inclined treadmill grades at constant speed (23). The dashed line for pedaling is an interpolation from zero load to the smallest load that allowed normal pedal movements. Relative load is expressed as a percentage of the maximum force applied to the wheel run, and 100% load corresponds to a power output of 63 watts.

Why then, does not the efficiency of runners increase as they run faster and develop an aerodynamic drag load? For example, if a runner could load his muscles enough to achieve an efficiency of 0.2, he could run 10 km at a pace faster than that at which a champion sprinter covers 100 m.

The answer to this question depends ultimately on knowing the speeds of contraction of the muscles involved, where contraction consists of shortening or lengthening while exerting tension. During a cycle of limb movement, the power output of a contracting muscle could vary continuously, being positive at one time if the muscle were shortening, and negative at another if the muscle were lengthening. The integral over time of this power curve is the work output of the muscle during the cycle of limb movement. This work output could be used in various ways. It could be transferred to the environment to overcome aerodynamic drag, used to overcome internal friction and viscosity, used to increase the kinetic or potential energy of the body and limbs, used to stretch an elastic element or used to stretch another muscle. When running occurs on a level surface at constant speed, the net changes in kinetic, potential and elastically stored energy during a limb cycle are zero. Any work done on another muscle is accounted for in the integral of the power curve for that muscle, so that the sum of the work outputs for all the muscles used in level running at constant speed on a hard non-skid surface is the work done to overcome air resistance, and internal friction and viscosity. Neglecting the latter, the summated work output divided by the duration of the limb cycle is the mean power output, or the numerator of equation 6.

The denominator of equation 6 estimates the mean power input to the muscles, which is the sum of the integrals over time of the power input curves for the various muscles, again divided by the duration of the limb cycle. The power input of a contracting muscle can be calculated from the power output if the efficiency of the muscle is known. The efficiency of muscles depends primarily on the speed of contraction where speed is positive for shortening and negative for lengthening (1,7, 9) (Fig.7). Thus, the efficiency of a transport process as defined by equation 6 is a function of the efficiencies of the component muscles, and these efficiencies in turn depend on the speeds at which the muscles contract.

In running and walking humans (2,12) and presumably in other walkers and runners, the muscles

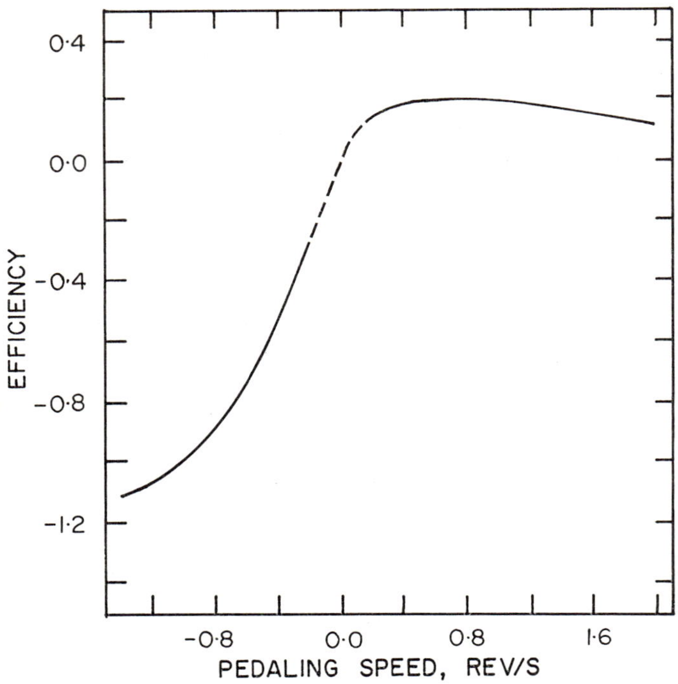

Figure 7 - Efficiency of pedaling a bicycle ergometer at various speeds. Positive speeds indicate that the feet move in the conventional direction and do work on the ergometer. Negative speeds indicate that the feet move in the opposite direction and the ergometer does work on them (1,4). The dashed line is an interpolation.

both shorten and lengthen while exerting tension. High efficiencies of transport result when the locomotor apparatus is arranged to permit muscle shortening and lengthening at speeds where the absolute values of muscle efficiency are high. Low efficiencies occur when the muscle shorten very rapidly, very slowly, or slowly lengthen while exerting tension.

Man can decrease the cost of transport by using mechanical devices such as gears or levers to change the speed of muscular contraction from that

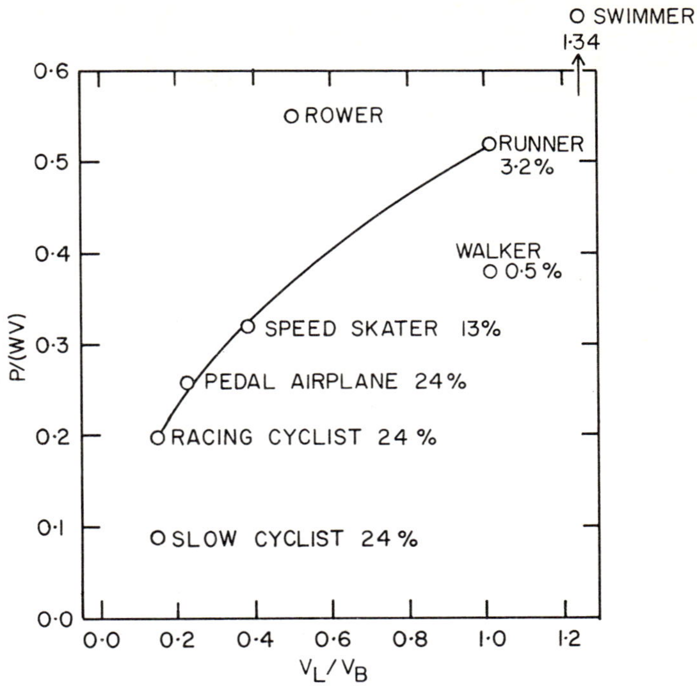

Figure 8 - Values of $P_i/(WV)$ and per cent efficiency for various forms of human locomotion at different values of V_L/V_B. The line is fitted to data for exhausting performances. The weight used for the calculations is body weight and is 687 newtons (70 kg) unless otherwise specified. The V_L/V_B value for swimming is unknown, but it is greater than 1, and the plotted value is a guess. The pedal airplane is a man carrying flying device powered by muscles turning bicycle pedals connected to a propeller. Data and sources as follows: Bicyclists (22), weight = 648 N; Pedal airplane (17) wing span = 18.3 m, aspect ratio = 11, altitude = 2 m above the ground, pedaling rate = 100/sec, 10% power loss through transmission and propeller; Rower, speed based on single scull record of 1 1/2 miles in 9 min, 7 sec (13), metabolic rate = 1670 watts; Swimmer, speed based on 1500 m in 17 min, 7.1 sec (14), metabolic rate = 1470 watts (10); Ice skating speed based on 5000 m in 7 min, 30.4 sec (14) metabolic rate = 1400 watts (11), drag coefficient = 1.02 (3), area = 0.18 m^2 (8), V_L = 4.5 sec; Running and walking, metabolic rates and speeds from (6), drag coefficient = 1.02, area = 0.54m^2(3).

during running. While running, the speed of extremities of the limbs (V_L) relative to the body must be as fast as the speed of the body (V_B) relative to the ground, at least during part of the movement cycle. Thus, the ratio V_L/V_B must be one at some time. The gearing of bicycles, for example, reduces this ratio to a typical value of 0.15. The efficiency of bicycle riding is high while the cost of transport is low, almost 1/4 the cost of transport for walking (Fig. 8). The figure shows that low values $P_i/(WV)$ for humans are related to the use of devices that decrease the V_L/V_B ratio.

REFERENCES

1. Abbott,B.C., Bigland,B. & Ritchie,J.M. (1952) The physiological cost of negative work. J.Physiol.117, 380-390.

2. Cavagna,G.A., Saibene,F.P. & Margaria,R. (1964) Mechanical work in running. J.Appl.Physiol.19, 249-256.

3. Dean,G.A. (1965) An analysis of the energy expenditure in level and grade walking. Ergonomics 8, 31-47.

4. Dickinson,S. (1929) The efficiency of bicycle-pedalling, as affected by speed and load. J. Physiol. 67, 242-255.

5. Gabrielli,G. & von Karman,T. (1950) What price speed? Mech.Engr. 72, 775-781.

6. Henry,F.M. (1953) The oxygen requirement of walking and running. Res.Quart.Amer.Assn.Hlth. Educ.Recr. 24, 169-175.

7. Hill,A.V. (1939) The mechanical efficiency of frog muscle. Proc.Roy.Soc.Lond.,ser.B,127, 434-451.

8. Hoerner,S.F. (1958) Fluid-dynamic Drag; Practical Information on Aerodynamic Drag and Hydrodynamic Resistance. 2nd Ed.,Midland Park,N.J.

9. Infante,A., Klaupiks,D. & Davies,R.E. (1964) Adenosine triphosphate: changes in muscles doing

negative work. Science 144, 1577-1578.

10. Magel,J.R. & Faulkner,J.A. (1967) Maximum oxygen uptake of college swimmers. J.Appl.Physiol. 22, 929-933.

11. Maksud,M.G., Wiley,R.L., Hamilton,L.H. & Lochart,B. (1970) Maximal V_{O_2} ventilation,and heart rate of Olympic speed skating candidates. J.Appl.Physiol. 29, 186-190.

12. Margaria,R. (1968) Positive and negative work performances and their efficiencies in human locomotion. Int.Z.angew.Physiol.ainschl.Arbeitphysiol. 25, 339-351.

13. Menke,F.G. (1950) The All-Sports Record Book. New York: A.S.Barnes & Co.

14. Menke,F.G. (1969) The Encyclopedia of Sports. 4th rev.ed., New York: A.S.Barnes & Co.

15. Pennycuick,C.J. (1969) The mechanics of bird migration. Ibis,111, 525-556.

16. Schmidt-Nielsen,K. (1972) Locomotion: energy cost of swimming, flying and running. Science 177, 222-228.

17. Sherwin,K. (1972) Man-powered flight as a sport. Nature, 238, 195-197.

18. Tucker,V.A. (1968) Respiratory exchange and evaporative water loss in the flying budgerigar. J.Exp.Biol. 48, 67-87.

19. Tucker,V.A. (1970) Energetic cost of locomotion in animals. Comp.Biochem.Physiol. 34, 841-846.

20. Tucker,V.A. (1972) Metabolism during flight in the laughing gull, Larus atricilla. Am.J. Physiol. 222, 237-245.

21. Tucker,V.A. Bird metabolism during flight: evaluation of a theory. J.Exp.Biol. (in press).

22. Whitt,F.R. (1971) A note on the estimation of the energy expenditure of sporting cyclists. Ergonomics 14, 419-424.

23. Workman,J.M. & Armstrong,B.W. (1964) A monogram for predicting treadmill-walking oxygen con-

sumption. J.Appl.Physiol. 19, 150-151.

Comparative Physiology, eds. L. Bolis, K. Schmidt-Nielsen and S. H. P. Maddrell

INVESTIGATION OF WING MOVEMENTS AND THE GENERATION OF AERODYNAMIC FORCES IN FLYING DIPTERA.

W. NACHTIGALL
Zoologisches Institut der
Universität des Saarlandes,
Saarbrücken, Germany

1. Introduction

Insect flight mechanics is a complexity of research, the main subject of which is external kinematics, i.e. wing movements. A knowledgeable understanding of this subject is necessary when dealing with such questions as aerodynamics, i.e. interaction between oscillating wings and the surrounding air, neuromuscular control and the mechanics of wing joints. Two major problems which arise, are that of measuring such wing movements and the interpretation of these measurements. The following communication offers an explanation of these problems through discussion of earlier publications and more recent, not yet published, investigations made on Calliphora erythrocephala and Phormia regina.

2. Stationary flight: wind tunnel techniques

The most simple flight condition is called "stationary, unaccelerated, horizontal, straight-ahead flight": summarized over a wing stroke period aerodynamic lift l (fig.1) equals body weight w; aerodynamic thrust t equals aerodynamic drag d of a flying insect. Insect flight analysis begins with this flight condition. A special, highly laminar wind tunnel (fig.2), was constructed in order to keep the insect in front of the analysing instruments. The wind velocity which must be opposite and equal to the insect's flying speed, may be controlled manually or directly by the insect itself via a

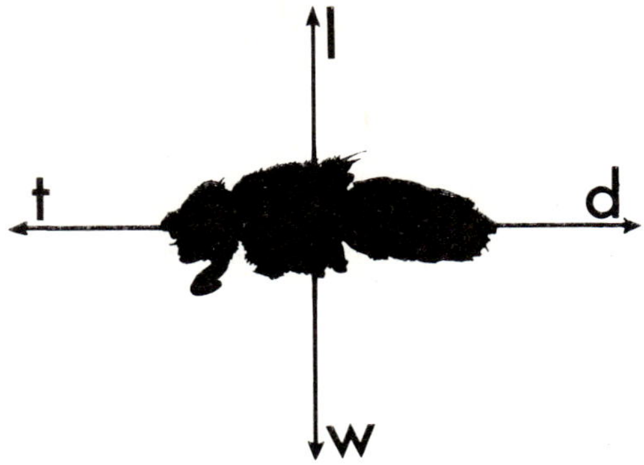

Figure 1 - Simplified diagram of balance of forces in a fly in flight (see text).

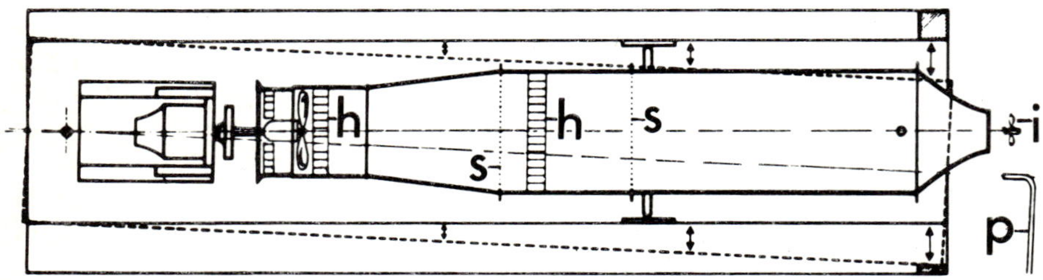

Figure 2 - Wind tunnel, h honeycombs, s screens. Turn-able from insect i to Pitot tube p. Diameter of nozzle 7.15 cm.

servo system. Under these conditions, the flying insect will remain steady. Systematic use of wind tunnel techniques was first made by Hollick (5) in analysing the flight of Diptera.

3. Measurement of stationary forces: aerodynamic balances

In order to know when stationary flight conditions have been reached, one requires the following data; weight w of the fly before experimentation and lift 1, drag d and thrust t during experimentation. When 1 = -w and t = -d, stationary conditions have been reached and high speed analysis can take place. The fly is glued at the tip of the abdomen to the rod r of a simple, non-automatic, aerodynamic balance (fig.3). This balance consists of 2 parts, a drag balance I with scale s_d supported by

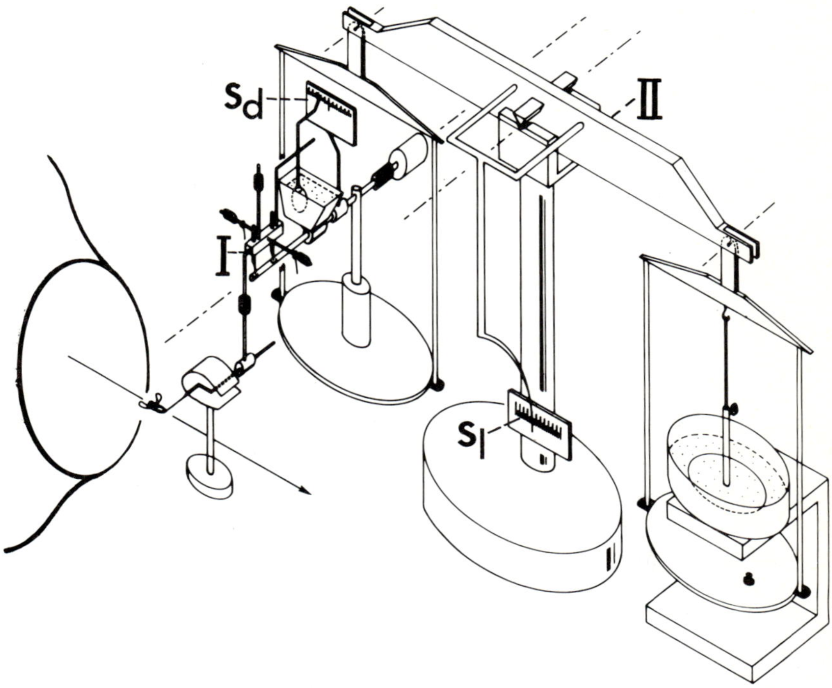

Figure 3 - Simple 2 component balance.
(See text)

a lift balance II with scale s_l. Both balances are adjusted to zero before flight. When they show zero again during flight, one has stationary flight conditions. A second, automatic balance which is used (fig.4) consists essentially of three parts, a compensated, torsion-moment balance with a horizontal rod perpendicular to the wind direction, thus allowing the fly to choose its own angle of attack between the longitudinal body axis and the direction of the streaming air. Balance I is supported by a lift balance II with a horizontal rod parallel to the wind direction; lift can be read off scale s_l. Balance II is supported by a drag balance III, with a vertical road which bears part of the servo-system on the upper side.

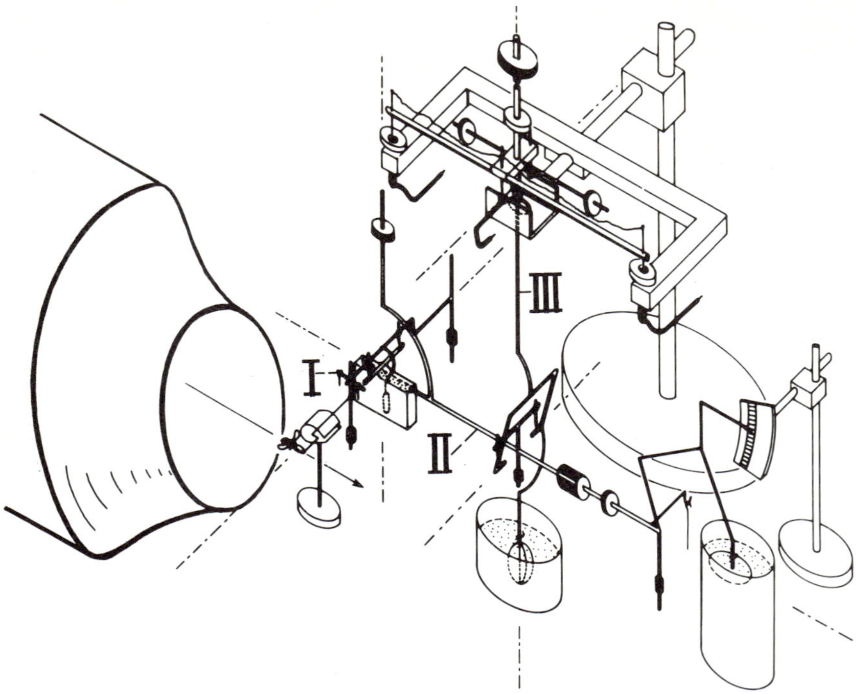

Figure 4 - Automatic 3 component balance. (see text)

4. Three-component high-speed technique

High speed stereo registration is necessary to analyse wing movements in space. Since the classical method of stereo-photography did not work for the dimensions of a fly, a 3-surface projection method was developed. A high speed camera hc (fig. 5) filmed the fly together with a tuning fork serving as a time control, against illumination coming from 4 low voltage light sources $l_1 - l_4$ and 4 mirrors, $m_1 - m_4$, placed above, behind and at the sides of the fly. A maximum of 8000 frames per second were used, so that, from 200 oscillations per second of a small fly, a single downstroke as well as a single upstroke was shown in approximately 15-20 frames.

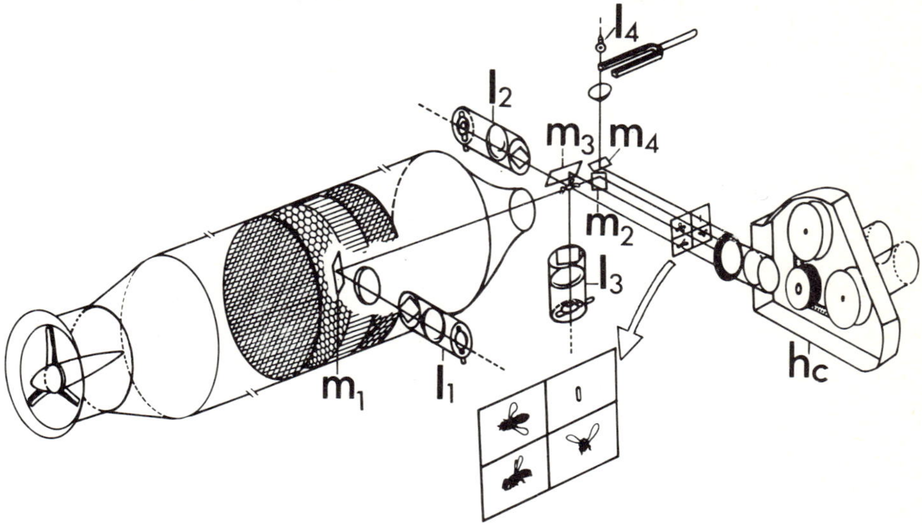

Figure 5 - Arrangement for filming insects from 3 directions on the same frame. See text.

5. Analysis of high speed films

Pictures were first redrawn by means of a movement-analysis-projector to get a qualitative impression (see fig.6). Wing rotation combined with twisting could be demonstrated at the upper and lower (fig.6) reversal phases as well as small-wing fluttering in the midst of the downstroke (fig.7, back view). The position of the longitudinal wing axis in the three projections was determined by 3 angles β , γ , δ (Fig.8), the time functions of which have been plotted (fig.9) and from which much kinematic data has been drawn. It has been shown that there is no simple sinusoidal oscillation. In stationary flight, single periods are kinematically the same. Downstroke lasts longer than upstroke. The beating wing oscillates at an average angle of 45° . Theories assuming sinusoidal oscillations (6) or constant velocity and/or wing movements perpendicular to the body axis (15) may not be valid for the flies investigated. There are turning movements of the wing, at least at the beginning of upstroke, which must be actively controlled.

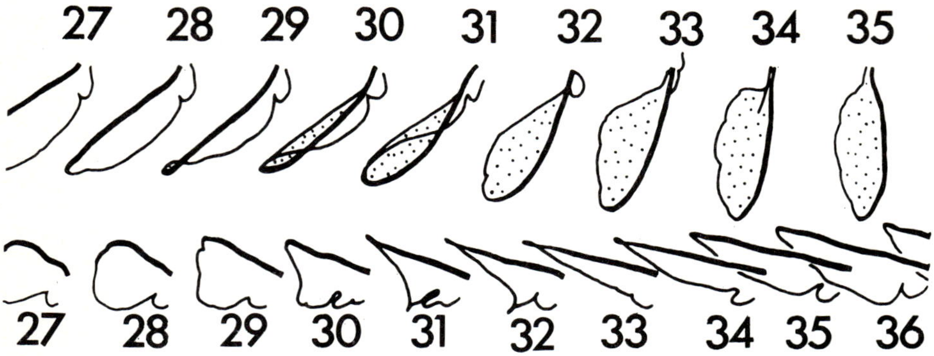

Figure 6 - Lower turning phase and beginning of upstroke, seen in direction of mean stroke plane (top) and perpendicular to it (bottom). Dotted area: morphological lower side is visible.

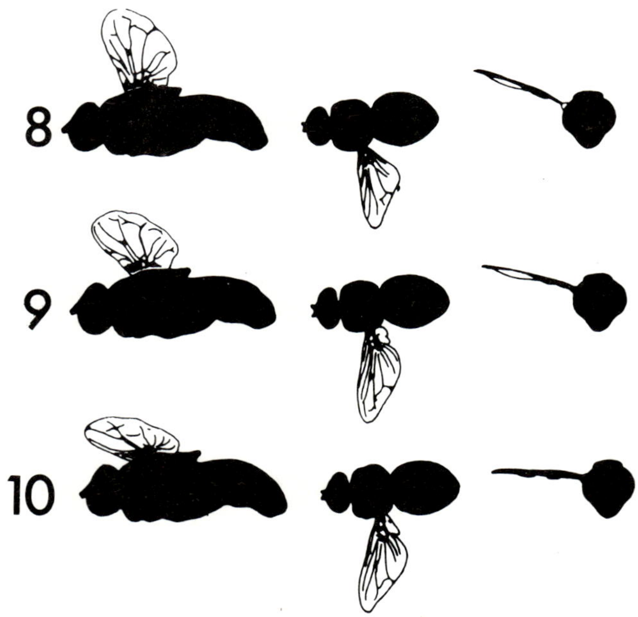

Figure 7 - Part of middle of downstroke, seen from 3 body axes. Time difference between two phases: 1/3200 sec.

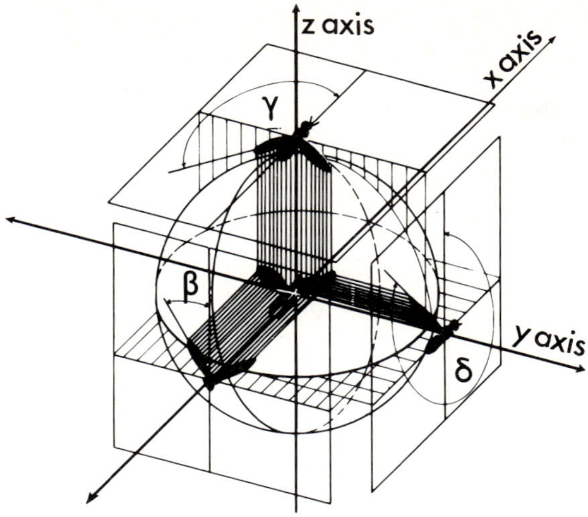

Figure 8- System of projection achieved with the arrangement of fig.5 (see text).

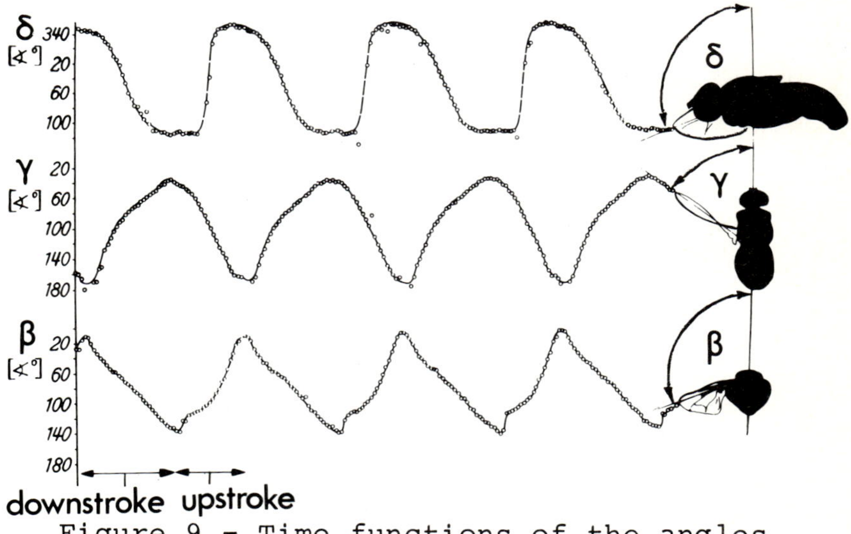

Figure 9 - Time functions of the angles of fig.8.

6. Wing-tip movements during a stroke period

Since a wing tip must move in a global fashion relative to the wing joint movements were reconstructed on a slate-globe (fig. 10) and then unfolded into a plane (fig.11). It can be seen that downstroke occurs more or less far ahead of upstroke. Relationships and transfer methods are described by Nachtigall (8).

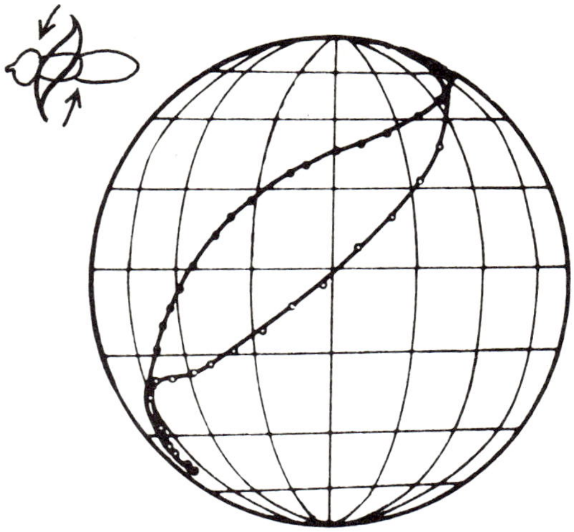

Figure 10 - Path of wing tip on a globe surface, the radius of which equals wing length. Head of the fly is left. Left: downstroke, right: upstroke.

7. Changes in angles of attack during a stroke period

The previously mentioned techniques permitted the wings to be shown in cross-section at any time during the beating cycle. Taking into account that the fly moves through the air (translation speed of the body equals wind speed from the tunnel) while its wings are moving relative to the body axes (fig.11) one can construct the wing-tip path relative to the surrounding air, i.e. relative to a "geocline" system of axes by adding the vector of translation. Therefore, the closed wing path of fig.11 is extended to a zig-zag like path in fig. 12. It can be seen that after a downstroke which is directed downwards and forward, the first two thirds of the upstroke are directed upwards and back, then the path bends upwards and forward. This means that during part of the upstroke, the wings move quicker backwards than the insect's body moves forward. The consequences of this behaviour are showns in parts 8 and 10.

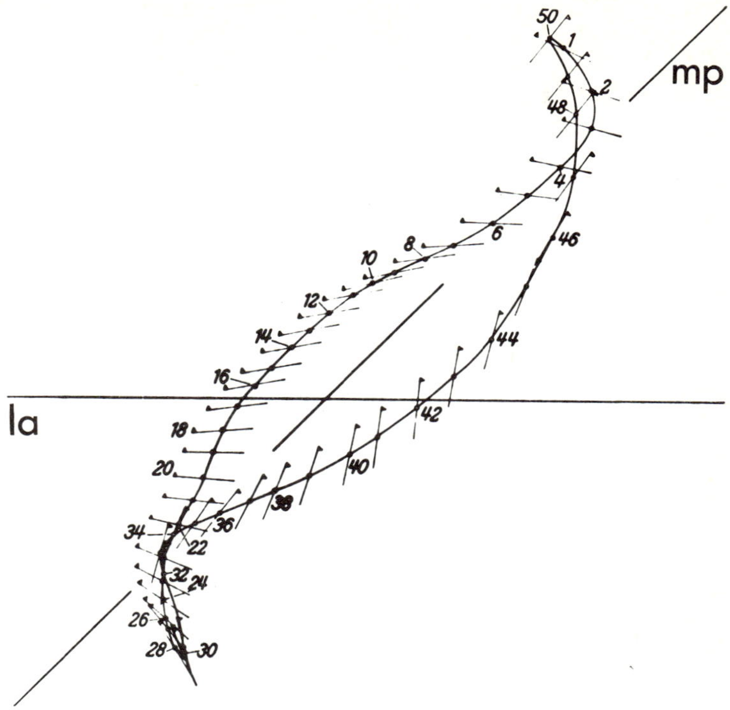

Figure 11 - Unfolding of fig.10.Numbers:film frames; time distance 1/6400 sec. 1a longitudinal axis (head left side), mp mean wing beat plane. The small triangles are located on the morphological upper side of the leading edge of the wing. Compare fig. 10.

A corrective step in the developing system has been omitted so that figs. 10 and 11 do not truly represent the projection used (Darnhoferdemar, personal communication). This deviance does not appear in the middle of a down- and upstroke, but distorts the ends. End-loops are therefore not so sharply defined and bending of the loops is not so strong. Since only the central parts of the strokes are relevant in the generation of thrust and lift forces and new experiments are in preparation,these 1966 figures have not been redrawn.

Figure 11 shows that there is a change in the

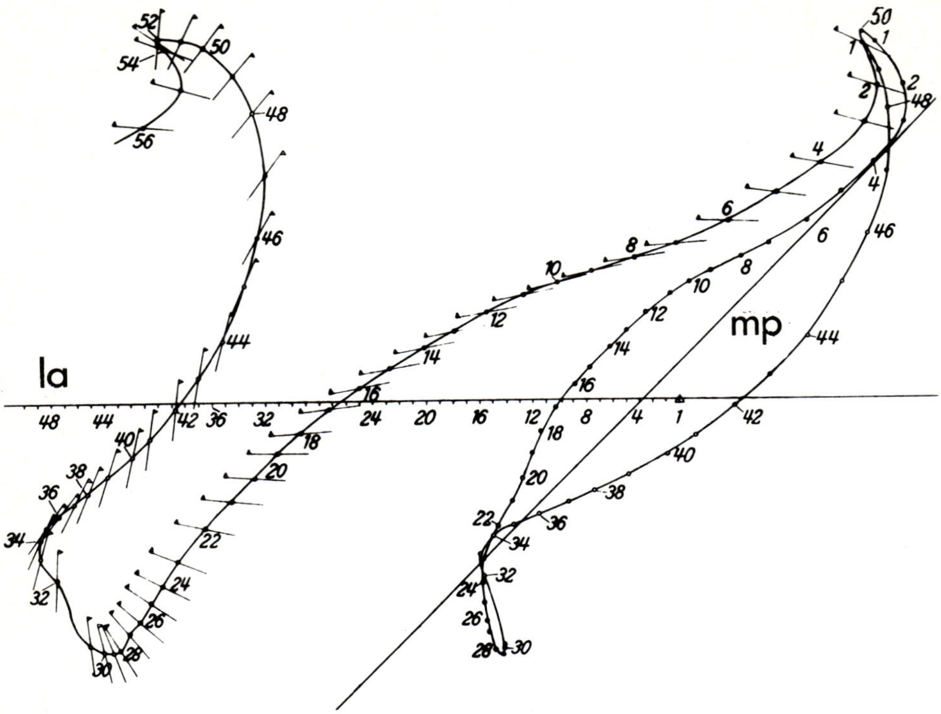

Figure 12 - Wing path in space constructed from fig.11 by adding translation. Compare fig.11.

angle of attack between wing cross-section (the outer third of the wing is used) and wing path as related to the insect's system of axes. At upper and lower reversal points there is a quick change of angle, i.e. a quick rotation of the wing, without or almost without any beating. This angle has been called the "geometrical angle of attack α_{geom}"; for exact definition see (8) fig. 15b.

The flight behaviour and wing-tip paths of a fly flown under wind tunnel ("quasi-natural") conditions differ greatly from those of a fly fixed to a simple rod without wind. In the latter case the "figure of 8 path" of the wing tip, as often mentioned in older textbooks, is usually seen (unphysiological conditions). Preliminary force constructions show that a fly cannot fly through the air with this coordination. Astonishingly, changes of

α_{geom} during a stroke period are quite similar in quasinatural and unphysiological conditions (fig. 13). This suggests that wing rotation is a process which is enforced by the combined movements of the wing joint and the surrounding wing sclerites, and that it is little influenced by aerodynamic forces. This point may be of particular interest for further analyses of joint kinematics.

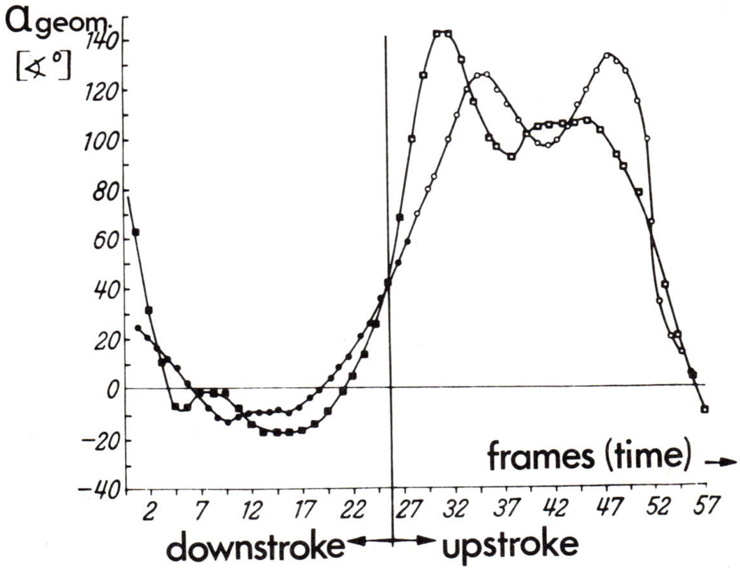

Figure 13 - Changes of α_{geom} as function of time during a stroke period. Circles: quasi-natural flight, squares: unphysiological conditions without wind.

Adding the vector of translation changes the angle α_{geom} to the "aerodynamic angle of attack α_{ae}, which is the angle between the more or less flat outer third (see part 13) of the wing and the direction in which the wing beats against the air or in which the air hits the wing.

In the upper turning phase (fram n.1 in fig. 12) α_{ae} is extremely high. The morphologically lower side of the wing is blown against (not dotted in fig.6).During the downstroke,the angle gradually decreases (frame 9) to the aerodynamically useful angles of 20-30° and further to 50°. There is certainly no stalling up to 20-50° in this range of

Reynolds numbers, (Re $\sim 2 \cdot 10^2 \div 10^3$, related to maximal wing chord), furthermore, these high angles are present for only a millisecond or so. In the lower turning phase there is a quick change in direction. During the first two thirds of upstroke (frame 35-45) the wing is blown against the morphologically upper side (dotted area in fig.6) at angles of up to 30° . Then comes a region of high angles when the wing is again blown against, on the morphologically lower side, and the wing rotates to a new downstroke in the upper turning phase (up to frame 52). Time function of the angle of attack is demonstrated in fig. 14.

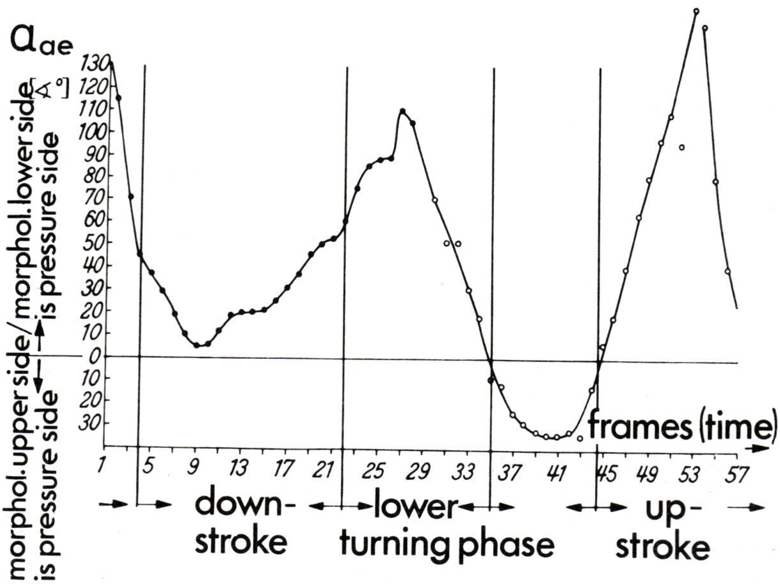

Figure 14 - Aerodynamical angle of attack α_{ae} as function of time during a stroke period. See text.

8. Preliminary assumptions on the generation of aerodynamic forces during a stroke period.

Discussion of the wing path and the aerodynamic angle of attack gives some preliminary ideas on how a fly's wing may generate aerodynamically essential force components. Two easily analyzed wing positions are discussed: wings fully spread in the midst of downstroke (frame 16, figs. 12-15) and in the midst of upstroke (frame 43, figs. 12-15). Downstroke: small α_{ae}-values give small drag d in direction of wing path and 2-3 times higher side forces perpendicular to the direction of movement. The resulting air force ae can be divided into a lift component 1 and a thrust component t, showing that the animal is lifted and moved forward during downstroke.Upstroke:wing moves upward and backward under small α_{ae}-values. (It does not matter that the morphologically upper side is now the pressure side, because the wing is not cambered). Forces d and s form an ae which is so located that it is divided into a big thrust component t and a small lift component l. During part of the upstroke due to the wing kinematics described, the fly is thrust forward and at the same time slightly raised. Since the velocities are approximately the same in frame 16 and 43 one can say: middle downstroke is the mean lifting phase, the first third of the upstroke is the mean thrusting phase. As far as force components can show, upper and lower turning phases do not generate useful aerodynamic forces, partly inducing small drag components and small negative lift components.

It is very important to measure accurately the lift and thrust components phasically, i.e. measuring the actual values from point to point during a wing beat cycle. This has been often attempted, but without success. One always measures many artefacts, e.g. mass forces due to the movements of the thorax system or oscillation characteristics of the transducer rods. With up to date equipment, it is possible to measure the phasic aerodynamic force components of larger, heavier insects. Zarnack (17) in my research group has succeeded in measuring the phasic lift in locusts. Details will be shown elsewhere.

Since a very quick change occurs in the angle of attack over a large field during the useful parts of down-and upstroke (mean values of $\Delta\alpha_{ae} \approx \frac{30°}{1\ \text{msec}}$) it can be assumed, that non-stationary aerodynamics play an important role in force generation, allowing for much higher lift – forces

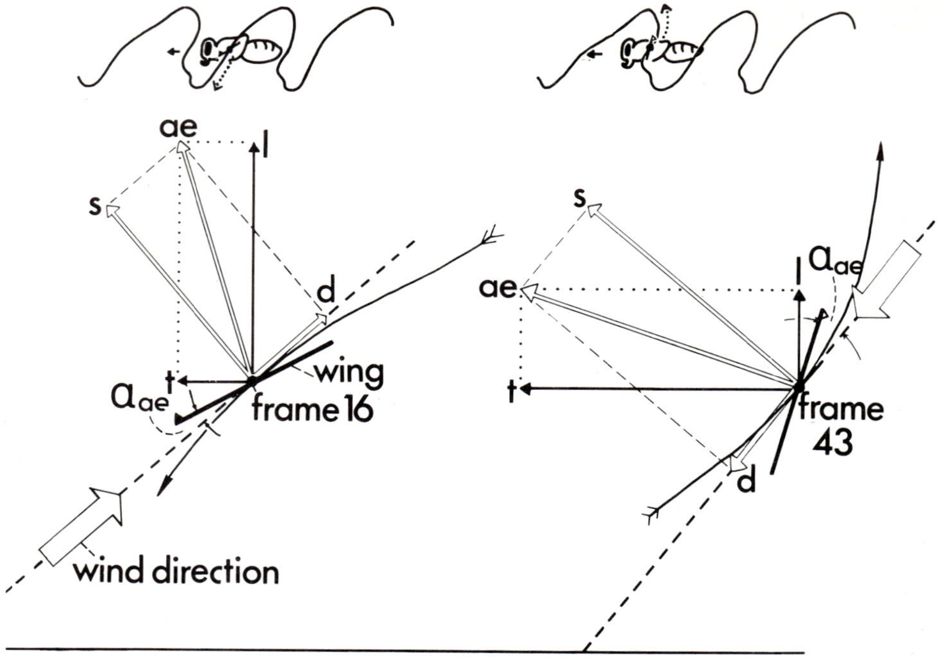

Figure 15 - Aerodynamic force constructions in the midst of downstroke (frame 16) and upstroke (frame 43). Compare fig.12 and the text.

than in conventional quasi steady-state conditions and preventing stall at high angles of attack, at which in steady flow stall usually occurs (14). Bennett (1), studying a big Melolontha wing model, pointed out that this might well be the case in natural wings, too. This conclusion is enforced by Wood's findings (16) that there is no stall in any wing position under quasi natural conditions of flight.

9. Wood's calculations of wing movements and air forces

Wood (16) working with Calliphora erythrocephala and Phormia regina in a wind tunnel, filmed the flies and measured the velocity distribution in the wake of the wings using the hot wire method. He found that changes in wing twisting and wing contours were qualitatively similar to my findings. His hot wire studies have shown that there is a low turbulence in quasi natural flight (wind tunnel), but very high turbulence, indicating stalling (wings no longer function as lift generators), when the insect is flying on a rod in still air. He concludes that α_{ae} especially during downstroke may be too high in the latter case. This is consistent with my kinematic analyses (compare Nachtigall (8), fig.33). Furthermore, he has shown that there is a high peak of the positive, vertical-velocity-component, presumably in the middle of the downstroke, and a smaller peak in the first part of the upstroke. This, too, is fully consistent with my kinematic constructions (fig.12) and my assumptions on aerodynamic force generation (fig.15). The latter statement - lift during upstroke - has been verified by the fact that paths are open and that the upstroke wing tip path lies posterior to the downstroke path (fig.11). Wood too, observed that open paths occurred much more frequently than figure of 8 paths (compare fig.11). His calculations however, have shown that the upstroke path lies anterior to downstroke path. I fixed the insect to the balance at the extreme tip of its abdomen, whereas Wood waxed the mesonotum to a fixed rod. To show whether the mode of fixation influenced wing path kinematics, studies of the wing movements of *Calliphora erythrocephala* in free flight were made (9,10).

10. Wing movements in free straight ahead flight

A glass flight cage 40 x 40 x 200 cm was used. Flies flying from one end to the other against a light source were filmed from above and from the side with the aid of two mirrors m_1 and m_2 using a Strobodrum-camera sc in stroboscobic light (fig.16). 1000 flashes per second gave 60 frames on a film strip of 150 cm length, mounted on the outside of the drum. Negatives were enlarged 1:1 and wing tip positions were measured. Although the time resolution had to be small with respect to the short length of the film, the side views under free flight conditions (fig.17) have shown that, in most cases, upstroke lies posterior to downstroke. This suggests that mounting flies at the tip of the ab-

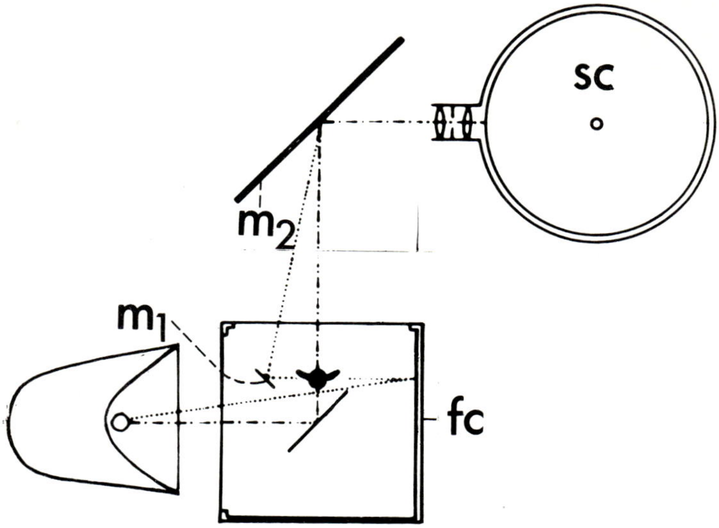

Figure 16 - Arrangement for filming flies in free flight from two directions with a drum camera and a stroboscope. (See text)

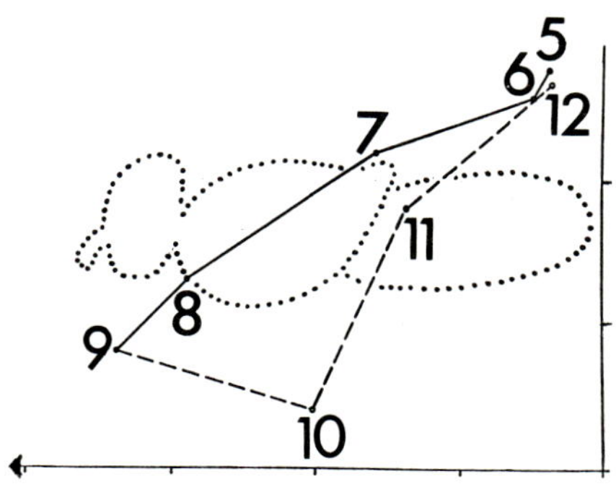

Figure 17 - Wing tip path of Calliphora in free flight, seen from the side. Frames 5 - 9 downstroke, frames 9 - 12 upstroke.

domen probably influences least of all the beating characteristics of their wings. Figure 18 shows the wing tip plot of an almost horizontal and steady free-flying *Calliphora erythrocephala* on a horizontal plane. Downstroke is represented by solid lines,

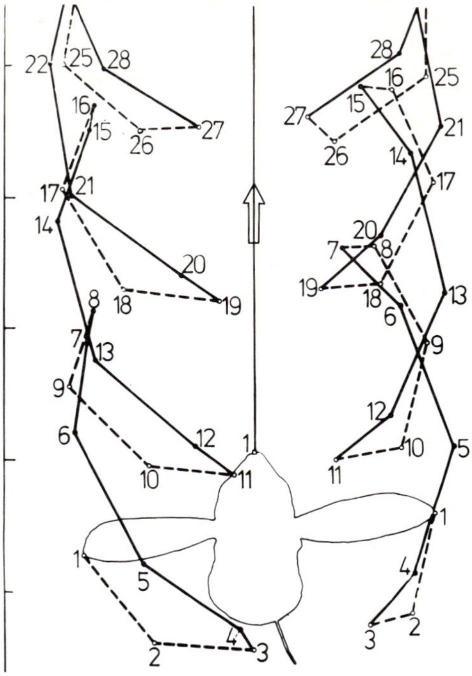

Figure 18 - Wing-tip path of Calliphora in free flight, seen from top. Full lines: downstroke; dashed lines: upstroke. Compare text.

upstroke by dotted lines. The backward direction of upstroke, which is the basis for the thrust-lift generation of the first one third of upstroke,is clearly seen.Note the mirror-similarity of right and left wings in this flight condition. More details will be published elsewhere.

11. Wing movements in free crabbing flight

Figure 19 shows a newly revealed flight behaviour in Calliphora.To get from point 13 to point 56 it need not fly with the body longitudinal axis parallel to line 13-56, but will do so by keeping the same angle between the body longitudinal axis and line 13-56: crabbing flight. Note strong asymmetries between right and lift wing-tip path.This probably means that neuromotor control of the two sets of left and right direct wing base muscles must be such that kinematic wing patterns are comple-

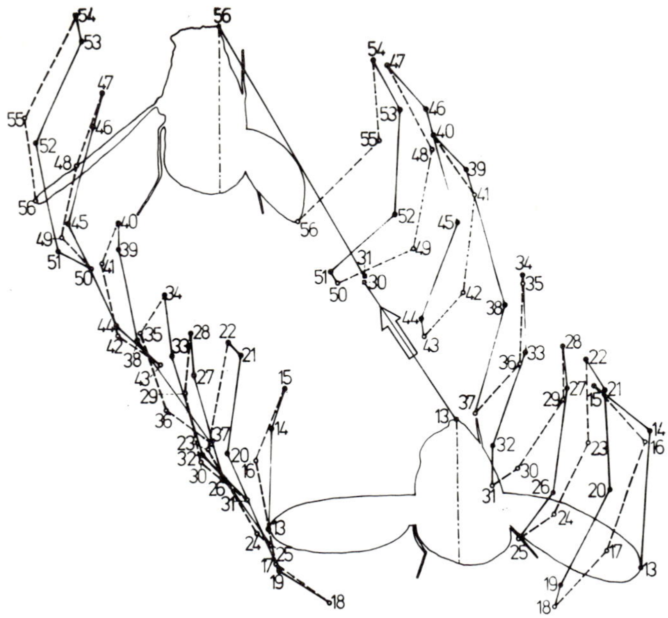

Figure 19 - Crabbing flight of a Calliphora. Head moves on a straight line from point 13 to point 56. Compare the text.

tely different in both wings, although both are beating with the same mean velocity and the body neither lifts nor falls, nor does it roll, yaw or gear. The result of these certainly complicated movements must be the generation of a simple side force without a change in mean thrust and mean lift. Details have still to be worked out.

12. Wing bending in quick acceleration

When a fly is disturbed in normal free flight it can quickly accelerate in any direction to which it turns its body axis. Figure 20 shows how this is accomplished. There is a preformed bending element in the wing (compare Ritter (13)). In extremely quick upstrokes, the wing can be bent passively by drag forces around this element, so that it works like an oar blade perpendicular to the airflow, generating strong thrust forces only.

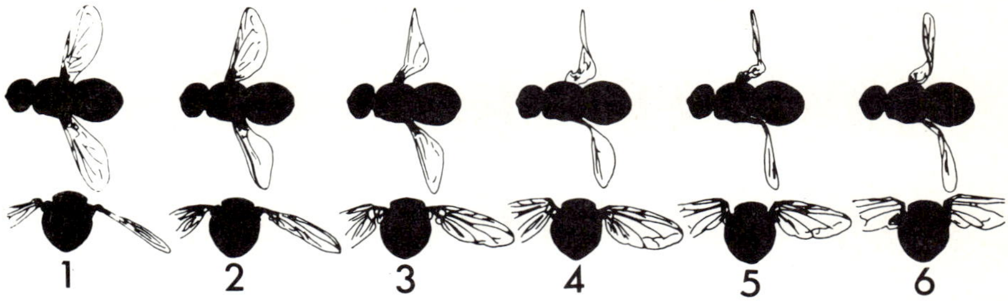

Figure 20 - Quick acceleration by wing bending. Seen from top and from rear side. Time distance between two phases approximately 1/3300 sec.

13. Final remarks

Many beating and steering possibilities are still unknown, thus offering an interesting research project for high speed filming with subsequent correlation to muscle activity and hinge mechanics on the one side and to aerodynamics on the other side. Although some advance has recently been made towards understanding the neuromuscular function of the steering system in "natural flight" (12, 3,4) and although some older, fine papers on the functional morphology of the Musca wing joint (7) and joint movements (2) exist, the bulk of the work still remains to be done. The investigations discussed here present a basic understanding of how the wings of a fly are set in motion and how they function. It must be said, however, that our knowledge is no more than a crude approach to the truth as shown, for example, by the fact that one assumed the outer one third of the wing to be a flat plate throughout the beat, which is only partly true. Kinematics are described in a very simplified manner and direct aerodynamic measurements were unsuccessful. Our methods have proved to be satisfactory as well as the experimental set up which provided the physiological flight conditions. It is therefore hoped that these experiments provide a good basis for further research on, and lead to a better understanding of, how a fly flies.

REFERENCES

1. Bennet,L. (1970) Insect flight: lift and rate of change of incidence. Science 167, 177-179.
2. Boettiger,E.G. & Furshpan,E. (1952) The mechanics of flight movement in diptera. Biol.Bull. 102, 200-211.
3. Heide,G. (1968) Flugsteuerung durch nicht-fibrilläre Flugmuskeln bei der Schmeissfliege *Calliphora*. Z.vergl.Physiol. 59, 456-460.
4. Heide,G. (1971) Die Funktion der nicht-fibrillä re Flugmuskeln von *Calliphora*. Teil II. Muskuläre Mechanismen der Flugsteuerung und ihre nervöse Kontrolle. Zool.Jb. Physiol. 76,99-137.
5. Hollick,F.S.J. (1940) The flight of the dipterous fly *Muscina stabulans* Fallen. Phil.Trans. Roy.Soc. B 230, 357-390.
6. Holst, E.v. & Küchemann,D. (1941) Biologische und aerodynamische Probleme des Tierflugs. Naturwiss. 29, 348-363.
7. Mihalyi,F. (1935/36) Untersuchungen über Anatomie und Mechanik der Flugorgane an der Stubenfliege. Arb.ungar.biol.Forsch.Inst.Tihany 8, 106-119.
8. Nachtigall,W. (1966) Die Kinematik der Schlagflügelbewegungen von Dipteren. Methodische und analytische Grundlagen zur Biophysik des Insektenflugs. Z.vergl.Physiol. 52, 155-211.
9. Nachtigall,W. (1969) *Gläserne Schwingen*. Moos, München. English edition in press.
10. Nachtigall,W. Flügelbewegung von *Calliphora* im freien Flug. - In press.
11. Nachtigall,W. & Weigold,H.P. Schiebeflug bei *Calliphora*. - In press.
12. Nachtigall,W., Wilson,D.M. (1967) Neuro-muscular control of dipteran flight. J.Exp.Biol. 47, 77-97.
13. Ritter,W. (1911) The flying apparatus of the blow-fly. Smithson.misc.Collns. 56 (12), 1-79.

14. Silverstein,A. & Joyner,V.T. (1939) Experimental verification of the theory of oscillating airfoils. NACA techn.Rep. 673, 619-623.

15. Walker,G.T. (1927) The flapping flight of birds. II. Jl.R.Aeornaut.Soc. 31, 337-342.

16. Wood,J. (1970) A study of the instantaneous air velocities in a plane behind the wings of certain diptera flying in a wind tunnel. J.Exp. Biol. 52, 17-25.

17. Zarnack,W. (1972) Flugbiophysik der Wanderheuschrecke (*Locusta migratoria* L.) I. Die Bewegungen der Vorderflügel. Z.vergl.Physiol. 78, 356-395.

Comparative Physiology, eds. L. Bolis, K. Schmidt-Nielsen and S. H. P. Maddrell

BODY DRAG AND EFFICIENCY OF SWIMMING IN MAN [1]

P.E. DI PRAMPERO [2] & D.W. RENNIE
Department of Physiology,
School of Medicine,
State University of New York at Buffalo,
Buffalo, New York, U.S.A.

SUMMARY

In order to determine the actual drag experienced by a swimmer, known drag loads have been added to (or subtracted from) swimmers moving at constant speed. This was obtained by means of a system of pulleys, fixed to a platform moving with the swimmer, which transformed the vertical pull of a weight into a force acting horizontally along the direction of movement. The steady state oxygen consumption was simultaneously measured. It has been found that, at constant speed, the swimmer's energy expenditure increases linearly with the added drag: the extrapolation of this function to resting oxygen consumption is assumed to yield the drag experienced by an unloaded swimmer. This amounts to 3.3 ± .9 (S.D.) kg at .55 m/s and to 7.2 ± 2.9 (S.D.) kg at .90 m/s, i.e. about twice the values observed on passively towed subjects. These values are reduced by about 10 to 20% when

---

(1 - These studies were aided by Contract N00014-68-A-0216, (NR 101-722), between the Office of Naval Research, Department of Navy, and the State University of New York at Buffalo.

(2 - On leave of absence from the Centro di Studio sulla Fisiologia del Lavoro Muscolare del C.N.R. (Italian National Research Council) c/o Istituto di Fisiologia Umana dell'Università di Milano. P.E.di Prampero is a Dr.Henry C. and Bertha H. Buswell Fellow.

considering that part of the oxygen consumption above basal during exercise is due to processes not directly related to progression. From the so obtained drag values and the simultaneously measured energy expenditure, the speed being known, the swimming efficiency can be calculated: it amounts to 2.7 $\pm$ .7 (S.D.) % at .55 m/s and to 6.0 $\pm$ 2.5 (S.D.) % at .90 m/s.

Introduction

During constant speed swimming in man the speed oscillations at each stroke are negligible (3,5) so that the average force exerted by the swimmer (thrust) is equal and opposite to the water resistance at the speed considered (drag). Hence the mechanical power output ($\dot{W}$) is given by the product of drag (D) times velocity of progression (v):

$$\dot{W} = D\ v \qquad 1)$$

Studies on the energetics of swimming in man have always been restricted by the absence of data on the actual drag experienced by the swimmer. This has generally been assumed to be equal to that measured on passively towed subjects (1,7,8). From the assumed values of drag and the simultaneously measured energy expenditure, if the speed is known, the efficiency can be calculated with the aid of equation 1). This procedure however is likely to yield minimal values of efficiency as, presumably, the swimmer, because of head, limbs, trunk movements, changes in buoyancy due to respiratory cycles, etc., experiences a drag higher than that observed on passively towed subjects.

The present report deals with an attempt to determine the actual drag of a swimmer, by adding (or subtracting) extra drag loads to (or from) swimmers moving at known predetermined speeds. The extra drag can be easily measured and related to the swimmer's energy expenditure in order to calculate the drag experienced by an unloaded swimmer. A similar approach has been applied on rainbow trout by P.W. Webb (11).

Methods and experimental procedure

The experiments have been performed on 10 male subjects,competitive swimmers of medium class in good training conditions (age 15 -28 years, weight 66 - 114 kg, height 165-188 cm, surface area 1.73-2.40 m^2).

The experiments were performed in a ring shaped swimming pool (depth 2.8 m, width 2.8 m, circumference at center of ring 58.6 m). The swimmer was paced by a platform moving at constant speed 50 cm above the level of the water. From the 4th to the 6th minute of exercise, when a steady state was attained, the expired air was collected in Douglas bags placed on the platform. From its volume and the O_2 and CO_2 concentration the steady state O_2 consumption was calculated. The energy expenditure of the exercise was never such as to exceed the subjects' maximal O_2 consumption.

Known weights (from a few hundred grams to 1.5 kg) were connected to the subject via a system of pulleys fixed to the platform which transformed the vertical pull into a force acting horizontally along the direction of movement. Since the platform was moving at the same speed as the swimmer, by positioning the weights behind or ahead of the swimmer, the force could be applied in either sense, fig.1, thus hindering, or helping, the swimmer's own activity. The weight applied to the swimmer can then be regarded as a known equivalent of drag added to (positive added drag, $+D_A$) or subtracted from (negative added drag, $-D_A$) the drag of the body.

The force applied to the subject was never such as to interfere with the swimming mechanics, as judged by an experienced observer walking along the edge of the pool.

Resting O_2 consumption was measured as the subject was lying motionless in water, supported by a harness. In following trials the energy expenditure, as given by O_2 consumption, was measured during constant speed swimming, with and without added drags. The order in which the various drags were applied was varied according to a table of random numbers. Different speeds were studied in successive days, also according to a randomized procedure.

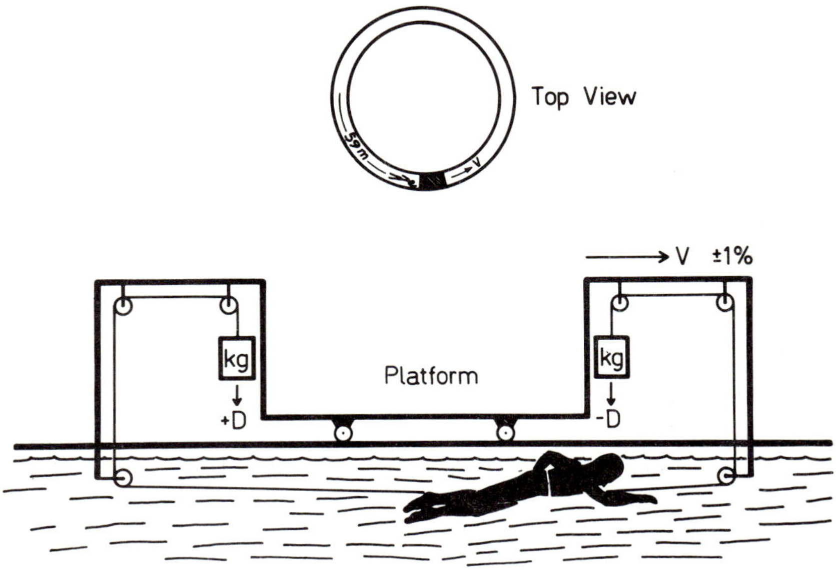

Figure 1 - Experimental set-up. The swimmer is paced by a platform which is moving at constant speed about 50 cm above the water surface. A Douglas bag placed on the platform allows expired air collection (not shown in figure). The propulsive force exerted by the subject during swimming can be increased (respectively decreased) by connecting the swimmer, via a system of pulleys fixed to the platform, to the known weight +D (respectively -D). The force +D (respectively -D) which acts horizontally, i.e. along the direction of progression, is therefore equivalent to an increase (respectively decrease) of the water resistance by the corresponding amount; it will be referred to as added drag.

Only two speeds were systematically investigated: .55 and .90 m/s.

The drag due to the face mask worn by the subject and the breathing apparatus (mouthpiece, valve system and hose) was determined by towing the subject in water with and without equipment. From this procedure the drag of the passively towed subject (passive drag, D_p), was also obtained.

In this study main attention was given to the crawl stroke.

Results and discussion

In figures 2 and 3 the O_2 consumption above resting (net $\dot{V}_{O_2}$) has been plotted for two subjects as a function of the added drag (D_A). The relationship between these two variables is linear, being displaced at a higher level the higher the speed.

In table 1 the average relationship between net $\dot{V}_{O_2}$ and D_A has been summarized for all subjects at the two investigated speeds. Speed of progression; overall number of observations (n); slope ± S.D., correlation coefficient for a straight line regression (r ± S.D.); net O_2 consumption at zero D_A, $\dot{V}_{O_2}^{fs}$ (i.e. net $\dot{V}_{O_2}$ when swimming naturally, "free swimming") are reported in table 1, together with swimming efficiency (η) as calculated from equation (4; extrapolated $-D_A$ at zero net $\dot{V}_{O_2}$ (D_b), and passive drag (D_p).

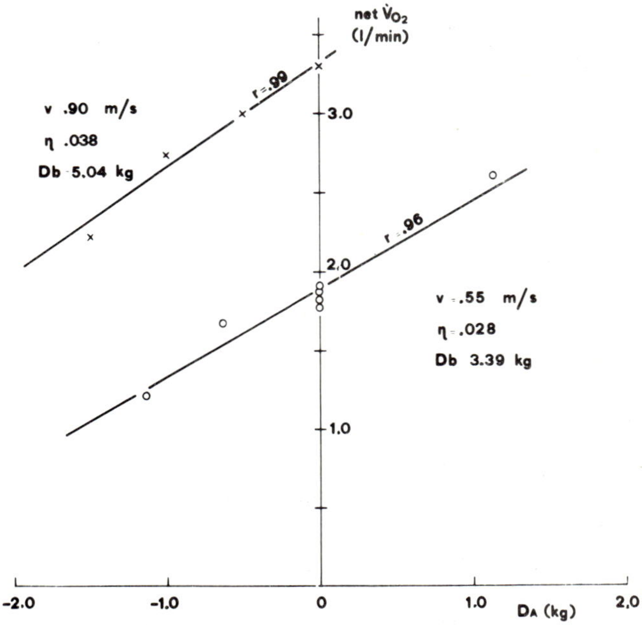

Figure 2

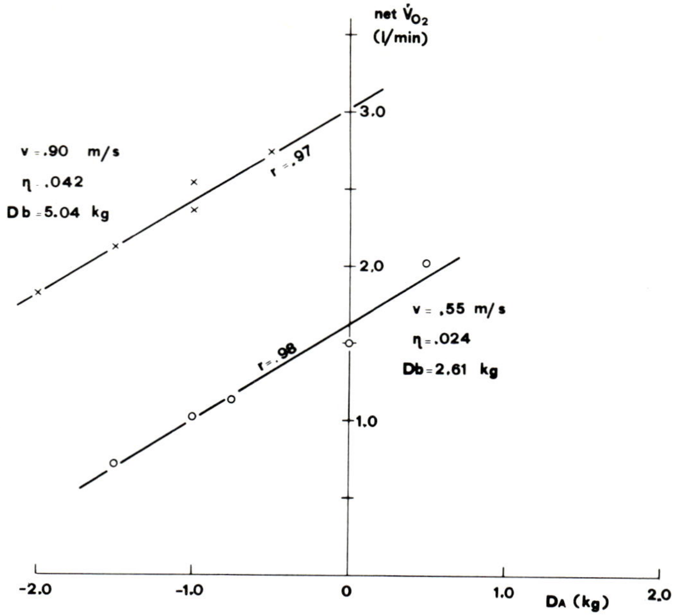

Figure 3

Figures 2 & 3. - Oxygen consumption above resting (net $\dot{V}_{O_2}$, 1/min) as a function of the added drag (D_A, kg) in two subjects. Swimming velocity .55 (o) and .90 (x) m/s.

Body drag

The relationship between net $\dot{V}_{O2}$ and D_A can be considered linear within the investigated range (figs. 2 and 3). On the assumption that this holds also beyond these limits, the extrapolation of this regression to net $\dot{V}_{O2} = 0$ is a measure of the force which, if applied to the swimmer in the direction of movement, in the forward sense ($-D_A$), would allow him to move at the given speed without any extra O_2 consumption above resting. This may then be assumed to indicate the average thrust exerted by the swimmer at that speed. Moreover, since the speed is constant, it should also be equal (and opposite) to the water resistance to progression (drag). This has been corrected for the drag due to the face mask and breathing equipment and is given in fig.4. In all cases it is sensibly higher than the passive drag obtained directly on the same subjects, D_p (table 1 and fig.4). It is also higher than the values calculated from Karpovich's formula (6) and by Alley (2) and Counsilman (4), whose values at the velocity investigated in this study are very close to Karpovich's data.

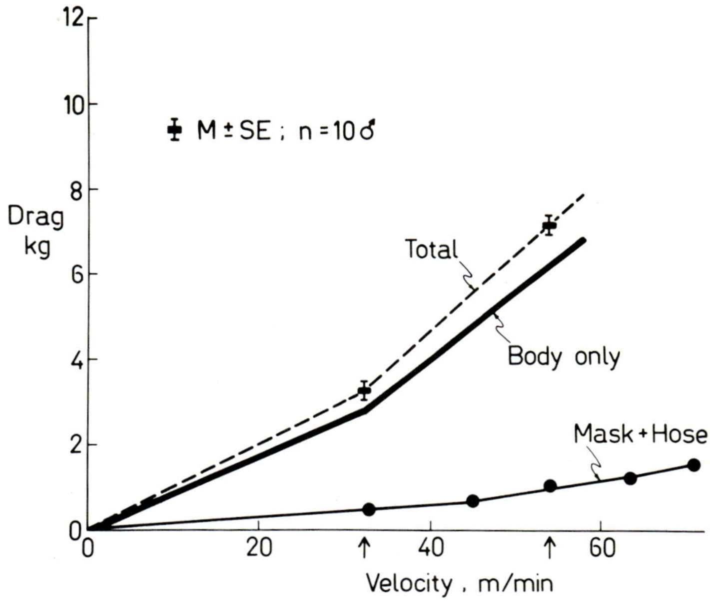

Figure 4 - Total drag (kg) as a function of the speed of progression (m/min). Once this is corrected for the drag due to mask and hose (lower thin line) the body drag only is obtained (thick continuous line). Mean ± standard error.

The difference between D_b, once corrected for the drag due to mask and breathing equipment, and D_p (see fig.4) may depend upon the O_2 consumption value selected as reference for the extrapolation procedure. Since this was the resting O_2 consumption (subject lying in water), the following assumptions were implicitly made:

a) that energetic requirements for maintenance metabolism are unaffected by muscular activity, and

b) that energy expenditure not directly useful for progression is nil.

Table 1

v (m/s)	n	slope ($1 min^{-1} kg^{-1}$)	r	$\dot{V}_{O_2}^{fs}$ ($1 \ min^{-1}$)	η %	D_b (kg)	D_p (kg)
.55	90	.61 ± .17	.89 ± .03	1.92 ± .46	2.7 ± .7	3.33 ± .93	1.73 ± .51
.90	53	.49 ± .17	.90 ± .12	3.12 ± .59	6.0 ± 2.5	7.16 ± 2.9	3.68 ± .51

Relationship between O_2 consumption above basal and added drag for all ten subjects at the two investigated speeds. Speed of progression (v); overall number of observations (n); mean slope (average of the slopes of the single experimental regression on all subjects); correlation coefficient for a straight line regression (r) (average of the single correlation coefficients of all regression on all subjects); net oxygen consumption at zero D_A ($\dot{V}_{O_2}^{fs}$); swimming efficiency (η); extrapolated $-D_A$ at zero net $\dot{V}_{O_2}$ (D_b); and passive drag (D_p).
Mean values ± S.D.

The influence of muscular activity on maintenance O_2 consumption is difficult to assess. However, because of the large increase in overall O_2 consumption due to exercise (up to 8-10 times), the assumption of a constant maintenance metabolism can represent only a minor source of error.

On the other hand some energy is presumably wasted in processes not directly related to propulsion such as:
i) the O_2 consumption of the respiratory muscles and of the heart;
ii) the work done against gravity alternately lifting and lowering the limbs and
iii) the work done in accelerating and decelerating the limbs at each stroke.

The term under i) increases with the overall O_2 consumption: it should not therefore significantly affect the intercept on the abscissa of the $\dot{V}_{O2}/D_A$ relationship, although it may affect its slope.

Since the specific gravity of the body is not far from 1.0, as a first approximation the work done against gravity (ii) can be considered limited to that part of the stroke during which the arms are lifted above the water surface; it should then be of minor importance since the overall center of gravity of the arm is moved by only a few centimeters in the vertical direction at each stroke.

The energy spent in accelerating and decelerating the limbs (iii) is presumably important only when the movement is rapidly changing direction, as in the case of the crawl kick. In these experiments however, the speed being relatively low, this component is presumably of minor importance.

Furthermore it can be calculated from the data of table 1 that, in order to account for the difference between extrapolated (D_b) and passive (D_p) drag, on the basis of a linear regression, non-propulsive $\dot{V}_{O2}$ should amount to about 50% of the free swimming O_2 consumption ($\dot{V}_{O2}^{fs}$). This value seems unrealistic, thus supporting the view that the drag experienced in actual swimming is higher than that measured on passively towed subjects.

As a first approximation non-propulsive $\dot{V}_{O2}$ can be assumed to be of the order of 10-20% of

$\dot{V}_{O_2}^{fs}$. In fact, when the negative added drag ($-D_A$) is increased to very high values, the subjects' O_2 consumption does not decrease below 10-20% of $\dot{V}_{O_2}^{fs}$, the $\dot{V}_{O_2}/D_A$ relationship becoming then a flat line (unpublished observation). If it is assumed therefore that this level of net $\dot{V}_{O_2}$ represents the non-propulsive energy expenditure and that this, in percentage of the net $\dot{V}_{O_2}^{fs}$, is not affected by the speed, the $\dot{V}_{O_2}/D_A$ regression should be extrapolated to this value of O_2 consumption in order to obtain a measure of the actual thrust exerted by the swimmer. The drag actually experienced during swimming should then be in all cases about .8-.9 D_b (i.e. about 1.5 times the passive drag values).

Swimming efficiency

When the subject's body drag is increased by known amounts, as done experimentally in this study, at constant speed, the extra mechanical power performed by the swimmer ($\Delta\dot{W}$) is given by the product of the extra added drag (ΔD_A), times the velocity of progression (v):

$$\Delta\dot{W} = \Delta D_A \quad v \qquad (2$$

The extra O_2 consumption ($\Delta\dot{V}_{O_2}$) required to increase the total power output by the amount W is given by:

$$\Delta\dot{V}_{O_2} = \dot{W}/\eta^p \qquad (3$$

where $\eta^p = \Delta\dot{W}/\Delta\dot{V}_{O_2}$ is the partial mechanical efficiency of this type of work. From (3 and (4:

$$\eta^p = \Delta D_A \ v/\Delta\dot{V}_{O_2} \qquad (4$$

Hence, at any given speed, if the partial efficiency of the exercise is independent of the added drag, the V_{O_2}/D_A relationship should be a straight line, as experimentally found; table 1 and fig.2 and 3.

Moreover, since the partial efficiency η^p is independent of the absolute drag value, it should be also equal to the overall efficiency of swimming under natural conditions, i.e. $\eta^p = \eta$. If the speed is constant and known, η can then be calculated from the slope of regressions such as those reported in figs. 2 and 3, with the aid of equation (4. The average values observed at the two speeds for all subjects are reported in table 1. The swimming efficiency increases with increasing speed from a percentage value of 2.7 $\pm$ .22 at .55 m/s to 6.0 $\pm$ .78 at .90 m/s. No indication that a maximal value has been reached can be derived from these data.

As previously indicated, the slope of the V_{O2}/DA relationship, which at constant speed is a function of η, is presumably increased by the O_2 consumption of the respiratory muscles and of the heart, this being roughly proportional to the overall $\dot{V}_{O2}$. Furthermore also the stroking frequency increases slightly with D_A, this leading to an increase in O_2 consumption due only to the increase in the speed of arm motion, independently of an increase in thrust.

It follows that the efficiency values, as measured by this method,are indicative of the "whole body" efficiency. The muscle efficiency itself is presumably higher, although only to a minor extent, since in all cases the O_2 consumption of the respiratory muscles and of the heart represents only a small fraction of the total (about 10% at the highest exercise level) and the increase in frequency is of the order of 10% over the whole range of D_A.

However, in order to provide information about the swimming technique, the effects of training, etc., the values of η as measured seem to be more important than a pure estimate of the muscle efficiency.

The efficiency values obtained by the present approach are about 30% higher, on the average,than those reported in the literature (1,7,8), this presumably reflecting the higher values of drag, and thus of mechanical work, obtained by the present indirect approach.

On swimming fish (trout) Webb (12) obtained efficiency values up to 20%. Closer values to the present study's (2-5%) were obtained by Prange and Schmidt-Nielsen (10) on swimming ducks.

The large variability of efficiency among different swimmers (table 1) emphasizes the fact that the actual swimming performance depends greatly on the technical skill of the individual swimmer. This is in contrast to what happens in running, where the energy expenditure per unit body weight and per unit distance is only about 5% higher in sedentary subjects than in long distance runners (9), so that the performance is practically entirely determined by the subjects' muscular power.

REFERENCES

1. Adrian,M.J., Singh,M. & Karpovich,P.V. (1966) Energy cost of leg kick, arm stroke, and whole crawl stroke. J.Appl.Physiol. 21, 1763-1766.
2. Alley,L.E. (1952) An analysis of water resistance and propulsion in swimming the crawl stroke. Res.Quart.23, 253-270.
3. Altever,R.J.G. (1958) A natographic study of various swimming strokes. Master's Thesis, Springfield College,Springfield,Massachusetts.
4. Counsilman,J.E. (1955) Forces in two types of crawl stroke. Res.Quart. 26, 127-139.
5. de Vries,H.S. (1959) Cinematographic analysis of dolphin swimming stroke. Res.Quart. 30,413-421.
6. Karpovich,P.V. (1933) Water resistance in swimming. Res.Quart. 4, 21-28.
7. Karpovich,P.V. & Pestrecov,K. (1939)Mechanical work and efficiency in swimming crawl and back stroke. Arbeitsphysiol. 10, 504-514.
8. Klissouras,V. (1968) Energy metabolism in swimming the dolphin stroke. Arbeitsphysiol. 25, 142-150.
9. Margaria,R., Cerretelli,P., Aghemo,P. & Sassi, G. (1963) Energy cost of running. J.Appl. Physiol. 18, 367-370.
10. Prange,H.D. & Schmidt-Nielsen,K. (1970) The metabolic cost of swimming in ducks. J.Exp.Biol. 53, 763-777.
11. Webb,P.W. (1971) The swimming energetics of trout. I.Thrust and power output at cruising speeds. J.Exp.Biol. 55, 489-520.

12. Webb,P.W. (1971) The swimming energetics of trout. II. Oxygen consumption and swimming efficiency. J.Exp.Biol. 55, 521-540.

Comparative Physiology, eds. L. Bolis, K. Schmidt-Nielsen and S. H. P. Maddrell

QUICK ESTIMATES OF FLIGHT FITNESS IN HOVERING ANIMALS, INCLUDING NOVEL MECHANISMS FOR LIFT PRODUCTION (Abstract).

T. WEIS-FOGH
Department of Zoology,
Cambridge, England.

Simple analytical expressions are derived for the average lift coefficient C_L, Reynolds number (Re), the aerodynamic power P_a, the moment of inertia of the wing mass $\underline{I}$, the bending moments caused by inertia (Q_i) and aerodynamic forces (Q_a), and for the dynamic efficiency η in animals which perform normal hovering with horizontally beating wings and on the assumption that steady-state aerodynamics prevails. It is shown that the majority of hovering animals, including large Lamellicorn beetles and Sphingid moths, depend mainly on normal aerofoil action. However, in some groups with wing loadings less than 10 N m^{-2} (1 kgf m^{-2}), non-steady aerodynamics must play a major role, namely in very small insects at low Reynolds number, in true Hover flies (Syrphinae), in large dragonflies (Odonata) and in many butterflies (Lepidoptera Rhopalocera).

The specific aerodynamic power ranges between 1.3 and 4.7 W N^{-1} (11 to 40 cal h^{-1} gf^{-1}) but power output does not vary systematically with size *inter alia* because the lift/drag ratio deteriorates at low Reynolds number. Comparisons between metabolic rate, aerodynamic power and dynamic efficiency show that the majority of insects require and depend upon an effective elastic system in the thorax which counteracts the bending moments caused by wing inertia.

The free flight of a *very small* Chalcid wasp *Encarsia formosa* was analysed by means of slow-motion films. At this low Reynolds number (10-20), the high calculated lift coefficient of 2 or 3 is not possible and the wasp must depend almost entirely on non-steady flow patterns. The wings of *Encarsia* move almost horizontally during hovering and there are three unusual phases in the wingstroke, a *clap*, a *fling* and a *flip*.

It is shown that the *fling* represents a novel mechanism for creating useful circulation *prior* to the wing movement itself. The fling corresponds to the normal pronation before the downstroke begins. The calculated and the observed wing velocities at which lift equals body weight are in agreement, and lift is produced almost instantaneously from the beginning of the downstroke and without any Wagner effect. The fling mechanism seems to be involved in the normal flight of butterflies and possibly of *Drosophila* and other small insects. Dimensional and other considerations show that it could be a useful mechanism also in birds and bats during take-off and in emergencies.

The *flip* mechanism is also believed to be novel but is less well understood and occurs in mid air, in contrast to the fling which depends on two opposing wings being flung open like a book. An analysis of freely flying Hover flies strongly indicates that the Syrphinae (and Odonata) depend almost entirely on the flip mechanism when hovering. Also, in this case, a transient circulation is supposed to be created *before* wing translation through the air, namely during the rapid pronation and supination which involve the stiff anterior but not the soft posterior part of the wing. However, vortices of the opposite sense must be shed and a Wagner effect present.

In some hovering insects, the wing twistings occur so rapidly that the speed of propagation of the elastic torsional wave from base to tip plays a significant role and appears to introduce beneficial effects.

Non-steady periods particularly flip effects, are present in all flapping animals and they will modify and become superimposed upon the steady-state pattern as described by the mathematical model presented here. However, the accumulated evidence indicates that the majority of hovering animals conform reasonably well with that model. Many new types of analysis are indicated in the text and are now open for future theoretical and experimental research.

Comparative Physiology, eds. L. Bolis, K. Schmidt-Nielsen and S. H. P. Maddrell

PROBLEMS OF CONTROL OF BREATHING IN FISHES

P. DEJOURS
Laboratoire de Physiologie
Respiratoire, C.N.R.S.
Strasbourg, France.

Respiratory Ecology of Aquatic Animals

The physical and chemical properties of the aquatic milieu that fish can survive in are extremely variable. The temperature of the environment can range from a little less than 0°C in arctic and antarctic waters to more than 40° C in the tropics. Some fresh waters are almost pure water; on the other hand salinity which is around 35‰ in the open ocean goes up to 41‰ in the Red Sea, and all intermediates between fresh and sea water are measurable in brackish waters.

The partial pressures of oxygen and free carbon dioxide in water (Pw_{O_2}, Pw_{CO_2}) may vary considerably. At sea level, water equilibrated with air has an O_2 tension around 155 torr; but O_2 consumption by living organisms can cause P_{O_2} to drop almost to zero. On the other hand, oxygen production by photosynthesis may raise P_{O_2} to 400 or 500 torr. Circadian variations of P_{O_2} from 10 to 500 torr have been observed between 4 a.m. and 2 p.m. on a sunny day in June in fresh water ponds in a temperate climate (17; see also 11). Partial pressures of CO_2 from near 0 to 20-30 torr may be encountered. As far as we know, circadian variations of P_{CO_2} have not been studied, but variations of pH have been recorded (see 20).

As for O_2 and CO_2 concentrations in water, many factors come into play. For O_2, the problem is relatively simple since its concentration depends on its solubility, which varies with temperature and salinity. For free CO_2 the same factors exist. But as is shown in Table 1, changes of temperature

Table 1 - Solubilities of O_2 and CO_2 in pure water at 0 and 30° C.

(This table shows that (1 CO_2 is more soluble than O_2; (2 an increase of temperature decreases the solubility of both gases, but of CO_2 more than O_2)

	α_{O_2} ml STPD/L	α_{CO_2}	$\alpha_{CO_2}/\alpha_{O_2}$
0° C	48.9	1713	35
30° C	26.1	665	25.3
30/0	0.53	0.39	

and salinity do not alter O_2 and CO_2 solubilities by the same factor. In any case, CO_2 is much more soluble in water than O_2 and consequently the CO_2 pressure in fish is always low (or, more exactly, differs little from P_{CO_2} of inspired water) (25). Furthermore, the problem of CO_2 change is more complicated than that of O_2 change, because the concentration of carbonate and bicarbonate in water is so extremely variable. In particular in sea water well equilibrated with air (i.e. with a CO_2 tension of 0.25 torr), the total CO_2 concentration is 2.03 mM, of which 0.23 mM is $CO_3^=$, 1.79 mM is HCO_3^- and a very small amount is dissolved CO_2. When an animal produces CO_2 and consequently H_2CO_3, some of the carbonic acid is buffered by the system $CO_3^=$ - HCO_3^-, so that the amount of carbonate decreases while the amount of bicarbonate increases (7). For example, an increase of 2 torr of P_{CO_2} in a non-carbonated water corresponds to an increase of 0.5 torr in sea water. The buffering value of some fresh or brackish waters may be much higher than that of sea water (20).

We conclude from this rapid survey of respiratory ecology of the aquatic milieu that no two natural waters have the same composition or undergo the same changes when submitted to a given O_2 extraction and a given CO_2 addition due to fish respiration. This contrasts with the air milieu which is remarkably homogeneous throughout the world; addition or subtraction of a given amount of CO_2 or of O_2 is accompanied by identical changes of partial pressure, the variation of barometric pressure being a relatively simple factor to deal with.

How is the breathing of fish affected by ecological variations in the environmental respiratory factors?

Technical difficulties

Data concerning the control of breathing in teleost fish is much less abundant than for air-breathers, particularly mammals. There may be several reasons for this, but certainly the technical difficulty in obtaining reliable data on fish represents an important drawback (15). First of all, anesthesia may considerably disturb breathing; in particular the most commonly used fish anesthetic, MS 222, provokes a prolonged increase of breathing. After anesthesia, as after manipulation, animals may take days to return to normal breathing.

Visual observation of respiratory movements has some great advantages, but except for measuring the frequency of breathing it is not very quantitative. In addition when the amplitude of ventilation is low, it may no longer be possible to see respiratory movements.

Mechanograms obtained with various mechanical or mechano-electrical sensors require a close confinement of the animals. Impedance operculograms (4) or electromyograms of certain buccal or opercular muscles (3) may be recorded permanently, up to four weeks in trout (9). These methods give exact information concerning the frequency, and also an index of amplitude, of the repiratory movements.

To measure water flow rate, two techniques are available.

1) Collection of expired water according to the technique of Van Dam (31); a rubber membrane is fixed around the snout to separate the expired and inspired water streams. This technique needs great delicacy in its use because manipulation shocks the

animal and the membrane may disturb breathing. In the technique of Davis and Cameron (6), the membrane is stitched around the margin of the fish's mouth.

2) Saunders' method (29). Our modification of Saunders' method will be described in some detail, as it was used to obtain the data reported here. Figure 1 shows the principle of the method. The O_2 consumption is measured from a knowledge of the flow rate of the water pumped through the container, and by measuring the O_2 concentrations of the ingoing water (I) and the outgoing water (E').Through a permanent catheter, passed behind the cleithrum into the opercular cavity, expired water, E, is

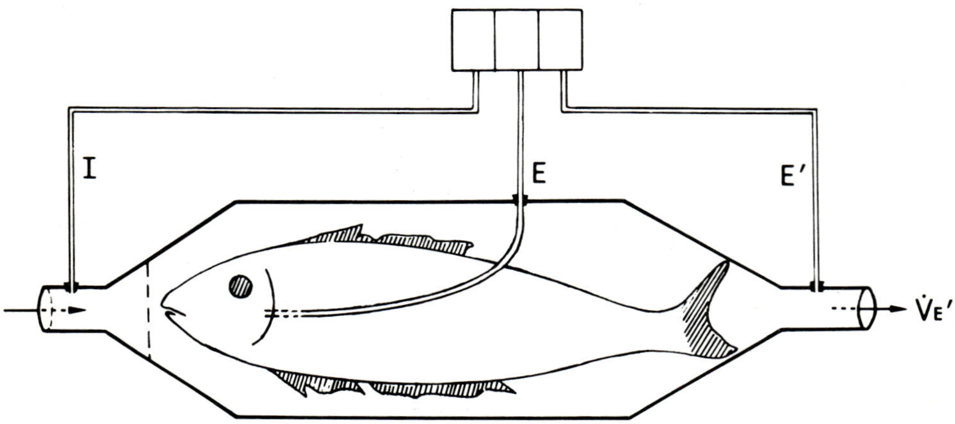

$$\dot{M}_{O_2} = \dot{V}_{E'} (C_{I_{O_2}} - C_{E'_{O_2}}) = \dot{V}_E (C_{I_{O_2}} - C_{E_{O_2}})$$

Figure 1 - The fish is placed in a lucite cylinder through which water flows at a rate $\dot{V}_{E'}$. Three pumps sample inspired water, I; expired water, E (collected through a permanent catheter in the opercular cavity); and the water E' at the exit of the cylinder. The ventilatory flow rate $\dot{V}_E$ of the animal may be calculated from the equations written according to the Fick principle. To prevent rebreathing, the cylinder flow rate, $\dot{V}_{E'}$, must be much higher than the ventilatory flow rate $\dot{V}_E$. For an evaluation of this method, see text.

collected very slowly over periods of several minutes. The O_2 concentration in the various waters are calculated from polarographic measurements of P_{O_2}.

In writing the equations of the Fig. 1 according to the Fick principle, we make four assumptions:

a) the animal is in a steady state;

b) there is no cutaneous respiration, i.e. all the O_2 uptake takes place in the gills and none through the skin (this assumption is of course false to some extent);

c) there is no rebreathing (if rebreathing were to occur, the inspired CO_2 must be known);

d) the collected expired water, E, is exactly representative of the mixed expired water. Garey (10) showed that this assumption is not valid for large carp, but as far as we know it is nearly valid for small fish (100-300 g).

Assumptions b and d are not negligible. However, the method is so easy to apply and gives a quantity of results so quickly, that it seems the most appropriate for estimating the respired water flow rate on the teleosts we studied. It may be added that in any case no one method seems entirely satisfactory.

Action of changes of water P_{O_2}, P_{CO_2} and pH on breathing

In Table 2, the collected data of many authors show that:

1) from the many investigations on the effect of hypoxia on ventilation, it is quite clear that hypoxia increases ventilation. Similar conclusions may be drawn from other studies on teleosts (6,12);

2) the effects of hyperoxia have been little studied, although, as we have seen, hyperoxia is not an uncommon situation. From the few studies that have been done, it appears that hyperoxia decreases ventilation.

3) hypercapnia generally leads to small increases of ventilation, and there are species differences in the responses to breathing CO_2. Any experiment where P_{CO_2} is made very high (for exa-

Table 2 - Action of hypoxia, hyperoxia, hypercapnia and decrease of pH on ventilation, V, or on respiratory frequency, f, in sea water and fresh water teleosts.

	Species	Method	Hypoxia	Hyperoxia	Hypercapnia	Decrease of pH
Babak 1907	*Misgurnus fossil.*	visual observation	increases $\dot{V}$	decreases $\dot{V}$	little effect	
	several Cyprinidae		"	"	"	
Oltoff 1934	*Perca fluviatil.*	visual observation			increases $\dot{V}$	nil
	Phoxinus laevis				"	"
Meyer 1935	*Uranoscopus*	direct meas. of insp.water	increases $\dot{V}$		increases $\dot{V}$	small increase
Van Dam 1938	*Anguilla vulg.*	collection of expired water	increases $\dot{V}$		little effect	
	Salmo shasta		"		increases $\dot{V}$	
Saunders 1962	*Catostomus com.*	Saunders' method	increases $\dot{V}$		small increase of $\dot{V}$	
	Ictalurus neb.		"		"	
	Cyprinus carpio		"		"	
Randall 1963	*Tinca tinca*	buccal mechanogram	increases $\dot{V}$		small increase of f	
Peyraud 1964	*Cyprinus carpio*	opercular mechanogram	increases $\dot{V}$	decreases $\dot{V}$	little effect	
Hughes 1970	*Salmo gairdneri*	Saunders' method	increases $\dot{V}$			
Saunders 1971	*Hemitripterus americanus*	opercular barogram	increases $\dot{V}$			
Belaud 1972	*Anguilla anguilla*	impedance operculogram		decreases $\dot{V}$		
Dejours 1972	several Cyprinidae	modified Saunders' method	increases $\dot{V}$	decreases $\dot{V}$	little effect	nil
	Salmo trutta		"	"	increases $\dot{V}$	"

mple by bubbling 10% CO_2 through the water) cannot lead to valid conclusions as to the normal control of breathing;

4) changes in the pH of water (at constant P_{CO_2}) cause little or no ventilatory response.

We present below some of our own observations on the ventilatory reactions of carp, tench, goldfish and trout to changes in the respiratorily significant conditions of the milieu, together with blood values, a class of data which as far as we know has not previously been reported in the literature.

Ventilatory flow rate was determined by the modification of the Saunders' method described above. Breathing fequency was counted when possible. We measured the oxygen consumption, $\dot{M}_{O_2}$; and calculated the ventilatory flow rate, $\dot{V}_E$.

Figure 2 shows the variation of ventilation in a tench for moderate changes of inspired P_{O_2}. It may

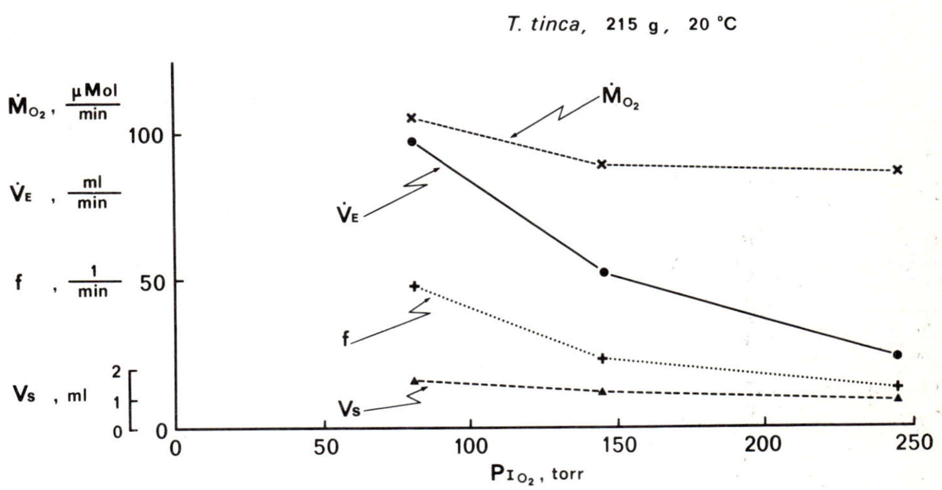

Figure 2 - Oxygen consumption $\dot{M}_{O_2}$, ventilatory flow rate $\dot{V}_E$, respiratory frequency f and ventilatory stroke volume Vs as functions of partial pressure of O_2 in inspired water, $P_{I\,O_2}$, in the tench. Similar data have been obtained in carp, goldfish and trout.

be seen that ventilation increased with hypoxia and decreased with hyperoxia. Similar observations were made in goldfish, carp and trout. The ventilatory requirement decreases by a factor of about 10 when P_{O2} is increased from 50 to 500 torr.

A decrease of ventilation due to hyperoxia must be accompanied by an increase of P_{CO2} in the opercular water; the hyperoxic depression of ventilation therefore implies that CO_2 is not a strong stimulus of respiration. This is what we have observed. In goldfish, an increase of P_{CO2} (6, 12 or 24 torr) did not lead to any obvious increase of the ventilatory requirement. Trout are different; in normoxia or hypoxia, the addition of CO_2 to water ($P_{I_{CO2}}$ 8 torr instead of 1 torr) increased ventilation by a factor of 2 or 3, but in hyperoxia ventilation was unchanged when the water was made hypercapnic.

If water of constant P_{CO2} has its pH changed by addition of bicarbonate, or by displacement of bicarbonate by <u>slow</u> addition of a strong acid, HCl, what is the effect on ventilation? The results are clear cut; the trout's ventilation does not change with changes of water pH at constant P_{CO2}, whereas change of P_{CO2}, accompanied by change of pH, has a strong action in a normoxic trout, although not in a hyperoxic one.

Changes of acid-base status of blood accompanying the ventilatory reactions to changes of aquatic O_2 and CO_2 concentrations (8).

Large carp (2-3.5 kg) were studied at 15°C. An indwelling catheter introduced through the mouth into the dorsal aorta was passed out through the upper jaw and fixed. The experimental animal was placed in a trough where it stayed quietly for many days or weeks. Hematocrit, which was measured from time to time, remained constant at more than 20%. A complete report of this work is in preparation but here, briefly, are our observations, made on many animals.

Hypoxia, which stimulates ventilation, caused moderate decreases of blood P_{CO2}, $[HCO_3^-]$ and $[H^+]$. Hyperventilation produced by hyperoxia caused increases of P_{CO_2}, HCO_3^- and $[H^+]$. Figures 3 and 4 illustrate these statements. It is remarkable that between hypoxia and hyperoxia, blood P_{CO_2} passed from 3 to 12 torr and pH from 8.0 to 7.7, whereas in normoxia P_{CO_2} was 4 torr and pH 7.93.

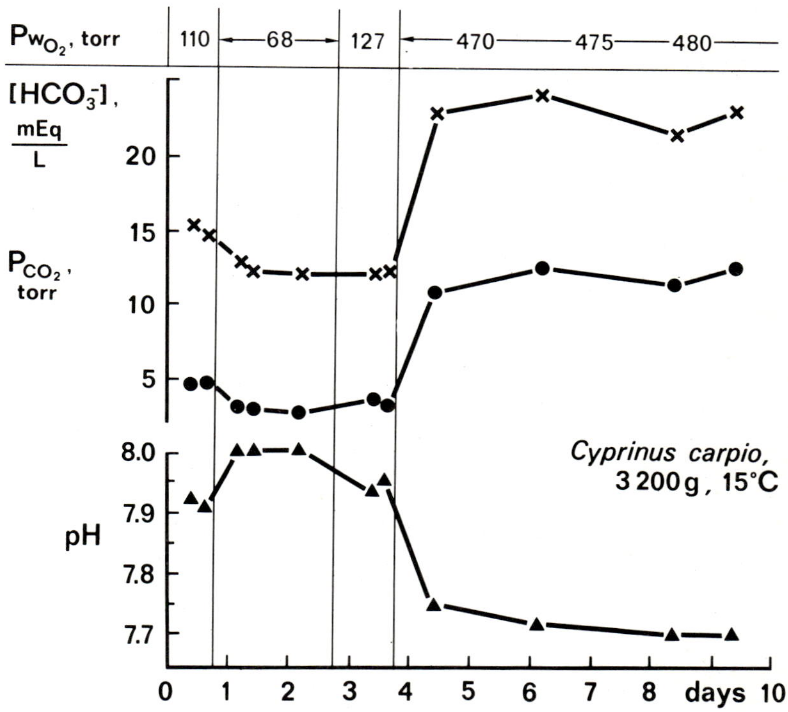

Figure 3 - Bicarbonate concentration, partial pressure of CO_2 (P_{CO_2}) and pH of arterial blood sampled in the dorsal aorta via an indwelling catheter in a large carp at 15° C. Abscissa, time in days. The partial pressure of O_2 in water, Pw_{O_2}, is indicated at the top of the figure.

Figure 4 shows that the acidosis due to the depression of ventilation in hyperoxia was partially compensated. The pH change is almost completed within 24 hours.

Hypercapnia (with water P_{O_2} of about 150 torr) leads to a pure respiratory acidosis, i.e. the aci-

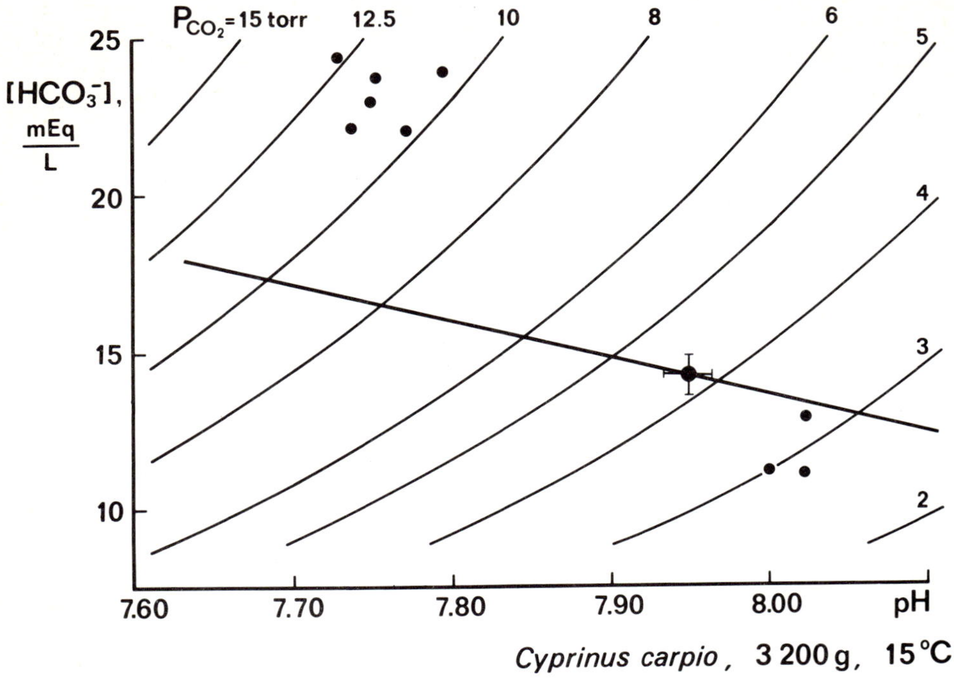

Figure 4 - Diagram of bicarbonate concentration as a function of pH of arterial blood. Same experiment as fig. 3. The straight line is the buffer line of the arterial blood at $p_{O_2} \simeq 120$ torr. (HCO_3^- = 14.3 $mEq \cdot L^{-1}$; pH = 7.95 $\pm$ 1 S.E.; N = 8). Three points lower right, hypoxia. Six points upper left, hyperoxia.

dotic and hypercapnic points remain on the buffer line determined under control conditions.

Variations of pH due to slow addition of HCl or to addition of $NaHCO_3$ to the aquarium water do not lead to any change of blood acid-base status.

What is known of the mechanism adapting breathing to changes of CO_2 and O_2 concentrations in water?

The isolated brain stem of goldfish, removed from the skull and placed on a glass slide in Ringer's fluid, exhibits a periodic activity which is considered to be respiratory (1). Such an endogenous rhythm of the respiratory centers suggests that the respiratory activity may be regulated by blood chemical factors: oxygen, carbon dioxide, hydrogen ions, catecholamines or metabolites, acting

1) either directly upon the brain stem (central chemical drive),

2) or indirectly, reflexly, by afferent stimulation originating in diffusely located receptors, or in receptors precisely grouped, for example near the vascular system, as in the carotid bodies of mammals and birds (chemoreflex drive).

Powers and Clark (24), studying the action of hypercapnia on breathing, reported that section of the 9th nerves stopped breathing and concluded that control of breathing is dominantly reflex. However, Saunders and Sutterlin (30) observed that bilateral transection of the olfactory nerves together with the 9th and 10th nerves did not alter the breathing response to hypoxia. But the possibility remains that receptors innervating the lips, the mouth, the pharynx and the cephalic part of the lateral line are sensitive to change in aquatic O_2 or CO_2 tensions. It is not easy to test this hypothesis, because the 5th and 7th nerves are mixed nerves innervating respiratory muscles. Laurent (18, 19) reported electrophysical evidence that the pseudobranch possesses a chemoreceptor function, a possibility which is discussed by Saunders and Sutterlin (30).

Recently, Eclancher (9) tried to determine the time course of the ventilatory reaction to sudden changes of water P_{O_2}. In unanesthetized tench or trout, the inflowing normoxic water was suddenly replaced by hypoxic or hyperoxic water over a 2 or 3 second period. Pharyngeal water was continually monitored by a fast O_2 electrode and recorded together with the electromyogram of a respiratory muscle. After a delay of the order of one second, sudden hypoxia or sudden hyperoxia of the pharyngeal water were accompanied, respectively, by an increase and a decrease of ventilation. The brevity of the delay between the changes of pharyngeal P_{O_2} and the ventilatory responses excludes the possibility that the mechanism may be due to variation of some blood metabolites in relation with the change of tissue oxygenation. It suggests, indeed, that the mechanism is reflex. Just where the receptors are remains a problem.

DISCUSSION

The ventilatory activity of teleosts is extremely dependent on the oxygen pressure in water. In *Cyprinidae* in hyperoxic water at 15° or 20° C, it is almost impossible to detect the respiratory movements, whereas their breathing is very active in hypoxic water. Certainly the ventilatory flow rate which we measured by Saunders' method is a good estimate of the breathing activity.

Since the oxygen consumption ($\dot{M}_{O_2}$) does not change much between normoxia and hyperoxia, the ventilatory flow rate ($\dot{V}_E$) and the inspired-expired O_2 concentration (and pressure) difference are inversely related, as may be seen in the Fick equation applied to external respiration:

$$\dot{M}_{O_2} = \dot{V}_E \; (C_{I_{O_2}} - C_{E_{O_2}})$$

In hypoxia, the same relationship holds with the complication that $\dot{M}_{O_2}$ increases in response to the extra work involved in the intense breathing in hypoxia, and it may decrease when P_{O_2} falls below the critical point. Finally the inspired-expired oxygen difference is very high in hyperoxia and very small in hypoxia; the lower the ventilation, the higher the inspired-expired O_2 differences, and vice-versa.

The inspired-expired CO_2 concentration difference necessarily increases when the ventilation is depressed. Thus the strict dependency of ventilation on water oxygenation has the corollary that it is relatively independent of CO_2 in water. There probably are differences between species concerning their respiratory sensitivity to hypercapnia. Nevertheless we continually find that all hyperoxic animals are insensitive to hypercapnia, and hyperoxia induces a very important respiratory acidosis which is barely compensated.

In trout, as in some other species (Table 2), hypercapnia in normoxia (and in hypoxia) leads to hyperventilation. One may wonder about the physiological meaning of such an observation in a water breather. Indeed, whereas one can see that it is an advantage to breathe more in hypoxia (the passage of more water over the gills limits the extraction of oxygen from the water, and the expired oxygen concentration is not intolerably low), hyperventilation cannot be a protection against hypercapnia, because the difference between inspired and expired water is always very small (14). On the contrary, in hypercapnic water, hyperventilation results in an increased CO_2 production related to the significantly greater CO_2 production by the respiratory muscles. It may be that, in natural waters, hypercapnia always occurs together with hypoxia (we do not know of any study of the simultaneous variations of O_2, CO_2 and pH in natural waters),and hypercapnia and hypoxia would then act together to stimulate breathing.

CONCLUSION

In various groups of teleost fish, hypoxia increases ventilation, hyperoxia decreases it; between hypoxia and hyperoxia the ventilatory flow rates may vary in the proportion of 10 to 1. Depression of breathing by hyperoxia leads to a considerable respiratory acidosis (e.g. blood pH will pass from 7.90 in normoxia to 7.65 in hyperoxia while P_{CO_2} will rise from 3 to 12 torr). The implication that hyperoxic animals are insensitive to hypercapnia is experimentally confirmed. Some mechanism exists by which breathing may change within one second after the fall or the rise of O_2 tension in the inspired water.

In recent years, attention has been focused on the remarkable relation existing between blood pH and temperature in poikilotherms. According to the concept of Rahn (13, 26, 27) the regulation of the H_2CO_3-HCO_3^- system is such that the "relative alkalinity" of the blood (i.e. at a given temperature, the difference between the blood pH and the neutral pH of pure water) is maintained approximately constant whatever the temperature. Consequently there may be a tendency to consider that what is regulated in respiration is mainly an adequate temperature-dependent pH by adjustment of the H_2CO_3-HCO_3^- system. It should not be overlooked, however, that the control of ventilation aims at maintaining a satisfactory oxygenation of the body and that changes of the ambient conditions of oxygenation lead to very important variations of the acid-base status of blood in fishes.

REFERENCES

1. Adrian,E.D. & Buytendijk,F.J.J. (1931) Potential changes in the isolated brain stem of the goldfish. J.Physiol.,London 71, 121-135.
2. Babak,E. & Dedek,B. (1907) Untersuchungen über den Auslösungsreiz der Atembewegungen bei Süsswasserfischen. Pflügers Arch. 119, 483-529.
3. Ballintijn,C.M. & Hughes,G.M. (1965) The muscular basis of the respiratory pumps in the trout. J.Exp.Biol. 43, 349-362.
4. Belaud,A.,Peyraud-Waitzenegger,M., & Peyraud,C. (1971) Techniques originales de prélèvement de paramètres physiologiques sur le Congre (*Conger conger* L.)libre de ses déplacements. I.Electrocardiogramme et activité respiratoire. Hydrobiologia 38, 49-59.
5. Belaud,A. & Barthelemy,L. (1972) Modifications de quelques paramètres physiologiques chez le Poisson en conditions hyperbares. J.Physiol., Paris 65, 346A.
6. Davis,J.C. & Cameron,J.N. (1971) Water flow and gas exchange at the gills rainbow trout,*Salmo gairdneri*. J.Exp.Biol. 54, 1-18.
7. Dejours,P., Armand,J. & Verriest,G. (1968) Carbon dioxide dissociation curves of water and gas exchange of water-breathers. Respir.Physiol. 5, 23-33.
8. Dejours,P. (1972) Action des changements de P_{O_2} et de P_{CO_2} de l'eau sur la ventilation et l'équilibre acide-base du sang chez quelques poissons téléostéens. J.Physiol.,Paris 65, 386A.

9. Eclancher,B. (1972) Action des changements de P_{O2} de l'eau sur la ventilation de la truite et de la tanche. J.Physiol.Paris 65, 397A.

10. Garey,W.F. (1967) Gas exchange, cardiac output, and blood pressure in free-swimming carp (*Cyprinus carpio*). Dissertation for Ph.D., University of New York at Buffalo, p.123.

11. Garey,W.F. & Rahn,H. (1970) Gas tension in tissues of trout and carp exposed to diurnal changes in oxygen tension of the water. J.Exp. Biol. 52, 575-582.

12. Gerald,J.W. & Cech,J.J. (1970) Respiratory responses of juvenile catfish (*Ictalurus punctatus*) to hypoxic conditions. Physiol. Zoöl. 43, 47-54.

13. Howell,B.J., Baumgardner,F.W., Bondi,K. & Rahn, H. (1970) Acid-base balance in cold-blooded vertebrates as a function of body temperature. Am.J.Physiol. 218, 600-606.

14. Hughes,G.M. (1964) Fish respiratory homeostasis. Symp.Soc.Exp.Biol. 18, 81-107.

15. Hughes,G.M. (1970) A comparative approach to fish respiration. Experientia 26, 113-122.

16. Hughes,G.M. & Saunders,R.L. (1970) Responses to the respiratory pumps to hypoxia in the rainbow trout (*Salmo gairdneri*). J.Exp.Biol. 53, 529-545.

17. Jones,J.D. (1961) Aspects of respiration in *Planorbis corneus* L. and *Lymnaea stagnalis* (Gastropoda Pulmonata). Comp.Biochem.Physiol. 4,1-29.

18. Laurent,P. (1967) La pseudobranchie des Téléostéens: preuves électrophysiologiques de ses fonctions chémoréceptrice et baroréceptrice. C.R. Acad.Sci.,Paris 264, 1879-1882.

19. Laurent,P. & Rouzeau,J.D. (1972) Afferent neural activity from pseudobranch of teleosts. Effects of P_{O2}, pH, osmotic pressure and Na^+ ions. Respir.Physiol. 14, 307-331.

20. Livingstone,D.A. (1963) Chemical composition of rivers and lakes. In *Data of Geochemistry*. Geo-

logical survey professional paper 440-G. Washington D.C.: United States Government Printing Office, Chapter G., p.64.

21. Meyer,H. (1935) Die Atmung von Uranoscopus scaber in ihrer Abhängigkeit vom Sauerstoffdruck, vom pH und von der Temperatur im Aussenmedium. Z.vergl.Physiol. 22, 435-449.

22. Olthof,H.J. (1934-35) Die Kohlensäure als Atemreiz bei Wassertieren, insbesondere bei Süsswasserfischen. Z.Vergl.Physiol. 21, 534-544.

23. Peyraud,C. & Serfaty,A. (1964) Le rythme respiratoire de la carpe (Cyprinus carpio L.) et ses relations avec le taux de l'oxygène dissous dans le biotope. Hydrobiologia 23, 165-178.

24. Powers,E.B. & Clark,R.T. (1942) Control of normal breathing in fishes by receptors located in the regions of the gills and innervated by the IXth and Xth cranial nerves. Am.J.Physiol. 138, 104-107.

25. Rahn,H. (1966) Aquatic gas exchange: theory. Respir.Physiol. 1, 1-16.

26. Rahn,H. (1967) Gas transport from the external environment to the cell. In Development of the lung, pp.3-23, ed. de Reuck,A.V.S. & Porter,R. A CIBA Foundation Symposium, London, J.& A. Churchill.

27. Rahn,H. & Baumgardner,F.W. (1972) Temperature and acid-base regulation in fish. Respir. Physiol. 14, 171-182.

28. Randall,D.J. & Shelton,G. (1963) The effects of changes in environmental gas concentrations on the breathing and heart rate of a teleost fish. Comp.Biochem.Physiol. 9, 229-239.

29. Saunders,R.L. (1962) The irrigation of the gill in fishes. II. Efficiency of oxygen uptake in relation to respiratory flow activity and concentrations of oxygen and carbon dioxide. Can.J.Zool. 40, 817-862.

30. Saunders,R.L. & Sutterlin,A.M. (1971) Cardiac and respiratory responses to hypoxia in the sea raven, Hemitripterus americanus, and an investi

gation of possible control mechanisms. J.Fish. Res. Board Canada 28, 491-503.

31. van Dam,L. (1938) On the utilization of oxygen and regulation of breathing in some aquatic animals. p.143, Groningen, Volharding.

Comparative Physiology, eds. L. Bolis, K. Schmidt-Nielsen and S. H. P. Maddrell

THE PHYSIOLOGY OF THE SWIMBLADDER

R. FÄNGE
Department of Zoophysiology
University of Göteborg, Sweden

Structural variability

The swimbladder besides its hydrostatic function may be adapted for several other functions like respiration, pressure reception, and reception and production of sounds. Different modes of controlling the gas volume have developed. In physostomes the swimbladder communicates with the oesophagus by a pneumatic duct, while in physoclists it is closed. Usually the physoclist swimbladder consists of two parts used for secretion and resorption respectively. Fishes with such a swimbladder are often termed euphysoclists.

In cyprinids (carp fishes), the swimbladder consists of two chambers separated by a smooth muscle sphincter. The anterior chamber has elastic walls and connects with the inner ear via the Weberian ossicles. It serves sensory functions, while the posterior chamber with a gas secretory apparatus and a pneumatic duct is presumably mainly hydrostatic. The gas pressure is high (about 60 mm Hg), which may impede blood circulation as the pressure in the swimbladder probably exceeds the arterial blood pressure (19). The wall consists of dense connective tissue and smooth muscles (1,19). In cobitids (*Misgurnus*) the anterior chamber is enclosed within a bony capsule, while the minute posterior chamber has a gas secretory apparatus but no pneumatic duct (34). In characinids (*Lebiasina*, *Erythrinus*) the posterior chamber has a wide pneumatic duct and is both respiratory and hydrostatic in function (63). Cyprinids, cobitids and characinids are grouped within the order Ostariophysi (Cypriniformes). Within this group the swimbladder may vary much in size and shape but regularly has a Weberian apparatus (1).

Gymnarchus of the order Mormyroidea has three swimbladders, two closed "auditory bladders" and one posterior bladder which is respiratory and hydrostatic. The two "auditory bladders" are very small but nonetheless have complete gas secretory devices (72). In the marine fishes Makaira (45) and Hemiramphus (own observation) the swimbladder consists of numerous closed chambers.

The list of fishes possessing swimbladder modified in different ways could be made very extensive. So far physiological investigations have concerned only few forms. Physostomes like cyprinids and salmonids, and euphysoclists, have been the main objects of research. Much work has concerned physical aspects of swimbladder function (1,13) and the mechanism of gas secretion.

Gas gland

The gas secreting apparatus of the swimbladder consists of a gas gland supplied by a counter current vascular system (rete mirabile). In the simplest case as in Coregonus, a salmonid, the gas gland may consist only of a rather unspecialized inner epithelium (22). More advanced conditions are found in the eel (Anguilla anguilla) and the toadfish (Opsanus tau) where gland-like cells form a single-layered epithelia (25). In euphysoclists the gas gland is often many-layered with duct-like clefts and occasional giant gas gland cells (28,81). In the fast-secreting gland of the bluefish (Pomatomus saltatrix) naked capillaries are situated on the surface of the gas gland in contact with the gas phase (78). Gas gland cells have a characteristic ultra-structure. The cytoplasm is rather empty but contains large amounts of minute glycogen granules. Smooth and rough endoplasmic reticulum are sparingly present and a Golgi apparatus may be found. The mitochondria are small with few cristae. They occur mainly at the base of the cells, where the membrane is greatly folded. At the apical pole a few microvilli project into the gas phase (8,16, 35). Lipid particles or vacuoles are abundant and histological signs of a cyclic secretory activity have been described (24,33) (Fig. 1).

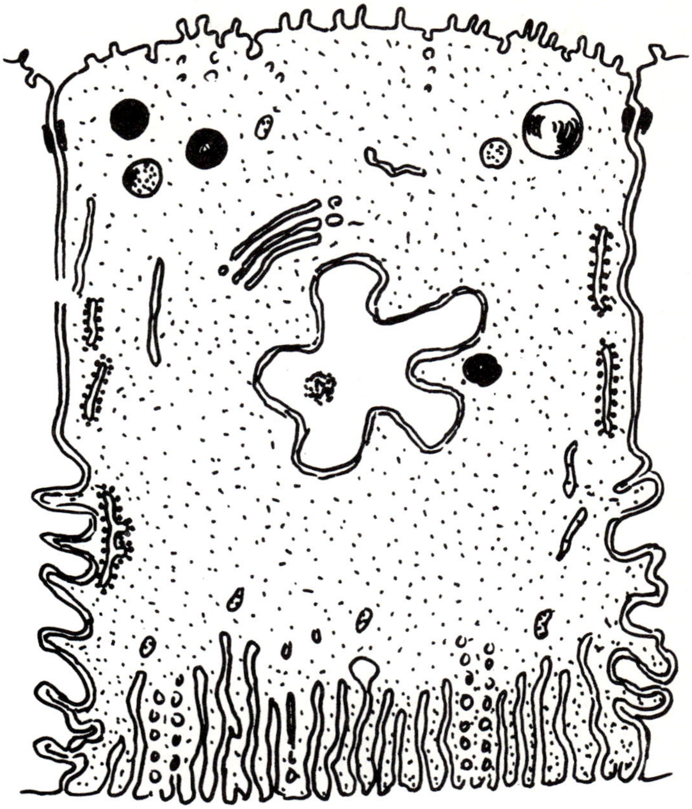

Figure 1 - Gas gland cell.
Diagram based on Dorn (1961), Copeland (1964), Jasinski and Kilarski (1969).

Rete mirabile

A rete mirabile may be defined as a blood vessel which abruptly ramifies into a system of parallel vessels before an ordinary capillary network is formed. J.Müller in 1840 surveyed the different kinds of rete mirabile to be found in animal organs and classified them into types (Table 1) (50).

Müller made the important statement that the rete mirabile of the swimbladder, whether unipolar or bipolar is always double, i.e. consists of arteries and veins with blood flowing in opposite directions.

Table 1

Type	Occurrence
I Simple unipolar	
II Double unipolar	Swimbladder; stomach and intestine of *Alopecias vulpes* (elasmobranch)
III Simple bipolar	Carotid rete mirabile in certain mammals; intercostal retia mirabilia of dolphins; pseudobranchs
IV Double bipolar	Swimbladder of the eel (*Anguilla*); teleostean choroidal gland; liver of tuna; arteria coeliaca of tuna; liver of certain elasmobranchs etc.

The functional importance of the rete mirabile in the swimbladder is underlined by the fact that it is strongly developed in fishes with efficient gas secretion (48), but only weakly developed in fishes with a slow gas secretion (22). Haldane about 50 years ago suggested that in the rete mirabile carbon dioxide is transported from venous into arterial capillaries. The secretion of gas into the swimbladder could be due to liberation of oxygen from oxyhemoglobin under influence of carbon dioxide (30). Hypotheses implying that the rete mirabile functions as a counter current multiplying device were already expressed by Jacobs and by Koch around 1930 (33,38). Similar hypotheses have been further developed by later investigators (2,3,13,18,41,42, 43,46,47,67,68). Species with a bipolar rete (*Anguilla anguilla*, *Sphyraena barracuda*) have been used for detailed studies of the function of the rete mirabile. Analyses of blood samples from veins connecting the gas gland and the rete mirabile have shown a low pH and large amounts of lactic acid (18,42,70). In the rete it is probable that carbon dioxide, lactate, hydrogen ions, oxygen and inert gases diffuse from the venous to the arterial capillaries.

The rate of blood flow appears to be most important for the function of the counter current multiplying system of the swimbladder (51,64). The kinetics of the oxygen-hemoglobin binding are such that oxygen released from hemoglobin takes a relatively long time to recombine and this enables oxygen to diffuse from the venous into the arterial capillaries of the rete (3). A filtration of fluid from arterial into venous capillaries may influence the concentration gradients across the rete (38). The enzyme carbonic anhydrase probably occurs in the walls of the blood vessels of the rete mirabile and may have an accelerating influence on diffusion of carbon dioxide (24,47). Contractile arterioles in the beginning of the rete mirabile may act to influence the blood flow (27).

Other counter current systems

The choroidal gland offers a parallel to the gas secretory apparatus of the swimbladder.It is a bipolar rete mirabile situated outside the retina in the choriocapillaris of the eye of certain teleosts (50).The fish retina like that of other nonmammalian vertebrates is avascular and gets oxygen by diffusion from the choriocapillaris. The choroidal gland probably multiplies the oxygen liberating effect of a metabolic acid produced by choroidal or retinal cells.In the large eyes of certain fast-swimming predators (*Pomatomus saltatrix*, *Echeneis naucrates*) tensions of oxygen which are 20 times or more of that of the arterial blood have been measured (77). As in the swimbladder, the enzyme carbonic anhydrase is necessary for the oxygen concentrating effect (23,12). According to Müller (50) the choriocapillaris in all vertebrates is a rete mirabile, but only in teleosts is this bipolar. The mesenteries of certain teleosts contain small arteries closely associated with veins (own observation), and similar arrangements have been described in the dura mater of the mammalian brain (60). Such vascular bundles are rather similar to the primitive rete mirabile in the swimbladder of certain physostomes. The importance of such weakly developed counter current systems does not seem to have been investigated. Some better known systems in various animals function as heat exchangers (66,68).

Permeability to gases

The rate at which gases penetrate the wall of the swimbladder is different in different parts of the bladder and also depends upon blood circulation. With no circulation the wall of the swimbladder may be practically impermeable. Thus the isolated swimbladder of a deep water fish may keep its high oxygen concentration for days (4), and the same is observed as regards the excess intrinsic pressure of the isolated cyprinid swimbladder (19). On the other hand, for example in the swimbladder of living physoclists which migrate upwards,outward diffusion of gases, mainly oxygen, is probably relatively rapid.

The swimbladder of the deep water fish, *Argentina silus*, in its anterior part has a silvery appearance caused by guanine crystals (26,79); the posterior resorbent chamber has no such guanine layer (21). Experiments on the eel (*Anguilla anguilla*) and the Conger eel (*Conger conger*) show that in these species the "silvery layers" are about 100 times less permeable to gases than is ordinary connective tissue (14,44). In organs other than the swimbladder, such as the eye, guanine-containing layers may also constitute barriers against diffusion (14). In the swimbladder dense layers of collagen fibrils probably contribute to the impermeability of the wall to gases.

Diffusion through the resorbent part of the swimbladder in a living fish follows the same laws controlling the diffusion of gases in other tissues. Carbon dioxide moves faster than oxygen, which diffuses faster than nitrogen (39). This relation may explain the occasional findings of very high concentrations of nitrogen in the swimbladder of certain species of fish (58,74), but other explanations have also been suggested (1). For resorption of gases to take place from the swimbladder or any other closed pocket of gases enclosed by living tissue, it is necessary that pressure gradients exist for each gas. The intrinsic total gas pressure always exceeds the sum of gas pressures in the blood, because oxygen is consumed by tissue respiration. Even nitrogen is finally resorbed, although this takes a long time (58,59).

In the living physoclists, gases are continuously lost by diffusion in the resorbent mucosa, but this is counterbalanced by gas secretion. The constant volume of the swimbladder in a living fish is a steady state controlled by autonomic nerve reflexes. The secretory mucosa not only secretes gases but also functions as a lock against diffusion losses, because of its counter current circulation. Furthermore the sometimes rather thick connective tissue membranes of the secretory mucosa may constitute a barrier against diffusion of gases. Movements of the impermeable secretory mucosa is used by the fish to regulate the area available for resorption of gases (cfr. Fig. 2).

Nervous control of gas elimination in physoclists

The availability of the swimbladder wall to diffusion is controlled by tonus changes of the muscularis mucosae and the blood vessels. Because smooth muscles change tonus slowly, the states of contraction of the secretory mucosa and the resorbent mucosa in euphysoclist swimbladders may be studied in newly killed fishes (24). In _Ctenolabrus rupestris_ the boundary between the two parts of the mucosa is a transverse diaphragm with a central hole. A jelly-like submucosa separating the inner and outer layers of the swimbladder makes large mucosal movements possible (Fig. 2), so that the diaphragm can glide forwards or backwards. With the diaphragm in an anterior position the swimbladder is in an absorbing state. When the same structure is situated in the posterior end of the swimbladder, gas secretion may take place.

In the cod (_Gadus morhua_), the perch (_Perca_

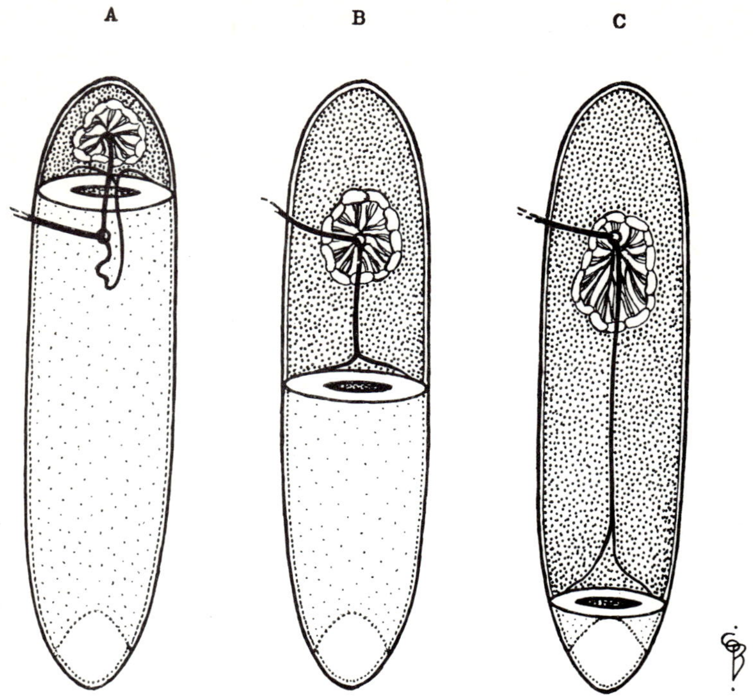

Figure 2 - *Ctenolabrus rupestris.* The appearance of the swimbladder during different conditions. A. Anterior position of the diaphragm produced by asphyxiation, or application of adrenaline, or electric stimulation of the nervus vagus. B. Middle position of the diaphragm, usual situation. C. Posterior position of the diaphragm as found during active gas secretion. This position of the diaphragm is also produced by section of the nervus vagus or by administration of α-adrenergic blocking agents etc. (Diagram drawn by Cynthia O'Brian).

fluviatilis) and many other species, there is an oval-shaped mucosal fold in the dorsal inner wall which resembles the diaphragm. The oval opens and closes much like the pupil of an eye (53,80). Experiments with different drugs have clarified the role of autonomic nervous control of the smooth muscle systems of the swimbladder. (Most studies have concerned adrenergic or cholinergic substances. No studies seem to have been made on the possible effects of 5-hydroxytryptamine on the swimbladder).

The control of the muscularis mucosae seems to be exerted exclusively by adrenergic vagal fibres, the secretory mucosa showing alpha-adrenergic activity and the resorbent mucosa beta-adrenergic activity (27,52). Blood vessels supplying the two parts of the mucosa seem to be controlled in a similar way (54,73).

A state of gas resorption can be induced experimentally by administration of adrenergic drugs, or by stimulation of the vagus nerve. Section of the vagus nerve or administration of adrenergic blocking agents inhibits gas resorption. Several kinds of external stimuli may induce a gas resorption reflex, for example asphyxiation (24).

Nervous control of gas elimination in physostomes

The eel (*Anguilla anguilla*), which is a physostome, eliminates gases by resorption (24), but physostomes may also release gases to the exterior via the pneumatic duct. For instance salmonids give off bubbles if the external pressure is slowly lowered to a critical level. This is called a gas-spitting reflex (15,28). They also "sound", i.e. release gas bubbles and dive, if startled and frightened (32). Both kinds of response require opening of the pneumatic sphincter, and contraction of the swimbladder muscles and body wall muscles. The structure of the sphincter guarding the entrance of the pneumatic duct into the oesophagus has been examined in only one species (19). It may contain both smooth and striated muscles (as does the sphincter of the mammalian urinary bladder). Results of nerve sectioning indicate that the nervus vagus has a tonic relaxing influence on the sphincter (28). The innervating fibres are probably adrenergic, because release of gases is facilitated by adrenergic stimuli and inhibited by adrenergic blocking agents (32,69). Noradrenergic varicose fibres innervate the smooth muscles of the swimbladder of the rainbow trout (*Salmo irideus*) (20).

Stretch receptors in the wall of the swimbladder may be responsible for initiation of the gas-spitting at lowered external pressure. Sensory nerve endings have been found in the swimbladder of cyprinids (62), but the inner ear and other sense organs may also be involved in the afferent reflex arch (1,15).

A few, mainly tropical, physostomes use the swimbladder for respiration. Air is taken up and expelled through the pneumatic duct. Smooth muscles of the swimbladder, or various systems of striated muscles may produce the respiratory movements, but few details are known. Phylogenetically interesting parallels may be drawn with the pulmonary muscles of lower vertebrates. In the lizard, *Uromastix*, smooth muscles contract spontaneously in the isolated lung (65), and in the frog, *Rana*, smooth muscles may assist in emptying the lung (61). Turtles have striated pulmonary muscles.

A reflex centre for gas-spitting is probably situated in the brain, but little is known about this. There are obvious similarities between the gas-spitting reflex and the gas resorbing (deflatory) reflex of physoclists (24,27).

Nervous control of gas secretion

Bohr (1894) discovered that in the cod (*Gadus morhua*) gas secretion is blocked by vagotomy (5). The result has been confirmed in this and other species (24,27). The fibres of the vagus nerve controlling secretion may not be identical with the adrenergic fibres which cause resorption of gases. The presence of acetylcholine esterase in the gas gland (2) and the fact that gas secretion is blocked by atropine (24) may indicate that the secretory fibres are cholinergic. A vagal reflex in the Indian fish *Ophiocephalus* produces bradycardia and increased concentration of oxygen in the bladder (11). This observation gives some additional although indirect support for the hypothesis of cholinergic secretory fibres. Nerve cells constituting a diffuse gas gland ganglion (24) could represent parasympathetic neurons innervating the gas gland cells.

Dreser (1892) reported that repeated injections

of pilocarpine into pike (Esox lucius) produced an increase of oxygen in the swimbladder (17).However, in my own experiments with Gadus morhua, Ctenolabrus rupestris and Esox lucius no significant increase of oxygen in the swimbladder was observed after varying doses of pilocarpine (unpublished results). Negative results were also obtained with mecholyl, another parasympathicomimetic drug (24). A significant increase of oxygen and carbon dioxide however, was obtained with alpha-adrenergic blocking drugs like yohimbine (24).

Table 2
Effects of yohimbine - Gadus morhua

Yohimbine	Vagotomy	CO_2%	O_2%	Oval
NO	NO	1 - 4.6	13.8-19.1	Open
NO	YES	1.5- 5.5	14.2-22.0	Closed
YES	YES	2.0- 7.9	16.2-23.0	Closed
YES	NO	2.4-12.8	14.8-50.0	Closed

(From experiments by R.Fänge and S.Nilsson,1972).

The effect of yohimbine disappears after vagotomy (Table 2) indicating the loss of tonic secretory vagal influence. The yohimbine effect is probably due to a block of resorption, and not to an increase in gas secretion. The secretory fibres of the vagus nerve may cause an increased metabolic activity of the gas gland cells. The mechanism of this action may be similar to that of, for example, mammalian chorda tympani fibres, which activate salivary gland cells.

It may be concluded that the gas content of the swimbladder increases as a result of 1) inhibition of adrenergic outflow in the vagus nerve, and 2) activation of secretory fibres in the vagus nerve. Both processes may be governed by a gas secretory (inflatory) reflex centre of the brain. Afferent impulses leading to gas secretion may come from the inner ear or other sense organs (24). The vasodilatation which takes place in the active gas gland (24) could be due to an inhibition of alpha-adrenergic vasoconstrictor influence possibly combined with the effects of substances produced by the metabolizing gas gland cells (carbon dioxide, lactic acid, peptides?, organic phosphates?).

The metabolism of the gas gland

During gas secretion glycogen disappears from the gas gland cells (24,47), and lactic acid appears (70). Studies on the enzymes of the gas gland may elucidate the metabolism of the gas gland cells. A rapid formation of lactic acid in the presence of oxygen was observed *in vitro* in the gas gland of the scup (*Stenotomus chrysops*)(2a), which has a very fast rate of gas secretion *in vivo* (78). Gas gland tissue from various species of marine fishes showed a high activity of the lactic dehydrogenase isoenzyme of the anaerobic type, also found in skeletal muscle (Table 3). Cytochrome oxidase activity

Table 3

Lactate Dehydrogenase Activity in Tissues

Species	Gas gland LDH	Gas gland p ratio	Cardiac musc. LDH	Cardiac musc. p rat.	Skelet.musc. LDH	Skelet.musc. p rat.
GADUS MORHUA	134	0.8	138	1.8	104	0.6
G.AEGLEFINUS	228	0.6	291	1.8	176	0.5
G.POUTASSOU	185	0.8	59	1.4	53	0.4
G.VIRENS	305	0.7	279	1.5	60	0.7
RANICEPS RAN.	52	0.7	127	1.7	39	0.8
CTENOLABRUS R.	270	0.7	517	1.5	81	0.9
LABRUS BERGG.	184	0.9	279	1.1	326	0.8
ANGUILLA ANG.	46	0.8	165	1.7	59	0.6
MEAN VALUES	185	0.78	279	1.69	74	0.67

The pyruvate ratio (p ratio) is defined in the original paper (29). (In the original table the headings Cardiac muscle and Skeletal muscle were confused). (From Gesser and Fänge, 1971).

was low (29). Other studies have shown that the

Krebs cycle is not a major pathway of glucose catabolism in the gas gland (6,9). The patterns of activities of some metabolic enzymes have recently been investigated in the gas gland of the cod (6) and other marine fishes. The activities of the same enzymes were also measured in another tissue with a high glycolytic rate, white skeletal muscle of the cod, and the activity ratios of representative enzymes in gas gland in relation to muscle were compared. The methodological approach was adopted mainly from Pette (56), who investigated enzyme ratios and metabolism in different types of skeletal muscle. The studies, which are not yet finished have given rather clear results. Activities of hexokinase, lactic dehydrogenase and hexosemonophosphate shunt enzymes (glucose-6-phosphate dehydrogenase and 6-phosphogluconate dehydrogenase) are higher in the gas gland than in the muscle. On the other hand phosphofructokinase, and the Krebs cycle enzymes fumarase and cytochromoxidase, are significantly less active in the gas gland than in the muscle (Table 4).

Table 4

Enzyme Ratios Gas Gland/White Muscle

Species	HK	G6PDH	6PGDH	PFK	GPDH	PK	LDH	FUM	COX
GADUS MORHUA	+	+	+	-	-	+-	+	-	-
GADUS VIRENS	+	+	+	-	-	-	+	-	-
GADUS POUT.	+	+	-	+	+	-	+	-	-
MOLVA MOLVA	+	+	-	-	-	+	+	-	-
MOLVA BIRKEL.	+	+	-	-	+	-	+	-	-
CORYPH.RUP.	+	+	+	-	-	+	+	-	-
BROSME BROS.	+	+	-	-	+	-	-	-	-
UROPHYCIS BL.	+	+	+-	-	-	+-	+-	-	-
ARGENTINA S.	+	+-	-	-	-	-	-	-	-

\+ high activity in gas gland
\- low activity in gas gland
+- equal activities in gas gland/muscle
(From Boström,Fänge,Johansson,1972,in preparation)

The predominance of the enzymes of the hexosemonophosphate shunt and the low activity of phosphofructokinase indicate that glycolysis proceeds mainly along the hexose monophosphate pathway. The allosteric enzyme phosphofructokinase is inhibited by high oxygen, but the glycolysis is "shunted around" this point of trouble. This may explain why the gas gland shows no Pasteur effect, i.e. lactic acid production is not blocked by oxygen as it is in other tissues. In fishes with a fast rate of gas secretion like certain deep water fishes the partial pressure of oxygen in contact with the gas gland cells may be very high (4,37,67). In fact the oxygen pressure would be high enough to cause oxygen poisoning, but, fortunately for the fish, only the swimbladder and the blood are exposed to the high oxygen. The cells of the swimbladder are metabolically adapted to high concentrations of oxygen (10). It is interesting that in the liver of rats exposed for prolonged periods to hyperbaric oxygen, an enzyme adaptation towards a high activity of the hexosemonophosphate pathway has been demonstrated (29a).

The complete oxidation of glucose-6-phosphate via the hexosemonophosphate shunt produces CO_2 and NADPH. The latter substance is much used by living cells for the synthesis of lipids. It is of interest therefore that the gas gland cells contain considerable amounts of lipid material (24). This is particularly so in the swimbladder of the deep sea fish, *Coryphaenoides* (rat-tail or grenadier). Analyses of lipid material from the swimbladder of this fish show that the lipids consist mainly of cholesterol and phospholipids (55,57). It is not known if the lipids formed by the gas gland are an excretory product, or if they have some special function. Gases are highly soluble in lipids. Thus to some extent gases may be transported into the bladder dissolved in the lipids. Also the surface tension lowering capacity of lipids may play a role in the liberation of gases from solution into the gas phase.

Because the Krebs cycle enzymes are poorly represented, carbon dioxide in the swimbladder is probably formed largely by the activity of the hexosemonophosphate shunt (6,9). However, a considerable part of the carbon dioxide in the swimbladder may also come from blood bicarbonate (78). Glucose-6-phosphate is probably generated in the gas gland cells both from stored glycogen and from glucose continuously transported into the gas gland cells by hexokinase, which has a high activity in these cells (6).The basal foldings and the mitochondria may be needed for uptake of glucose and for generation of ATP used for the transport. Very possibly the enzyme machinery of the gas gland cells is influenced by various hormones (insulin, glucagon, corticoids?), but this does not seem to have been investigated.

The gas gland region of the swimbladder has a remarkably high activity of the enzyme carbonic anhydrase. This occurs not only in the gas gland cells and in the erythrocytes of the blood passing the gas gland but probably also in the vascular walls of the rete mirabile. No carbonic anhydrase is found in the fish blood plasma, which appears to contain an inhibitor of the enzyme (47). The enzyme is probably needed, to facilitate transport of carbon dioxide across the rete mirabile, the metabolism of the gas gland cells, and processes of the blood leading to a Bohr effect. Without carbonic anhydrase the Bohr effect would be too slow to operate in the secretion of oxygen (71).

Figure 3 shows a hypothetical outline of the metabolism of the gas gland cell.

Gas secretion without counter current circulation

A release of gas into intracellular vacuoles is known to occur in the protozoan *Arcella*. The gas is supposed to consist mainly of oxygen formed by decomposition of hydrogen peroxide (7). Liberation of oxygen, associated with transformation of glycogen into lipids, has been observed in *Chironomus* larvae during anaerobiosis (31). When larvae of another insect, *Eristalis*, are enclosed without access to air, gas accumulates in the tracheal system and causes the larvae to float up (40). The gas was thought to be carbon dioxide. In the swimbladder of larvae of the fish *Hemichromis*, gas secretion has been observed at a stage before any

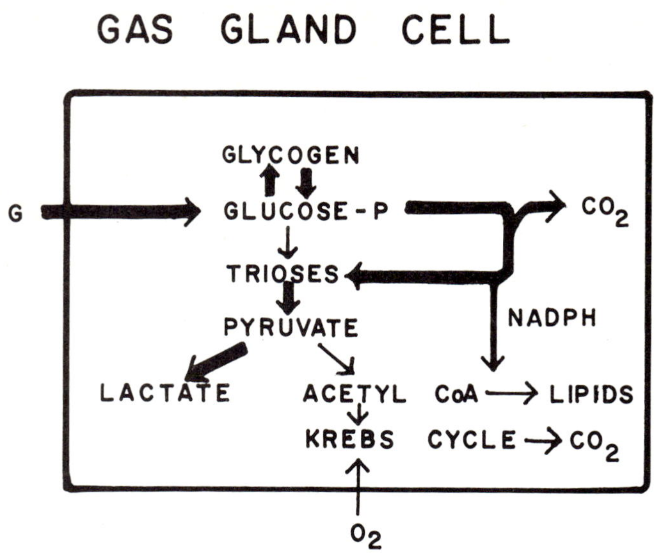

Figure 3 - (Modified from a diagram in Tepperman,1966 (74) used to illustrate effects of insulin on the metabolism of the fat cell).

rete had developed and when no pneumatic duct was present (49). A similar situation has been noted in another fish species (36). Gas secretion,however, was inferred from morphological evidence which may be difficult to interpret.

A most remarkable kind of gas secretion ought to be mentioned here, namely that observed in the pneumatophores of siphonophores. The gas which accumulates consists largely of carbon monoxide;it is thought to be formed from folic acid derivatives (76). The biochemistry of the gas gland of siphonophores is probably unique. These examples show that there exist, in nature, gas liberating processes in which no counter current multiplying effect is needed. Perhaps such a process in the fish swimbladder cooperates with the rather well established, lactic acid dependent,counter current multiplying mechanism.

CONCLUDING REMARKS

An attempt to summarize processes thought to

take place in the swimbladder during gas secretion is found in Fig.4. This is a diagram modified from previous similar diagrams (27,47). No account is taken of any cellular gas secretion acting without a rete mirabile, because such a process in the swim bladder still seems unproved. The Bohr (or Root) effect is thought to play an important role.

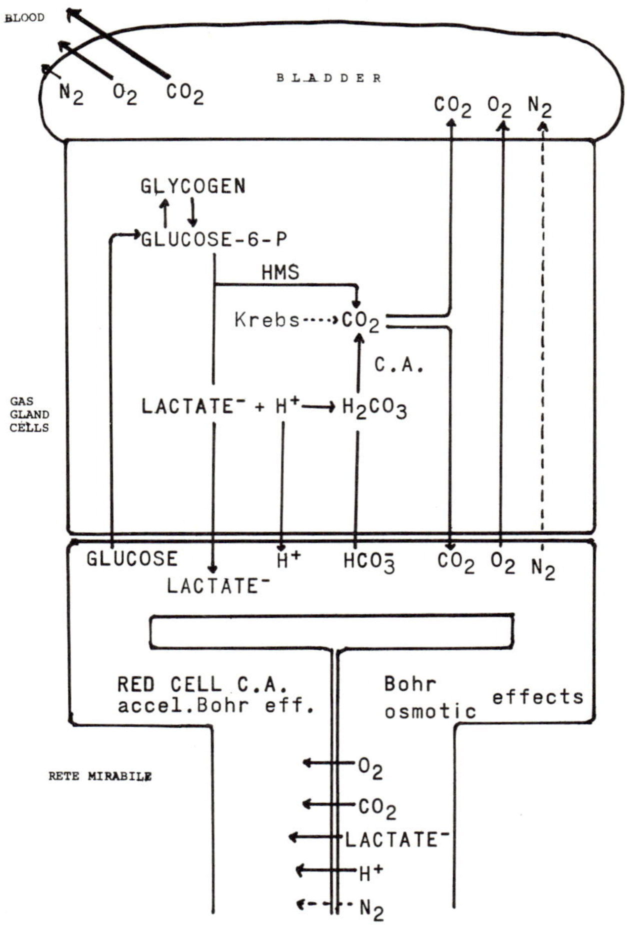

Figure 4 - (C.A. = Carbonic Anhydrase; HMS = Hexosemonophosphate Shunt).

Organic phosphates (2,3-DPG or ATP) in the erythrocytes probably participate in this effect. It is not known whether the activity of the gas gland cells may increase the concentrations of organic phosphates within the blood cells. Lactic acid may also exert an osmotic (salting out) effect. The possibility that gas gland cells in addition to lactic acid produce some non-acid, low molecular-weight substance, which may decrease the solubility in the blood of nitrogen, has been discussed (1).

The nature of gas secretion is only one aspect of the physiology of the swimbladder. Numerous problems concerning the nervous and humoral control of the varying functions of this organ remain to be investigated.

REFERENCES

1. Alexander,R.McN. (1966) Physical aspects of swimbladder function. Biol.Rev. 41,141-176.

2. Augustinsson,K.B. & Fänge,R. (1951) Innervation and acetylcholine splitting activity of the air-bladder of fishes. Acta Physiol.Scand.22, 223-230.

2a. Ball,E.G.,Strittmatter,C.F. & Cooper,O. (1955) Metabolic studies on the gas gland of the swimbladder. Biol.Bull. 108, 1-17.

3. Berg,T., & Steen,J.B. (1968) The mechanism of oxygen concentration in the swim-bladder of the eel. J.Physiol. 195, 631-638.

4. Biot,J.B. (1807) Mémoire sur la nature de l'air contenu dans la vessie natatoire des poissons. Mém.Phys.Chim.Soc. d'Arcueil 1, 252-281.

5. Bohr,C. (1894) The influence of section of the vagus nerve on the disengagement of gases in the airbladder of fishes. J.Physiol. 15,494-500.

6. Boström,S.L., Fänge,R., & Johansson,R.G. (1972) Enzyme activity patterns in gas gland tissue of the swimbladder of the cod (*Gadus morhua*). Comp.Biochem.Physiol. 43B, 473-478.

7. Cicak,A.J.,McLaughlin,J.A. & Wittenberg,J.B. (1963) Oxygen in the gas vacuole of the rhizopod protozoan, *Arcella*. Nature 199, 983-985.

8. Copeland,D.E. (1969) Fine structural study of gas secretion in the physoclistous swim bladder

of Fundulus heteroclitus and Gadus callarias and in the euphysoclistous swim bladder of Opsanus tau. Z.Zellf. 93, 305-331.

9. D'Aoust,B.G. (1970) The role of lactic acid in gas secretion in the teleost swimbladder. Comp. Biochem.Physiol. 32, 637-668.

10. D'Aoust,B.G. (1971) Natural resistance to oxygen poisoning. In Underwater Physiology,pp. 23-34, ed.Lambertsen,C.J., New York & London Academic Press.

11. Dehadrai,P.V. (1962) Observations on certain physiological reactions in Ophiocephalus striatus exposed to air. Life Sci. 11,653-657.

12. Dehadrai,P.V. (1966) Mechanism of gaseous exophthalmia in the Atlantic cod, Gadus morhua L. J.Fish.Res.Bd.Canada 23,909-914.

13. Denton,E.J. (1969) The buoyancy of fish and cephalopods. In Progress in Biophysics, vol.11, pp. 177-234, ed Butler,J.A.V. & Katz,B., New York Pergamon Press.

14. Denton,E.J., Liddicoat,J.D. & Taylor,D.W.(1972) The permeability to gases of the swimbladder of the conger eel (Conger conger). J.Mar.Biol.Ass. U.K. 52,727-746.

15. Dijkgraaf,S. (1950) Über die Auslösung des Gasspuckreflexes bei Fischen. Experientia 6, 188.

16. Dorn,E. (1962) Über den Feinbau der Schwimmblase von Anguilla vulgaris . Z.Zellf.55,894-912.

17. Dreser,H. (1892) Notiz über eine Wirkung des Pilokarpins. Arch.Exp.Path.Pharmak. (Leipzig) 30, 159-160.

18. Enns,T., Douglas,E. & Scholander,P.F. (1967) Role of the swimbladder rete of fish in secretion of inert gas and oxygen. Advances Biol. Med.Physics. 11, 231-244.

19. Evans,H.M. & Damant,G.C.C. (1828-29) Observations on the physiology of the swimbladder in cyprinoid fishes. Brit.J.Exp.Biol. 6,42-55.

20. Fahlén,G. (1965) Monoamines in the swimbladder of <u>Gadus callarias</u> and <u>Salmo irideus.</u> Acta Physiol.Scand. <u>64</u>, 119-126.

21. Fahlén,G. (1965) Histology of the posterior chamber of the swimbladder of Argentina. Nature <u>207</u>, 94-95.

22. Fahlén,G. (1967) Morphology of the gas bladder of <u>Coregonus lavaretus</u> L.. Acta Universitatis Lundensis Sect.II, n.28, 1-37.

23. Fairbanks,M.B., Hoffert,J.R. & Fromm,P.O. (1969) The dependence of the oxygen-concentrating mechanism of the teleost eye (<u>Salmo gairdneri</u>) on the enzyme carbonic anhydrase. J.Gen.Physiol. <u>54</u>, 203-211.

24. Fänge,R. (1953) The mechanism of gas transport in the swimbladder of euphysoclists. Acta Physiol.Scand. <u>30</u>, suppl. 110.

25. Fänge,R. & Wittenberg,J.B. (1958) The swimbladder of the toadfish (<u>Opsanus tau</u> L.). Biol.Bull. <u>115</u>, 172-179.

26. Fänge,R. (1958) The structure and function of the gas bladder in <u>Argentina silus.</u> Quart.J. Microscop. Sci. <u>99</u>, 95-102.

27. Fänge,R. (1966) Physiology of the swimbladder. Physiol.Rev. <u>46</u>, 299-322.

28. Franz,G. (1937) Über den Reflex des Gasspuckens bei Fischen und die Funktion des Weberschen Apparates. Z.Vergleich.Physiol. <u>25</u>, 193-238.

29. Gesser,H. & Fänge,R. (1971) Lactate dehydrogenase and cytochrome oxidase in the swimbladder of fish. Inter.J.Biochem. <u>2</u>,163-166.

29a. Gorman,R.R., Jordan,J.P., Simmons,J.B. & Clarkson,D.P. (1971) Biochemical adaptation in rat liver in response to marginal oxygen toxicity. Biochem.J. <u>125</u>, 439-447.

30. Haldane,J.S. (1922) <u>Respiration.</u> New Haven, Yale Univ.Press. (New edition, Oxford Clarendon Press,1935).

31. Harnisch,O. (1958). Leben ohne Sauerstoff na-

mentlich bei euroxybionten Süsswassertieren. Gewässer und Abwässer Heft 20, 7-12.

32. Harvey,H.H. (1963) Pressure in the early life history of Sockeye salmon. Ph.D.thesis Univ. Brit. Columbia, pp.1-267.

33. Jacobs,W. (1930) Untersuchungen zur Physiologie der Swimmblase der Fische. I. Über die "Gassekretion" in der Schwimmblase von Physoklisten. Z.Vergleich.Physiol. 11, 565-629.

34. Jasinski,A. (1964) Weberian apparatus and swimbladder of the pond-loach (*Misgurnus fossilis* L.). Acta Biol.Cracov.Ser.Zool. 7, 11-20.

35. Jasinski,A. & Kilarski,W. (1971) The gas gland in the swimbladder of the burbot (*Lota lota* L.) and stone-perch (*Acerina cernua* L.), its macro- and microscopic structure based on observations of electron microscopy. Acta Biol.Cracov.7, 111-125.

36. Johnston,R.S. (1953) The embryonic development of the swimbladder of the large-mouth bass *Micropterns salmoides salmoides* (Lac.). J. Morph. 93, 45-67.

37. Kanwisher,J. & Eberling,A.(1957) Composition of the swimbladder gas in bathypelagic fishes. Deep-Sea Res. 4,211-217.

38. Koch,H. (1934) L'émission de gaz dans la vésicule gazeuse des poissons. Rev.Quest.Sci.(4), 26, 385-409.

39. Krogh,A. (1919) The rate of diffusion of gases through animal tissues, with some remarks on the coefficient of invasion. J.Physiol.52, 391-408.

40. Krogh,A. (1943) Some experiments on the osmoregulation and respiration of *Eristalis* larvae. Entomol.Medd. (Jubilae umsbind) Copenhagen 23, 49-65.

41. Kuhn,W. & Kuhn,H.J. (1961) Multiplikation von Aussalz- und anderen Einzeleffekten für die Bereitung hoher Gasdrucke in der Schwimmblase. Z.f.Elektrochemie 65, 426-439.

42. Kuhn,H.J., Moser,P. & Kuhn,W. (1962) Haarnadelgegenstrom als Grundlage zur Erzeugung hoher Gasdrücke in der Schwimmblase von Tiefsee-Fischen. Arch.ges.Physiol. 275, 231-237.

43. Kuhn,H.J., & Marti,E. (1966) The active transport of oxygen and carbon dioxide into the swimbladder of fish. J.Gen.Physiol. 49, 1209-1220.

44. Kutchai,H. & Steen,J.B. (1971) The permeability of the swimbladder. Comp.Biochem.Physiol.39A, 119-123.

45. Lamonte,F.R. (1958) Note on the alimentary, excretory, and reproductive organs of Atlantic Makaira. Bull.Amer.Musc.Nat.History 114,396-401.

46. Lesslauer,W., Burger,F., Kuhn,H.J. & Kuhn,W. (1966) Modellversuch zur Gaskonzentrierung in der Schwimmblase der Fische. Pflügers Arch. 290, 56-69.

47. Maetz,J. (1956) Le rôle biologique de l'anhydrase carbonique chez quelques Téléostéens. Bull.Biol.France Belg. (Paris) Suppl.40,1-129.

48. Marshall,N.B. (1965) Systematic and biological studies of the Macrourid fishes (Anacanthini, Teleostei). Deep-Sea Res. 12, 299-322.

49. McEwen,R.S. (1940) The early development of the swimbladder and certain adjacent parts in *Hemichronis bimaculata.* J.Morphol. 67, 1-58.

50. Müller,J. (1840) Über Nebenkiemen und Wundernetze. Arch.f.Anat.Physiol. pp. 101-142.

51. Niesel,W. & Röskenbleck,H. (1963) Die Bedeutung der Stromgeschwindigkeit in den Gefässsystemen der Niere und der Schwimmblase für die Aufrechthaltung von Konzentrationsgradient. Arch. ges.Physiol. 277, 302-315.

52. Nillson,S. & Fänge,R.(1967) Adrenergic receptors in the swimbladder and gut of a teleost (*Anguilla anguilla*). Comp.Biochem.Physiol. 23, 661-664.

53. Nilsson,S. (1971) Adrenergic innervation and

drug responses of the sphincter in the swimbladder of the cod (Gadus morhua). Acta.Physiol. Scand. 83, 446-453.

54. Nilsson,S. (1972) Autonomic vasomotorinnervation in the gas gland of the swimbladder of the teleost (Gadus morhua). Comp.Gen.Pharmacol. (in press).

55. Patton,S. & Thomas,A. (1971) Composition of lipid from foam from swimbladders of two deep ocean fish species. J.Lipid Res. 12, 331-335.

56. Pette,D. (1971) Metabolic differentiation of distinct muscle types at the level of enzymatic organization. In Muscle Metabolism During Exercise pp.33-49, ed. Pernow,B. & Saltinl,B. Plenum Press.

57. Phleger,C.F. (1971) Pressure effects on cholesterol and lipid synthesis by the swimbladder of an abyssal Coryphoenoides species. Amer. Zool. 11, 559-570.

58. Piiper,J., Humphrey,H.T. & Rahn,H. (1962) Gas composition of pressurized, perfused gas pockets and the fish swimbladder. J.Appl.Physiol. 17, 275-282.

59. Piiper,J. (1965) Physiological equilibria of gas cavities in the body. In Handbook of Physiology, Sect. Respiration (3), vol.3, pp. 1205-1218.

60. Pool,J.L., Nason,G.J. & Forbes,H.S. (1934) Cerebral circulation. III. The effect of nerve stimulation and various drugs on the vessels of the dura mater. Arch.Neurol. 32, 1202.

61. Price,J.A. & Pierce,J.A. (1963) The physical properties of the frog lung. Proc.Soc.Expl.Biol. Med. 112, 538-541.

62. Quotob,Z. (1962) The swimbladder of fishes as a pressure receptor. Archs.neérl.Zool. 15, 1-67.

63. Rowntree,W.S. (1907) On some points in the visceral anatomy of the Characinidae, with an enquiry into the relations of the ductus pneumaticus of the Physostomi generally. Trans.Linn. Soc.London, Ser.2, Zool. 9, 47-81.

64. Röskenbleck,H. & Niesel,W. (1962). Die "Sauerstoffsekretion" in die Schwimmblase von Tiefseefischen. Naturwiss. 49, 114-115.

65. Saalfeld,E. v. (1933) Die nervöse Regulierung der Atembewegungen bei *Uromastix.* Pflügers.Arch. 233, 449-468.

66. Schmidt-Nielsen,K. (1972) *How animals work*. London & New York Cambridge, University Press.

67. Scholander,P.F. & van Dam,L. (1953) Composition of the swimbladder gas in deep sea fishes. Biol.Bull. 104, 75-86.

68. Scholander,P.F. (1958) Counter current exchange. A principle in biology. Hvalradets Skrifter. Norske Videnskaps-Akad. Oslo No.44, 1-24.

69. Srivastava,V.M.B. (1964) Some functions of the swimbladder and its ducts in Atlantic and Pacific Herring. Ph.D.thesis Univ.Brit.Coulumbia (Diss.Abstr.28, 1736-B-1967-8).

70. Steen,J.B. (1963) The physiology of the swimbladder of the eel *Anguilla vulgaris.* III. The mechanism of gas secretion. Acta Physiol.Scand. 59, 221-241.

71. Steen,J.B. (1970) The swimbladder as a hydrostatic organ. In *Fish Physiology* vol. IV,pp. 413-443, ed.Hoar,W.S. & Randall,D.J., New York & London Academic Press.

72. Stipetiĉ,E. (1939) Über das Gehörorgan der Mormyriden. Z.Vergleich.Physiol. 26, 740-752.

73. Stray-Pedersen,S. (1970) Vascular responses induced by drugs and by vagal stimulation in the swimbladder of the eel *Anguilla vulgaris.* Comp.Gen.Pharmacol. 1, 357-364.

74. Sundnes,G., Bratland,P. & Strand,E. (1969) The gas content in the coregonid swimbladder. FiskDir.Skr.Ser.HavUnders. 15, 274-278.

75. Tepperman,J. (1966) *Metabolic and Endocrine Physiology*. Chicago,Year Book Med.Publ.

76. Wittenberg,J.B., Noronha,J.M. & Silverman,M. (1962) Folic acid derivatives in the gas gland of *Physalice physalis* L.. Biochem.J.85, 9-15.

77. Wittenberg,J.B. & Wittenberg,B.A. (1962) Active secretion of oxygen into the eye of fish. Nature 194, 106.

78. Wittenberg,J.B., Schwend,M.J. & Wittenberg,B.A. (1964) The secretion of oxygen into the swim-bladder of fish. III. The role of carbon dioxide. J.Gen.Physiol. 48, 337-355.

79. Voit,C. (1865) Über die in den Schuppen und der Schwimmblase von Fischen vorkommenden irisierenden Krystalle. Z.Wiss.Zool. 15, 515-521.

80. Woodland,W.N.F. (1911) On some experimental tests of recent views concerning the physiology of gas production in teleostean fishes. Anat. Anz. 40, 225-242.

81. Woodland,W.N.F. (1911) On the structure and function of the gas glands and retia mirabilia associated with the gas bladder of some teleostean fishes, with notes on teleostean pancreas. Proc.Zool.Soc.London 1, 183-248.

Comparative Physiology, eds. L. Bolis, K. Schmidt-Nielsen and S. H. P. Maddrell

GAS EXCHANGE IN AVIAN LUNGS: MODELS AND EXPERIMENTAL EVIDENCE

J.PIIPER and P.SCHEID
Abteilung Physiologie,
Max-Planck-Institut für
experimentelle Medizin
Göttingen, Germany

The aim of this report is to evaluate some of the recent morphological and physiological data pertaining to gas exchange in birds' lungs. In particular the experimental results on unidirectional respiratory air flow and the evidence for cross-current gas exchange will be analyzed in detail. A complete review of the recent literature will not be attempted. On the contrary, the selection of materials will be rather subjective, with emphasis placed on our own experimental results and on the morphological data obtained by Duncker (6,7,8).

Models for gas exchange organs in vertebrates

There are three basic types of external gas exchange organs in vertebrates, each of them corresponding to a different functional model (13) (Fig. 1).

(1) Mammalian lungs:uniform pool model. The voluminous air spaces of the mammalian lung contain a gas mixture - alveolar gas - of nearly uniform composition. Pulmonary capillary blood equilibrates with the alveolar gas with respect to P_{CO_2} and P_{O_2}. Under idealized conditions (diffusion resistances and "functional inhomogeneity" being wholly absent) the partial pressures in end-capillary or arterial blood reach the corresponding values in the alveolar gas. In reality this goal is not achieved:arterial P_{O_2} is lower and arterial P_{CO_2} is higher than the alveolar values.

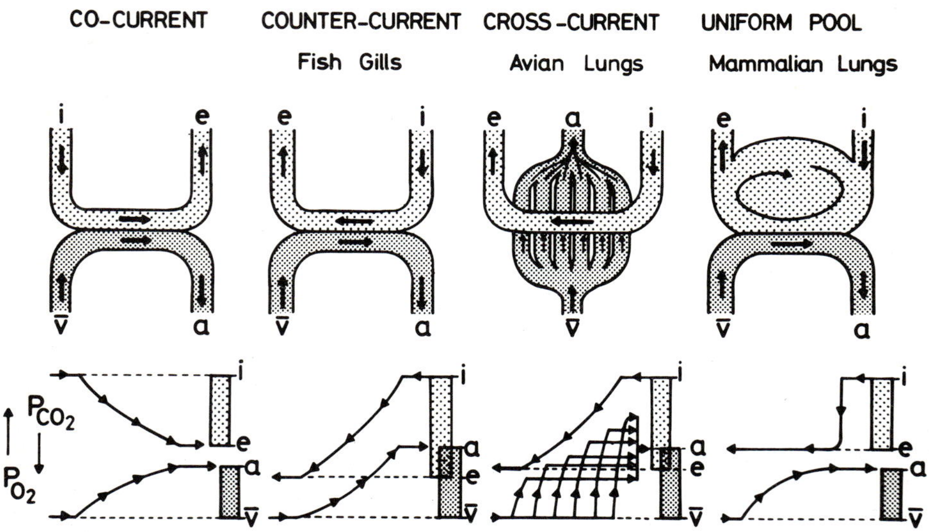

Figure 1 - Models for gas transfer in external gas exchange organs of fishes, birds and mammals. The symbols i,e,$\bar{v}$ and a refer to medium entering and leaving, and to blood entering and leaving the gas exchange organ, respectively. Lower panel: schematic representation of P_{CO_2} and P_{O_2} changes in medium and blood during passage through the gas exchange organ.

(2) Fish gills: counter-current model. In fish gills water in the interlamellar spaces and blood in the secondary lamellae flow in opposite directions. The blood leaving the gills, i.e. arterial blood, has had its last gas equilibration contact with water that has just entered the interlamellar space, i.e. with inspired water.Conversely, water leaving the system,i.e. expired water,potentially equilibrates with mixed venous blood.Thus arterial blood P_{CO_2} and P_{O_2} may come close to inspired values, and expired P_{CO_2} and P_{O_2} approach the values in mixed venous blood. Thus, if the resistance to diffusion is sufficiently low, an "overlapping of

water and blood P_{CO_2} and P_{O2} would be expected to occur in the manner that the arterial value lies between the inspired and the expired values, and the expired value is between the arterial and the venous values. Such "overlapping", which implies a high level of gas transfer efficacy, cannot occur in mammalian lungs and therefore their maximum gas transfer efficacy is lower than that of fish gills.

(3) Bird lungs: hypothesis of cross-current model. The structure of avian lungs differs fundamentally from that of mammalian lungs. There are no dead-end alveolar ducts and alveoli. Gas exchange takes place in the walls of the parabronchi (more precisely, in the air-blood capillary network contained in the walls of the parabronchi) which open into the bronchial system at both end (see Figs. 3,4 and 10). Therefore, the gas composition of the parabronchi must change considerably from the inspiratory to the expiratory end. According to the cross-current model first introduced by Zeuthen (21)and later employed by Scheid and Piiper (15) for analysis of gas exchange in avian lungs, parabronchial capillary blood flows at right angles to the long axis of the parabronchus. Accordingly, capillary blood equilibrates with parabronchial gas of widely differing P_{CO2} and P_{O2}, the values of which depend on the distance from the inspiratory end of the parabronchus, and the arterial blood is formed by mixing streams of end-capillary blood of widely varying gas composition. Since the blood perfusing the inspiratory end of the parabronchus has a chance to equilibrate with gas entering the parabronchus, it is perfectly possible that the mixed arterial blood leaving the lung reaches a higher P_{O2} and lower P_{CO2} than the gas leaving the lung, i.e. that there is some degree of "overlapping".

Efficacy of gas transfer in models

There are numerous ways of characterizing the efficiency or efficacy of a gas exchange system. A simple way is to consider the inflowing medium and blood as constant, of given gas composition, and to view the composition of the effluent medium and blood in relation to these (13). One such possibility is to consider the ratio x,

$$x = \frac{P_e - P_a}{P_i - P_{\bar{v}}}$$

where i refers to medium entering the organ, e, to medium leaving the organ, $\bar{v}$, to blood entering the organ, and a, the blood leaving it.

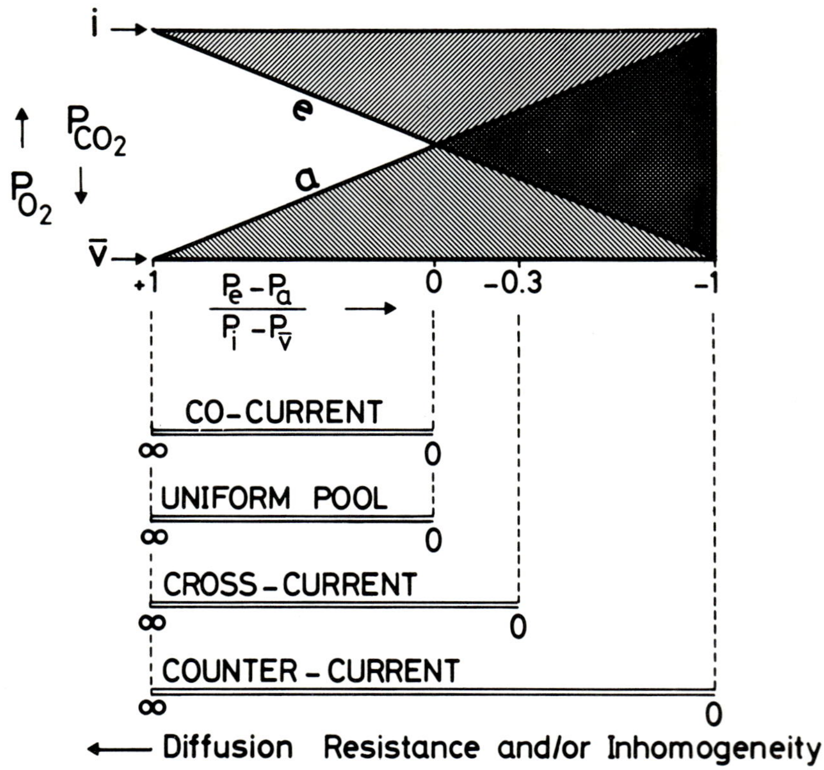

Figure 2 - Efficacy of gas transfer models. $(P_e - P_a)/(P_i - P_{\bar{v}})$, index for efficacy of CO_2 or O_2 exchange, with possible range from -1 to +1. The range for each model is indicated by a double line, the right end of which marks the maximum efficacy.

Negative values of x signify high efficacy (Fig. 2). The theoretical possible limit x = -1 can be approached by a counter-current system lacking diffusional limitation under a certain combination of ventilation and perfusion conductances (13). The cross-current system can attain x = -0.3 under optimal conditions. With the uniform pool model, and also with the co-current model, the lower limit for x is zero.

These are the upper limits for efficacy. Of course, in real systems there are several reasons why these limiting values are not reached. The more important ones are (1) deviation from the above-mentioned matching of medium and blood flow conductances; (2) presence of a resistance to diffusion between medium and blood (which functionally includes resistances due to rate-limiting chemical reactions); (3) inhomogeneity of the system with respect to dimensions, flow rates, diffusion resistances etc. Therefore, a low value of the overall efficacy ($x > 0$) can occur with all models. Only high efficacy values can be helpful in excluding some models. In particular negative values for x cannot be explained on the basis of co-current and uniform pool models.

Morphology and direction of air flow

The gross anatomy of the respiratory system in birds is schematically represented in Fig. 3. The trachea divides into two main bronchi whose intrapulmonary portions are called mesobronchi. Each mesobronchus gives rise to a more proximal group of secondary bronchi, the ventrobronchi, and to a more distal group of secondary bronchi, the dorsobronchi and the laterobronchi, before terminating in the abdominal air sac. The main intrapulmonary airways are the parabronchi which give rise to the peripara bronchial air capillary network where gas exchange with blood capillaries takes place. The parabronchi are long narrow tubes of uniform diameter (about 0.2 mm to 1 mm, depending on species) typically connecting the ventrobronchi, which lie on the ventral-medial surface of the lungs, with the dorsobronchi situated on the lateral-dorsal surface of the lungs.

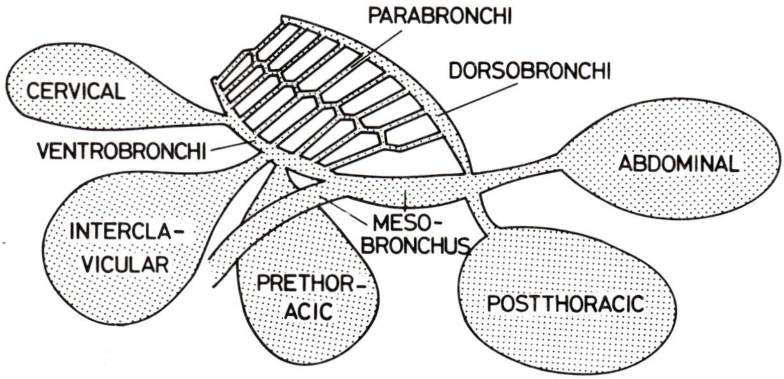

Figure 3 - Schema of the lung-air-sac system of birds.

The parabronchi are arranged in a regular parallel pattern and show some degree of anastomosing, particularly in the central plane of the lung between the ventrobronchi and the dorsobronchi (see also Fig. 9).

The lungs, i.e. the parabronchi, are relatively rigid structures. The changes in airway volume during the respiratory cycle seem to be exclusively confined to the voluminous air sacs. Usually there are 4 pairs of air sacs and an unpaired sac (the interclavicular sac). Two groups of air sacs should be discerned on the basis of their anatomical connections and with regard to their physiological role in respiration. The cranial air sacs (cervical, interclavicular and prethoracic) are connected to the ventrobronchi, whereas the caudal air sacs (postthoracic and abdominal) communicate with the mesobronchus in the region where the dorsobronchi and the laterobronchi take their origin.

This description holds for the major, and probably phylogenetically older part of the lung designated as paleopulmo by Duncker (6,7). In addition, another parabronchial network, termed neopulmo by Duncker, may be developed to a variable extent between secondary bronchi and the air sacs (see below).

The pattern of air flow through this rather complex system which appears to offer alternative pathways has been a challenging problem to which a variety of tentative solutions has been offered (a short historical review has recently been presented by Bretz and Schmidt-Nielsen (5)). A selection is presented in Table 1.

Table 1. - Hypotheses on direction of gas flow in the respiratory system of birds.

A. Flow bidirectional (inspiration: from ventrobronchi via parabronchi to dorsobronchi; expiration: from dorsobronchi via parabronchi to ventrobronchi)

Zeuthen (1942)

B. Flow unidirectional (from dorsobronchi via parabronchi to ventrobronchi)

I. Flow during one respiratory phase

a. Flow during inspiration only

Portier (1928)
Dotterweich (1936)

b. Flow during expiration only

Vos (1935)

II. Flow during both respiratory phases

Bethe (1925)
Dotterweich (1930)
Hazelhoff (1943)

Since all efforts to find morphological valves in the bronchial system had failed, the assumption of Zeuthen (21) postulating a reversal of air flow during expiration as compared to inspiration in all parts of the respiratory system, including parabronchi, seemed to be the most straight forward and logical. However, many findings such as the differences in gas concentrations between cranial and caudal air sacs, sites of preferred deposition of coal dust and the behaviour of hydrodynamic models led the majority of the authors to believe that air flow through the parabronchi was unidirectional from the dorsobronchi through to the ventrobronchi. As far as the qualitative flow pattern and not its mechanisms are concerned, the differences between the hypotheses of the various authors reside in assuming air flow to take place during inspiration only, or during expiration only, or during both respiratory phases (Table 1).

In the last few years three groups working independently have succeeded in directly measuring the air flow in bronchi. Bretz and Schmidt-Nielsen (5) used a heated thermistor probe introduced via tracheostomy into primary or secondary bronchi of ducks. Scheid and Piiper (16) implanted a heated wire anemometer into dorsobronchi in ducks. Brackenbury (3,4) measured air flow in the dorsobronchi of geese by an implanted Pitot type flowmeter. The results of these studies showed beyond any doubt that the parabronchial gas flow was from the dorsobronchi via the parabronchi to the ventrobronchi both during inspiration and expiration, in accordance with the hypotheses propounded by Bethe (1) and later by Hazelhoff (9). The same conclusion was reached by Bouverot and Dejours (2) who experimentally analyzed the time course of CO_2 concentration changes in various air sacs following administration of CO_2 into inspired gas or into air sacs. The air flow patterns during inspiration and expiration based on measurements performed by Scheid *et al.* (17) on duck lungs *in vivo* and on reduced duck lung preparations are illustrated in Fig. 4.

Possible relationships between air flow and blood flow

The parabronchi are surrounded by a layer of intermingling air and blood capillary networks (cf. 6). The blood capillary net seems to receive blood inflow at all levels of the lenght of a parabronchus, and there appear to be outflow channels distributed over the whole length of the parabronchus. Thus the most probable pathway of a blood element would be that offering least resistance, i.e. the shortest one connecting inflow and outflow vessels inside the same segment of the parabronchus (or of several adjacent parabronchi). This would mean that

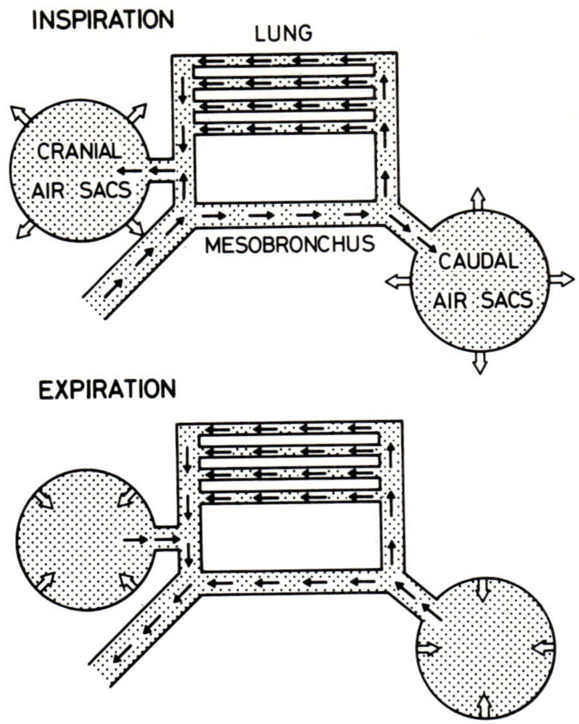

Figure 4 - Functional model of the lung-air-sac system of birds. The arrows indicate direction of gas flow during inspiration and during expiration. After Scheid, Slama and Piiper (17).

the air-blood gas exchange relationships are comparable to those of the cross-current model (Fig. 1). However, this anatomical evidence is not conclusive, since the relationships between the inflow and outflow vessel have not been investigated in such detail as to exclude the possible dominance of one direction of blood flow parallel to the long axis of the parabronchus.

If the blood flow had a preferential longitudinal flow direction component the system would be comparable either to a co-current model or to a counter-current model depending on the reciprocal relations between the gas flow in the parabronchial lumen and the longitudinal blood flow in the periparabronchial tissue layer. As we have explained above, there is a large difference between the co-current and counter-current models in the maximum gas transfer efficacy that they can achieve (Figs. 1 and 2). Therefore, some elucidation of the situation may be expected from experimental analysis of gas exchange in birds.

In a series of experiments performed in unanesthetized domestic fowl spontaneously breathing normal air or room air, or 3% CO_2 and 13% O_2 in N_2, several respiratory and circulatory parameters were measured simultaneously including P_{CO_2} and P_{O_2} in expired gas, in arterial blood and in mixed venous blood (12, 15). The mean values are plotted in Fig. 5 in the same manner as in Fig. 1.

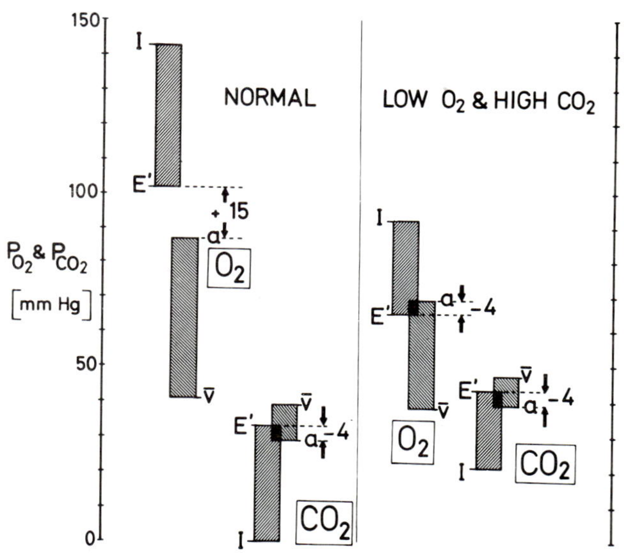

Figure 5 - Gas-blood CO_2 and O_2 equilibration in domestic fowl. I, E', a and $\bar{v}$ refer to inspired, end-expired, arterial and mixed venous P_{CO_2} and P_{O_2} values, respectively. The differences between $P_{E'}$ and P_a are indicated, their values are negative for gas-blood "overlapping". After Piiper, Drees and Scheid (12) and Scheid and Piiper (15)

In both cases the gas P_{CO_2} and blood P_{CO_2} ranges show a definite "overlapping". For P_{O_2} the "overlap" is present when low O_2 air is inspired but there is no "overlap" with normal room air. Comparison with Figs. 1 and 2 shows that these data are compatible with the cross-current and the counter-current models, but not with the co-current model. When the fowl is breathing normal air, it is probable that "functional inhomogeneities" of the kind well known to reduce the gas exchange efficiency of mammalian lungs (unequal rates of ventilation and perfusion, venous admixture) prevent the attainment of the "overlapping" condition.

Certain difficulties for quantitative evaluation of these experimental data in terms of different models arise from the fact that inspired gas and end-expired gas do not adequately represent the gas entering and leaving the lungs, respectively. Since the caudal air sacs, which provide the gas entering the lungs during expiration, contain more CO_2 and less O_2 than inspired gas (see below) the average gas entering the lungs must on average have a higher P_{CO_2} and a lower P_{O_2} than inspired gas. Furthermore, the gas leaving the lungs could have a somewhat higher P_{CO_2} and lower P_{O_2} than the exhaled end-expired gas because this also contains gas expelled from the caudal air sacs via the mesobranchus during expiration. Consideration of these factors reveals that the extent of "overlapping" of gas and blood P_{CO_2} and P_{O_2} ranges in the lung is almost certainly greater than that obtained from measurements of inspired and end-expired gas.

"Overlapping" is compatible with both the cross-current and the counter-current models. In order to differentiate between these two possibilities a further specific experimental test is required.

Experiments with reversal of parabronchial air flow

If in a counter-current model one of the flows is reversed a co-current system with decreased gas transfer efficacy must result. In the cross-current system, however, reversal of one of both flows should produce no effect. It appeared to us to be worth while to design an experiment on this basis

(18,19). Since reversal of the blood flow seemed not to be feasible, it was sought to reverse the air flow through the lungs (Fig. 6).The technical

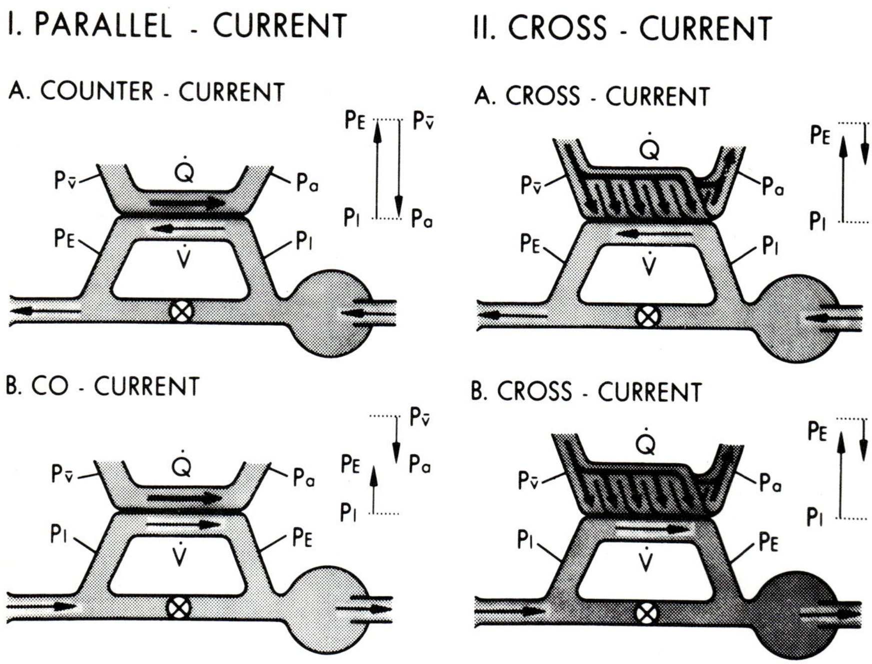

Figure 6 - Theoretical basis for experiments with reversal of air flow in duck lungs. Left side: a counter-current system is converted to a co-current system, or vice-versa, with large change in efficacy of gas transfer. Right side: a cross-current system remains unchanged with reversal of flow. From Scheid and Piiper (19).

problem was to make sure that the gas flow through the lungs proper was quantitatively reversed,given the possibilities of variable "shunting" of gas through the mesobronchus. For this purpose we attempted to block the mesobronchus between the orifices of the ventrobronchi and those of the dorsobronchi by means of an inflatable balloon.

In 3 experiments on domestic ducks we succeeded in placing a balloon catheter in the desired position (Fig. 7). Cannulae were introduced into

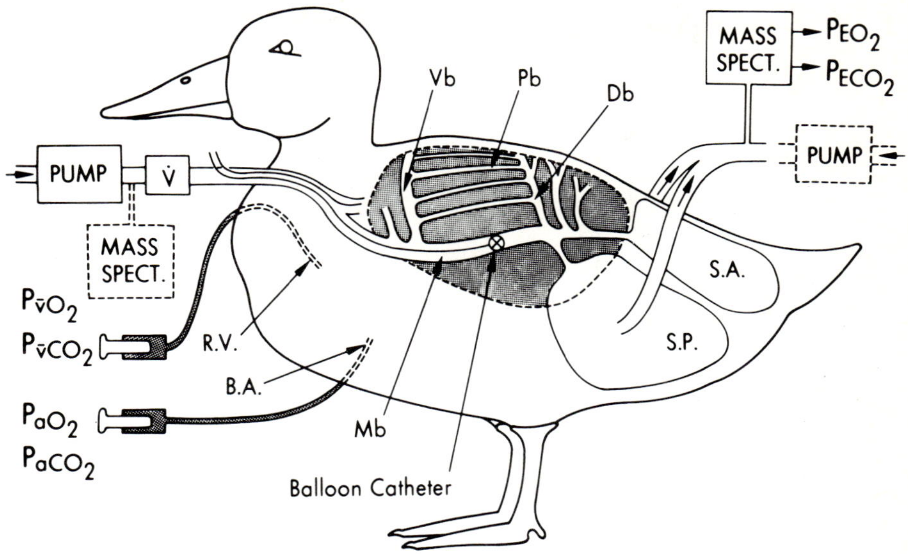

Figure 7 - Schema of experimental set-up for experiments with reversal of air flow in duck lungs. Mb, mesobronchus. Vb, ventrobronchi. Db, dorsobronchi. Pb, parabronchi. R.V., right heart ventricle. B.A., brachial artery. S.P., post-thoracic air sac. S.A., abdominal air sac. Mass spect., mass spectrometer. V̇, flow measured by pneumotachograph. From Scheid and Piiper (19)

the trachea and into the postthoracic air sacs. Thus a steady air stream could be led from the postthoracic air sacs through the lungs in the "normal" direction (from dorsobronchi via parabronchi to ventrobronchi) or from the trachea through the lungs in the "reversed" direction (from ventrobronchi via parabronchi to dorsobronchi). The air flow rate was measured by means of pneumotachograph. P_{CO_2} and P_{O_2} in inspired and expired gas (which in this condition should be identical with gas entering and leaving the lungs, respectively) were monitored by a mass spectrometer. Samples of arterial and mixed venous blood were taken for measurement of P_{CO_2} and P_{O_2} using standard electrode techniques.

The mean results are summarized in Fig. 8. It

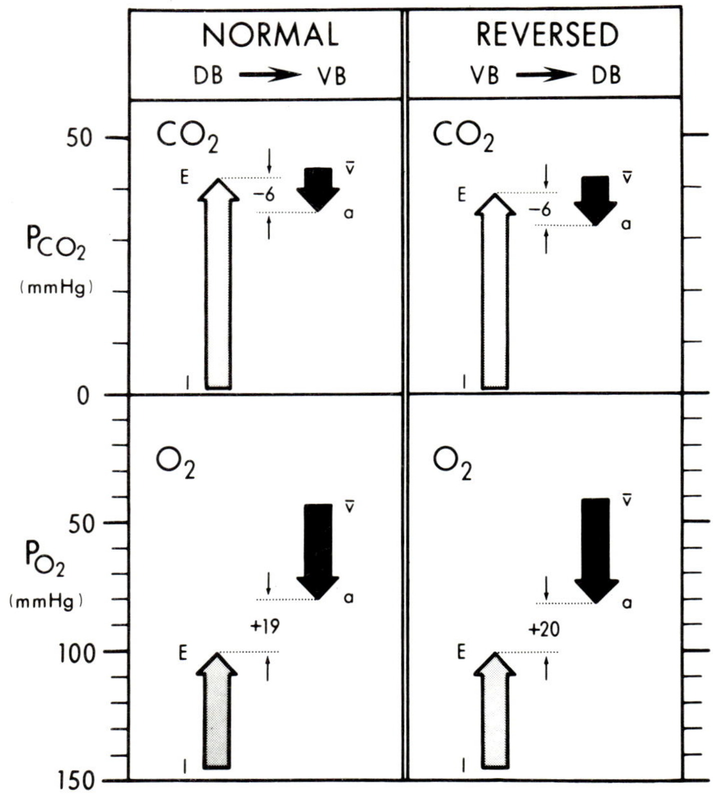

Figure 8 - Mean results of experiments with reversal of air flow in duck lungs. I, E, a and $\bar{v}$ denote inspired, expired, arterial and mixed venous P_{CO_2} and P_{O_2}, respectively. Representation analogous to Fig. 5. Normal: air flow direction from dorsobronchi via parabronchi to ventrobronchi. Reversed: air flow direction from ventrobronchi via parabronchi to dorsobronchi. From Scheid and Piiper (19).

is evident that the P_{CO_2} and P_{O_2} values and their relationships remained unchanged upon reversal of air flow through the lungs. The gas and blood P_{CO_2} ranges show a considerable extent of "overlap", whereas for O_2 the "overlap" is absent. These relationships are very similar to those previously found in the domestic fowl (Fig. 5).

If a counter-current system had been present, it should have been transformed into a co-current system when the flow of air through the lung was reversed (Fig. 6). This transformation would be expected to reduce the overall gas exchange efficiency, in particular it should have abolished the "overlapping" of gas and blood P_{CO_2} ranges. Thus the experimental results are not compatible with the counter-current model for gas exchange, but are in good agreement with the cross-current model, which allows "overlapping" of blood and gas partial pressures with both gas flow directions.

Physiological significance of unidirectional parabronchial gas flow

What are the advantages of the unidirectional gas flow when compared to the simpler model of Zeuthen (1942) with bidirectional gas flow?

A change in flow direction implies a transitory interruption of gas flow, albeit of short duration. The experimental records of the unidirectional gas flow in the dorsobronchi showed no check in the gas flow, although its rate decreased toward the end of expiration considerably at low respiratory frequencies (3, 16). At high respiratory frequencies (panting) the gas flow became much more steady, displaying but relatively small oscillations with the respiratory phase.

A check in the ventilatory air stream would cause a fall of P_{O_2} in the gas exchanging airways at a rate dP_{O_2}/dt which is proportional to the momentary O_2 uptake, $\dot{V}_{O_2}$, and inversely proportional to the volume of the airways, V,

$$- \frac{dP_{O_2}}{dt} = f \cdot \frac{\dot{V}_{O_2}}{V}$$

(f is a conversion factor depending on temperature, pressure and water vapor pressure).

The total volume of the lung-air sac system has been determined, by wash-out techniques, to average 170 ml in hens of 1.6 kg body weight (14). According to Duncker (7) only 11.2% of this volume, or about 20 ml, is located in the lungs. The O_2 uptake in 1.6 kg hens has been found to average 24.3 ml/min (12,15). From these figures it turns out that the initial P_{O_2} fall rate, if ventilation ceased, would be 18 torr/sec, which would lead to severe hypoxia in a few seconds. In mammals an apnea of short duration has a much lesser effect because of the relatively much larger alveolar space. For man, with V_{O_2} = 300 ml/min and an alveolar volume of 3000 ml, the initial P_{O_2} fall during apnea turns out to be only 1.4 torr/sec. These calculations show the importance of maintaining an uninterrupted air flow through avian lungs.

The unidirectional air flow pattern predicted by Bethe (1) and confirmed by direct recording of air flow appears to have another fundamental advantage with respect to efficiency of gas exchange during the expiratory phase. In Zeuthen's model gas from the caudal air sacs, which at least in part must have passed through the parabronchi and therefore must have acquired an increased CO_2 content and a decreased O_2 content, will enter the lungs during expiration, which would decrease the oxygenation of the blood traversing the capillaries. In the unidirectional flow model the caudal air sacs should contain essentially inspired gas (except for re-inspiration of end-expired gas from the relatively small dead space) and thus would provide fresh air for lungs during expiration. Unfortunately, however, the situation is not quite so simple as this because the caudal air sac turns out to contain gas somewhat different in composition from inspired gas.

Caudal air sac gas and neopulmo

Numerous measurements have revealed that the gas of caudal air sacs has a higher P_{CO_2} and lower P_{O_2} than inspired gas, although to a considerably lesser degree than the gas in cranial air sacs which is similar to end-expired gas. For example, Bouverot and Dejours (2) found the following mean values in unanesthetized domestic fowl,
abdominal air sac P_{CO_2}, 27.9 torr,

interclavicular air sac P_{CO2}, 45.2 torr,

end-tidal gas P_{CO_2}, 43.4 torr.

The origin of this CO_2 excess and of O_2 deficit in the caudal air sacs with respect to inspired gas is of considerable interest in the understanding of the gas exchange function of the respiratory system in birds. Three explanations appear to be theoretically feasible.

First, CO_2 may diffuse directly from surrounding tissues and blood vessels into the air sacs. Because of the relatively high ventilation rate of the caudal air sacs and because diffusion conditions are much less favorable than in the parabronchi, this source is usually considered not to be of importance.

A second explanation results from a particular feature of the hypothesis of Hazelhoff (9). He assumed that gas entering the ventrobronchi from the parabronchi during inspiration recirculated via the mesobronchus into the dorsobronchi and into the caudal air sacs. However, measurements in fixed lung preparations (17) showed air flow from the mesobronchus to the cranial air sacs during inspiration (Fig. 4). Furthermore, preliminary measurements in the mesobronchus distal to the orifices of the ventrobronchi have yielded P_{CO_2} and P_{O_2} values close to those in inspired gas during inspiration (Piiper, unpublished observations). These results are not in accordance with the "recirculation loop" postulated by Hazelhoff.

According to the third explanation the major part of CO_2 found in caudal air sacs originates in gas exchange in the parabronchial system connecting the mesobronchus and the dorsobronchi (and the laterobronchi) to the posterior air sacs (Fig. 9). This part of the lungs is apparently of more recent origin and, therefore, has been termed "neopulmo" by Duncker (6,7), in distinction from the "paleopulmo" which constitutes the major part of the lungs and which up to now has been exclusively considered in this paper. The neopulmo is absent in primitive groups (e.g. penguins) and reaches its highest degree of development in fowl-like birds and in song-birds (in which it extends also to the ventrobronchi), but it never exceeds 20% of the total lung tissue mass (6).

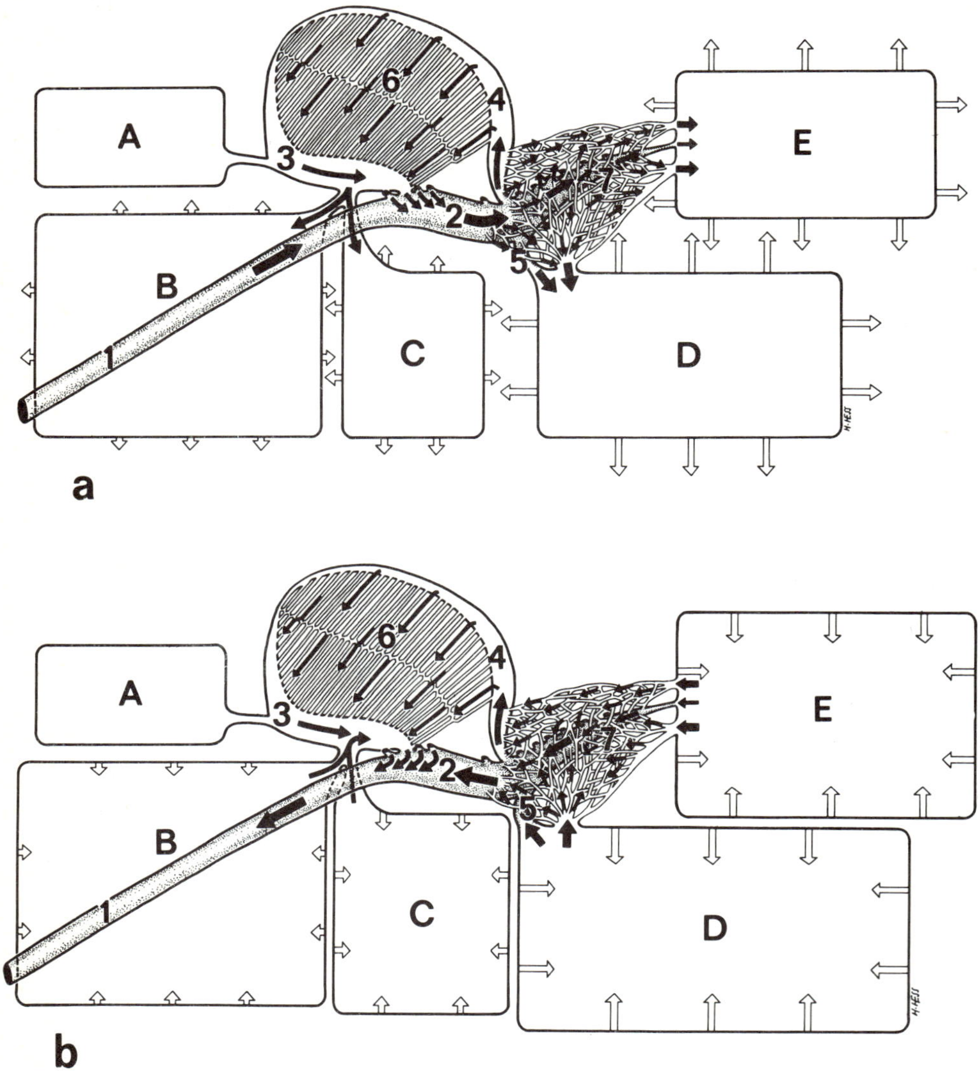

Figure 9 - Schematic representation of air flow through paleopulmonic (6) and neopulmonic parabronchi (7) during inspiration (a) and during expiration (b). A,B,C,D,E, cervical, interclavicular, prethoracic, postthoracic, and abdominal air sacs, respectively. 2, mesobronchus. 3, ventrobronchi. 4, dorsobronchi. 5,laterobronchi. From Duncker (6).

It is highly probable that the neopulmonic parabronchi are bidirectionally ventilated, viz. in the direction from mesobronchus or dorsobronchi towards caudal air sacs during inspiration, and in the opposite direction during expiration (Fig. 9). This becomes a necessity with the neopulmonic parabronchi connecting the mesobronchus to the abdominal air sac in song-birds, in which the distal portion of the mesobronchus, leading to the abdominal air sac, is reduced to the size of a single parabronchus thus leaving the net of neopulmonic parabronchi virtually the only connecting path between mesobronchus and abdominal air sac (6,7).

Thus the higher CO_2 and lower O_2 concentrations of the caudal air sacs, as compared to inspired gas, can be explained by gas exchange of that part of the inspired air which enters the caudal air sacs by way of the neopulmonic passageways. It should be noted that during expiration,by the same token of neopulmonic gas exchange, the gas leaving the caudal air sacs is even enriched in CO_2 and depleted of O_2. Thus the paleopulmo which during expiration is ventilated by gas originating in caudal air sacs receives gas with even higher concentration of CO_2 and lower concentration of O_2 than in caudal air sacs. Functionally,paleopulmo and neopulmo are arranged in parallel during inspiration, with inspired air entering the parabronchi of both types, whereas during expiration both types of parabronchi, at least in part, are arranged in series, with expired air in part ventilating both types of parabronchi consecutively.

It is evident that the presence of the neopulmo very much complicates any analysis of gas exchange in bird lungs. Therefore, it may prove to be profitable to use species with no,or poorly developed,neopulmo for experimental work on pulmonary gas exchange.

Since the neopulmonic parabronchi seem to be bidirectionally ventilated, it appears that the manner of air flow, unidirectional or changing direction with the respiratory phase, is not of prime importance for gas transfer in the parabronchi.

Gas exchange between parabronchial gas and blood

Attempts to determine the overall conductance for diffusional gas transfer between parabronchial gas and blood have been undertaken in the domestic fowl using techniques worked out for mammalian lungs. By a CO rebreathing technique Piiper, Pfeifer and Scheid (11) estimated the diffusing capacity for CO, D_{CO}, at 0.80 $ml \cdot min^{-1} \cdot torr^{-1}$ in domestic fowl of 1.4 kg average body weight. Later Scheid and Piiper (15) analyzed O_2 and CO_2 transfer in domestic fowl of 1.6 kg average body weight, breathing high CO_2 and low O_2 mixtures. The diffusing capacity for O_2, D_{O2}, was estimated at 1.1.to 1.5 $ml \cdot ml \cdot min^{-1} \cdot torr^{-1}$. Recently Duncker (7,8) has published some results of his morphometric determinations of the blood-gas exchange surface area in avian lungs. In Table 2 the relationships between the diffusing capacity for O_2 (D_{O2}), the resting O_2 uptake ($\dot{V}_{O2}$) and the pulmonary gas exchange surface area (S) for the domestic fowl and for the dog are compared.

It is clear that D_{O2} per unit surface area is higher in the bird, but the ratio $D_{O2}/\dot{V}_{O2}$ is higher in the dog. The exchange surface area per resting $\dot{V}_{O2}$ is lower in the hen. This is possibly explainable by the higher gas exchange efficacy at a lower D_{O2}. It must be stated that this comparison should be considered more as a program for future work than as a valid result, mainly due to problems in the interpretation of the D_{O2} values determined by physiological methods. Thus, in spite of considerable efforts, determination of D_{O2} in mammalian lungs is still problematic (e.g. 10).

As has been described above, gas exchange between gas and blood takes place in the network of air and blood capillaries which constitutes the bulk of the periparabronchial tissue. The transfer of O_2 from the parabronchi and the transfer of CO_2 into the parabronchi has to be accomplished by diffusion in the periparabronchial airways which comprise the atria, the infundibula and the air capillaries (6). Due to lack of specific morphometric data the resistance to diffusion cannot be estimated at present. But one can predict that if

Table 2 - O_2 diffusing capacity, D_{O_2} and exchange surface area, S, of lungs in a bird and in a mammal
Domestic hen: gas exchange data (15), morphometric data (7)
Dog: gas exchange data (10), morphometric data (20).

	Domestic fowl	Dog
Body weight	1.6 kg	23 kg
Exchange surface area, S	2.88 m^2	72 m^2
Resting O_2 uptake, $\dot{V}_{O_2}$	22 ml/min	144 ml/min
O_2 diffusing capacity, D_{O_2}	1.3 $\frac{ml/min}{torr}$	22 $\frac{ml/min}{torr}$
D_{O_2}/S	0.45 $\frac{ml/min}{torr \cdot m^2}$	0.31 $\frac{ml/min}{torr \cdot m^2}$
$D_{O_2}/\dot{V}_{O_2}$	0.059 $torr^{-1}$	0.15 $torr^{-1}$
$S/\dot{V}_{O_2}$	0.13 $\frac{m^2}{ml/min}$	0.50 $\frac{m^2}{ml/min}$

a gradient due to diffusion resistance were to exist, its role in limiting the overall gas transfer from parabronchial gas to blood would be diminished by the counter-current-like arrangement of the air and blood capillaries, based on anatomical evidence (Fig. 10).

According to Duncker (6) the inflow into pulmonary blood capillaries occurs at the periphery of a "parabronchial unit" considered to include the periparabronchial air-blood capillary network. By contrast, the collecting venules are located at the parabronchial lumen. If a certain O_2 diffusion

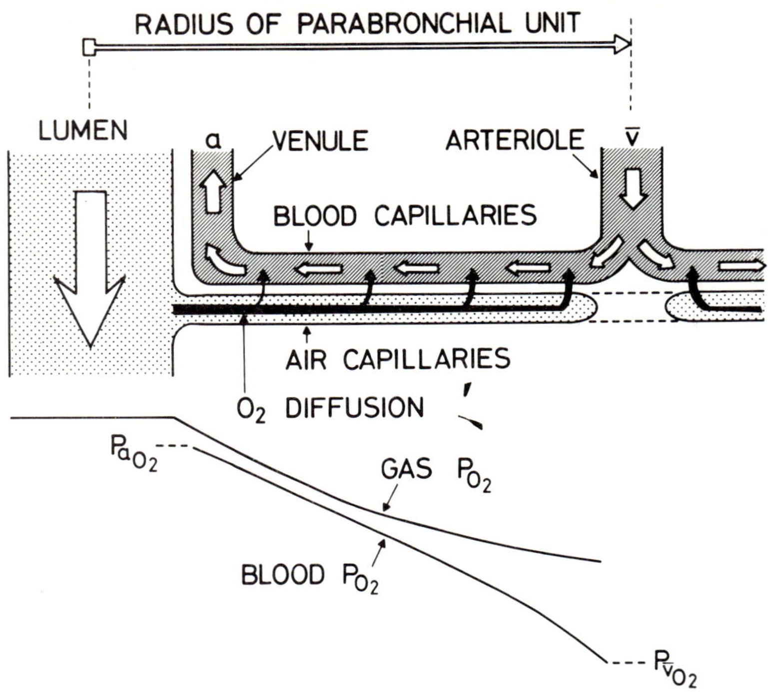

Figure 10 - Schema of the counter-current-like effect enhancing O_2 transfer between air and blood capillaries. Blood flows from the periphery towards the lumen of the parabronchus, O_2 in the air capillaries diffuses in the opposite direction. Lower panel: P_{O_2} profiles in gas and blood.

gradient in the small periparabronchial airways is present, the capillary blood equilibrates with increasing P_{O_2} along its course and enters the venules after contact with gas whose P_{O_2} is close to that in the parabronchial lumen. The ensuing P_{O_2} profiles resemble those occurring in a counter-current model (Fig. 1). It is evident that this counter-current-like mechanism is expected to contribute to the effectiveness of CO_2 transfer in the same manner.

It should be pointed out, however, that - in contrast to the true counter-current arrangement - this counter-current like mechanism is not capable of producing an "overlapping" of gas and blood P_{O_2} and P_{CO_2} ranges, leading to negative values of the ratio

$$\frac{P_e - P_a}{P_i - P_{\bar{v}}}$$

as discussed above. All it can do is to enhance equilibration of capillary blood with parabronchial gas when there is a diffusion gradient of P_{O_2} and P_{CO_2} along the small airways of the parabronchial unit.

REFERENCES

1. Bethe,A. (1925) Atmung: Allgemeines und Vergleichendes. In Handbuch der normalen und pathologischen Physiologie, pp. 1-36, ed. Bethe, A., Bergmann,G.V., Embden,G. & Ellinger,A., Berlin, Springer-Verlag.

2. Bouverot,P. & Dejours,P. (1971) Pathway of respired gas in the air sacs - lung apparatus of fowl and ducks. Respir.Physiol. 13, 330-342.

3. Brackenbury,J.H. (1971) Airflow dynamics in the avian lungs as determined by direct and indirect methods. Respir.Physiol. 13, 319-329.

4. Brackenbury,J.H. (1972) Physical determinants of air flow pattern within the avian lung. Respir.Physiol. 15, 384-397.

5. Bretz,W.L. & Schmidt-Nielsen,K. (1971) Bird respiration: flow patterns in the duck lung. J.Exp.Biol. 54, 103-118.

6. Duncker,H.R. (1971) The lung air sac system of birds. Erg.Anat.Entwickl.Gesch. 45 (6),1-171.

7. Duncker,H.R. (1972) Structure of avian lungs. Respir.Physiol. 14, 44-63.

8. Duncker,H.R., Der quantitative Aufbau des Lungen-Luft-sacksystems der Vögel. Verh.Anat. Ges. 67: Anat.Anz. (in press).

9. Hazelhoff,E.H. (1943) Bouw en functie van de vogellong. Versl. gewone Vergad.Afd.Natuurk. Kon.Ned.Akad.Wet. 52, 391-400. English translation: (1951) Structure and function of the lung of birds. Poultry Sci. 30, 3-10.

10. Piiper,J., Huch,A., Kötter,D. & Herbst,R.(1969) Pulmonary diffusing capacity at basal and increased O_2 uptake levels in anesthetized dogs. Respir.Physiol. 6, 219-232.

11. Piiper,J., Pfeiffer,K. & Scheid,P. (1969) Carbon monoxide diffusing capacity of the respiratory system in the domestic fowl. Respir. Physiol. 6, 309-317.

12. Piiper,J., Drees,F. & Scheid,P. (1970) Gas exchange in the domestic fowl during spontaneous breathing and artificial ventilation. Respir.Physiol. 9, 234-245.

13. Piiper,J. & Scheid,P. (1972) Maximum gas transfer efficacy of models for fish gills, avian lungs and mammalian lungs. Respir. Physiol. 14, 115-124.

14. Scheid,P. & Piiper,J. (1969) Volume, ventilation and compliance of the respiratory system in the domestic fowl. Respir.Physiol. 6, 298-308.

15. Scheid,P. & Piiper,J. (1970) Analysis of gas exchange in the avian lung: theory and experiments in the domestic fowl. Respir.Physiol. 9, 246-262.

16. Scheid,P. & Piiper,J. (1971) Direct measurement of the pathway of respired gas in duck lungs. Respir. Physiol. 11, 308-314.

17. Scheid,P., Slama,H. & Piiper,J. (1972) Mechanisms of unidirectional flow in parabronchi of avian lungs: measurements in duck lung preparations. Respir.Physiol. 14, 83-95.

18. Scheid,P. & Piiper,J. (1972) Experimental evidence for cross-current gas exchange in avian lungs. The Physiologist 15/3, 258.

19. Scheid,P. & Piiper,J. Cross-current gas exchange in avian lungs: effects of reversed parabronchial air flow in ducks. Respir.Physiol. (in press).

20. Siegwart,B., Gehr,P., Gil,J. & Weibel,E.R. (1971) Morphometric estimation of pulmonary diffusion capacity. IV. The normal dog lung. Respir.Physiol. 13, 141-159.

21. Zeuthen,E. (1942) The ventilation of the respiratory tract in birds. Kgl.danske Vid.Selsk. biol. Medd. 17, 1-51.

Comparative Physiology, eds. L. Bolis, K. Schmidt-Nielsen and S. H. P. Maddrell

COMPARATIVE VERTEBRATE VENTILATION AND HETEROGENEITY

G.M. HUGHES
Research Unit for Comparative Animal Respiration,
The University,
Bristol, England.

When making a comparative survey of a particular function at least two main attitudes may be adopted. The first involves emphasis on phylogeny, the information being used to show relationships between different groups of animals. We all know some of the dangers involved in arguing from the structure of living forms to that of the ancestors that constitute the phylogeny. This was emphasised, for example, in relation to the possible role of cutaneous respiration during the origin of land vertebrates (27). At Göttingen many of us enjoyed a lecture by Romer (45) in which he indicated how the balance of palaeontological evidence did not support such a view. And, as he pointed out, it is equally important that we should remember limitations of the fossil record, especially when dealing with the functioning of soft parts of the animal such as are involved in gaseous exchange.

A second approach is what Sir James Gray describes as 'taking the animals as we find them' and to investigate mechanisms whereby they carry out a particular function and their adaptation to the needs of the animal. Links between different animals are here made because of similarity of mechanism and do not necessarily indicate a common ancestry. I suppose most comparative physiologists generally adopt this view, but those who have come via zoology tend to retain an interest in phylogeny which is greater than that found in those mainly trained in mammalian physiology.

All comparative physiologists know the importance of choosing the right animal and there are many examples from comparative vertebrate ventilation, none of which illustrates this better than recent investigations on birds. Several workers (Brackenbury (3), Bretz and Schmidt-Nielsen (6), Scheid and Piiper (47)) have found the duck lung ideally suited to such studies because of the accessibility of particular bronchial structures through which the flow direction is critical to decide between two major theories. Once more we must remember the dangers of such an approach because of the tendency to over-generalise from information gained with such favoured animals. Consequently it is my view that at the same time as such detailed studies are in progress, it is important to survey other related organisms so as to judge the validity of such generalisations.

In surveying present knowledge of ventilation among vertebrates, three main themes will be emphasised:

1. The value of modern pressure recording techniques, together with gas analysis of samples from similar sites. Flow measurements will also be mentioned.

2. The concept of heterogeneity among vertebrate respiratory systems has developed during the last few years. This concept is discussed with respect to a variety of phenomena, brought together here because previously each had been considered as homogeneous.

Earlier descriptions of ventilation mechanisms are usually regarded as applicable to the whole respiratory surface of a given animal. Further investigations show that this is not always true and have led to more detailed and comprehensive studies. This has been closely paralleled by morphometric studies which also emphasise regional differences within respiratory organs. Furthermore, corresponding changes are also being recognised on the blood side of the exchanger (46, 11).

3. All studies have given increasing evidence for the value of detailed studies of the morphological relationships and dimensions of relevant parts of ventilatory systems.

<u>Aquatic vertebrates</u>. - Primitive vertebrates

probably had gill slits that were homogeneous, all gill filaments being of equal dimensions and equally ventilated. Nothing is known about the ventilatory mechanisms in such animals. Among living vertebrates the cyclostomes are somewhat specialised and as yet studies have not been sufficiently detailed to indicate any heterogeneity. It is generally thought that all pouches have the same basic structure and are ventilated equally. However, only some of the pouches are ventilated under certain circumstances (55).

Among modern elasmobranchs, some of the more primitive members have a larger number of gill slits (e.g. Hexanchus, Heptranchias) but the common dogfish and sharks are characterised by the possession of 5 pairs of slits. This was one of the earliest mechanisms to be studied using pressure transducers and even at that stage indications were noted that the 5 pouches do not function in precisely the same way (26). However, the stability of the equipment then available was not good enough to allow a precise investigation. Another indication of heterogeneity from previous studies was the separateness of the paths of water entering the mouth and spiracle and their exit through different gill slits; probably due to the presence of laminar flow at the low Reynolds number operating throughout the system, which functions at relatively low velocities.

Recently we have carried out more detailed pressure recording studies at Plymouth using unanaesthetized, unrestrained fish. Polyethylene cannulae were inserted into the oro-branchial (OB) cavity and one or more of the parabranchial (Pb) cavities. Cannulae inserted above or below the slits are referred to as dorsal or ventral. Only rarely was a single cavity cannulated in more than one position in order not to disturb the system. Many simultaneous recordings have been made of pressure changes in the OB and Pb cavities and the differential pressure between them using Sanborn manometers (36). For analysis the height and areas of the different phases of the pressure cycles were measured.

The dogfish ventilatory cycle consists of a contraction phase in which the volume of the OB and Pb cavities is reduced and results in an increased positive pressure. This is succeeded by relaxation of the musculature; elasticity of the branchial skeleton and its ligamentous connexions produces an expansion phase during which pressure in these cavities falls below that of the outside water. Water enters through the mouth and spiracles but cannot enter through the external gill slits because of the passive flap valves.

Before considering the pressure changes, it is necessary to be clear about the likely pathways of water flow and their resistance (Figure 1). The well developed septum must increase that part of the gill resistance (R_{sept}) relative to that of teleosts (30) but will be lessened because of the septal channel (40). The resistances of the internal (R_a) and external (R_{valve}) slits are clearly important and differ among different elasmobranchs and for different slits of the dogfish. The pressure changes are biphasic, the relative sizes of the positive and negative phases varying between slits. The negative phase of the first three is similar but that of the 4th and 5th slits is smaller (Figure 3A). The time course of its return to zero is also more rapid in these two slits - indicative of a lower gill resistance, assuming similar elastic or restoring forces for all slits.

The differential pressure across the gills has three phases, the first peak (I) being due to the pressure pump phase, usually succeeded by a brief reversal (II) and finally a parabranchial suction phase (III). A brief pause occurs before the next active cycle commences. The numbering of the phases used here differs from that adopted in 1960, which was based upon a direct comparison with the teleost system (37). In view of the fuller understanding of the ventilatory mechanism derived from electromyography (33) the new nomenclature is more appropriate.

Figure 2 shows typical pressure recordings from the OB cavity and third parabranchial cavity and the differential pressures between them, also note the differences between dorsal and ventral recording from Pb_3. The relationship between the different phases is shown in Figure 3B for the first slit. Thus with an increase in the total diffe

A

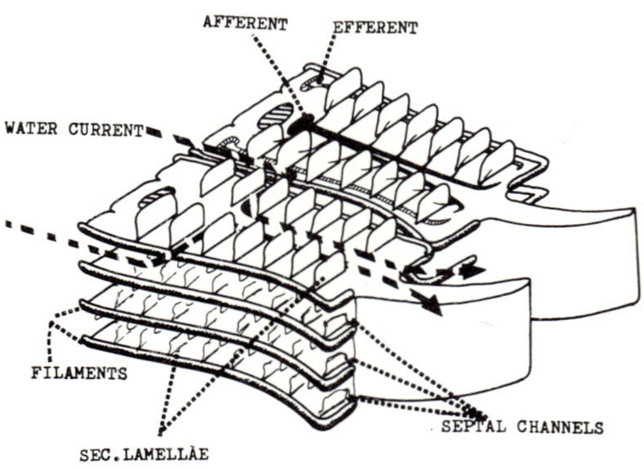

B

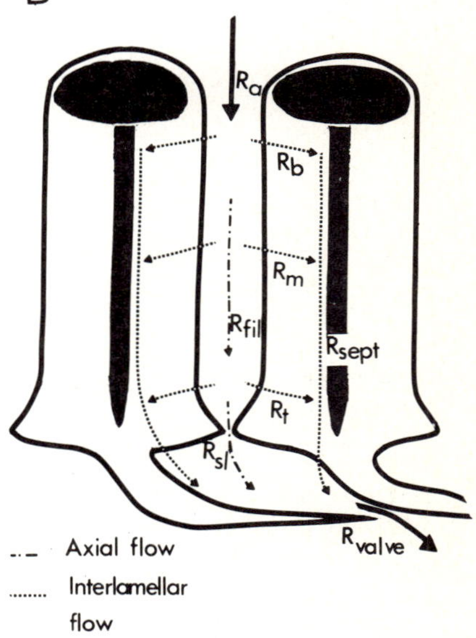

Figure 1 - A. Diagram to show path of water across the secondary lamellae of a dogfish. B. Diagram indicating subdivisions of the resistance to water flow during its passage from the oro-branchial cavity to the exterior. All water enters between a pair of gill arches (R_a) and leaves through the external valve (R_{valve}). The current (dotted) which passes between the secondary lamellae (R_b,R_m,R_t) and along the septal channels (R_{sept}) are differentiated from that which passes directly between the filaments (R_{fil}) and their tips (R_{s1}) before entering the parabranchial cavity.

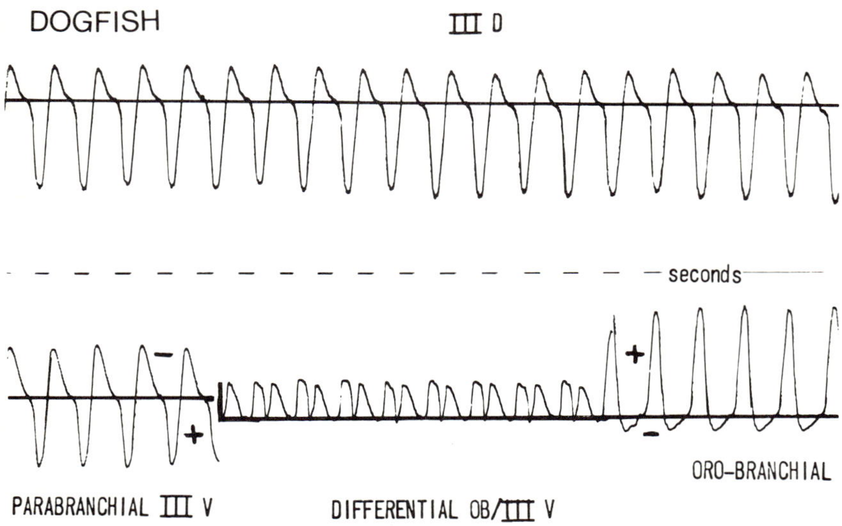

Figure 2 - Scyliorhinus canicula. Simultaneous recording of pressure changes in the oro-branchial cavity and dorsal (IIID) and ventral (IIIV), of the third parabranchial cavity. The differential pressure between the oro-branchial and third ventral parabranchial is also recorded. Notice the larger positive phase in the dorsal region but greater negative phase in the recording of the ventral Pb_3. Zero pressure is the pressure outside the fish. Calibration pressure of + 1 cm water.

rential pressure area, presumably reflecting an increased ventilation volume, there is a relatively greater increase in phase I, i.e. the oro-branchial pressure pump. This is to be expected in view of the passive nature of the expansion phase. Comparisons between slits were made with respect to slit 3 which was taken as the typical member because of its central position. Figure 4 summarises the relative size of the negative phase of the parabranchial pressures with respect to total differential pressure.

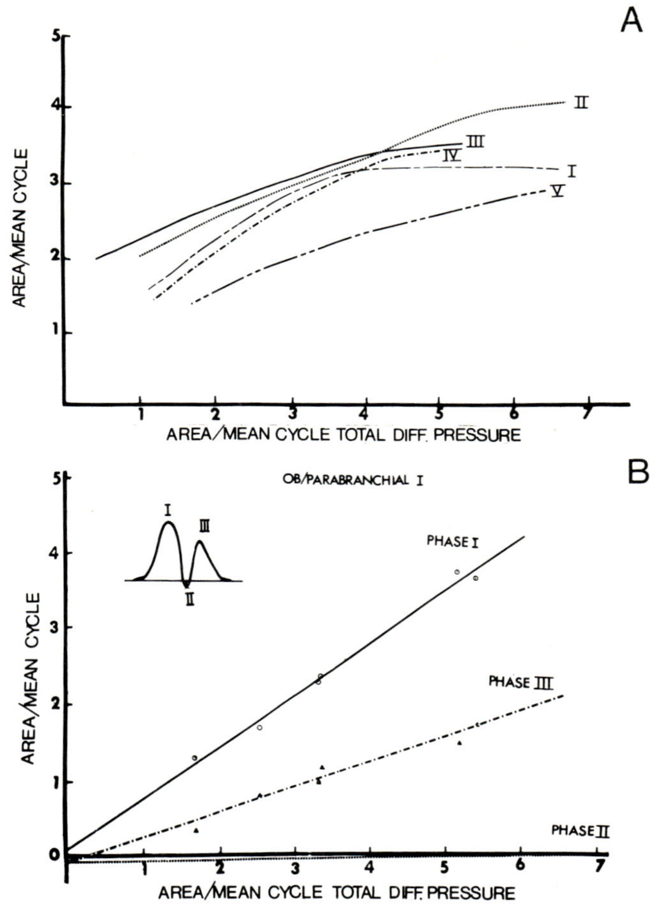

Figure 3. - A. Area of the negative phase of the parabranchial pressure in the five pouches of a dogfish plotted against total area of the differential pressure. Plots indicate a gradual increase in size of the negative phase with total differential area. B. Plot showing the increased area of the three phases of the differential pressure between the oro-branchial and 1st parabranchial cavity in relation to the total area. Phase I (oro-branchial pressure pump) increases more rapidly than the other two phases.

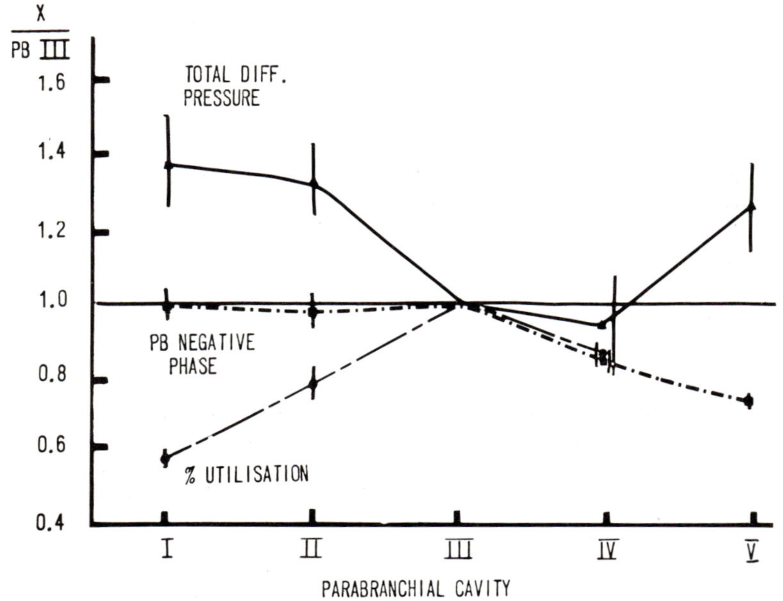

Figure 4 - Scyliorhinus canicula. Graph summarising comparison between different parabranchial cavities, all data is plotted relative to the third pouch. Mean values are given for each point ± standard error.

Comparison with O_2 tension of water samples from different regions of parabranchial cavity. Figure 4 also shows the percentage utilisation of O_2 determined by means of similar cannulae. There seems to be a correlation between the relative size of the differential pressure and utilisation, the latter being less where the total area (I + III - II) of the differential pressure is greatest. This relationship agrees with many other studies showing a reduction in overall percentage utilisation as ventilation volume increases.

Measurements on both sides of a fish were more or less equal, which was a necessary control for comparisons between the dorsal and ventral parts of parabranchial cavities with respect to their pressure and utilisation (Table 1). Utilisation measured using dorsal cannulae is greater than from ventral regions and perhaps indicates a greater flow of water into the ventral part of the parabranchial cavities. Differential pressures also indicate a flow of water from dorsal to ventral within a given cavity.

TABLE 1 - The % utilisation of oxygen found by sampling water from different parabranchial cavities compared using Student's T test. The significance of differences between dorsal and ventral regions of single cavities and between corresponding cavities on the left and right side were tested. Results indicate that % U is greater for dorsal samples and that the left and right sides are not significantly different.

P for identity	(No.)	% U for samples from PARABRANCHIAL CAVITY	P for identity	(No.)
<0.01	(13)	I	<0.7	(3)
<0.001	(9)	II	<0.9	(6)
<0.01	(11)	III	-	
<0.3	(15)	IV	<0.01	(8)
D > V			L = R	

As yet it has not been possible to make sufficiently detailed comparisons of the results of the physiological and morphological studies. Table 2 summarises some results which indicate a fairly even distribution of gill area among the five slits. The 5th slit only has one hemibranch, the consequent reduction in gill resistance correlates with the parabranchial pressure waveform. Differences in % utilisation probably reflect differences in relative flows through individual slits but so far these have not been measured. Such heterogeneity also has implications in relation to estimates of diffusing capacity; there must be variations in the O_2 tension difference across the water/blood barrier, which also show regional differences.

TABLE 2 - Comparison of some features of the gills from different slits of a dogfish. % utilisation are averages obtained from dorsal and ventral samples from several specimens.

	1	2	3	4	5
% of total gill area	22.9	23.3	23.4	22.1	8.3
Ave.sec.lam./mm both sides of filament	14.72	14.68	14.46	14.34	14.26
Ave.% utilisation of oxygen	44.4	59.0	65.4	61.1	

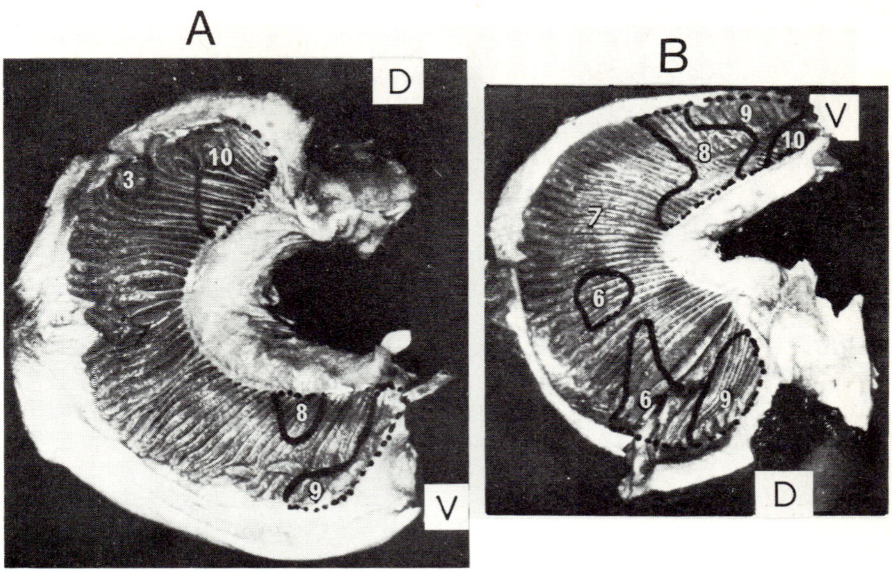

Figure 5 - Scyliorhinus stellaris - Photographs of two hemibranchs from the left side. A. Posterior hemibranch of 1st arch. B. Anterior hemibranch of 4th arch. The frequency of secondary lamellae on one side of the filaments is indicated by superimposed contours. Note the higher frequency towards the two ends of each hemibranch in the regions dorsal and ventral to the internal gill slit.

One structural aspect of the elasmobranch parabranchial system is the way many gills have dorsal and ventral regions which are probably relatively unventilated. The secondary lamellae at the two ends of the branchial arch are also more closely spaced (Figure 5). In the very active sharks with large internal and external slits, these parts are reduced to a minimum, and the gill sieve is more homogeneous. Comparable situations are also found among bony fish where tunas illustrate one extreme in which the whole sieve must be very well and uniformly ventilated as the fish swims along with open mouth and opercular slits, the flow of water through the system being controlled by the degree of opening of these two apertures

(10). Ventilation of tuna gills results in a very high percentage utilisation of oxygen (50) and contrasts with some of the low figures obtained for active freshwater fish such as the trout (44). The method used for trout, cannulation through the operculum, has many disadvantages relative to direct measurements of ventilation (35) which was used in the tuna experiments, and perhaps accounts for some of these relatively low figures. The value of direct methods has also been emphasised by Davis and Cameron (14) in relation to the trout and they obtained higher % U which was of the same order as the 50-75% found by van Dam (13). On the other hand, the whole gill system is constructed quite differently in some other teleosts and in these the same cannulation technique gave much higher values e.g. values over 80% in the trigger fish Balistes (28). In the latter, and also the sunfish (Mola mola) the gill system is reminiscent of that found particularly in the bottom-living elasmobranchs which have a relatively small mouth and a small external opening to the parabranchial cavities. Furthermore, the portion of the total length of the gill arches through which water can pass directly from the buccal to the opercular cavities is less than half of their total length (1). In such systems uneven ventilation of the gills is highly probable even at low Reynolds numbers.

On the perfusion side of the exchanger there are also differences, and fish seem to be able to regulate the blood flows to different portions according to circumstances. Flattened elasmobranchs (rays and skates) provided useful preparations from this point of view, and it has been shown (46) that blood flow to the anterior two slits has a more rapid rising phase than that to the three posterior afferent branchials. During exercise the increase in flow to the posterior arches is greater than that to the anterior regions. Similar evidence for regulation of blood flow to the slits was found in the stingray (Dasyatis sabina) but there was no evidence for a regulation of water flow through the different slits when the blood flow to them was altered experimentally (11). It is clear that much needs to be learned of these relationships between the overall organisation of gill systems and their differential ventilation.

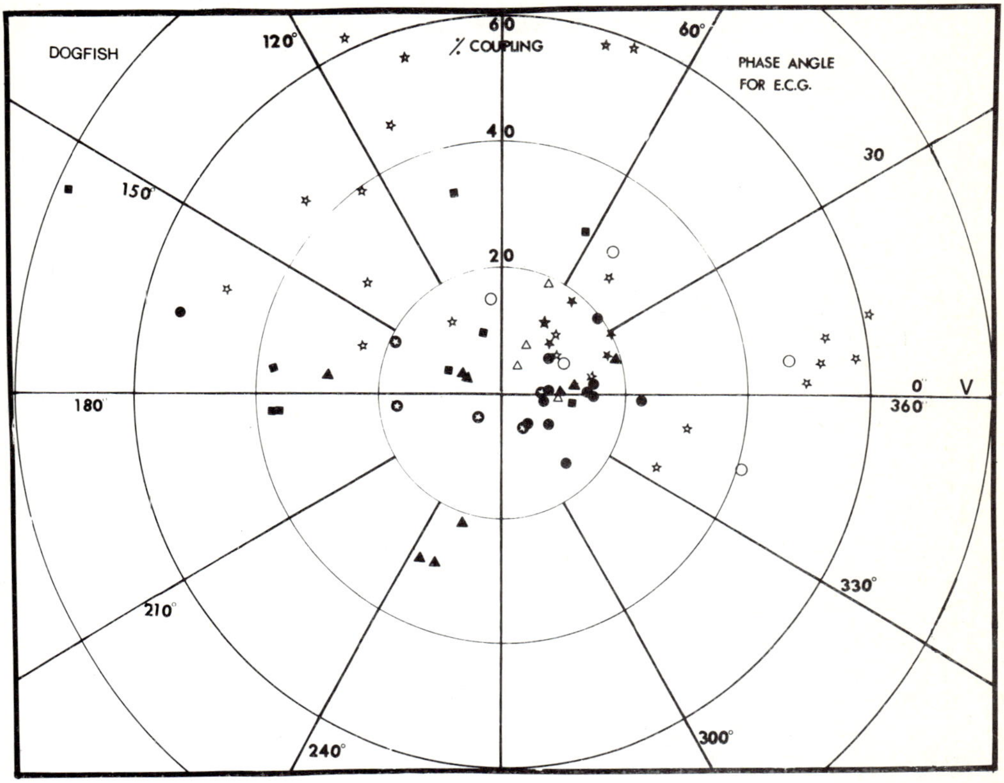

Figure 6 - Diagram summarising the distribution of occurrence of the ECG within ventilatory cycles for several dogfish. The phase angle between the two rhythms together with their % coupling are plotted (From Hughes (31)).

Related to these findings are recent studies on the coupling between the ventilatory and cardiac (29,31) pumps which have indicated that there is not always so rigid a co-ordination between the two rhythms as has sometimes been supposed. The precise coupling between the two rhythms varies and also the phase relationship (Figure 6).Results suggest that we cannot yet define a simple overall pattern of detailed ventilation/perfusion of the gill system which is universally applicable (31). Individual fish vary but always the impression is gained that the precise relationship between ventilation and perfusion at the level of the secondary lamella is more precisely controlled. Measurements that have so far been possible are too gross to indicate some of these adaptations. Thus although the pressure measurements reported here give far more detailed information about pressure distribution during the ventilatory cycle, they have also indicated the presence of many variations in the overall pattern. The observations were made on unanaesthetised and unrestrained dogfish which had been acclimated to the experimental circumstances for some time, and were relatively unhindered by the operative procedures (31). When the animal is placed under stress e.g. hypoxia or anaesthesia, or during exercise, it appears that the system becomes more constant and easier to define. Thus, although the ray experiments seem to indicate that the fish are unable to regulate the overall distribution between the slits, nevertheless they may regulate the flow within possible channels through the gill resistance. There are,however, differences in flow between slits, as indicated by differences in the path of water entering the spiracle or mouth, and also in some sharks where water enters through the first gill slit (24).

Ventilation in air-breathing vertebrates. A great variety of different mechanisms have been suggested including for example the use of the eyes in the mudskipper *Periophthalmus* (56) and the liver as a piston in crocodiles (19).

Living vertebrates showing a transition between aquatic and aerial respiration include lungfish and many teleosts with accessory air-breathing organs.

Ventilation in all three genera of living lungfish has been investigated in recent years (2, 23,38,41). The operculum is poorly developed and contrasts with the well-developed opercular bones of Devonian dipnoans (Figure 7). The apparently relatively more developed opercular mechanism of *Neoceratodus* is probably a secondary adaptation (53). Pressure recordings from *Protopterus* indicate a dominance of the buccal pressure pump during gill ventilation which is modified to inflate the lung during air breathing (42). In these cases, there does not seem to be any suction pump although the animals possess ribs in the body region.

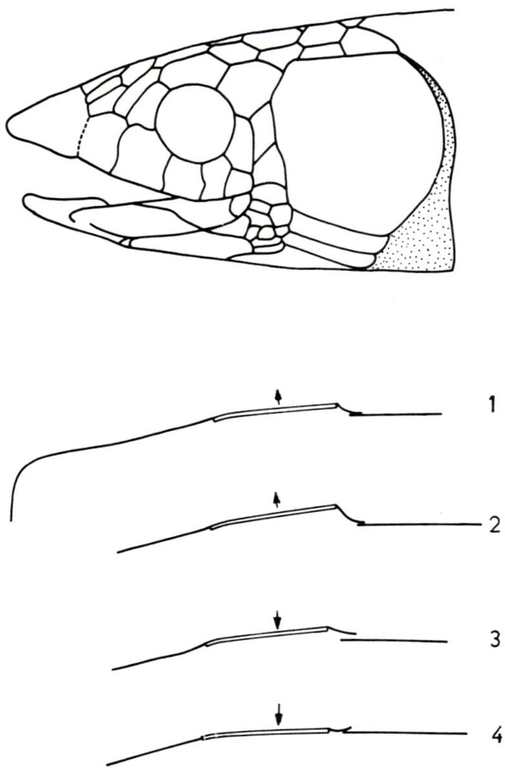

Figure 7 - Head of Dipterus valenciennesi to show the well developed opercular bones (from Westoll (54), together with four suggested stages in a complete cycle of its movements. (From Thomson (53)).

Movements of the 'cranial ribs' are associated with lung-filling as has been revealed in X-ray and electromyographic studies (41). The ventilation cycle begins with a compression phase, during which most of the main muscles are active, and serves to empty the buccal cavity of water before the mouth is opened. Expansion involves the action of elastic restoring forces together with activity in a part of the constrictor hyoideus.

For several reasons lungfish may be considered something of a red herring; from many physiological points of view their adaptations do not seem as close to those of tetrapods as those found in some air-breathing tetrapod ancestors. Thus the O_2-carrying capacity of dipnoan blood is low and its buffering mechanism is not so well developed as in air-breathing fish such as _Anabas_ or _Clarias_.

In these transitional forms the gills are the main sites of CO_2 release (27); the accessory organ being more important for O_2 uptake. The skin is an important secondary site of CO_2 liberation although some forms e.g. _Anabas_, have no cutaneous respiration. Other aspects of their respiration and acid-base balance would suggest that such a form could become exclusively air-breathing without the need for secondary surfaces for CO_2 release to evolve, e.g. skin.

Measurements of the O_2 and CO_2 tension of the gas in the suprabranchial chamber (Figure 8) show that when the next air breath takes place, the P_{CO_2} is as high as 30-35 mm.Hg. (49). Pressure changes within these chambers have given some insight into the mechanism of ventilation which seems to include an active suction component.

These transitional forms, then, illustrate very well the importance of heterogeneity in ventilation mechanism in that CO_2 and O_2 are not exchanged equally at the different surfaces. There are large differences in the construction and functioning of a lung, a gill and skin. It is remarkable how the particular advantages accruing from aquatic respiration (e.g. high solubility of CO_2) are used together with the greater availability of oxygen from the air in such a variety of forms from many different groups. Morphometric studies show how different respiratory surfaces increase in size at different stages of the life cycle. In some cases the change is related to the environmental conditions. For example, some air-breathing fish hatch out in the rainy season when O_2 levels in the water are high. During these early stages of development, the fish mainly relies upon the gills for gaseous exchange and they increase in size relatively rapidly. At a later stage, which normally coincides with worsening conditions for aquatic respiration, the growth rate of the air-breathing organs increases as they become increasingly important.

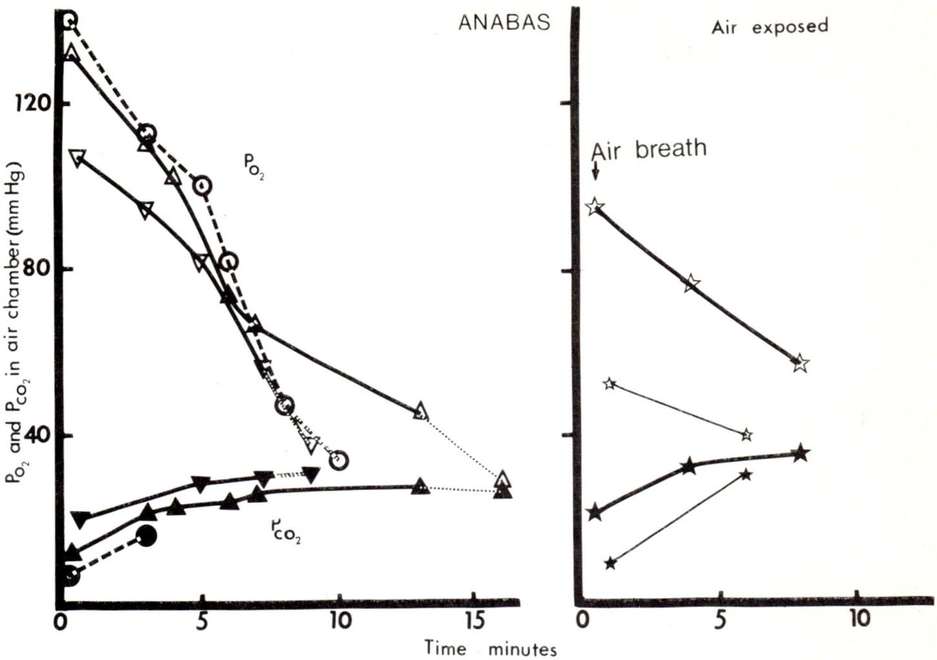

<u>Figure 8</u> - The climbing perch (<u>Anabas testudineus</u>), graph showing change in composition of gas in the suprabranchial chambers following an air breath, with the fish kept in water but allowed to surface, and when out of water. Mean values are shown for samples taken at the various times after an air breath and are then extrapolated (dotted line) to the instant at which the next air breath is taken, and thus estimates obtained for the composition of the gas just before an air breath. (From Singh & Hughes (49)).

In modern Amphibia such as the frog, recent studies have emphasised further the importance of the same three sites of gaseous exchange at particular phases of the life cycle. The mechanism of pulmonary ventilation has generally been considered to involve a large degree of mixing within the buccopharynx of spent gas from the lungs with fresh air inspired through the nares. More recent studies are thought to support the view that such mixing does not occur - gas from the lung being 'jetted' directly through the buccopharynx and out of the nares. Such a pathway is difficult to prove without the experimental procedures interfering too greatly with the process being analysed. But such a mechanism provides an excellent illustration of the way in which the use of modern techniques of rapid gas analysis (mass spectrometry) and of single ventilatory cycles could improve our insight of differences within systems which previously appeared homogeneous (Figure 9). The role of the buccopharyngeal

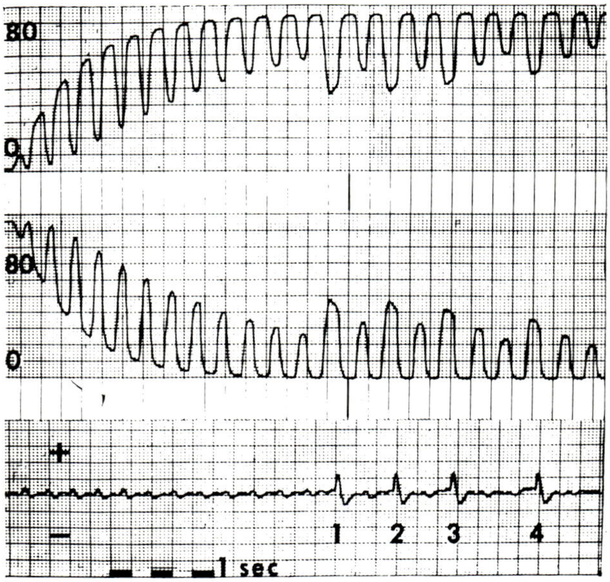

Figure 9 - Changes in expired argon and nitrogen composition upon switching breathing mixtures from 80% argon and 20% oxygen to air. Top, nitrogen; centre, argon; bottom, pressure in atmospheres. Record of buccal pressure shows buccopharyngeal cycles and four pulmonary cycles (labelled 1-4). Note that shapes of the gas concentration curves for the buccopharyngeal and pulmonary expirations are different, the high initial concentration during the latter being produced by the jet effect. The pressure record has been horizontally displaced to compensate for the difference in time delays. (From Gans, De Jongh and Farber, (20)).

movements in mixing the contents of the buccopharynx between successive pulmonary expulsion phases is of respiratory significance as was shown using mass spectrometry (39), after the system had been filled with argon. However, the recordings during pulmonary deflation did not indicate a complete 'jetting' effect (Fig.9) and suggested that 24-75% of the gas remaining in the buccal cavity is derived from the pulmonary efflux. Calculations based on a simple mixing system gave comparable results (20).

Reptile ventilation. - The wide range of habits among reptiles invites investigation of their ventilation mechanisms but when one looks into detail, there is still a need for further research. However, the application of pressure recording and electromyographic techniques during the past 7 or 8 years has greatly improved our knowledge. Surprisingly enough, ventilation in 'typical' reptiles i.e. lizards, is still in great need of study although it seems clear that the basic mechanism involves the costal muscles. It has been shown that some of these muscles fire during expiration and others during inspiration (52, 15). The movements are triphasic as suggested by Boelaert (7), lung inflation being biphasic. As is typical of many reptiles, the lung remains partially inflated during periods of apnoea which separate periods of repeated ventilatory cycles. The lungs of some Lacertilia show good heterogeneity in development of respiratory surfaces but as yet there is no good evidence for different degrees of ventilation.

Chelonians are of special interest because of the fusion of ribs with the carapace and the adoption of aquatic and semi-aquatic habits. The lungs are well developed and in some forms lie closely connected to the dorsal carapace. Ventilation results from changes in volume of the perivisceral cavity, the pressure changes imposed upon the lung being co-ordinated with opening and closing of the glottis to produce ventilation. The volume changes are due to inward movements of the limbs and girdles (Figure 10B) as has been shown by electromyographic recording in *Testudo graeca* (34), and later in aquatic forms by Gaunt and Gans (22). Though the movements are triphasic, the actual inflation and deflation of the lung is biphasic and

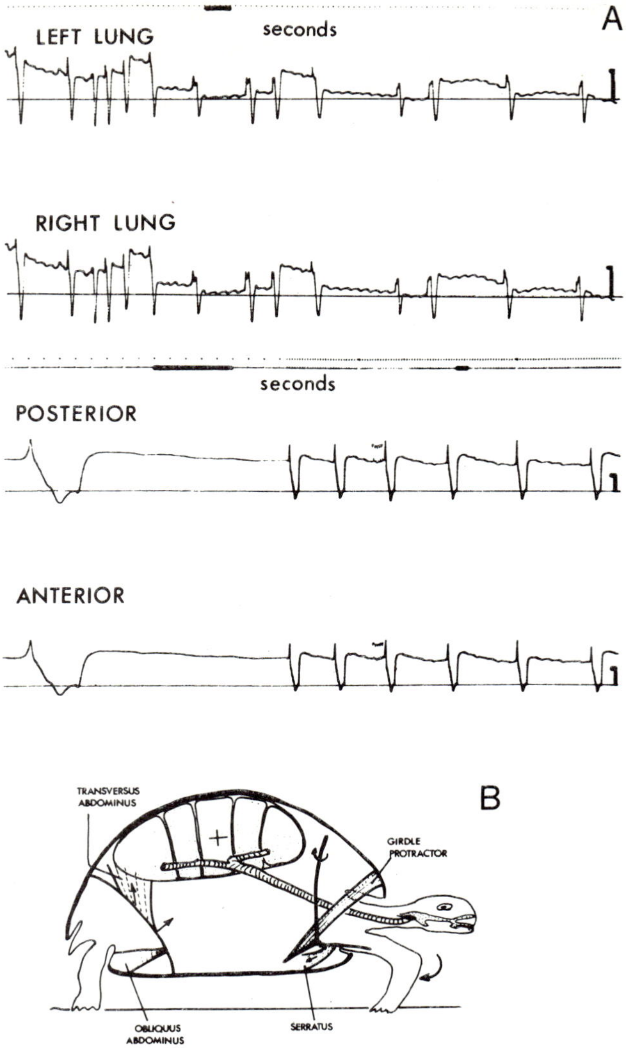

Figure 10 - *Testudo graeca*. A. Simultaneous recordings of intrapulmonary pressure from cannulae in a single lung, and from the left and right lungs. (From Gans and Hughes (21)). B. The mechanism of lung ventilation as deduced from pressure recording and electromyography. Diagram shows decrease in volume of perivisceral coelom by activity in transversus abdominus and pectoralis muscles resulting in increased intrapulmonary pressure. The antagonistic muscles obliquus abdominus and serratus are active during inspiration.

it seems that the final so-called exhalation is largely passive, resulting from the return of the viscera to its resting position. A number of structures, muscular in nature and close to the lung have been variously described as 'diaphragms' but they are certainly not homologous, and only partly analogous to the mammalian diaphragm.

The relatively large spaces within reptile lungs suggests that diffusion would equalise the concentration of gas in different parts. Samples from Testudo graeca obtained via cannulae inserted through the carapace, indicate an absence of any major differences in O_2 and CO_2 content, and although pressure changes within these cavities remain essentially the same (Figure 10A) small differentials were recorded which may be important in further analyses. In other reptile lungs, and also in amphibians and Lepidosiren, there are regional differences in development of the alveoli and it has often been suggested that ventilation of these parts also varies. I do not believe that proof of this supposition has yet been obtained.

In crocodiles electromyography and pressure recording have shown that both inspiration and expiration are active and that individual ventilatory movements are biphasic except for the later ones in a series where the passive expiratory phase of Boelaert leads to the maintenance of the lung in a moderate degree of inflation between bursts of ventilatory cycles (43). Boelaert (8) concluded that the final partial expiration resulted from air being forced out of a closed glottis following relaxation of the diaphragm and not because of visceral recoil alone, as in lizards. Among crocodilians he noted also that the abdominal and thoracic breathing movements are in the same direction, not opposite to one another as observed in lizards. He also attributed this difference to the functioning of the muscular diaphragm. Further experimentation using whole body plethysmography gave results not entirely in agreement with Boelaert's view. In particular, the partial expiration was not present in these recordings (Figure 11) showing that it results from a redistribution of the total body volume, as there was no overall change. When the animals were excited, the final

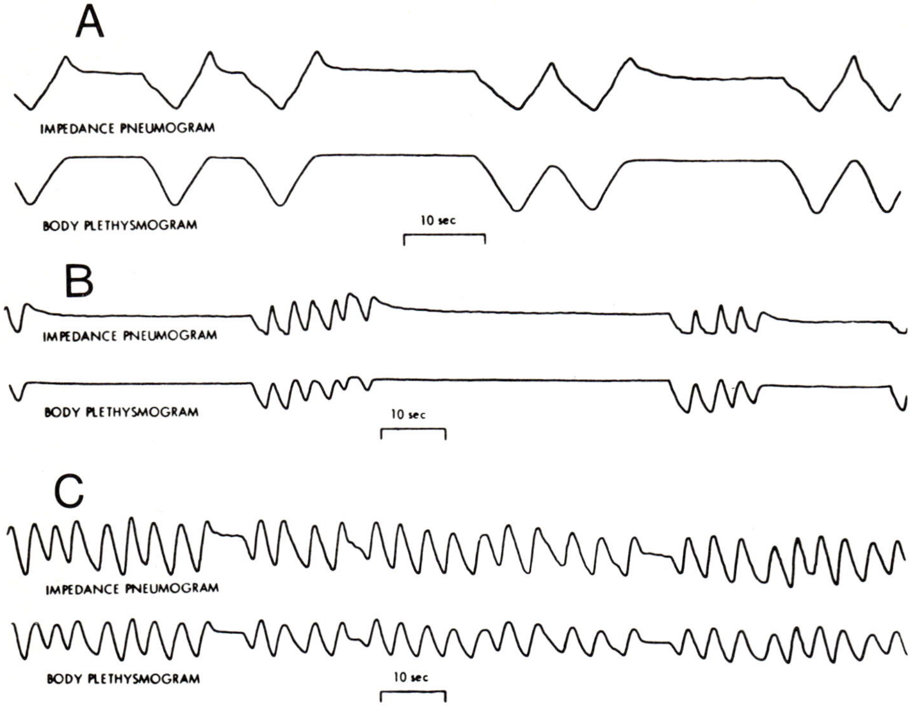

Figure 11 - Simultaneous recordings using impedance pneumograph and whole body plethysmographs of crocodilians. A. Recordings from caiman show triphasic waveforms particularly in the final cycle of an individual group. Body plethysmograph indicates absence of any air loss during the final expiratory phase (E_2). B. Similar recordings made on a calm alligator. C. Recording from an alligator during a period of excitement shows actual loss of air during the E_2 phase. (From Naifeh, Huggins and Hart (43)).

expiratory phase included some passive escape of air. Thus in the crocodile, activity in the abdominal transversus muscle produces inward movement of the abdominal wall during expiration. Presumably this pushes the abdominal contents up against the diaphragm thus assisting in the evacuation of air from the lungs. Activity in the diaphragm itself occurs during inspiration and probably functions as Boelaert believed to force the abdominal viscera down and so produce a small negative pressure which assists lung filling.

It may be concluded that although much has recently been learnt about reptile ventilation there is little indication of functional heterogeneity except perhaps in a temporal sense, i.e. variations in ventilatory rhythms. Morphometric observations indicate heterogeneity of respiratory surfaces and suggest that more detailed studies will reveal corresponding differences in physiology.

The avian system. - A great deal of information has been gained during recent years about ventilation of this very highly developed system. Expiration is always active but during quiet breathing inspiration is a passive process. Electromyographic studies emphasise temporal overlaps in the timing of different inspiratory and expiratory muscle groups (18). Studies in a number of different laboratories (9,3,6,17,18) have resolved the problem as to which direction air passes through the parabronchi of the bird lung. It now seems clear that the system suggested by Hazelhoff (25) and others, that air passes in a caudo-cranial direction (Figure 12) during both inspiration and expiration, is found in at least the major portion of bird lungs.

The parabronchial lung of the bird contrasts with alveolar lungs of other vertebrates in being a relatively rigid, constant volume structure. The ventilatory movements producing pressure changes within the body cavity are mainly transmitted to the air sacs and have scarcely any effect on the lung itself. It has generally been supposed that pressure changes within different air sacs are identical but it is only recently that they have been recorded in detail. During eupnea the pressure waveforms are characterised by an exponentially decaying expiratory phase of longer duration than the inspiratory phase (Figure 13A). These waveforms are suggestive of a constant muscular effort during expiration and show an initial rapid pressure change at the beginning of inspiration which is similar in all sacs and probably due to expiratory recoil. Differential pressures (Figure 13B) recorded between the posterior and anterior sacs amount to 0.5-2 mm.H_2O (5). Simultaneous measure-

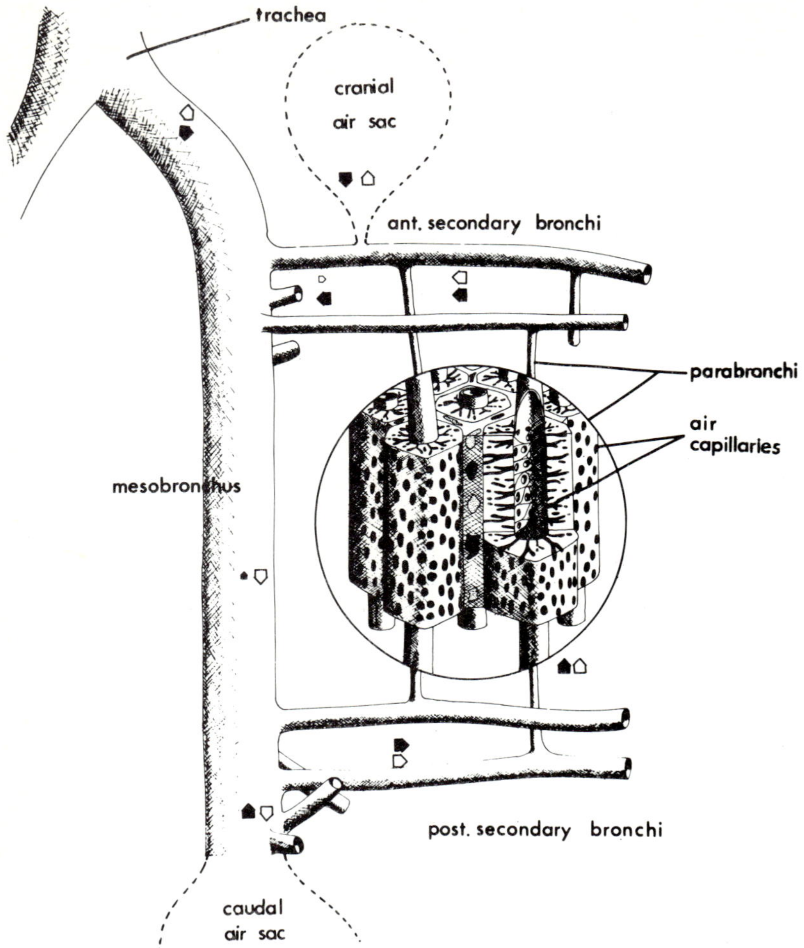

Figure 12 - Diagram of main airways within the bird lung and connexions with anterior and posterior air sacs. Directions of air flow are indicated by arrows (white - inspiration; black - expiration). (After Hughes (32)).

ments of air flow showed a great increase in caudo-cranial flow in the posterior dorsal secondary bronchi during the inspiratory phase. Initially expiration seems to depress the flow. Differences in pressure between air sacs might be expected in view of differences in resistance of pathways between the outside air and the individual sacs. An analysis of the system on this basis has very recently been carried out by Brackenbury (5).

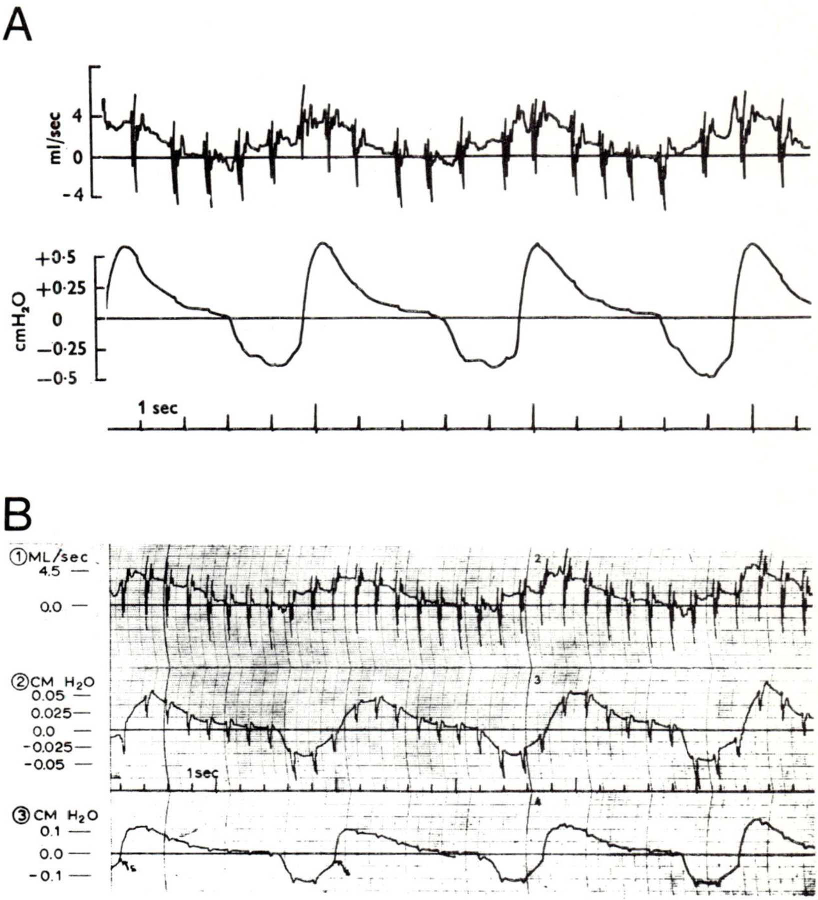

Figure 13 - A. Pressure recordings from the right posterior thoracic sac of a goose during normal ventilation, together with the respiratory flow in a posterior dorsal secondary bronchus (From Brackenbury (3)). B. Respiratory airflow in the right posterior dorsal secondary bronchus (1), differential pressure between posterior thoracic sac and interclavicular sac (2), and pressure differential between posterior thoracic sac and trachea 3 cm anterior to the syrinx (3). Caudo-cranial flow above the line in (2). (From Brackenbury (4)).

He has concluded that the passage of air through the parabronchi during inspiration mainly results from the aspiratory action of the anterior sacs and during expiration the posterior sacs force air forwards through the parabronchi. Because of differences in resistance to air flow in different pathways it is possible to think of a sort of antagonism between different air sacs for the entering air. The resulting flows will be determined by the different resistances and as a consequence pressure differentials may be recorded between the air sacs.

Most of the experiments carried out so elegantly in recent years on flow through the parabronchi have made use of the palaeopulmo which forms the whole of the parabronchial lung in many primitive birds but in more advanced groups such song birds the neopulmo (17) may constitute as much as 25% of the total parabronchial volume (Figure 14B). This system seems to be in a position analogous to the recurrent parabronchi shown in Zeuthen's diagram (Figure 14A) and it is almost

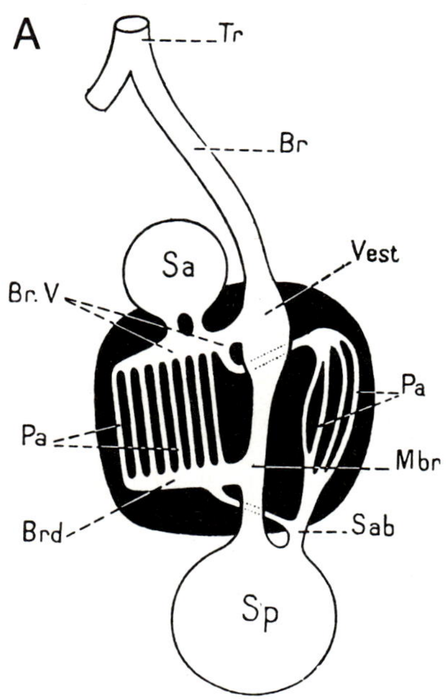

Figure 14A - Diagram to show connexions between the main airways in the bird lung (After Zeuthen (57)).

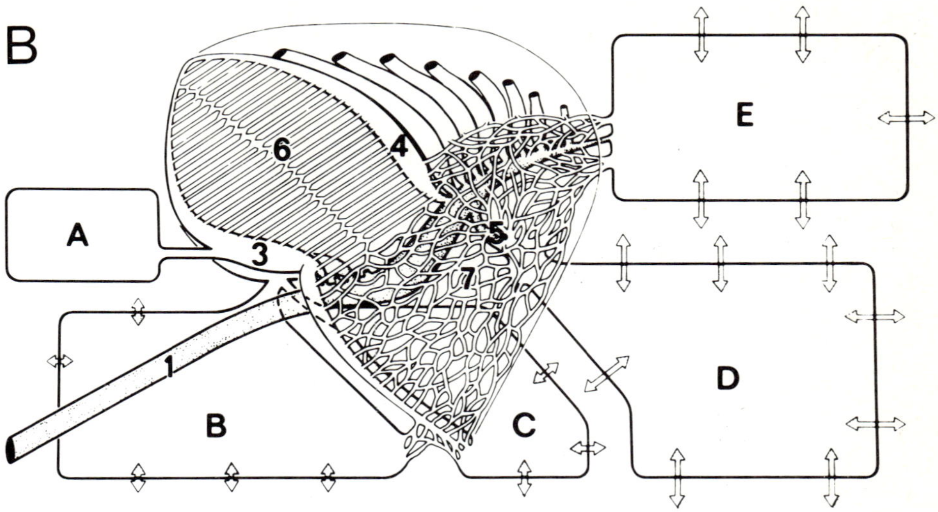

Figure 14B - Diagram of the airways from the bird lung showing the palaeopulmo (6) to illustrate the position of the parabronchial net of the neopulmo (7) occupying the whole latero-ventral part of the lung. In addition to connexions between the primary bronchus (1) and initial parts of the dorsobronchi (4) and the posterior thoracic (D) and abdominal (E) sacs, connexions to the first ventrobronchus (3) are developed. (From Duncker (17)). Notice that the neopulmo corresponds to the recurrent parabronchial system of the Zeuthen's diagram.

certain that air flow through these parabronchi will reverse during inspiration and expiration.

Mammals. - Many studies on the functioning of the mammalian lung have emphasised differences between regions of the lung, especially in relation to their ventilation/perfusion ratio. Theoretical treatments have been carried out using models of varying degrees of complexity (12), the more compartments which are considered in general the closer the model approximates to the actual conditions. The mammalian physiologists generally use the term 'inhomogeneity' to indicate the uneven distribution of alveolar ventilation, pulmonary perfusion and of diffusing capacity.Although these regional differences can be of great functional importance, nevertheless the overall structure and function of the mammalian lung is surprisingly homogeneous and hence 'inhomogeneity' is appropriate. This discussion suggests that the differences are far greater than in mammals and justifies the use of the term 'heterogeneous'.

Within mammals, however, some lungs do show quite marked differences in structure (32) including for example the proportion of the non-respiratory and gas exchange areas. A particularly good example is shown by diving mammals, and recently studies (16) have indicated a correlation with differences in physical properties of the lung which are important during diving. The pressure/volume characteristics of the lungs of sea lions, for example, indicate that they empty far more completely during compression than does the lung of a dog. Support has been given to Scholander's view that cartilaginous reinforcement of the small airways provides some protection from decompression sickness among marine mammals. In this way the gas remaining in the lung during a dive is removed from contact with the alveolar epithelium.

CONCLUSIONS

There is now sufficient evidence for the existence of differences in structure and functioning within vertebrate respiratory organs for it to be regarded as a general feature. The assumption which has often been made that all parts of a given respiratory system function in the same way must clearly be abandoned. Advances in instrumentation have largely contributed to these findings but they have also emphasised the need for equally elegant investigations of the morphological relationships of the system concerned. There is thus a great need for an integrated morphological and physiological approach to vertebrate ventilation, and clearly there is abundant material for comparative studies.

SUMMARY

1. Studies on the pressure changes and gas tensions sampled from different regions of the respiratory system of the dogfish indicate differences in the conditions for ventilation and gas exchange within the system. Relationships between these and the morphology of the system are discussed. Heterogeneity is perhaps related to differences on the blood side of the exchanger as revealed by studies of the cardioventilatory coupling, and of the ventilation/perfusion ratio of the anterior and posterior parabranchial pouches.
2. Within the air-breathing fish and amphibians, different respiratory surfaces are used and their degree of development and function is related to the habits and environmental conditions of the animal.
3. Ventilation of the air-breathing organs of fish involve modifications of the mechanisms found in typical water-breathing species. The buccal pressure pump is predominant in the present-day dipnoans and suggests the type of mechanism found in the earliest amphibians.
4. Similar heterogeneity of ventilatory mechanisms is shown among Amphibia and includes the possibility of spatially-separated flows within the single bucco-pharyngeal cavity during lung emptying.
5. The lungs of reptiles are characterised by the presence of non-respiratory regions and the large interconnections presumably allow equalisation of gas distribution. During activity heterogeneity is possible at a physiological level and may be related to the known differences in morphometry.
6. The avian system illustrates differences in regional distribution and the importance of the resistances within different airways. These result in differential pressures between air sacs.
7. Differences in ventilation/perfusion ratio within the mammalian lungs are well known. Notable differences between respiratory and non-respiratory portions of the lung are shown by diving mammals.

ACKNOWLEDGEMENTS

I wish to thank the Natural Environment

Research Council and the Nuffield Foundation for their financial support.

REFERENCES

1. Adeney,R.J. & Hughes,G.M.. Observations on the gills of the sunfish (Mola mola). in preparations.

2. Bishop,I.R. & Foxon,G.E.H. (1968) The mechanism of breathing in the South American lungfish, Lepidosiren paradoxa; a radiological study. J.Zool.Lond. 154, 263-271.

3. Brackenbury,J.H. (1971) Pressure-flow phenomena within the avian respiratory system. J.Anat. 198, 609-610.

4. Brackenbury,J.H. (1971) Airflow dynamics in the avian lung as determined by direct and indirect methods. Respir.Physiol. 13, 319-329.

5. Brackenbury,J.H. (1972) Physical determinants of air flow pattern within the avian lung. Respir.Physiol. 15, 384-397.

6. Bretz,W.L. & Schmidt-Nielsen,K. (1971) Bird respiration: flow patterns in the duck lung. J.exp.Biol. 54, 103-118.

7. Boelaert,R. (1941) Sur la physiologie de la respiration des lacertiens. Arch.Intern. Physiol. 51, 379-437.

8. Boelaert,R. (1942) Sur la physiologie de la respiration de l'Alligator mississippiensis. Arch.Intern.Physiol. 52, 57-72.

9. Bouverot,P. & Dejours,P. (1971) Pathway of respired gas in the air sacs-lung apparatus of fowl and ducks. Respir.Physiol. 13, 330-342.

10. Brown,C.E. & Muir,B.S. (1970) Analysis of ram ventilation of fish gills with application to skipjack tuna (Katsuwonus pelamis). J.Fish. Res.Bd.Can. 27, 1637-1652.

11. Cameron,J.N., Randall,D.J. & Davis,J.C. (1971) Regulation of the ventilation-perfusion ratio in the gills of Dasyatis sabina and Squalus

sucklevi. Comp.Biochem.Physiol. 39A, 505-519.

12. Chinet,A., Micheli,J.L. & Haab,P. (1971) Inhomogeneity effects on O_2 and CO_2 pulmonary diffusing capacity estimates by steady-state methods. Theory. Respir.Physiol. _13_, 1-22.

13. van Dam,L. (1938) On the utilization of oxygen and regulation of breathing in some aquatic animals. Dissertation, Groningen.

14. Davis,J.C. & Cameron,J.N. (1970) Water flow and gas exchange at the gills of rainbow trout, _Salmo gairdneri_. J.exp.Biol. _54_, 1-18.

15. Dawson,W.R. & Templeton, J.R. (1963) Respiration in the lizard _Crotaphytus collaris_. Physiol.Zool. _36_, 104-121.

16. Denison,D.M., Warrell,D.A. & West,J.B. (1971) Airway structure and alveolar emptying in the lungs of sea lions and dogs. Respir.Physiol. _13_, 253-260.

17. Duncker,H.R. (1972) Structure of avian lungs. Respir.Physiol. _14_, 44-63.

18. Fedde,M.R., Burger,R.E. & Kitchell,R.L. (1961) Anatomic and electromyographic studies of the costo-pulmonary muscles in the cock. Poultry Sci. _43_, 1177-1184.

19. Gans,C. (1970) Strategy and sequence in the evolution of the external gas exchangers of ectothermal vertebrates. Forma et Functio _3_, 61-104.

20. Gans,C., De Jongh,H.J. & Farber,J. (1969) Bullfrog (_Rana catesbiana_) ventilation: how does the frog breath? Science _163_, 1223-1225.

21. Hughes,G.M. & Gans,C. (1967) The mechanism of lung ventilation in the tortoise _Testudo graeca_ Linnei. J.Exp.Biol. _47_, 1-20.

22. Gaunt,A.S. & Gans,C. (1969) Mechanics of respiration in the snapping turtle, _Chelydra serpentina._ J.Morph. _128_, 195-228.

23. Grigg,G.C. (1965) Studies on the Queensland lungfish, _Neoceratodus forsteri_ (Krefft). III. Aerial respiration in relation to habits. Aust.J.Zool. _13_, 413-421.

24. Grigg,G.C. (1970) Water flow through the gills of Port Jackson sharks. J.exp.Biol. 52,565-568.

25. Hazelhoff,E.H. (1951) Structure and function of the lung of birds. Poultry Sci. 30, 3-10.

26. Hughes,G.M. (1960) The mechanism of gill ventilation in the dogfish and skate. J.exp.Biol. 37, 11-27.

27. Hughes,G.M. (1966) Evolution between air and water. CIBA Foundation Symposium on Development of the Lung, pp.64-80, ed. De Rueck,A.V.S. & Porter,R., London, Churchill Ltd.

28. Hughes,G.M. (1967) Experiments on the respiration of the trigger fish (*Balistes capriscus*). Experientia 23, 1077.

29. Hughes,G.M. (1971) The cardiac and ventilatory rhythms in the dogfish (*Scyliorhinus canicula*). Experientia, 27, 721.

30. Hughes,G.M. (1972) Morphometrics of fish gills. Respir.Physiol. 14, 1-26.

31. Hughes,G.M. (1972) The relationship between cardiac and respiratory rhythms in the dogfish *Schyliorhinus canicula*. J.exp.Biol. 57,415-434.

32. Hughes,G.M. (1973) The Vertebrate Lung. Oxford Biology Readers, O.U.P. (in press).

33. Hughes,G.M. & Ballintijn,C.M. (1965) The muscular basis of the respiratory pumps in the dogfish (*Scyliorhinus canicula*). J.exp.Biol. 43, 363-383.

34. Hughes,G.M. & Gans,C. (1965) Electromyographic analysis of respiratory movements in *Testudo graeca*. Am.Zool. 6, 272-273.

35. Hughes,G.M. & Knights,B. (1968) The effect of loading the respiratory pumps on the oxygen consumption of *Callionymus lyra*. J.exp.Biol. 49, 603-615.

36. Hughes,G.M. & Roberts,J.L. (1970) A study of the effect of temperature changes on the respiratory pumps of the rainbow trout. J.exp.Biol. 52,177-192.

37. Hughes,G.M. & Shelton,G. (1958) The mechanism of gill ventilation in three freshwater teleostean fishes. J.exp.Biol. 35, 807-823.

38. Jesse,M.J., Shub,C. & Fishman,A.P. (1967) Lung and gill ventilation of the African lungfish. Respir.Physiol. 3, 267-287.

39. de Jongh,H.J. & Gans,C. (1969) On the mechanism of respiration in the bullfrog, *Rana catesbiana* A re-assessment. J.Morph. 127, 259-290.

40. Kempton,R.T. (1969) Morphological features of functional significance in the gills of the spiny dogfish, *Squalus acanthias*. Biol.Bull. 136, 226-240.

41. McMahon,B.R. (1969) A functional analysis of the aquatic and aerial respiratory movements of an African lungfish, *Protopterus aethiopicus* with reference to the evolution of the lung-ventilation mechanism in vertebrates. J.exp. Biol. 51, 407-430.

42. McMahon,B.R. (1970) The relative efficiency of gaseous exchange across the lungs and gills of an African lungfish, *Protopterus aethiopicus* J.exp.Biol. 52, 1-15.

43. Naifeh,K.H., Huggins,S.E. & Hoff,H.E. (1970) The nature of the ventilatory period in crocodilian respiration. Respir. Physiol. 10, 338-348.

44. Randall,D.J., Holeton,G.F. & Stevens,E.D.(1967) The exchange of oxygen and carbon dioxide across the gills of rainbow trout. J.exp.Biol. 46, 339-348.

45. Romer,A.S. (1972) Skin breathing - primary or secondary. Respir.Physiol. 14, 183-192.

46. Satchell,G.H., Hanson,D. & Johansen,K. (1970) Differential blood flow through the afferent branchial arteries of the skate, *Raja rhina*. J.exp.Biol. 52, 721-726.

47. Scheid,P. & Piiper,J. (1970) Analysis of gas exchange in the avian lung: theory and experiments in the domestic fowl. Respir.Physiol. 9, 246-262.

48. Scheid,P., Slama,H. & Piiper,J. (1972) Mechanisms of unidirectional flow in parabronchi of avian lungs: measurements in duck lung preparations. Respir.Physiol. 14, 83-95.

49. Singh,B.N. & Hughes,G.M. Cardiac and respiratory responses in the climbing perch, *Anabas testudineus.* (In preparation).

50. Stevens,E.D. (1972) Some aspects of gas exchange in tuna. J.exp.Biol. 56, 809-823.

51. Taylor,E.W. & Butler,P.J. (1971) Some observations on the relationship between heart beat and respiratory movements in the dogfish (*Scyliorhinus canicula* L.). Comp.Biochem.Physiol. 39A, 297-305.

52. Templeton,J.R. & Dawson,W.R. (1963) Respiration in the lizard *Crotaphytus collaris*. Physiol.Zool. 36, 104-121.

53. Thomson,K.S. (1969) Gill and lung function in the evolution of the lungfishes (Dipnoi): an hypothesis. Forma et Functio 1, 250-262.

54. Westoll,T.S. (1949) The evolution of the Dipnoi In *Genetics, Palaeontology and Evolution*, ed. Jepsen, *et al.*, Princeton, Princeton University Press.

55. Wikgren,B. (1959) Osmotic regulation in some aquatic animals with special reference to the influence of temperature. Acta Zool.fenn. 71, 1-102.

56. Willem,V. & Boelaert,R. (1937) Les manoeuvres respiratoires de Periophthalmus. Bull.de la Classe des Sci.Acad.roy.de Belg. 23,942-959.

57. Zeuthen,E. (1942) The ventilation of the respiratory tract in birds. K.K.dansk.Vidensk. Selsk.Biol.Med. 17, 1-50.

Comparative Physiology, eds. L. Bolis, K. Schmidt-Nielsen and S. H. P. Maddrell

FILTRATION STRUCTURES IN EXCRETORY SYSTEMS. A COMPARISON.

G. KÜMMEL
I. Zoologisches Institut
der Freien Universität Berlin,
Berlin, Germany.

INTRODUCTION

Filtration as a mechanism of urine formation was postulated for the first time in 1844 by Ludwig for the kidney of vertebrates. Since then it has been proved, or at least become probable for a large number of animals, that the primary urine has the character of a filtrate or of an ultrafiltrate (36,44,31). Since there are such clear physiological indications of a filtration process, it is advisable to look also for morphological indications of such a process, i.e. for filtration structures. Filtration structures should be confined to the first segment of excretory systems, because if filtration takes place, it will be the first step in the over-all process of urine formation (36).

Examples of Filtration Structures

Configurations which can be interpreted as filtration structures were demonstrated for the first time (20,39,59) in the Bowman's capsule of the vertebrate kidney. (Fig. 1A & B). The epithelial cells lining the inner wall of the capsule - called podocytes - have long, arm-like processes ("arms"), from each of which small processes spring out, called pedicels. These processes surround the capillaries, and they are attached to a basal lamina. The pedicels interdigitate with each other just leaving narrow and meander-like slits (about 250 Å

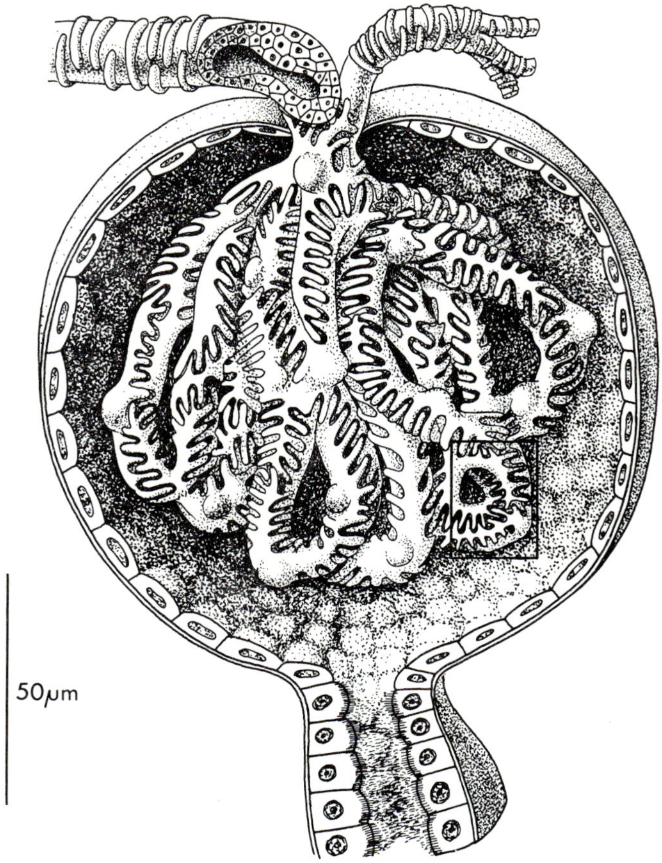

A

Figure 1A - Schematic representation of the renal corpuscle, cut away to show the lumen of the Bowman's capsule and podocytes overlying the capillaries of the glomerulus. (Redrawn and modified after Bergmann, 3).

wide) between adjacent pedicels. At the level of the basal lamina these slits are bridged by a very thin membrane (60 Å in thickness or less),the slit membranes or diaphragms. In grazing sections the diaphragms often appear to be dissected by an intermediate line (46). The plasma membranes of the pedicels are often associated with dense cytoplasmatic material at the attachment regions of the dia-

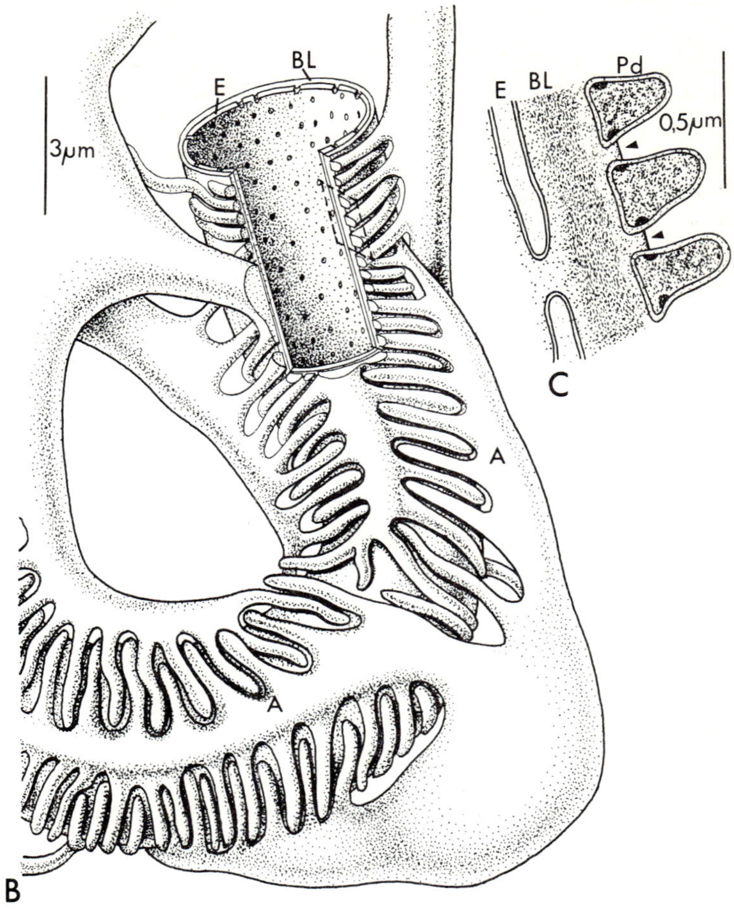

Figure 1B - A part of A (vid.rectangle in A) at a higher magnification, very schematically. Podocytes with arms and interdigitating pedicels surrounding a capillary. (Redrawn and modified after Ham,21). C: a part of B (vid.rectangle in B) at a higher magnification. Pedicels in cross section, with diaphragms (arrows) and dense cytoplasmic material at the attachment regions.

phragms (3). The basal lamina is three-layered and has a fibrillar appearance. Cytoplasmic filaments and microtubules extend from the cell body to arms and pedicels (35,3). A remarkable feature is the scarceness of mitochondria in the podocytes.

In nephridial excretory organs of arthropods (antennal and maxillary glands of Crustacea, coxal glands of Chelicerata, maxillary-and labial glands of Tracheata) ultrafiltration probably takes place through the wall of the coelomic sac (40,47,48,26). The epithelial cells of the wall of the coelomic sac again look like podocytes (30,32,50,52,1,22,23, 18,24,5,12,16,13). As examples the next two figures (Figures 2 and 3A-E) demonstrate podocytes from the antennal gland of two crustacean groups. The podocytes of different crustacean groups differ from each other only in nuances. The principal difference of the arrangement in the Bowman's capsule is the lack of an endothelium; moreover, in these podocytes so far only microtubules are described but no other cytoplasmic filaments.

In many molluscs there is also good evidence that ultrafiltration is the initial step in urine formation (43). Since the excretory organs usually drain the pericardium, probably in most cases ultrafiltration takes place through the wall of this coelomic cavity. In gastropods, except for pulmonates, and in mussels the most likely sites of ultrafiltration are the walls of either the ventricle or the auricles. In cephalopods on the other hand there is little doubt that the branchial heart appendages are the sites of filtration.

Recently Andrews and Little (2) have reported that the epithelial cells over the ventricle of a gastropod (Poteria) resemble podocytes (Figure 4); but apparently, diphragms are lacking.

Likewise in the branchial heart appendages of the cephalopods, podocyte-like cells have been detected (32,49,58), which at least in Octopus usually separate the pericardial cavity from a bloodsinus (Figure 5). The podocytes are attached with arms and pedicels to a fibrillar multi-layered basal lamina. The gaps between the pedicels are bridged by membrane-like elements (Fig. 5D), comparable to the diaphragms. These diaphragms again are contiguous with plasma membrane areas associated with dense cytoplasmic material. In grazing sections (Fig. 5 B,C) a periodical substructure with an intermediate line appears in the diaphragms.

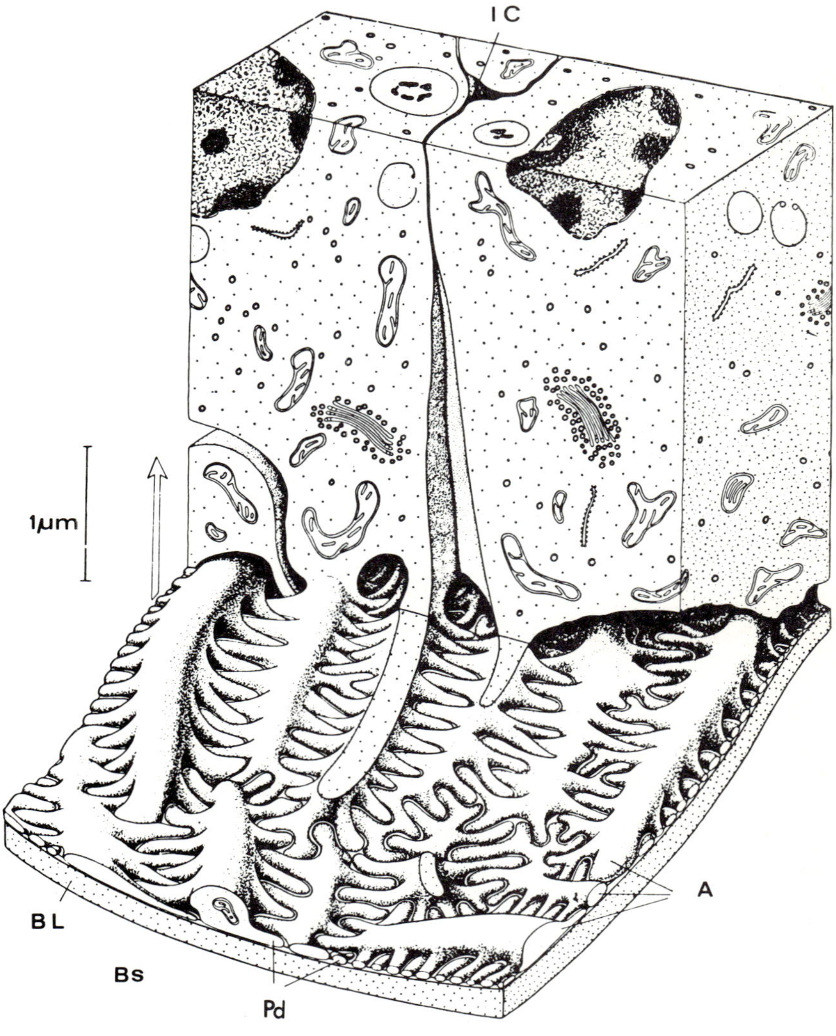

Figure 2 - Blockdiagram of podocytes of a crayfish (Orconectes, Decapoda, Crustacea). The arrow indicates the direction of the filtration from a bloodsinus to the lumen of the coelomic sac. The slits between the interdigitating pedicels are again bridged by diaphragms (from 32). Drawn by C.St. Friedmann.

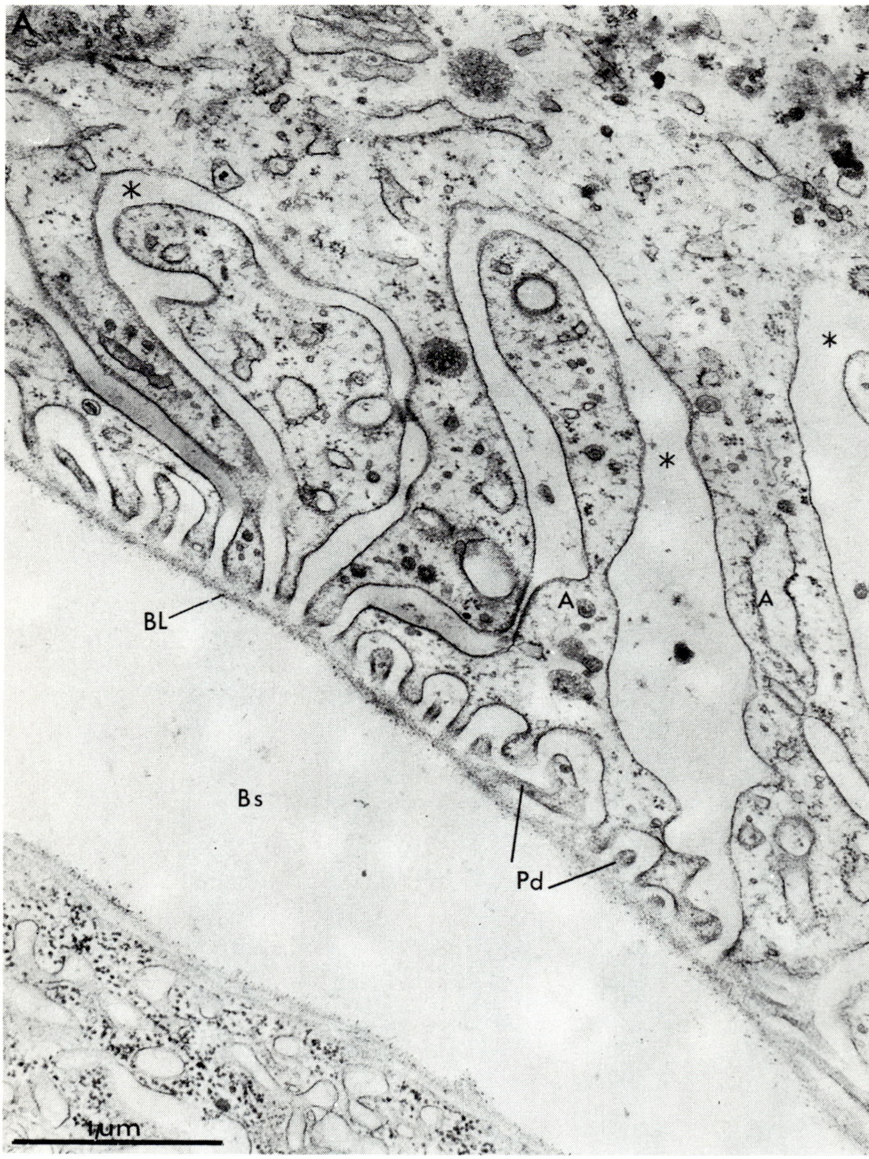

Figure 3 - Electron micrographs:Podocytes from the antennal gland of *Gammarus* (Amphipoda, Crustacea).
A. Longitudinal section through the basal part of podocytes. *:intercellular spaces, communicating with the lumen of the coelomic sac through intercellular channels.

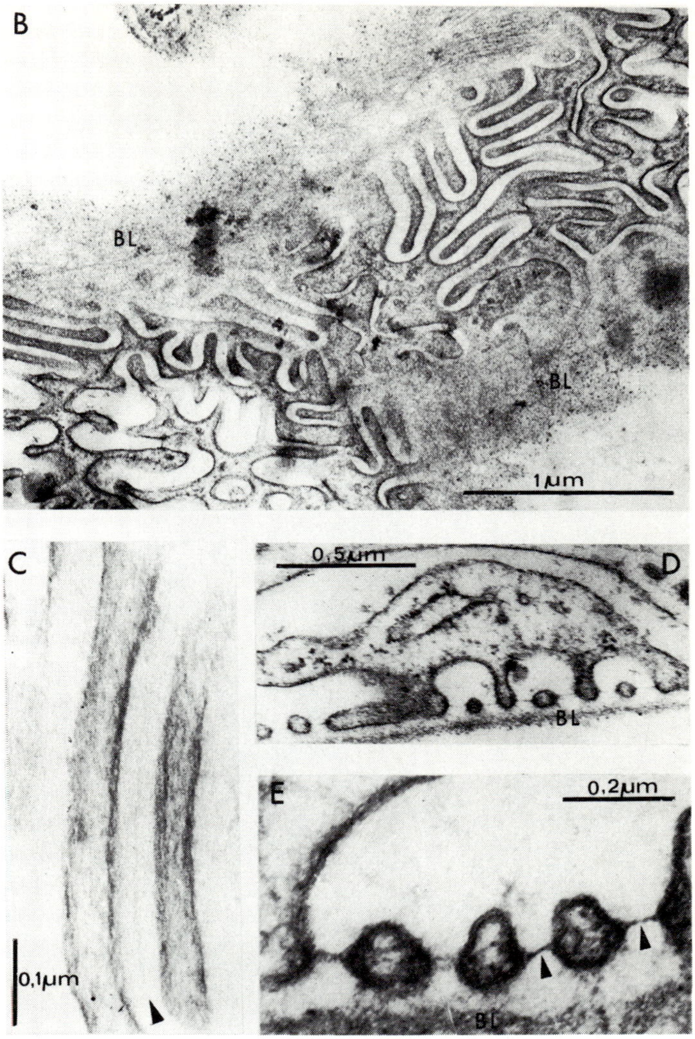

Figure 3 - Electron micrographs: Podocytes from the antennal gland of Gammarus (Amphipoda, Crusta-

B. Tangential section through pedicels, interdigitating with each other. Basal lamina has a fibrillar appearance. C: Grazing section through a diaphragm between adjacent pedicels. Arrow: intermediate line. D: Arms and cross sectioned pedicels. Diaphragms attached to plasma membrane areas associated with dense cytoplasmic material. E: Pedicels in cross section. Arrows: Diaphragms. (A-D, after Morelli,unpubl.;E, after Tobolla, unpubl.).

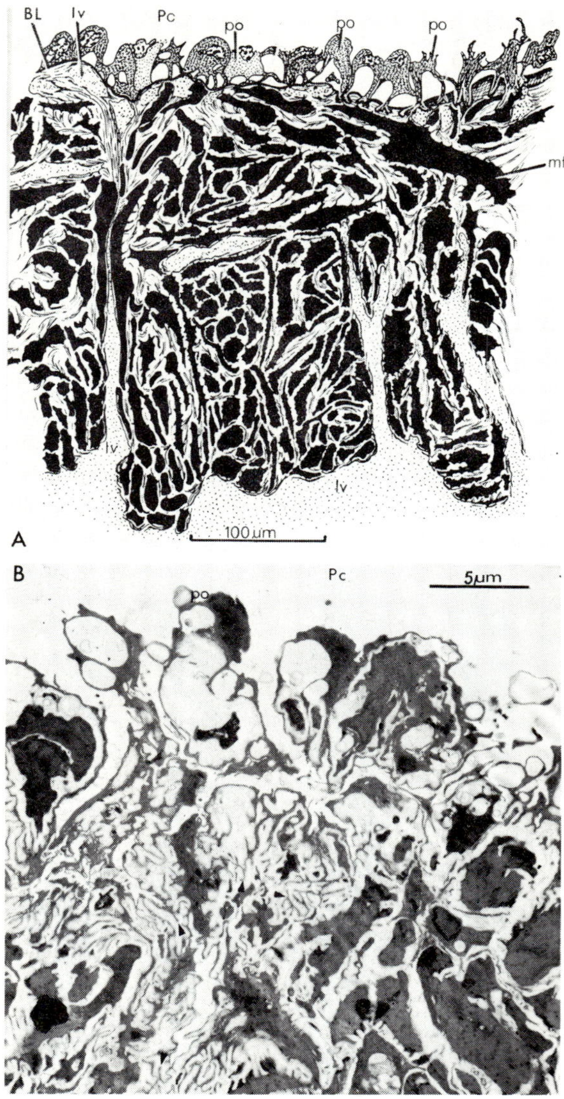

Figure 4 - Wall of the ventricle of Poteria (Prosobranchia Gastropoda). A: diagrammatic cross section through the wall. B: electron micrograph of the epithelium (= "podocytes") and part of the wall of the ventricle. Arrows: major and finer processes of the podocytes, tangentially sectioned (Courtesy of El. Andrews).

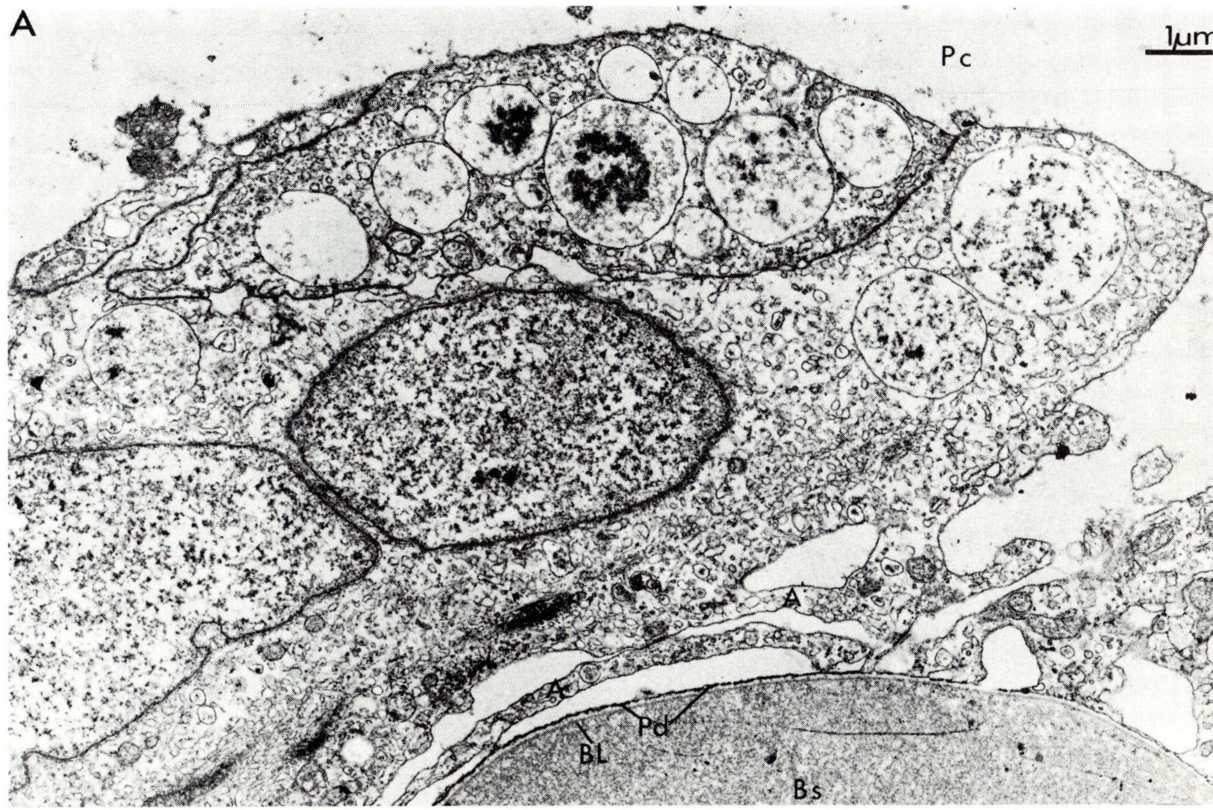

Figure 5 - Electron micrographs: podocytes from the branchial heart appendages of Octopus (Cephalopoda). A: Podocytes in a survey.

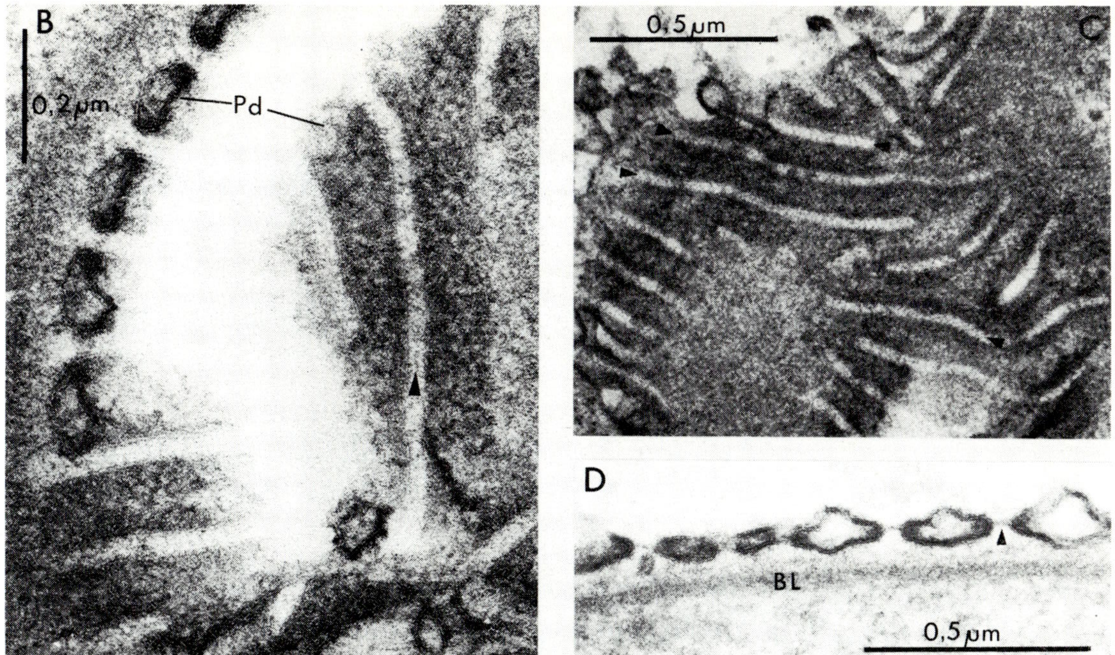

Figure 5 - Electron micrographs: podocytes from the branchial heart appendages of *Octopus* (Cephalopoda). B:Pedicels,obliquely and tangentially sectioned. Arrow:diaphragm with an intermediate line and a periodical substructure. C:Pedicels, tangentially sectioned.The diaphragms show a periodical substructure (arrows). D:Pedicels in cross section. Arrow:diaphragm (A-D, after Witmer, unpublished).

In electron micrographs it has been shown that the glomerulus epithelium of Glossobalanus minutus, an enteropneust, is also differentiated as podocyte-like cells (55, Fig. 6), with diaphrag-

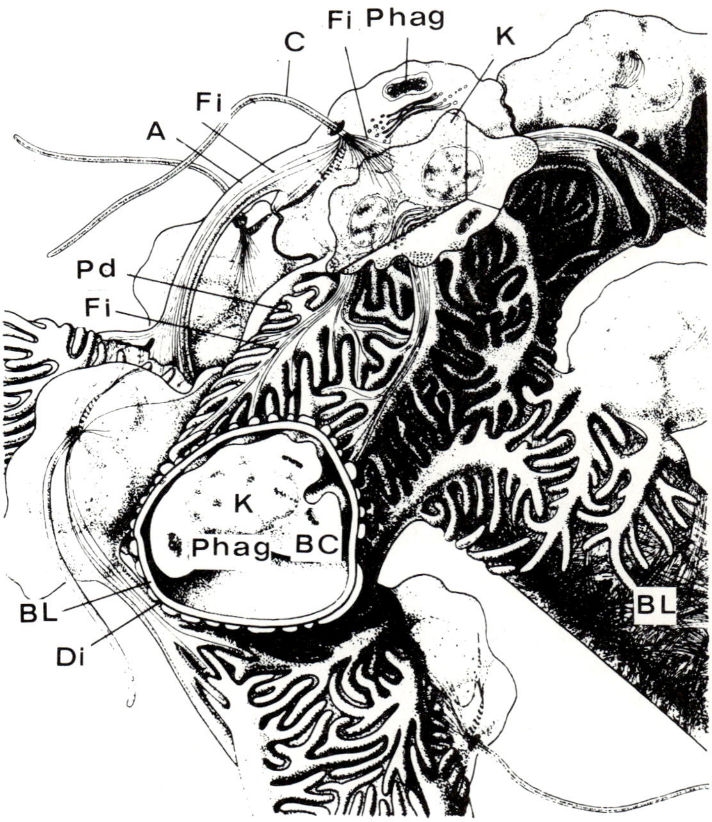

Figure 6 - Glomerulus of Glossobalanus (Enteropneusta). Podocyte-like cells, diagrammatically (from 55).

ms bridging the slits between the pedicels, again associated with dense cytoplasmic material.Current investigations will have to clarify whether the assumption of a filtration process is correct in this case.

Certainly the most common type of excretory systems are the protonephridia with the flame cell (or terminal cell) as first part. One function often ascribed to the flame cell is that it is a site of filtration of body fluids (28,41,42,9,10, 56,37). The presence of filtration structures in the region of the flame cell can be taken as morphological support of this function (28,7,45,37,57,25, 53,11). The tubular wall around the flame cilia is at least partly "disintegrated" in rods leaving long narrow slits between adjacent rods. The slits are always covered, usually with distinct membranes. Because the wall around the flame cilia looks like a weir, we have named the cell "cyrtocytes" (33). Figure 7 gives, as an example, pictures of the flame cell (= cyrtocyte) of the miracidium of the liver fluke.

DISCUSSION

This rather incomplete (vid. 27,8) record has made it clear that structures, which can be interpreted as filtration areas are regularly found in excretory systems in which the initial step of urine formation is a filtration process (Some exceptions are discussed in ref.5). What do filtration structures have in common? The cells possess flattened regions; these regions show narrow perforations, or they are at least partly disintegrated into rods, with narrow extended slits between the rods. Of course such a "grid" has in itself a certain sieve-effect. But the width of the slits (250 Å) is too coarse for an ultrafiltration. Therefore, the main function of this "grid" is to provide an adequate support for the filter membranes on the one hand and yet on the other hand to guarantee the largest possible filter area.However, the grid is probably not merely a passive holder for the filter-membranes: active transport processes (15,38) have often been described and movements of the pedicels have occasionally been described (35). Moreover, the podocytes play a role both in the formation and continual renewal of the basal lamina (34).

It is suggested that the basal lamina and/or the diaphragms function as filter-membranes. The fibrillar appearance of the basal lamina can be taken as support of their filter-like function. Some observations indicate that the basal lamina acts like a thixotropic gel (38).

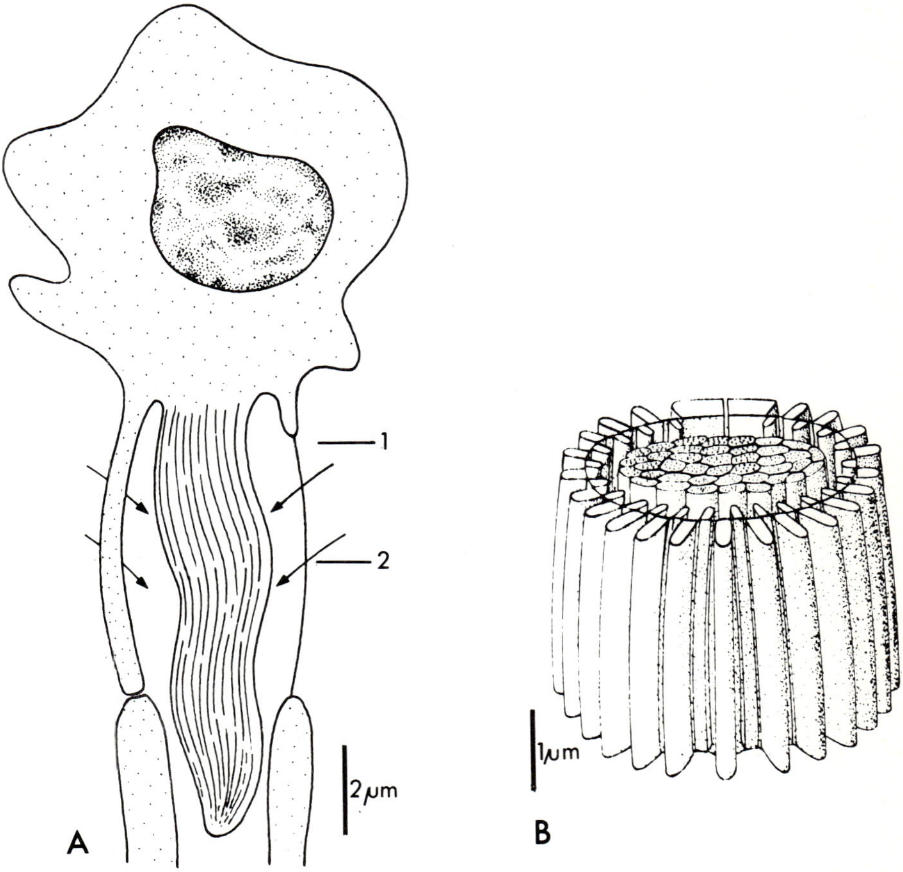

Figure 7 - Flame cell of the miracidium of the liver fluke Fasciola (Trematoda).
A: a highly diagrammatic longitudinal section through a flame cell. On the left side a rod of the weir,on the right side the membrane between the rods is sectioned. Arrows: direction of the assumed filtration from extracellular spaces into the weir lumen.
B: a view of a part of a transversely sectioned weir (very schematically), with the rods of the weir and the flame cilia in the weir-lumen. The lines (1,2) in A indicate the levels of the sections.

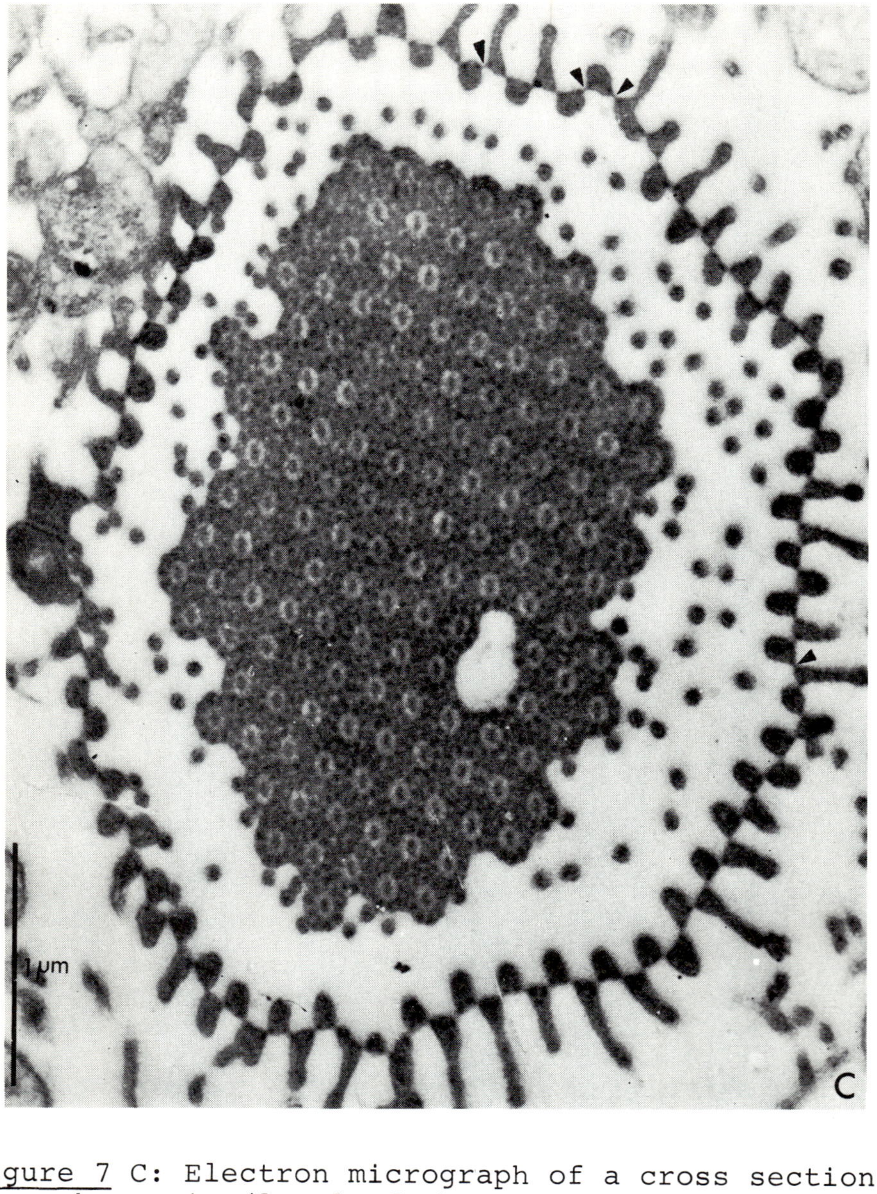

Figure 7 C: Electron micrograph of a cross section through a weir (level of the section line 1 in A). Arrows: membrane-like structures between adjacent rods; in the weir-lumen the cross sectioned flame cilia are clearly visible (from 31).

The mere structure of the diaphragms does not of course allow one to conclude that they function as a filter. The nature of the diaphragm is controversial: perhaps the various diaphragms are not all similar. Some authors have suggested a continuity with the outer leaflet of the unit membrane (1,51). Groniowsky _et al._ (19) were able to show that the slits of podocytes are covered by a surface coat continuous with the surface coat of the pedicels which appears to be a part of the outer leaflet of the unit membrane. Finally Farquhar _et al._ (14,15) have drawn attention to resemblances to desmosomes and adhering zonules: the association with dense cytoplasmic material and the intermediate line in the intercellular space. All these findings are perhaps consistent with the suggestion that all diaphragms are specialized regions of extra-or intercellular material.

The filter membranes are the only barrier between blood spaces and the lumen of excretory systems. The filtrate,therefore, does not need to cross cell membranes proper; after its passage through the filter membrane its onward flow is not impeded by junctional complexes.

Unfortunately,only in the vertebrate kidney has the significance of the diverse filter membranes been investigated (17,38,35). It would be very desirable if such investigations could be extended to other animal groups.

ABBREVIATIONS: A,arms; BC,blood cell; BL,basal lamina; Bs,blood sinus; C,cilium; Di,diaphragm; E,endothelium; Fi,filaments; IC,intercellular channels; K,nucleus; lv,ventricle-lumen; mf,muscle fiber; Pc,pericard cavity; Pd,pedicels; Phag,phago-lysosome; po,podocytes.

REFERENCES

1. Altner,H. (1968) Die Ultrastruktur der Labialnephridien von _Onychiurus quadriocellatus_ (Collembola). J.Ultrastruc.Res. 24, 349-366.

2. Andrews,El. & Little,C. (1971) Ultrafiltration in the gastropod heart. Nature 234, 411-412.

3. Bargmann,W. (1962) Histologie und mikroskopische Anatomie des Menschen. 4.Auflage,Stuttgart Thieme Verlag.

4. Bargmann,W. & Hehn,v.G. (1971) Über das Nephron der Elasmobranchier. Z.Zellforsch. 114, 1-21.

5. Berg,M. (1970) Die Feinstruktur der Labialniere von Schizophyllum sabulosum aimatopodium (Risso) (Diplopoda, Myriapoda) Diplomarbeit,F.U. Berlin.

6. Bonga,W.S.E. & Boer,H.H. (1969) Ultrastructure of the renopericardial system in the pond snail Lymnaea stagnalis L. Z.Zellforsch.94, 513-529.

7. Brandenburg,J. (1966) Die Reusenformen der Cyrtocyten. Eine Beschreibung von fünf weiteren Reusengeisselzellen und eine vergleichende Betrachtung. Zool.Beiträge N.F. 12, 345-418.

8. Brandenburg,J. & Kümmel,G. (1961) Die Feinstruktur der Solenocyten. J.Ultrastruct.Res. 5, 437-452.

9. Braun,G., Kümmel,G. & Mangos,J.A. (1966) Studies on the ultrastructure and function of a primitive excretory organ, the protonephridium of the rotifer Asplanchna priodonta. Pflügers Arch.ges.Physiol. 289, 141-154.

10. Clément,P. (1968) Ultrastructures d'un Rotifère, Notommata copens. I. La cellule-flamme. Hypothèses physiologiques. Z.Zellforsch. 89, 478-498.

11. Ebrahimzadeh,A.E. & Kraft,M. (1971) Ultrastrukturelle Untersuchungen zur Anatomie der Cercarien von Schistosoma mansoni II. Das Excretions system. Z.Parasitenk. 36, 265-290.

12. El-Hifnawi,E.S. & Seifert,G. (1971) Über den Feinbau der Maxillarnephridien von Polyxenus lagurus (Diplopoda, Penicillata).Z.Zellforsch. 113, 518-530.

13. Fain-Maurel,M.A. & Cassier,P. (1971) Différenciations cytoplasmatiques en relation avec la fonction excrétrice dans les reins céphaliques de Petrobius maritimus Leach (Insecta, Aptérygota). J.de Microscopie 10, 163-178.

14. Farquhar,M.G. & Palade,G.F. (1963) Junctional complexes in various epithelia. J.Cell Biol.17,

375-412.

15. Farquhar,M.G., Wissig,S.L. & Palade,G.E. (1961) Glomerular permeability. I.Ferritin transfer across the normal glomerular capillary wall. J.Exptl.Med. 113, 47-66.

16. François,J. (1972) Ultrastructure du rein labial céphalique de Campodea chardardi Condé (Diplura, Insecta). Z.Zellforsch. 127,34-49.

17. Graham,R.C. & Karnowsky,M.J. (1966) Glomerular permeability. Ultrastructural cytochemical studies using peroxidases as protein tracers. J.Exptl.Med. 124, 1123.

18. Groepler,W. (1969) Feinstruktur der Coxalorgane bei der Gattung Ornithodorus (Acari, Argasidae) Z.Wiss.Zool. 178, 235-257.

19. Graniowski,J., Biczyskowa,W. & Walski,M.(1969) Electron microscope studies on the surface coat of the nephron. 40, 585-601.

20. Hall,B.V. (1954) Further studies of the normal structure of the renal glomerulus. Proc.7th Annual Conf. on the Nephrotic Syndrome. National Nephrosis Foundation, New York.

21. Ham,A.W. (1965) Histology. 5th Edition. London, Pitman & Philadelphia,Lippincott.

22. Haupt,J. (1968) Zur Feinstruktur der Labialdrüse des Silberfischschens Lepisma saccharina L. Zool.Beitr. N.F. 15, 139-170.

23. Haupt,J. (1969) Zur Feinstruktur der Maxillarnephridien von Scutigerella immaculata Newport (Symphyla, Myriapoda). Z.Zellforsch. 101,401-407.

24. Hecker,H., Diehl,P.A. & Aeschlimann,R. (1969) Recherches sur l'ultrastructure et l'histochemie de l'organe coxal d'Ornithodorus mourbata (Murrag) (Ixodoidea; Argasidae) Actatrop. (Basel) 26, 346-359.

25. Howells,R.E. (1969) Observations on the nephridial system of the Cestode Monezia expansa. Parasitology 59, 449-59.

26. Kirschner,L. & Wagner,S. (1965) The site and permeability of filtration locus in the crayfish antennal gland. J.exp.Biol. 43, 385-395.

27. Koechlin,N. (1966) Ultrastructure du plexus sanguin périoesophagien; ses relations avec les néphridies de Sabella pavonina Savigny. C.R. Acad.Sci. (Paris), 262, 1266-1269.

28. Kümmel,G. (1958) Das Terminalorgan der Protonephridien. Feinstruktur und Deutung der Funktion. Z.Naturforschg. 13b, 677-679.

29. Kümmel,G. (1962) Zwei neue Formen von Cyrtocyten. Vergleich der bisher bekannten Cyrtocyten und Erörterung des Begriffes "Zelltyp". Z.Zellforsch. 57, 172-201.

30. Kümmel,G. (1964) Das Cölomsäckchen der Antennendrüse von Cambarus affinis Sag. (Decapoda,Crustacea). Zool.Beitr. N.F. 10, 227-252.

31. Kümmel,G. (1965) Druckfiltration als ein Mechanismus der Stoffausscheidung bei Wirbellosen. In: Funktionelle und morphologische Organisation der Zelle. Sekretion und Exkretion. 2. Wiss.Konferenz der Gesellsch.Deutscher Naturforscher und Ärzte, 1964, Berlin, Springer Verlag.

32. Kümmel,G. (1967) Die Podocyten. Zool.Beitr. N.f. 13, 245-263.

33. Kümmel,G. & Brandenburg,J. (1961) Die Reusengeisselzellen. Z.Naturforschg. 10b, 692-697.

34. Kurtz,M. & Feldman,J.D. (1962) Experimental studies on the formation of the glomerular basement membrane. J.Ultrastruct.Res. 6, 19-27.

35. Latta,H. (1970) The glomerular capillary wall. J.Ultrastruct. Research 32, 526-44.

36. Martin,A.W. (1957) Recent advances in knowledge of invertebrate renal function. In Invertebrate Physiology, ed. Scheer,B.T., pp.247-276, University of Oregon Publications.

37. McKanna,J.A. (1968) Fine structure of the protonephridial system in Planaria. I.Flame cells. Z.Zellforsch. 92, 509-523.

38. Menefee,M.G. & Mueller,C.B. (1967) Some morphological considerations of transport in the glomerulus. In: Ultrastructure of the kidney, ed. Dalton,A.J. & Haguenau,F., New York & London, Academic Press.

39. Pease,D.C. (1955) The fine structure of the kidney seen by the electron microscopy. J. Histochem.Cytochem. 3, 295-301.

40. Peters,H. (1936) Über den Einfluss des Salzgehaltes im Aussenmedium auf den Bau und die Funktion der Exkretionsorgane decapoder Crustacean (nach Untersuchungen an Potamobius fluviatilis und Homarus vulgaris). Z.Morph.Ökolog. Tiere 30, 355-381.

41. Pontin,R.M. (1964) A comparative account of the protonephridia of Asplanchna (Rotifera) with special reference to the flame bulbs. Bio.Zool. Soc.Lond. 142, 511-525.

42. Pontin,R.M. (1966) The osmoregulatory function of the vibratile flames and the contractile vesicle of Asplanchna (Rotifera). Comp.Biochem. Physiol. 17, 1111-1126.

43. Potts,W.T.W. (1967) Excretion in the Molluscs. Biological Reviews 42, 1-41.

44. Potts,W.T.W. & Parry,G. (1969) Osmotic and ionic regulation in animals. Pergamon Press.

45. Rees,G. (1967) The histochemistry of the cystogenus gland cells and cyst wall of Parorchis acanthus Nicoll and some details of the morphology and fine structure of the cercaria. Parasitology 57, 87-110.

46. Rhodin,J.A. (1962) The diaphragm of capillary endothelial fenestrations. J.Ultrastr.Res. 6, 171-185.

47. Riegel,J.A. (1971) Excretion - Arthropoda. In Chemical Zoology, ed. Florkin,M. & Scheer,B.T., vol.VI, pp. 249-277, New York, Academic Press.

48. Riegel,J.A. & Kirschner,L.B. (1960) The excretion of inulin and glucose by the crayfish antennal gland. Biol.Bull. 118, 296-307.

49. Schipp,R., Höhn,P. & Schäfer,A. (1971) Elektro-

nenmikroskopische und histochemische Untersuchungen zur Funktion des Kiemenherzanhanges (Pericardraldrüse) von _Sepia officinalis_. Z.Zellforsch. _117_, 252-274.

50. Schmidt-Nielsen,B.K.E., Gertz,K.H. & Davis,L.E. (1968) Excretion and Ultrastructure of the antennal gland of the fiddler crab _Uca mordax_. J.Morph. _125_, 473-496.

51. Strangways,-Dixon,J. & Smith,D.S. (1970) The fine structure of gill podocytes in _Palinurus argus_ (Crustacea).Tissue and Cell _2_, 611-624.

52. Tyson,G.E. (1968) The fine structure of the maxillary gland of the brine shrimp, _Artemia salina_: the end sac. Z.Zellforsch. _86_, 129-138.

53. Warner,F.D. (1969) The fine structure of the Protonephridia in the Rotifer _Asplanchna_. J.Ultrastruct.Res. _29_, 499-524.

54. Webster,L.A. & Wilson,R.A. (1970) The chemical composition of protonephridial canal fluid from the Cestode _Hymenolepis diminuta_.Comp.Biochem. Physiol. _35_, 201-209.

55. Wilke,U. (1972) Die Feinstruktur des Glomerulus von _Glossobalanus minutus_ Kowalewsky (Enteropneusta).Cytobiologie _5_, 439-44.

56. Wilson,R.A. (1967) The protonephridial system in the miracidium of the liver fluke _Fasciola hepatica_ L. Comp.Biochem.Physiol. _20_, 337-342.

57. Wilson,R.A. (1969) The fine structure of the protonephridial system in the miracidium of _Fasciola hepatica_. Parasitology _59_, 461-67.

58. Witmer,A. & Martin,A.W. - Z.Zellforsch. (in press).

59. Yamàda,E. (1955) The fine structure of the renal glomerulus of the mouse. J.Biophys. Biochem.Cytol. _1_, 551-566.

Comparative Physiology, eds. L. Bolis, K. Schmidt-Nielsen and S. H. P. Maddrell

SOME TRANSPORT PROPERTIES OF AMPHIBIAN AND MAMMALIAN NEPHRONS [1]

G.GIEBISCH
Department of Physiology,
Yale University School of Medicine,
New Haven, Connecticut, U.S.A.

INTRODUCTION

Although kidneys are present in all vertebrates and play an important role in osmoregulation they are not the sole organ involved in water and salt exchange. In fresh water fishes the gills aid the kidney in the regulation of volume and composition of extracellular fluid, and in marine fishes the gills carry out an even greater proportion of the osmoregulatory work than do the kidneys. In amphibians the charge of osmoregulation is distributed between the skin, urinary bladder and the kidney. In reptiles, and some marine birds, salt glands assist in the excretion of sodium chloride. In mammals the kidneys alone are responsible for osmoregulation (1,1a,2).

In most vertebrates the extracellular fluid composition is quite similar and ranges, with the exception of some fishes, from some 200 to 300 mOsm. Sodium and chloride make up some 80 to 90% of the total osmolarity. In most vertebrates, the make-up of body fluids is kept fairly constant even in the presence of large variations in the environment.

The kidneys are the only organs in vertebrates that can secrete large volumes of water and at the same time conserve solutes to such a degree that their urinary concentrations fall to less than 1%

---

1 - Work in the author's laboratory was supported by grants from the National Institutes of Health and the American Heart Association.

of their total level in the body fluids. Thus, they are the main organs for the excretion of water and the conservation of solutes. Despite the fact that salt glands and gills participate in the regulation of salt homeostasis, the kidneys assume the main responsibility of this function in higher vertebrates. It is of interest that the kidneys have remained in all species the main route for excretion of divalent ions. In addition, solutes derived as by-products of metabolism (urea, uric acid, ammonia, creatinine, etc.) may also be excreted by the kidneys. Finally, the latter do play a major role in acid-base regulation (1,1a,2,3).

General Properties of Nephrons.

Two types of nephrons have played a major role in the development of our concepts as to how the kidney functions. These are the amphibian and mammalian nephron on which the most direct work - on the single nephron level - has been carried out. Whereas a fascinating multitude of different morphological segments make up the nephron of both fresh and salt water fishes (4), far less is known about their precise functions and the underlying transport mechanisms although there are some notable exceptions (5,6,7). Similarly, the reptilian and bird nephron has received less attention but progress has been made (8,9).

Figure 1 provides a schematic summary of nephrons of different vertebrates (10). The basic functional unit of the kidney, the nephron, is made up of a vascular and a tubular part. The tubular structure is segmented into units of varying length, diameter and morphology. While in exceptional cases (Agnatha) one glomerulus may communicate with several tubules (see 2) or be entirely absent in some marine fishes (10), it is the common pattern that glomeruli of varying size and number connect with a single tubule.

It is the tubular structure which may vary considerably and show important adaptations to the particular functional requirements. The amphibian nephron, examples of which are shown in Figures 2 and 3, has all the segments of the typical vertebrate nephron (1). The glomeruli are joined to the

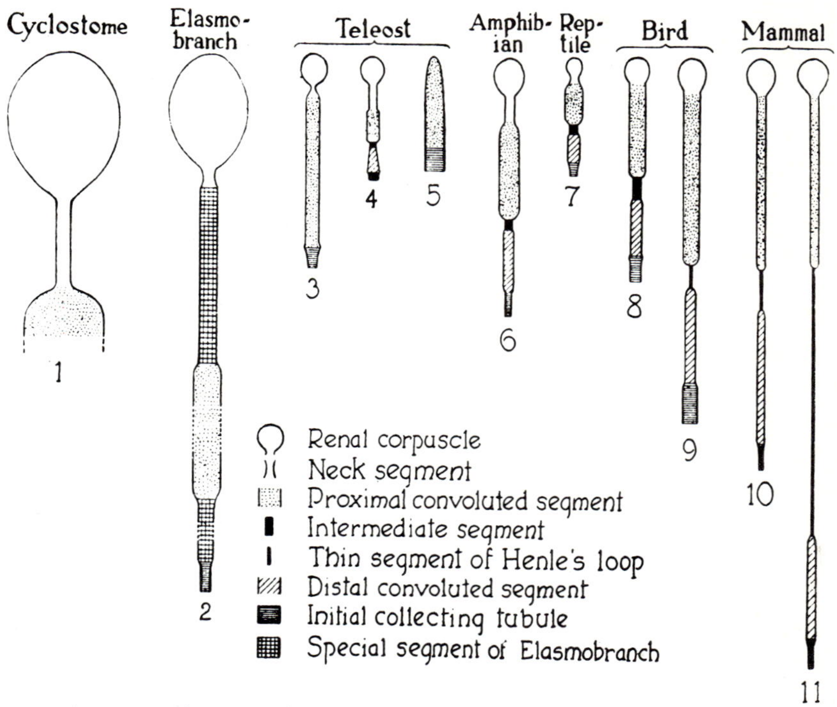

Figure 1 - Schematic representation of nephron of different vertebrates (10).

proximal tubule by a ciliated neck segment. The proximal tubule is lined with epithelial cells having a well-developed brush border. A small intermediate segment is present. Subsequently, the distal tubule forms several coils and drains into the collecting ducts. These run at right angles to the tubules and join the ureter. As shown in Figure 4, in reptiles the highly convoluted tubules are arranged in parallel to each other and also run at right angles to the collecting ducts. In contrast to the amphibian reptilian nephron, some proximal tubules in birds and all of them in mammals are followed by a loop-like structure,the loop of Henle; it dips into the medullary part of the kidney and runs at right angles to the collecting ducts. It is well established that the length

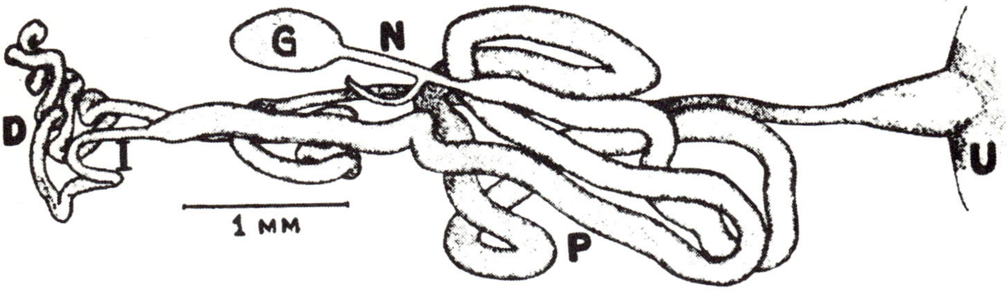

<u>Figure 2</u> - Nephron of Necturus, after Kempton (88). G: glomerulus, N:neck, P: proximal tubule, I: intermediate segment, D: distal tubule, U: ureter.

<u>Figure 3</u> - Frog kidney. VPR: renal portal vein, S.Coll.: collecting duct, C.W.: ureter, SBR: proximal tubule, C: neck segment, SGR: intermediary segment, SB: distal tubule, AO: renal artery, VCP: renal efferent vein. (From ref. 1).

of this hair-pin structure is functionally related to the kidney's ability to produce an osmotically concentrated urine (11,12). It is of interest that the bird kidney contains a mixture of nephrons similar to reptilian <u>and</u> mammalian nephrons. Braun

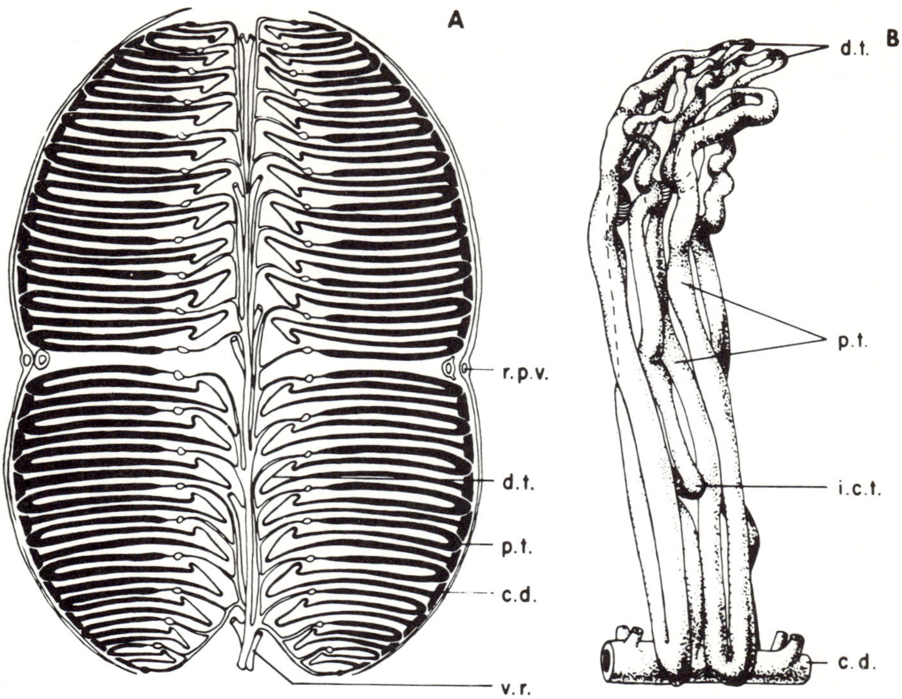

Figure 4 - Reptile Kidney. A: crocodile kidney lobule. V.R.: vena revehens, c.d.:collecting duct, p.t.:proximal tubule, d.t.: distal tubule. B Two crocodile renal nephrons. i.c.t.: initial collecting tubule (from ref. 1).

and Dantzler have studied the structure and function of mammalian-type and reptilian-type nephrons in kidneys of the desert quail and Figure 5 is taken from their work (8). Most nephrons are similar to reptilian nephrons, having a simple proximal and distal tubule without loops of Henle,and are arranged at right angles to collecting ducts. In addition to these, there are some nephrons having loops of Henle, arranged in parallel to both vasa recta and collecting ducts, and forming typical medullary cones. It appears safe to conclude that these latter structures function as countercurrent multipliers and play the key role in producing an osmotically concentrated urine. An impor-

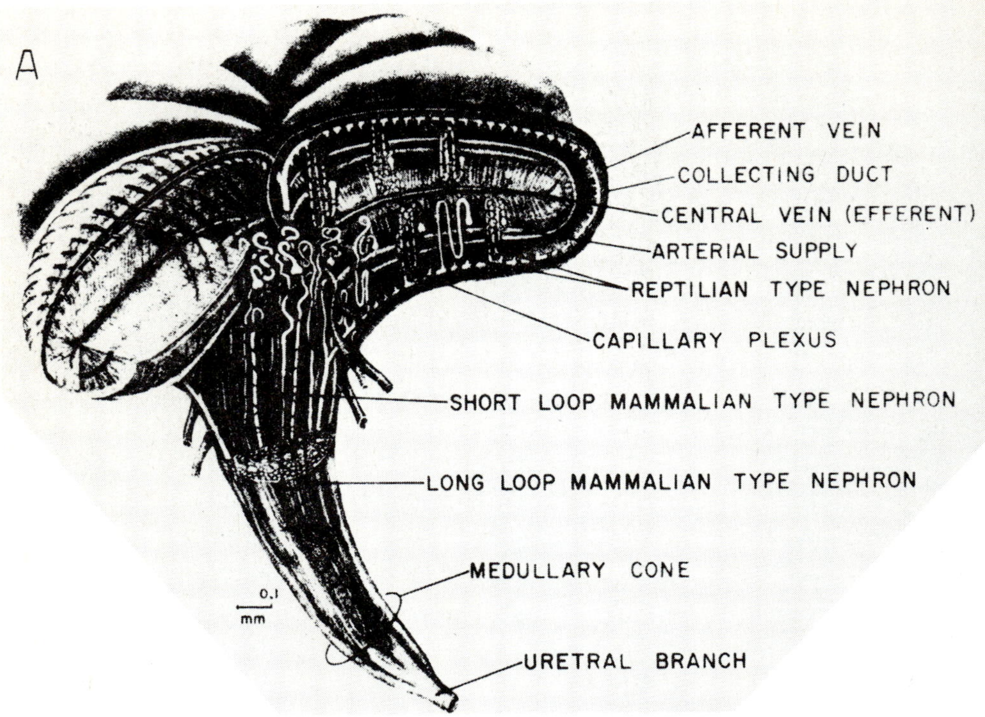

Figure 5A - Section of avian kidney showing arrangement of different types of nephrons. (From ref. 8).

tant functional heterogeneity was also revealed. With severe dehydration, it could be demonstrated that only reptilian-type nephrons decreased their filtration rate, thus effectively curtailing water loss, whereas the mammalian-type nephrons continued functioning and producing an appropriately concentrated urine (8).

In addition to differences with respect to the presence of Henle's loop, a consideration of the blood supply to the kidneys of different classes of vertebrates indicates large variations in the pattern of circulation. Figure 6 illustrates four representative types of blood supply to the nephron. In general the blood supply to the glomerulus is arterial, with afferent and efferent arterioles supplying and draining the glomerular capillary tuft. In mammalian kidneys the afferent capillaries form a network of peritubular capillaries and

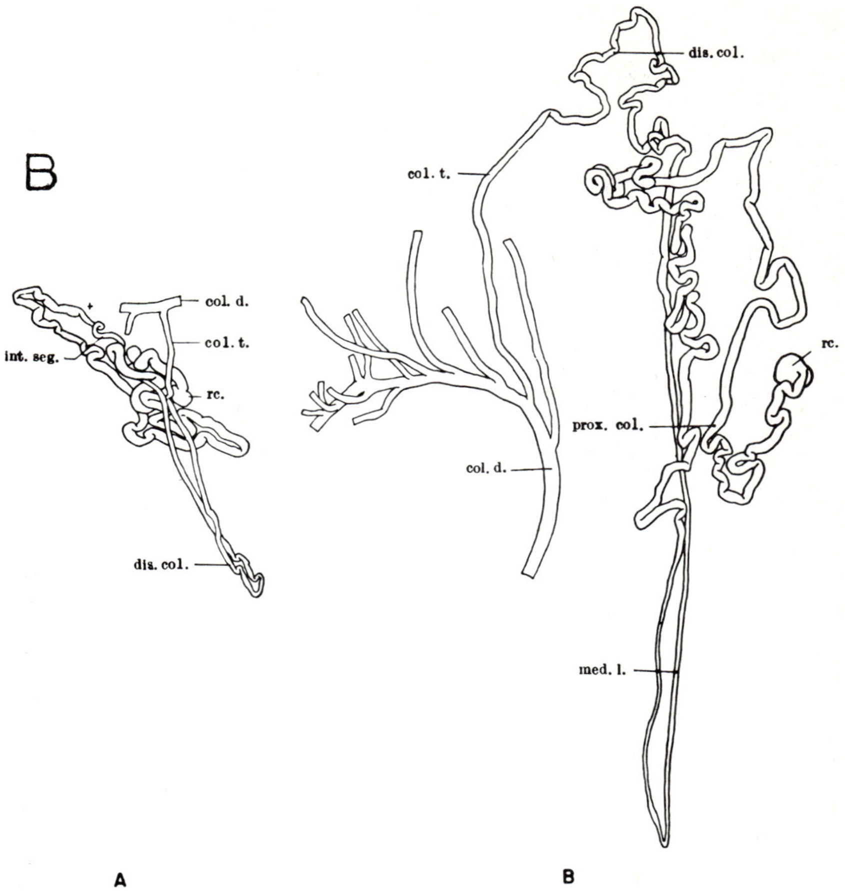

Figure 5B - Nephrons from bird kidney.
A: Single reptilian-type nephron from chicken kidney. v.c.: renal corpuscle, int.seg.: intermediary segment; dis. col.: distal tubule, col.t.:initial collecting tubule, col.d.: collecting duct.

B: A single nephron with a loop of Henle from a chicken kidney. v.c.: renal corpuscle, prox.col.: proximal tubule, med.l: ascending limb of the loop of Henle, dis.col.: distal tubule, col. t.: initial collecting tubule, col.d.: collecting duct. (From ref. 1).

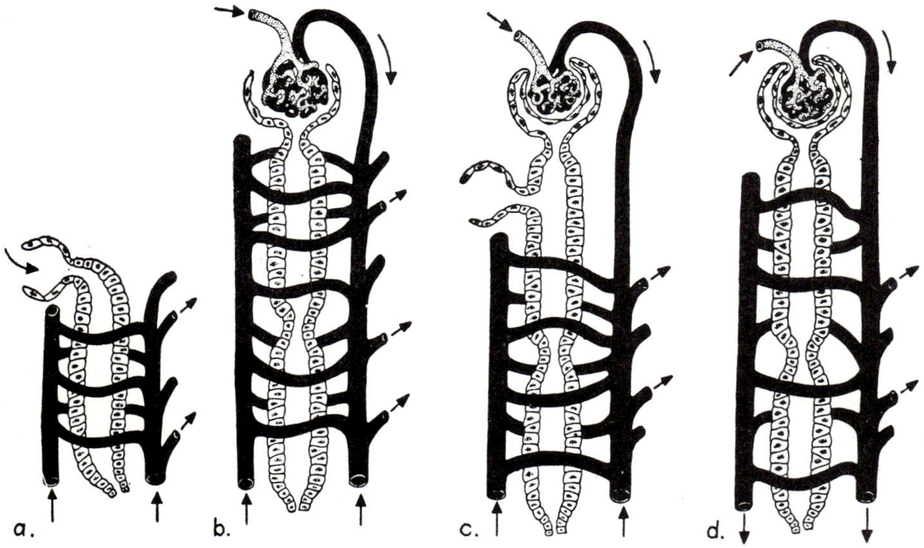

Figure 6 - Stages in the evolution of the vertebrate nephron.
(a) In protovertebrates the renal tubule drained the coelom by means of the coelomostome.

(b) The glomerulus in the earliest vertebrates was first only loosely related to the coelomostome.

(c) Later the glomerulus became sealed within the end of the tubule.

(d) In higher vertebrates, the coelomostome disappeared, the renal portal system persists through the fishes, amphibia, reptiles and birds (a to c) but disappears in mammals (d) leaving the tubules supplied only by postglomerular blood. (From 13).

vasa recta which supply the tubular tissue. In contrast, all amphibians, as well as birds, reptiles and most fishes, have a functioning renal portal system. Venous blood from the posterior regions of the body drains into the renal portal vein which splits into a capillary network. This supplies the tubules, thus joining the post-glomerular capillary network of vessels surrounding the nephrons. By the addition of marker substances to the renal portal circulation which enter the tubular fluid only via the glomerular route (i.e. by filtration) it can be demonstrated that at normal pressure distribution between the aortal and portal circulation circuits, in Necturus only a very small fraction, usually less than 5%, supplies the glomeruli (14).

The described differences in tubular blood supply have profound consequences. In those vertebrate classes with portal blood supply the threat of dehydration can be met by the dramatic reduction of the rate of filtrate formation (either by a fall in the number of functioning glomeruli or by a dramatic reduction of filtration in all glomeruli) because metabolic integrity of the tubular tissue is not jeopardized (12). The neurohypophyseal hormone arginine vasotocin (AVT) has a dual effect on the kidney; it depresses filtration rate and increases the renal tubular conservation of water by augmenting water permeability of the distal nephron and facilitating equilibration of the hypotonic tubular contents with the relatively hypertonic peritubular environment (see ref.2). In contrast,recent clearance and micropuncture studies on distal tubules of reptiles have shown that the water permeability of the distal tubule or collecting duct system in many reptiles is relatively insensitive to the state of hydration and relatively unaffected by injections of arginine vasotocin (1). It has recently been shown that the reptilian-type short nephrons of some bird kidneys do respond to dehydration by a profound reduction in nephron filtration rate (8). Obviously, such a regulatory mechanism in response to dehydration could not be tolerated by mammalian tubular structures since their metabolic demands depend critically on supply with post-glomerular arterial blood. The high rate of glomerular filtration in the mammalian kidney necessitates high tubular rates of reabsorptive sodium and water transfer and, thus, dependence upon a high rate of oxygen delivery. On the other hand, the relatively low and variable rates of filtrate formation, associated with similarly low rates of transepithelial sodium and fluid reabsorption and a low arterial pressure system in the amphibian permits continued tubular activity with portal renal venous blood supply. Indeed, the absence of oxidative enzymes such as cytochrome oxidase and succinic dehydrogenase in proximal tubules of *Necturus* which derive their blood supply mainly from the renal portal circulation underscores those principal differences in the metabolic structure of amphibian and mammalian nephrons (15).

The functional organization of the amphibian and mammalian nephron.

(1) *Distribution of function along the nephron.*

The functional properties of the single nephron have been most extensively studied in the kidney of *Necturus*, the frog and the rat. After the pioneering micropuncture studies of Richards,Walker, Oliver,Bott and their respective associates (16,17) investigations dealing with the localization of individual transport processes within the nephron, their magnitude and direction as well as their underlying mechanism have made great progress (16,17, 18,19,20,21,22,23).

The *proximal tubule* is the primary site of filtrate retrieval. Since the function of the amphibian kidney is primarily that of excreting excess of water - most amphibia live in fresh water habitat - the capacity of fluid reabsorption is less developed in this class of vertebrates (19,24,25). In contrast, the mammalian proximal tubule has at its disposal a more powerful mechanism for the conservation of fluid which aids in the conservation of fluid in a terrestrial environment. In addition to fluid conservation (18,23,27) the proximal tubule is also the main nephron site of sodium (18,23, 27), chloride (21), calcium (27), phosphate (28), potassium (29), and bicarbonate reabsorption (30). Furthermore, it is the sole site of glucose (19,18) and aminoacid (31) conservation. It secretes hydrogen ions (32,33) and, in the mammalian at least, ammonium ions (34,35) as well as some foreign anions (36).

As has been pointed out before, it is the loop of Henle - absent in all but avian and mammalian kidneys - which is responsible for the generation of an osmotically concentrated urine. This concentrating operation is achieved by the reabsorption and deposition of sodium within the papillary interstitium and subsequent osmotic equilibration of the contents of the collecting ducts with this hypertonic medium (37,38,39).

The *distal tubule* serves, in very general terms, two functions. It is sensitive to the action of hypophyseal peptides such as vasopressin and vasotocin which enhance the hydraulic conductivity of the distal tubular epithelium and allow its hypotonic contents to equilibrate osmotically with the extratubular medium (37,39). This leads to the production of at best an isosmotic urine. In addition to playing a role in water conservation, in species below the avian and mammalian level the distal tubule is probably the main tubular site of water conservation which is subject to regulation by the hypophyseal hormones: this nephron segment is also responsible for the continued reabsorption of salt from the tubular fluid. Both the distal tubular epithelium and the collecting ducts are characterized by a more variable ionic transport pattern than the proximal tubular epithelium (23). The fraction of the glomerular filtrate reabsorbed at this site is fairly small in the mammalian nephron, some 10% in the normal state of hydration, but micropuncture studies in *Necturus* under conditions of free-flow have shown that as much water can be reabsorbed by its distal tubule as by its proximal tubule (19,24, 25,40). In contrast to the proximal tubule, tubular fluid in both amphibian and mammalian species in the absence of antidiuretic hormones is rendered hypotonic by the gradual development of sizeable concentration gradients of sodium chloride. Normally, the concentration of sodium and chloride declines along the distal tubule, whereas the transport pattern of potassium is highly variable: both the mammalian as well as the amphibian distal tubule is capable of the massive addition of potassium by net secretion leading to the excretion of potassium moieties greatly in excess of that filtered (23,29,41,42,43,44). On the other hand, the distal tubular epithelium is also a nephron site where continued reabsorption of potassium may reduce the potassium excretion to less than 1% of the filtered potassium load (29,42). Thus, the distal tubule is the most important nephron site with respect to the regulation of urinary potassium excreticn.

The collecting duct epithelium is the nephron site where the steepest concentration differences with respect to sodium, chloride, potassium and hydrogen ions have been observed (42,32,33,45,46). Net transfer rates of electrolytes are fairly small compared to more proximal transport sites but important in the elaboration of a urine which can be rendered almost free of sodium chloride or potassium in states of salt deprivation (46), or during water diuresis (47). It should be noted that both the distal tubular epithelium as well as that of the collecting ducts are subject to the regulatory effects of adrenal mineralcorticoids; their presence has been shown to be necessary for the ability of these tubular segments to lower the concentrations of sodium and potassium in the tubular fluid (48,49). Finally, it is along the mammalian collecting ducts that a markedly hypertonic urine is formed (37,38,39).

(2) The proximal tubular epithelium.

(a) Fluid and salt reabsorption. The proximal tubule, as judged by the presence of the brush border, is the only nephron segment occurring in all classes of vertebrates (10), reabsorbing large quantities of fluid and electrolytes from the filtrate. Figures 7 and 8 provide quantitative information on the fractional reabsorption of fluid from the proximal tubule of *Necturus* and the proximal convolution of the rat, respectively. As judged by the concentration increase of inulin, some 30% of the filtrate are absorbed along the proximal tubule of *Necturus* whereas some 60% are transferred by the time tubular fluid has reached the end of the proximal convolution of the rat kidney. In view of the fact that the rat's filtered fluid load greatly exceeds that of *Necturus* (GFR in the rat: 300-

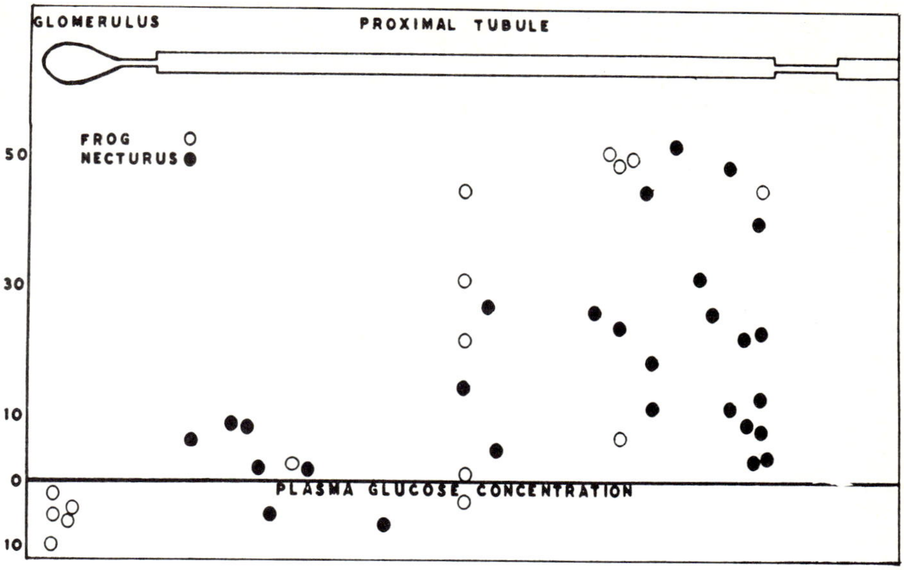

Figure 7 - Plot of differences in glucose concentration between blood plasma and fluid collected from various levels of proximal tubules of phlorhizinized Necturus and frogs. (From 19). Inulin concentration ratios are similar (From 19).

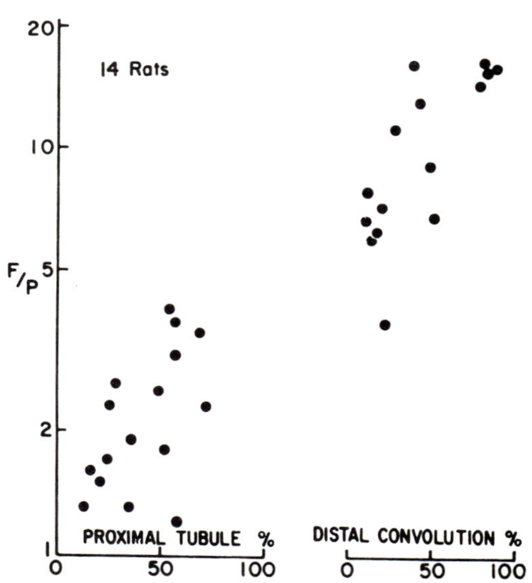

Figure 8 - Inulin fluid/plasma concentration ratios along superficial nephrons in the non-diuretic rat. (From 26).

400 ml/Kg.hr; GFR in _Necturus_: 8-15 ml/Kg.hr (90, 91,92)), the absolute differences in the rates of fluid transport along the proximal tubule are even greater than the fractional differences. The presence of cilia in the neck segment and the intermediate segment in the amphibian nephron (10,50) is probably related, to aid the lower filtration pressure in propelling fluid down the tubule. Cilia disappear coincidentally with the attainment of a high blood pressure and with the development of steeper pressure gradients along the nephron in the avian and mammalian kidney.

It is well established that the driving force for proximal fluid reabsorption is provided for by the tubules ability to transport sodium ions actively across the proximal tubular epithelium (23,51, 52). It is further believed that the active translocation of sodium ions establishes an osmotic concentration difference within the epithelium and thus provides an osmotic driving force for passive fluid movement. Experimental evidence supporting such a mode of transport is based on experiments in both single proximal tubules of _Necturus_ (52) and the rat (51). A representative example is shown in Figure 9. Summarized are results from stationary perfusion experiments on single proximal tubules in the rat. The perfusing fluids used in these experiments contained different concentrations of sodium chloride, isosmolarity in relation to plasma being maintained by addition of requisite amounts of nonelectrolytes. Hence, water movement could not be due to pre-existing osmotic forces acting across the tubular epithelium. It is apparent that water movement, measured on the basis of inulin concentration changes in the collected perfusate, depended upon the intratubular sodium concentration. In _Necturus_, fluid was still reabsorbed from solutions having a sodium concentration greater than 66 mEq/L in the rat above 95 mEq/L. Thus the mammalian proximal tubule (51,53) establishes somewhat higher (40-50 mEq/L) limiting transepithelial sodium concentration differences than do those amphibian proximal tubules (30 mEq/L) in which such studies have been carried out (52). Since in all these studies the collected perfusates had equal osmolalities to plasma but a lower sodium concentration it follows

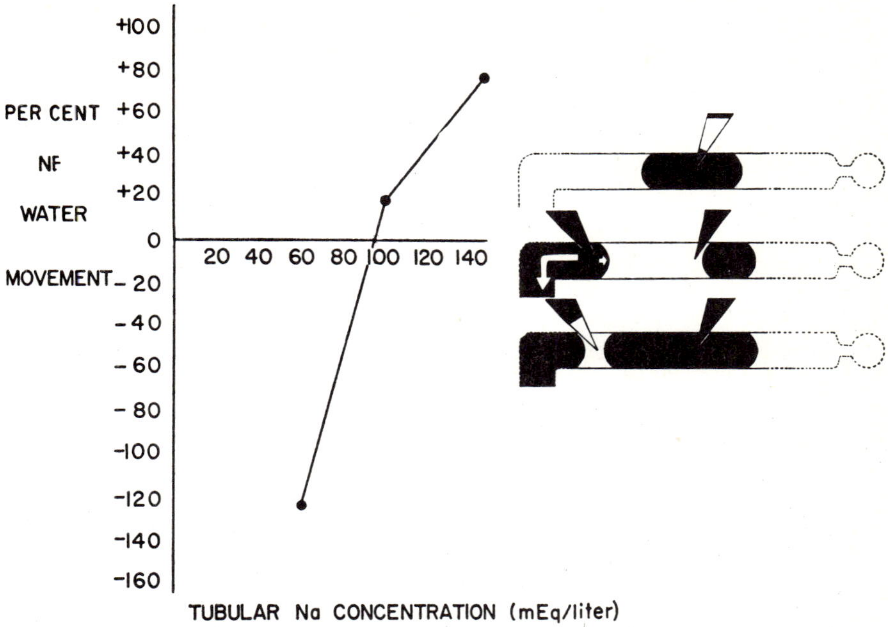

Figure 9 - Fractional net water movement as function of mean intratubular sodium concentration. Schematic drawing on right shows method of isolation of fluid column and of collection of perfusate Colored mineral oil is injected into a proximal tubule through a micropipette and split by injection of perfusion fluid. The position of the oil-fluid-oil assembly is stabilized by downstream injection of mineral oil. Collection is accomplished by injection of more oil through the proximally located micropipette and by aspiration of perfusion fluid through the more distally located micropipette. Plus sign: movement of fluid out of the tubular lumen; minus sign: movement of fluid into the lumen (From 23).

that sodium was reabsorbed against a concentration gradient with osmotically equivalent amounts of fluid following. Hence, water movement is osmotically coupled to the transfer of solutes. With the exception of quantitative differences these mechanisms are present in both the amphibian and mammalian nephron.

Measurements of osmotically induced water movement in perfused nephrons have provided information on the water permeability of the proximal tubular epithelium. The hydraulic permeability of *Necturus* proximal tubule is of the order of 0.15 x x 10^{-8} ml/sec.cm^2 (53) whereas that of the rat proximal tubule is some one hundred times higher (54). However, the oncotic pressure of proteins in the peritubular blood cannot directly account for more than a minor fraction of fluid reabsorption since the tubular wall is too tight (43,54).Nevertheless, it shall be pointed out that those factors responsible for the ultimate transfer of the tubular reabsorbate into the peritubular capillaries, i.e. the colloid-oncotic and hydrostatic pressure in peritubular capillaries, provide an important regulatory mechanism with respect to overall transtubular solute movement. Importantly though,their site of action is beyond the active transfer mechanism of sodium ions (55,56).

There seem to be some significant differences with respect to the site of tubular fluid acidification between amphibian and mammalian species. Studies on amphibian single proximal tubules have shown that no significant pH changes occur as tubular fluid passes along the proximal tubule (25,57). The fact that there occurs no pH change along the proximal tubule of the amphibian nephron should not be taken to indicate absence of net bicarbonate reabsorption. Since the tubular fluid pH does not increase, it follows that bicarbonate ions are reabsorbed in proportion to fluid movement.Although direct evidence is lacking it is tempting to speculate that, similar to the mammalian proximal tubule (32,33), hydrogen ion secretion is involved in the process of amphibian bicarbonate reabsorption. In contrast to the amphibian proximal tubule, measurements of the pH of tubular urine in mammalian kidneys have provided unequivocal evidence that proximal tubules decrease the pH of their fluid significantly below that of plasma (30,32,33). A comparison of the situation in proximal tubules of *Necturus* and the rat is given in Figures 10 and 11.

The phenomena are related to the differences in generating a significant pH difference across the proximal tubular epithelium in amphibian and mammalian proximal tubules. The first concerns the

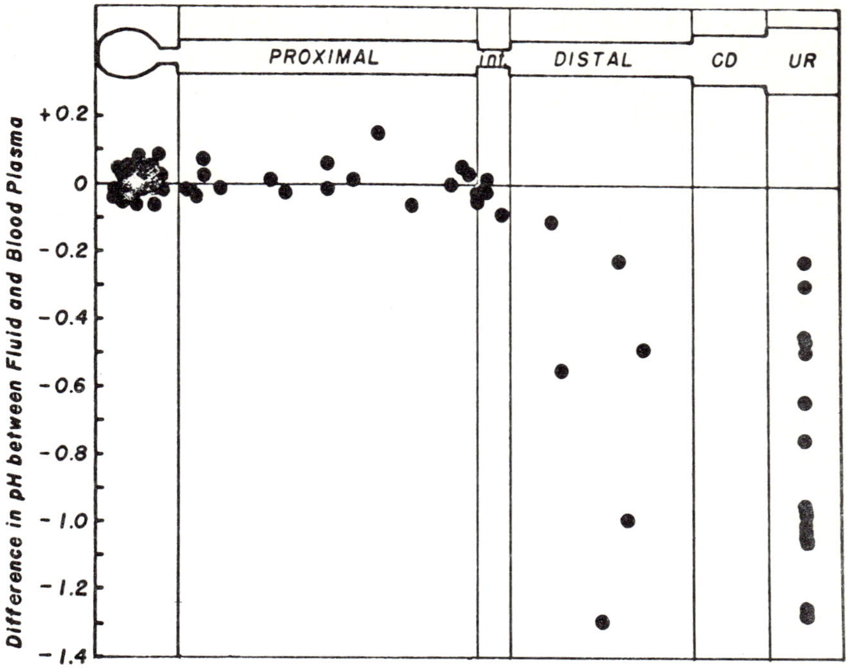

Figure 10 - Sites of acidification of the urine in the amphibian kidney. (From 57)

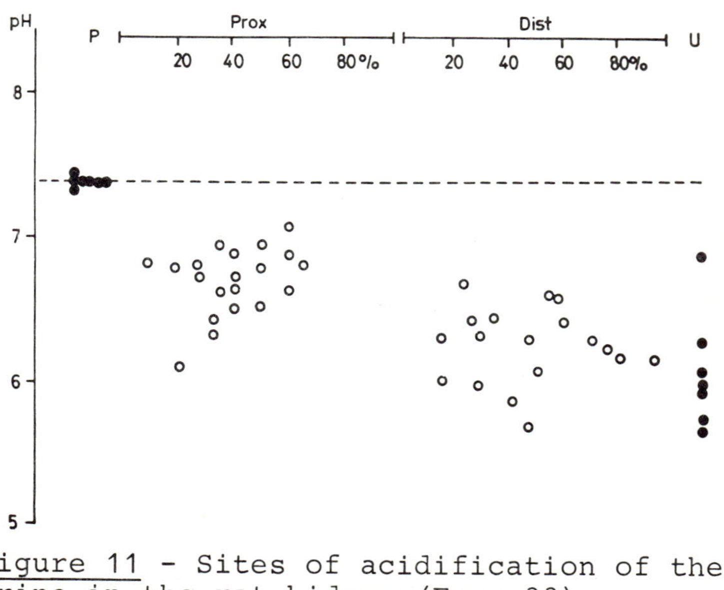

Figure 11 - Sites of acidification of the urine in the rat kidney. (From 33)

secretion of ammonia. In the amphibian nephron secretion of ammonia is restricted to the distal tubule (58), whereas in the proximal tubule of the rat ammonia secretion occurs along the whole length of the nephron (34,35). Indeed, a quantitative comparison of the contribution of various nephron segments in the rat to the total rate of urinary ammonium secretion has clearly demonstrated that the proximal tubular epithelium is the main site of ammonium secretion (34,35). Since it has been shown that ammonium ions are transferred into the tubular fluid by "diffusion-trapping", i.e. diffusion as NH_3 and subsequent conversion into the poorly permeant ionic form NH_4^+ (59), it is not surprising to find that the generation of a marked transepithelial pH gradient across the proximal tubular epithelium favors ammonium accumulation at the same nephron site.

A second consequence of the fall in pH along the proximal tubule is that it leads to preferential bicarbonate reabsorption as the latter ion is titrated by hydrogen ions to carbonic acid and subsequently decomposes into CO_2 and water (32,33). Since the proximal tubular fluid is isosmotic under all experimental conditions and since the extent of acidification leads to concurrent changes in bicarbonate concentration it follows that differences in net acidification affect the degree to which the tubular chloride concentration exceeds that of plasma. In the mammalian nephron of the rat the quite significant acidification of tubular fluid reduces the bicarbonate concentration to about 50 to 25% of its level in the glomerular filtrate (30-37-33). Concomitantly the chloride concentration is proportionately increased above plasma levels (18,21,22). The significant transepithelial chloride concentration difference provides a driving force for passive proximal tubular chloride reabsorption. Direct measurements of the tubular chloride conductance in a mammalian nephron have shown that the transtubular concentration difference indirectly generated by hydrogen ion secretion and preferential bicarbonate reabsorption is of adequate magnitude to account quantitatively for proximal tubular chloride reabsorption (60). In contrast, the concentration of chloride in proximal tubular fluid of those amphibian tubules where this has been studied is quite similar to plasma levels (25,61). It is of interest that the transepithelial electrical potential difference (lumen negative) is higher across the amphibian than the mammalian proximal tubule (62, see also Table 2). In the amphibian proximal tubule it is the electrical potential difference and not the transepithelial concentration difference which provides the main reabsorptive driving force for chloride movement.[2]

(b) Electrophysiology. In recent years, the application of electrophysiological techniques in kidney micropuncture has led to considerable progress in our understanding of both the functional organization of the proximal tubular epithelium as well as of some fundamental control mechanisms of proximal tubular solute and fluid movement. It is in this area that the amphibian nephron has been particularly useful. For the visualization of

---

[2] It is also of interest, and related to the different extent of acidification in amphibian and mammalian proximal tubules, that during split-drop experiments with isotonic raffinose solution after attainment of steady-state conditions the chloride concentration does not fall below plasma levels in the mammalian proximal tubule (63-64), but does fall below plasma levels in similar experiments carried out in *Necturus* (Diezi-Chomety, Diezi, Whittembury and Giebisch). The extensive removal of bicarbonate ions from the split-drop sample in the mammalian nephron leads to an increase in tubular chloride concentrations which fortuitously cancels the decrease which would be caused by the establishment of a limiting sodium concentration difference and the concomitant fall in chloride concentration. In the amphibian proximal tubule in which the absence of a pH gradient prevents the fall in bicarbonate concentration, the chloride concentration falls significantly and in proportion to the sodium concentration. In *Necturus*, the transepithelial chloride concentration difference is of similar magnitude as that expected from the electrical potential difference.

single cells, because of their large size, kidney preparations of Necturus maculosus or Amphiuma are of great value. Since many electrical parameters are dependent on chemical concentration gradients across elements of the tubular epithelium, the perfusion of the lumen and the peritubular capillaries with solutions of known composition is of considerable interest. Amphibian kidney preparations have the advantage of a dual blood supply, making possible the separate control of the luminal and peritubular fluid in doubly-perfused kidneys where separate solutions perfuse the aortal and portal circulation. In the mammalian kidney, the development of peritubular perfusion techniques has made it possible to control the peritubular fluid composition.

With respect to overall transepithelial electrical potential across the proximal tubule there are significant differences between the amphibian and mammalian nephron. Table 1 provides relevant information. With respect to the amphibian nephron, the existence of a significant transepithelial potential difference of some 10-20 mV (lumen negative) has recently been confirmed in Necturus (62, 65,66). In contrast, the potential difference under free-flow conditions across both the proximal tubule of the rat (67), of the dog (60) and across the perfused rabbit proximal tubule (68) are small, at best of the order of a few mV. This topic remains a controversial one since the absence of a significant potential difference, or even slightly positive potentials have been reported, although under somewhat artificial conditions (69,70,71). It is difficult, at this time, to decide on the true magnitude of the proximal transepithelial potential difference. On balance it seems preferable to suppose that the sodium transport mechanism, as in many other epithelia (71a) generates a transport potential and renders the lumen electrically negative. The electrical resistance of the proximal epithelium is exceedingly low (some 5 ohm cm^2) (60, 62,71,72). Accordingly, only an insignificant fraction of the transport potential will appear because of the attenuation by the extensive array of low resistance pathways in parallel with the epithelial cell structure (see below).

Table 1 - Comparison of transepithelial electrical characteristics of proximal tubules of different species (From 62).

		Necturus in vivo	Dog kidney perfused	Rat in vivo	Rabbit isolated tubule
P.d.	10^{-3} V	−15.26 ± 0.47 (141)	−1.97 ± 0.2 (135)	−0.5 ± 1.3 (92)	−0.7 ± 0.6 (7)
		−20 (25)			−3.8 ± 0.3 (16)
R_{input}	10^3 ohm	1031 ± 38 (101)	29.0 ± 2.0 (46)	33 ± 2.1 (12)	
				100 ± 25 (10)	
				32 ± 1.3 (14)	
				25 ± 1.5 (3)	
Length constant	10^{-4} cm	492 ± 33.48 (14)	93.6 ± 9.2 (16)	55	
				78 ± 8 (3)	
				97 ± 15 (8)	
				86 ± 3 (6)	
R_m	ohm·cm^2	69.87 ± 8.47 (14)	5.58	4.9—5.7	
G_m	10^{-3} ohm^{-1} cm^{-2}	14.3	179	189	
T_{Na}/T_{Cl}			1.38	1.58	
T_{Cl}			0.405 (2)	0.37	
G_{Cl}			72.5 (2)	69.8	

Abbreviations:

p.d. = transepithelial potential difference;
R_{input} = transepithelial input resistance;
R_m = transepithelial specific resistance;
G_m = transepithelial specific conductance;
T_{Na}/T_{Cl} = ratio of transport numbers for Na and Cl;
G_{Cl} = partial chloride conductance.
Values are means ± S.E.M. followed by the number of observations between parentheses.
The superscripts indicate the references from which the data are taken

Experiments on amphibian proximal tubule cells indicate a significant degree of asymmetrical cell polarization. The luminal cell membrane maintains a lower potential difference than the peritubular cell boundary and makes the lumen electrically negative with respect to peritubular fluid (14,73). Studies on single proximal tubule cells of Necturus carried out by experimental procedures schematically shown in Figure 12, indicate that the potential

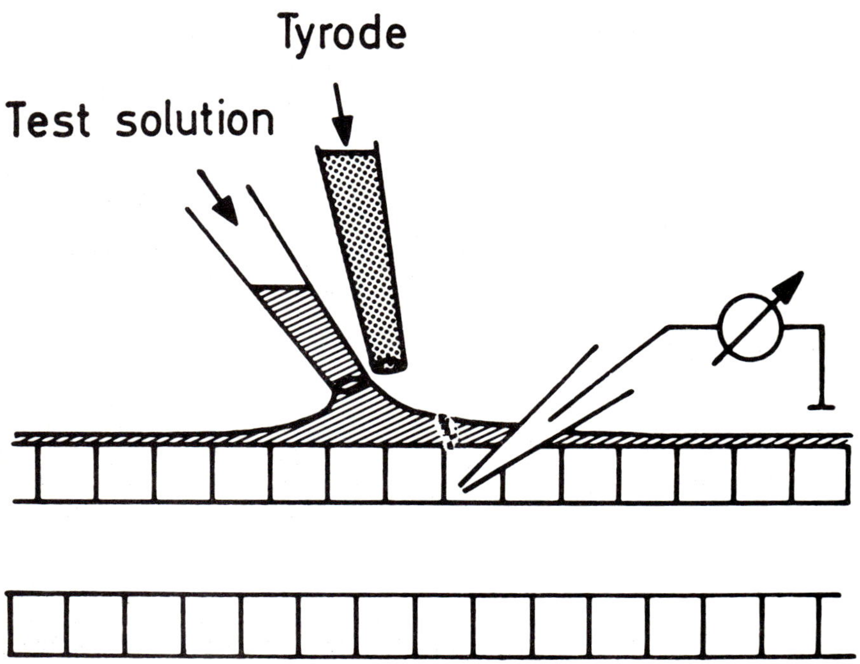

Figure 12 - Procedure to alter the peritubular fluid composition during continuous recording of cell potentials. (From 71).

difference across the peritubular cell membrane is mainly due to a potassium diffusion potential, although chloride ions also carry current across this cell boundary (14,62,71). Cell impalements of single mammalian tubule cells are technically more difficult because of the small cell size and because of interference from pulsatile and respiratory movement of the kidney. Some successful recent experiments, based on the response of the peritubular potential difference to changes in the external environment have clearly shown that the rat's proximal tubule cell shares many permeability properties with the amphibian tubule cell (71). Thus, its peritubular cell membrane is mainly permeable to potassium and chloride. An interesting finding was the observation of a significant bicarbonate permeability of the peritubular cell membrane of proximal tubule cells in the rat (71). A summary of these findings is given in Figure 13. Although this

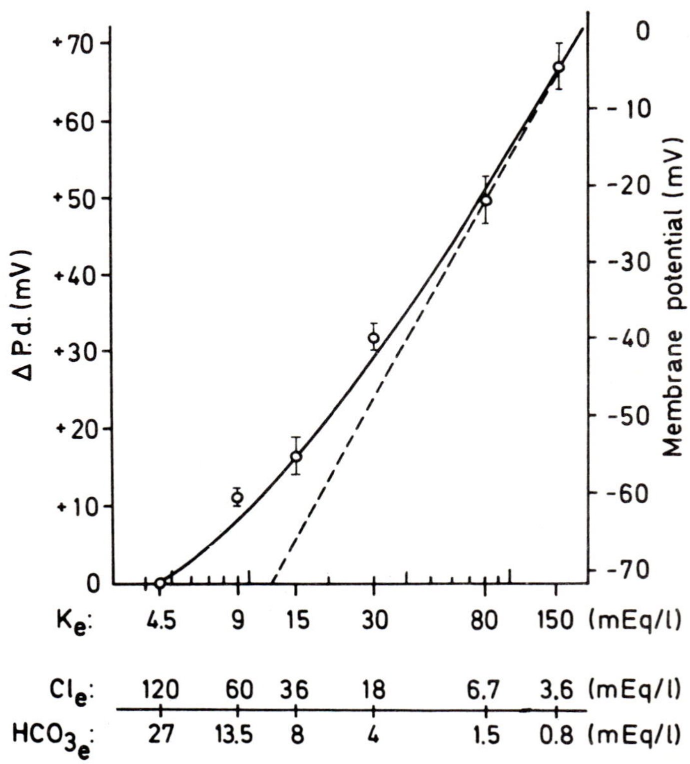

Figure 13 - Peritubular membrane potential as function of the peritubular concentrations of K, Cl and HCO_3. Left ordinate shows changes in potential difference; right ordinate gives the absolute values. (From 71).

problem has not been studied extensively in *Nectu-*

rus some evidence is available and supports the view of a significant bicarbonate permeability of the peritubular membrane of proximal tubule cells. This observation is of interest since it is relevant to the acidification operation of tubular tissue. In particular it might provide a functional link between changes in extracellular potassium concentration, intracellular pH and rate of proximal tubular hydrogen ion secretion (32). It is safe to conclude that peritubular cell membranes sharply distinguish between sodium and potassium. With respect to the permeability properties of the luminal cell membrane, the available information indicates that its sodium permeability exceeds that of the peritubular cell membrane, a property aiding the entry of sodium from the tubular lumen into the cell (62).

Most of the evidence in support of the view that there exists an important intercellular transport pathway linking the lumen directly to the peritubular fluid compartment (62,71) has been derived from work on amphibian proximal tubules but it is safe to conclude that the situation is similar in the mammalian tubule (60,62,71,72). Several lines of evidence, summarized in Table 2, can be

Table 2 - Supporting evidence for major intercellular ion transfer path.

1. Transepithelial specific resistance $\lll$ series resistance of tubular cell membranes.
2. Ranking and range of transepithelial ionic permeability coefficients different from those of tubular cell membranes (Electrophysiological evidence).
3. Magnitude of transepithelial P_{Na} and P_{Cl} $\ggg$ magnitude of cellular P_{Na} and P_{Cl} (Flux data).
4. Electrical coupling between peritubular and luminal potential changes.

listed in support of the recognition that low resistance pathways are a functionally important element in the organization of the proximal tubular nephron segment. The key arguments are (1) that the total transepithelial conductance greatly exceeds that to be expected on the basis of the resistive contributions of the bordering cell membranes. This suggests the presence of low resistance shunt paths in parallel with cell membrane resistances. Figures 14 and 15 show the methods used for measuring transepithelial and transmembrane resistances and some relevant results. At this time the method for measuring the intraepithelial length constant has only

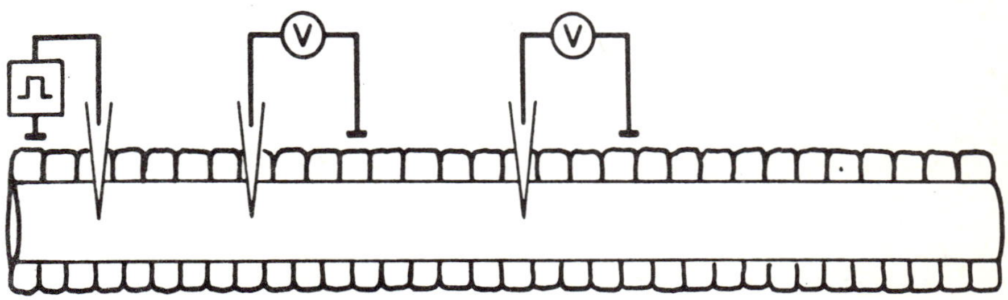

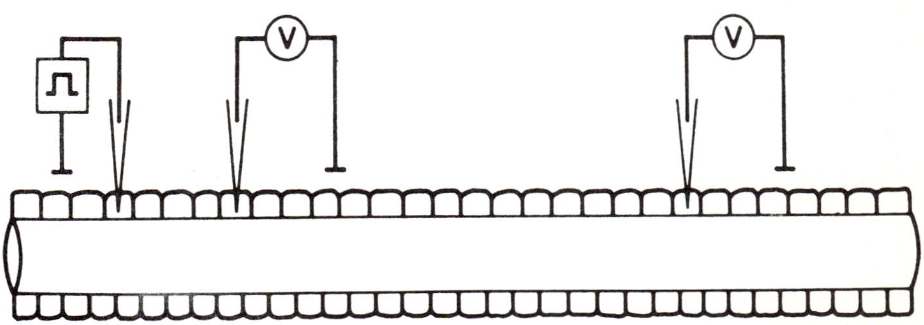

<u>Figure 14</u> <u>Top</u>: method for the measurement of the transepithelial length constant. Representative values: *Necturus* proximal tubule: 492 µm; dog proximal tubule: 93 µm; rat proximal tubule: 50 to 100 µm, rat distal tubule: 477 µm, rabbit collecting tubule: < 1000 µm. (From 62).
<u>Bottom</u>: Method for measuring transepithelial length constant (From 62).

been used on the amphibian proximal tubule. It is most likely that the anatomical routes of the low resistance paths are the exceedingly short junctional complexes and the intercellular fluid space between proximal tubule cells.

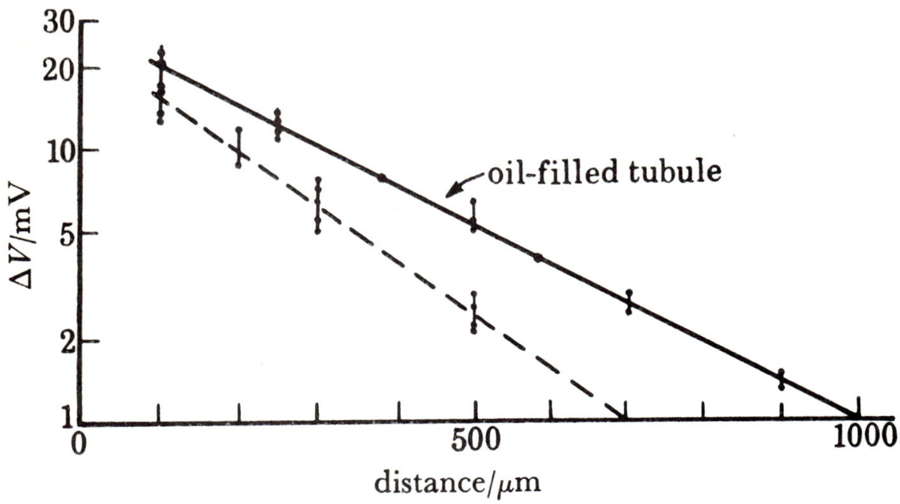

Figure 15 - Results from the application of the method illustrated on the bottom of Figure 14. Shown is the voltage attenuation within the proximal tubular epithelium (*Necturus*) for an oil-filled tubule (continuous line), and for a tubule filled with Ringer's solution (broken line). (From 62,74).

Figure 16, taken from the recent work of Whittembury *et al.* on *Necturus*, supports this view (75). (2) The second argument favoring an important role of the intercellular shunt pathway, a line of argument supported by work on both amphibian (62) and mammalian(60,71) proximal tubules, is that the ionic selectivity of the luminal and peritubular cell membrane of single tubule cells is entirely different from that of the epithelium. Thus, from the magnitude and direction of transepithelial potential differences it has become evident that the ranking within the separate cation or anion series is quite similar to that expected from the ionic mobilities in free solution. However, the finding in studies on the rat's proximal tubule that anion

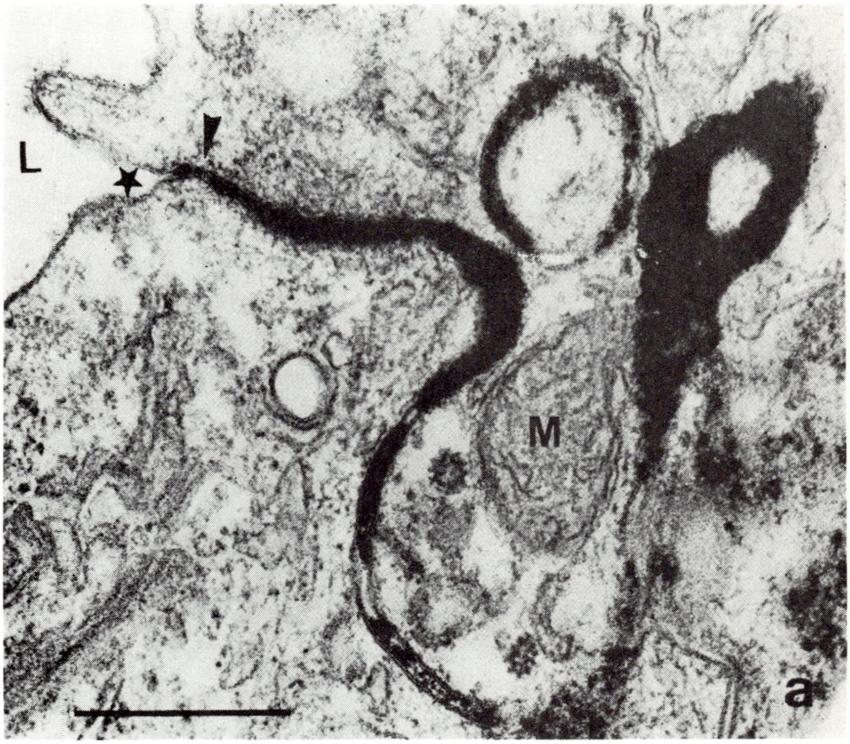

Figure 16 - Electron micrograph of the proximal tubule of Bufo marinus. The La $^{3+}$-containing solution bathed the tubular lumen (L) and the $SO_4{}^{2-}$-containing solution bathed the peritubular space. The precipitate is seen almost exclusively at the tight junctions and in the intercellular spaces. The bar represents 0.5 μ. (From 75).

movement is relatively reduced compared to that of cationic species is consistent with the view that there are negative, fixed charges within the low-resistance transepithelial fluid channels. Figure 17 summarizes a simplified equivalent circuit of a

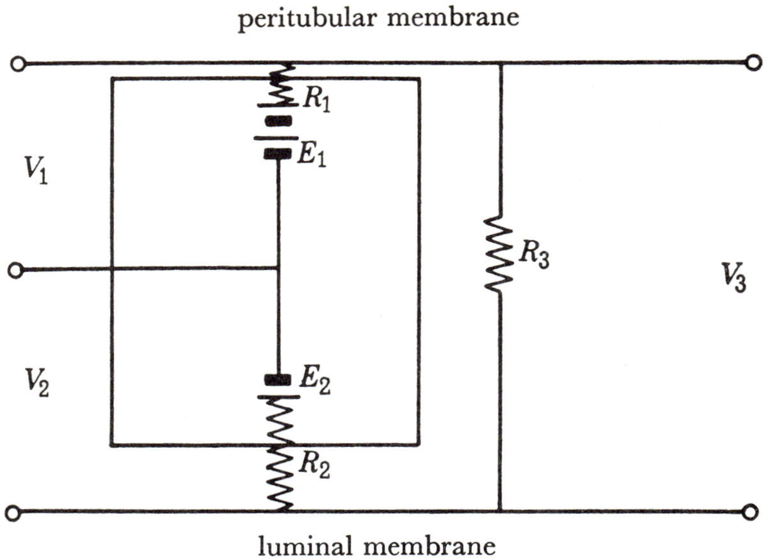

Figure 17 - Equivalent circuit of a proximal tubule cell. V_1: peritubular membrane potential, E_1: peritubular e.m.f., R_1: peritubular resistance; V_2: luminal membrane potential; E_2: luminal e.m.f., R_2: luminal resistance, V_3: transepithelial potential; R_3: "shunt" or paracellular resistance. (From 62).

proximal tubule cell (62). It should be noted that evidence in support of such an epithelial structure has also been more recently emerging from studies on the intestine (76) and the gall bladder (77).

It is the kidney, however, in which the described intercellular transport pathway has also been explored under conditions of altered proximal tubular net fluid and sodium transport (65). What has emerged from these experiments is the recognition that a variable backflux component along the described intercellular transport path can dramatically affect the efficiency of the proximal tubular sodium transport system. In particular, it appears that physical factors at the peritubular side potentially change the tightness of the junctional complex connecting individual tubule cells and thus affect the rate of net tubular sodium transport. A representative example is that of extracellular volume expansion by saline, a condition characterized by a decreased rate of proximal sodium reabsorption and by a diminished driving force (peritubular protein dilution, increase in peritubular hydrostatic pressure) for the transfer of reabsorbate from interspaces into the peritubular capillaries. It has been shown that during such decrements of proximal tubular sodium reabsorption, an increased backflux of sodium occurs. This is supported by the finding of a significant increase in the transepithelial conductance at a time when the cell membrane resistance of tubule cells has remained unchanged, and by the observation that the permeability of both sodium chloride and a non-electrolyte such as raffinose is significantly enhanced (65). Figure 18 shows experimental results demonstrating the increased leakiness of a proximal tubule to both sodium chloride and an extracellular marker such as raffinose in the volume expanded state. These observations are important because they underscore the important role of the extracellular shunt pathway in the regulation of net sodium transport. Since the tight junction is an inherently weak element in the functional organization of the proximal tubular epithelium it provides a variable leak, responsive to pressure changes, and affects the pump/leak ratio of the proximal tubular sodium transport system. Since this extracellular pathway is relatively unselective it is not surprizing to find that in conditions of an increased sodium and fluid leak other proximal transport mechanisms such as that of chloride, potassium, bicarbonate, calcium, phosphate and glucose are also rendered less effective.

It should be noted (see Figure 15) that the voltage difference resulting from intracellular current injections along the proximal tubular cell column declines logarithmically with distance. Such significant intraepithelial current spread demonstrates presence of low resistance coupling pathways between neighboring tubule cells (62,74) of the nature demonstrated also in other epithelial structures (80).

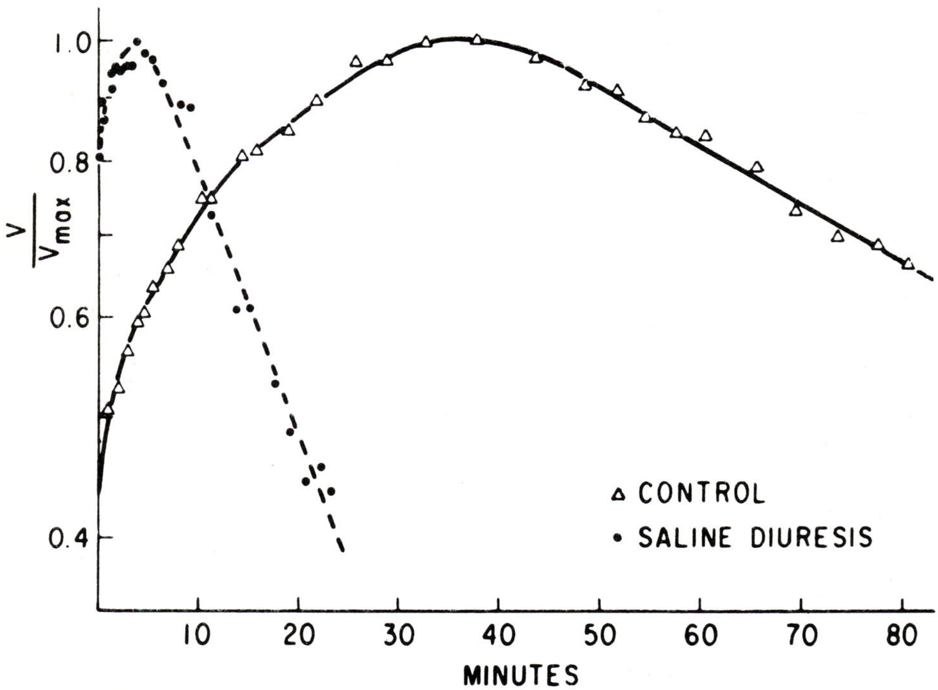

Figure 18 - Changes in volume of split-droplets of isotonic raffinose in proximal tubules of *Necturus*, expressed as a fraction of maximal volume and plotted against time. Triangles and continuous lines: control condition. Dots and interrupted lines: after volume expansion (From 65).

(c) Other transport mechanisms. - From the comparative point of view, an interesting difference between the amphibian and mammalian proximal tubule concerns the transport mechanism of para-amino hippurate (PAH). In the mammal, this substance is effectively cleared from blood by a powerful secretory mechanism. Active proximal tubular secretion was demonstrated by the application of micropuncture to the rat nephron (78a). In contrast, the studies of Kinter and Tanner (79,90,92) on the *Necturus* kidney have demonstrated a more variable behaviour. Of particular interest are studies on the doubly-perfused kidney of *Necturus* in which PAH could be shown to be actively reabsorbed from the proximal tubular fluid. When octanoate was present in the perfusion fluid, net secretion could be induced. This indicates that in *Necturus*, proximal tubular epithelium is endowed with the capacity of *both* reabsorbing *and* secreting PAH against sizeable concentration gradients. Whether simultaneous reabsorptive *and* secretory transport mechanisms for PAH reside within the proximal tubular epithelium of mammalian species cannot be ascertained from the presently available evidence.

Glucose is actively reabsorbed from the proximal tubule of *Necturus*, the frog and the rat (18, 19). Although there are marked quantitative differences, qualitatively the transport mechanism appears to be similar. Inadequate experimental information is available for a comparison between the amphibian and mammalian transport mechanisms of such substances as amino acids, phosphate, uric acid, calcium and sulfate. With respect to potassium ions, the available evidence indicates that the proximal tubule of *Necturus* is not endowed with the capacity of extensive potassium reabsorption whereas extensive tubular potassium reabsorption along the proximal nephron has been demonstrated in mammalian species (see 41). Finally, it has become apparent that species differences do exist with respect to the handling of urea. Whereas urea is absorbed along the mammalian proximal tubule (26), the whole nephron of *Rana catesbiana*, including the proximal tubular segment, has the capacity of urea secretion (81).

(2) Distal Tubules and Collecting Ducts.

(a) *Fluid movement.* - The fluid entering the mammalian distal tubule is rendered hypotonic with respect to plasma by the reabsorption of sodium chloride along the relatively water-impermeable ascending limb of Henle's loop (37,39). In water diuresis, i.e. in the absence of a high level of circulating antidiuretic hormone, the fluid remains hypotonic. Indeed, some sodium chloride abstraction along the collecting ducts in diabetes insipidus rats produces a final urine of greater hypotonicity than the distal tubule (47).

In contrast, in the dehydrated state or after administration of antidiuretic hormone, osmotic equilibration proceeds along the second half of the distal tubule of the rat and generates frequently an isosmotic urine although recent evidence suggests that the extent of osmotic equilibration may be less complete than previously thought (84).Also, there appear to be significant species differences between mammals. Despite dehydration or the presence of high levels of circulating antidiuretic hormone the distal tubular fluid of dogs (82) and monkeys (83) remains hypotonic during the elaboration of a markedly hypertonic urine. This indicates a lower hydraulic permeability along the distal tubule and a downward shift of osmotic equilibration into the collecting ducts.

From the evidence obtained by micropuncture of amphibian distal tubules it can be concluded that their contents are normally markedly hypotonic, the degree of hypotonicity increasing as sodium,chloride and potassium are progressively reabsorbed (20, 24). Studies in *Necturus* have shown that fluid distribution is about equally distributed between the proximal and distal tubule (24). The administration of arginine vasotocin diminishes the free-water clearance in amphibians (1,2) indicating that hypophyseal peptides increase the water permeability of some tubular segment. In the absence of relevant micropuncture studies it cannot be stated with certainty whether the distal tubule or the collecting ducts are involved.

(b) *Electrolyte movement.* - Micropuncture studies on mammalian and amphibian distal tubules and collecting ducts have clearly shown that this nephron segment is a tubule site at which sodium ions are actively reabsorbed against increasing transepithelial concentration gradient (23,42). Figures 19 and 20 provide data obtained on distal tubules of rats and of *Amphiuma*. With respect to the rat tubule (Figure 19), it is apparent that the sodium concentration declines along the distal tubule; an amount equivalent to some 8% of the filtered sodium load is normally reabsorbed along the distal tubule

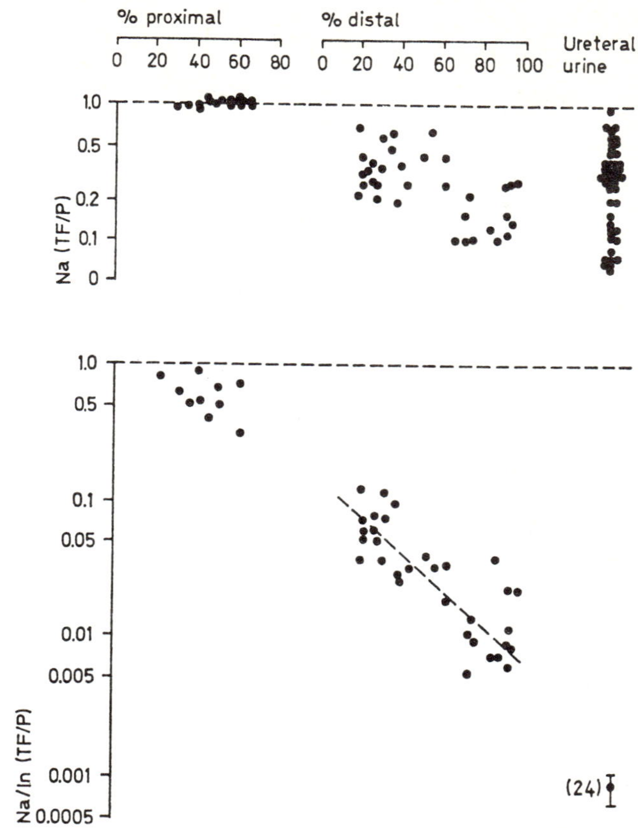

Figure 19 - Summary of proximal and distal sodium and sodium/inulin tubular fluid to plasma concentration ratios. Data were obtained in rats on a normal sodium intake (From 42).

(42). This is not at all a maximum value. With increased delivery of fluid and sodium to both the distal tubular and collecting duct epithelium their transport rate with respect to sodium ions can be shown to increase dramatically. Indeed, demonstration of a transport maximum (T_M) along the distal tubule is impossible and may only be approached at the collecting duct level under rather extreme loading conditions with isotonic or hypertonic sodium chloride (45, Khuri,R., Strieder,N., Wiederhold,M. and G.Giebisch, unpublished observations). Similarly, in the rat distal fractional reabsorption rates

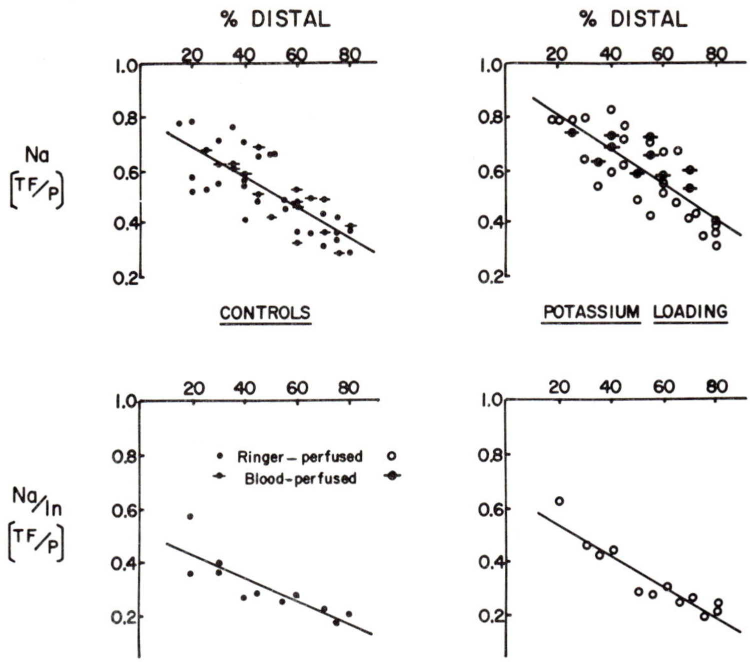

Figure 20 - Progression with tubule distance of distal sodium and sodium/inulin concentration ratios in non-perfused and perfused *Amphiuma* kidneys under control and potassium-loaded conditions. (From 85).

of sodium are dramatically increased after those diuretics which block sodium reabsorption at the level of the loop of Henle and at the proximal tubular level and thereby lead to the delivery of increased amounts of sodium to the distal tubular reabsorptive site.

Inspection of Figure 20 summarizes data on transepithelial sodium concentration differences and fractional reabsorption rates in an amphibian distal tubule (85). Similar to the mammalian distal tubule, the sodium concentration falls along the distal tubule. Since sodium ions move against a sizeable electrical and chemical potential difference this transport mechanism is active in nature. Maximum transepithelial concentration differences must be achieved at a site beyond the distal tubule, i.e. along the collecting ducts, since the sodium concentration in samples of collecting duct fluid were of the order of 1-5 mEq/L, i.e. considerably lower than end-distal samples. Compared to the mammalian distal tubule, a much larger fraction of the filtered sodium load, somewhat more than 50% compared to 8%,enters the amphibian distal tubule. This is due to the markedly lower rate at which more proximally located nephron segments reabsorb fluid and sodium in amphibian species. Some 90% of the filtered sodium load has been reabsorbed by the end of the distal tubule. Of those, some 40% are reabsorbed proximally while approximately 50% are abstracted from distal tubular fluid. It should be noted that tubular sodium transport is not significantly affected by potassium loading (see figure 20). This is of interest since the direction of net potassium transport is different in the two situations. Extensive net reabsorption obtains in control animals, whereas distal tubular net secretion could be induced in *Amphiuma* receiving an exogenous potassium load.

Direct investigations on both amphibian and mammalian nephrons have shown that the distal tubule and the collecting ducts determine the rate of urinary potassium excretion (41,43,44,45,46,85). With respect to potassium transport the distal tubular epithelium shows a wide range of variation.

In the rat, evidence obtained by micropuncture techniques indicates that net secretion of potassium along the distal tubule accounts for most of the excreted potassium, confirming Berliner's thesis of an essentially distal secretory origin of the excreted potassium moiety in the mammal. The distal tubular transport pattern of potassium is highly variable and depends critically upon the metabolic situation. Data on distal tubular excretion patterns of potassium in states of potassium conservation and of maximal potassium secretion in the rat are compared in Fig.21. Control data are also shown.Although the potassium concentration along the distal tubule increases both during potassium depletion and potassium loading,fluid abstraction accounts for the concentration increments in low-potassium animals whereas massive net secretion of

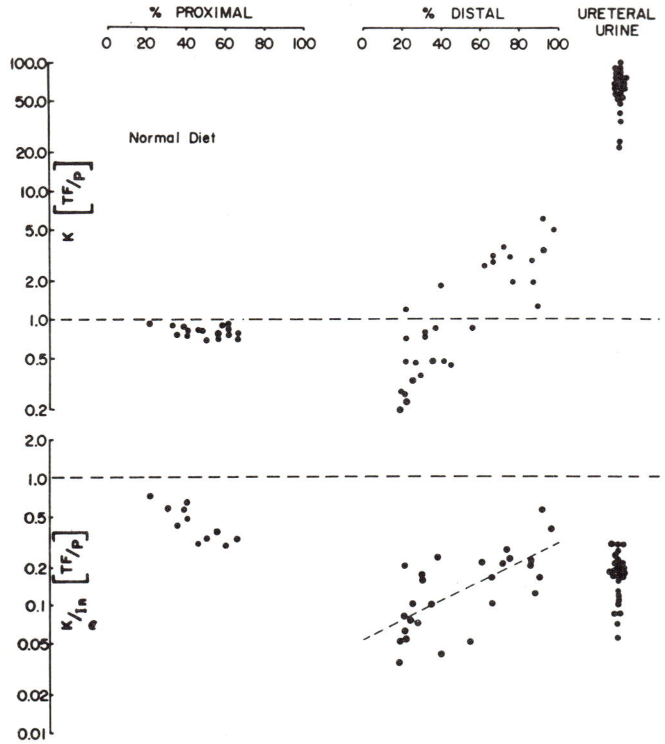

Figure 21A - Summary of potassium and potassium/inulin concentration ratios from rats on a control diet. (From 29).

potassium is a regular finding in those animals in which potassium excretion had been stimulated. In the mammal the collecting duct has been shown to have also the capacity of either net secretion or net reabsorption of potassium ions. On a normal potassium intake, net secretion can be shown to be the most frequently observed transport pattern along the mammalian distal tubule.

In some aspects the situation differs in the amphibian distal tubule. In those amphibians which live in a fresh water habitat and have been studied extensive potassium reabsorption is observed along the distal nephron. Figure 22 summarizes relevant micropuncture data (85). It is obvious that not only is the tubular potassium concentration at le-

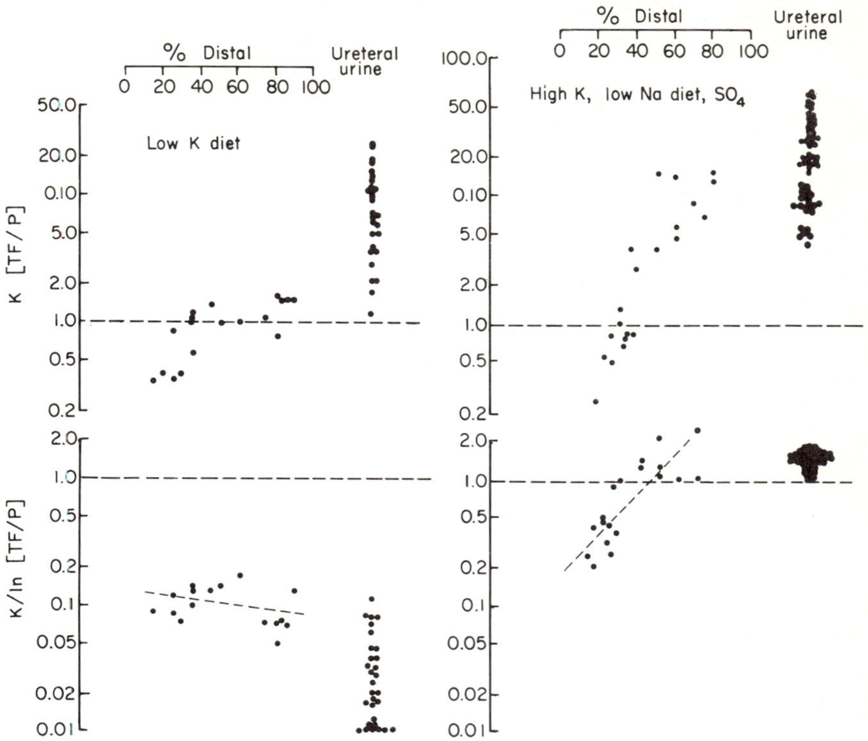

Figure 21B - Summary of distal tubular potassium and potassium/inulin concentration ratios. Left: data from rats kept for several weeks on a low-potassium diet. Right: data from rats in which potassium excretion was maximally stimulated by pretreatment with a low sodium high potassium diet and the infusion of potassium sulfate and dichlorphenamide. (From 42).

vels above that of plasma at the beginning of the distal tubule but it also declines as fluid passes along this nephron segment. It is also evident from the potassium/inulin TF/P ratio of unity at the early distal tubular level that the amount of potassium entering the distal tubule, compared to that filtered, has remained unchanged. This is due to the fact that the amphibian nephron lacks a loop of Henle, a site of significant sodium, chloride and potassium reabsorption in the mammalian nephron, and that only a short intermediate segment connects

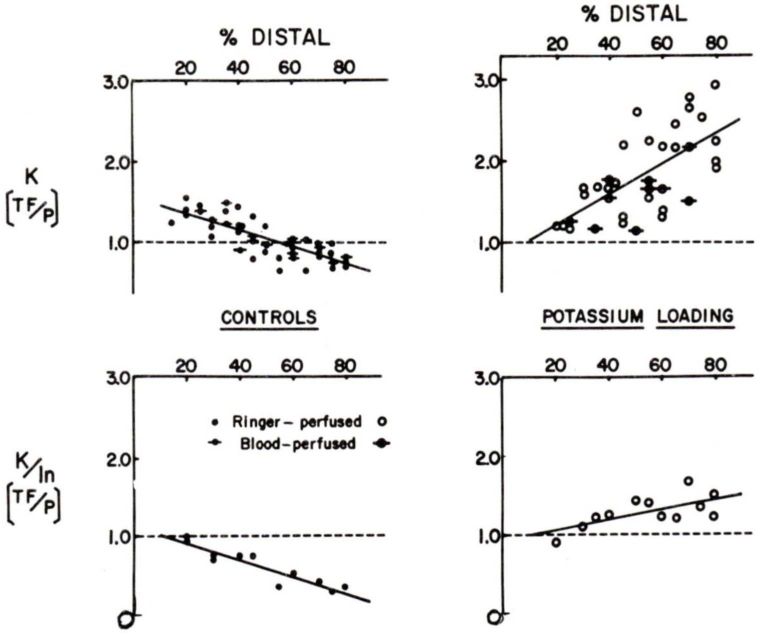

Figure 22 - Progression with tubule distance of distal potassium and potassium/inulin concentration ratios in *Amphiuma* kidneys under control and potassium-loaded conditions (From 85).

the proximal with the distal tubule. This lack of ability of the amphibian nephrons to reabsorb potassium at nephron sites prior to the distal tubule contrasts sharply with the extensive reabsorption in the mammal along similar segments of the tubule; some two-thirds of the filtered potassium is normally reabsorbed during passage of fluid along the proximal convolution and the loop of Henle.Only a quantity of potassium equivalent to some 5-10% of the filtered amount enters the distal tubule. In contrast,in the amphibian,modifications of the filtered potassium moiety are exclusively within the domain of the distal tubule.In *Amphiuma*, some 75% of the filtered potassium load is reabsorbed along the distal tubule.Quantitatively,the situation in this amphibian species is similar to the transport pattern of the rat distal tubule under conditions of maximal potassium conservation. In the amphibian, the fact that potassium is avidly reabsorbed from the distal tubular lumen may reflect its permanent state of maximum electrolyte conservation and renal tubular adaptation to a fresh water environment.

Inspection of figure 22 shows that similar to the mammalian distal tubular apparatus, the amphibian distal tubule also has the capacity of altering magnitude and direction of potassium transfer in response to metabolic stimuli. Subsequent to the exposure of <u>Amphiuma</u> to a high ambient potassium concentration (some 100 mEq/L) the potassium concentration along the distal tubule increases strikingly. Since fluid abstraction cannot account for the observed rise in potassium concentration, potassium secretion is indicated. As shown in Figure 23 a similar situation obtains in animals pretreated with the carbonic anhydrase inhibitor acetazolamide whereas in ouabain-treated kidneys potassium re-

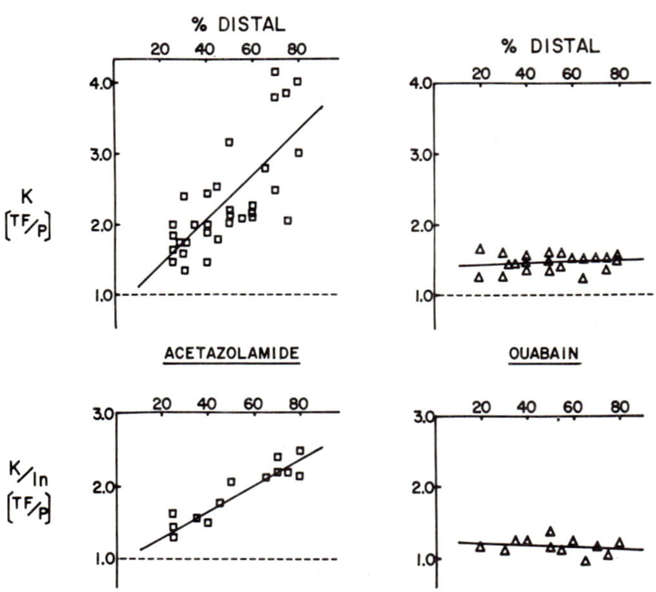

<u>Figure 23</u> - Progression with tubule distance of distal potassium and potassium/inulin concentration ratios in acetazolamide and ouabain-treated kidneys. Perfusion was carried out with 1.0×10^{-4}M acetazolamide (Diamox) and with 5.5×10^{-6}M ouabain. (From 85).

absorption is suppressed. This cardiac glycoside abolishes net potassium transport across the distal tubular epithelium (85). Ouabain also inhibits potassium reabsorption in the mammalian distal tubule (86).

In comparing the distal tubular transport system of potassium in amphibian and mammalian species it appears that the mammalian system is, at normal potassium intake levels, poised toward secretion, that of aquatic amphibia towards net reabsorption. A series of tracer flux studies carried out in both amphibian and mammalian distal tubules (87) has provided additional information on some apspects of these potassium transport system.

Figure 24 provides a summary of the situations

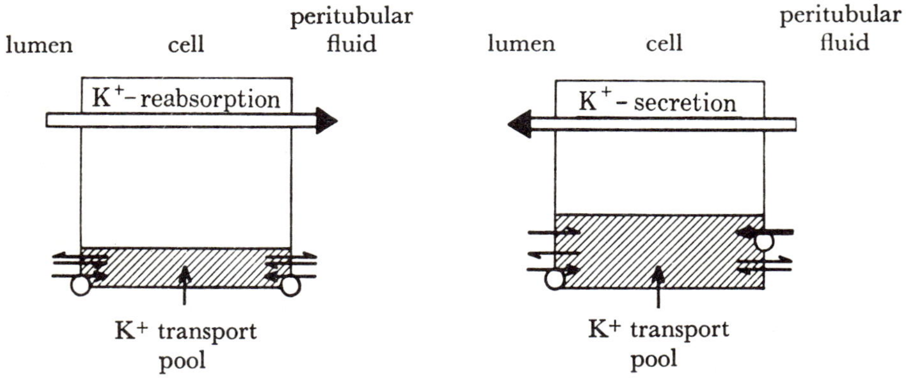

Figure 24 - Cell schema depicting distal tubular potassium transfer across the luminal and peritubular cell membrane during state of net reabsorption and net secretion in *Amphiuma* distal tubule. The transport pool of potassium is larger as potassium secretion is stimulated. This is the consequence of increased potassium uptake across the peritubular cell membrane. (From 85,89).

obtained during net reabsorption and during net secretion of potassium. An important finding was that only part of the cellular potassium is involved in net potassium transport. With the change from net reabsorption to that of net secretion (induced either by potassium loading or by giving a carbonic anhydrase inhibitor) the potassium transport pool increases. It could also be clearly shown that the increase was due to stimulation of potassium uptake across the peritubular cell border. Essentially similar findings were observed in tracer flux studies carried out on rat distal tubules with the notable exception that the potassium transport pool comprised a much larger fraction of cell potassium (87). Since there were no modifications of the luminal kinetic transport parameters it appears that changes in peritubular potassium uptake control the magnitude of the intracellular potassium pool.

These data provide a rational explanation of the differences in potassium transport in the mammalian and amphibian nephron. Based on data dealing with the electrochemical potential differences across individual cell membranes and the direction of net potassium movement, some properties of distal tubule cells were derived which are summarized in Figure 25. The key elements involve an active uptake step at the peritubular cell membrane which determines the magnitude of the amount and the concentration of potassium within the cell. Potassium entry across the luminal cell membrane is passive, driven by the electrical and chemical potential difference across this cell boundary, and opposed by an active reabsorptive mechanism. Importantly, the magnitude and direction of net potassium transfer depends in this pump-leak system on the relative magnitude of the driving forces acting on potassium ions. We propose that it is the difference in the magnitude of the potassium transport pools in amphibian and mammalian distal tubules - itself a reflection of the activity of the peritubular potassium pumping mechanism - which leads to the different direction of net potassium movement. In the mammalian distal tubule the relatively large magnitude of the intracellular potassium pool leads to potassium movement across the luminal cell membrane which exceeds active uptake from the lumen; net potassium secretion obtains. As the cellular potassium pool, and the effective intracellular potas-

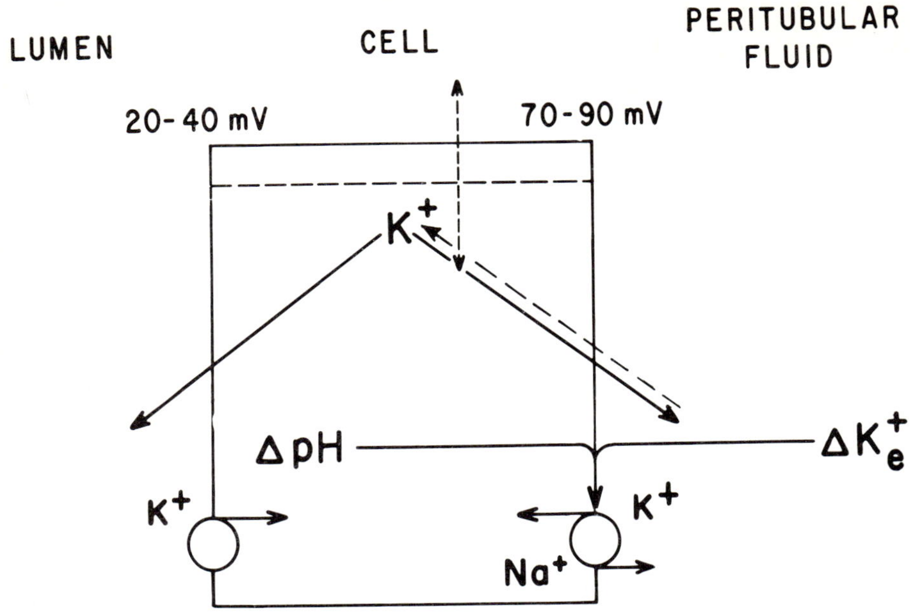

Figure 25 - Schematic presentation of some transport properties of distal tubule cell. From the available information the situation in the amphibian and mammalian distal tubule differ in that a larger fraction of cell potassium partakes in the mammalian transport system. (From 41, 89).

sium concentration falls in potassium deprivation, the active uptake component is largely unopposed; net reabsorption of potassium takes place. This is the normal situation in *Amphiuma* where the transport pool represents only a relatively small fraction of cell potassium. As a consequence, it is highly likely that the effective cellular potassium concentration is low and that the passive influx into the lumen falls below the unidirectional active uptake component. It is only with stimulation of peritubular potassium uptake that the situation is reversed. It is tempting to speculate that the low potassium transport pool in *Amphiuma* is a reflection of the need to conserve potassium maximally, an adaptation to the fresh water environment. It is with the higher potassium intake of terrestrial organisms that the distal tubule develops a progressively more potent secretory system by recruiting a larger fraction of cell potassium into the transport mechanism.

REFERENCES

1. Schmidt-Nielsen,B.M. & MacKay,W.C. (1972) Comparative physiology of electrolyte and water regulation, with emphasis on sodium,potassium, chloride, urea and osmotic pressure. In: Clinical Disorders of Fluid and Electrolyte Metabolism. Ed. Maxwell,M.H. & Kleeman,C.R., New York McGraw-Hill Book Company, pp. 45-93.

1a. Forster,R.P. Comparative vertebrate physiology and renal concepts. In Handbook of Physiology. Renal Function, ed. Orloff,J. & R.W. Berliner, Amer. Physiol.Society, Washington (in press).

2. Bentley,P.J. (1971) Endocrinology and Osmoregulation, Zoophysiology and Ecology, vol.1, New York, Heidelberg,Berlin, Springer-Verlag.

3. Smith,H.W. (1951) The Kidney, Structure and Function in Health and Disease. New York Oxford University Press.

4. Hickman,C.P. jr. & Trump,B.F. (1969)The kidney. In Fish Physiology, vol.1: Excretion, Ionic Regulation and Metabolism, ed. Hoar,W.S. & Randall,D.J., pp.91-239, New York & London Academic Press.

5. Forster,R.P. & Goldstein,L. (1969) Formation of excretory products. In Fish Physiology, vol. 1 Excretion, Ionic Regulation and Metabolism, ed. Hoar,W.S. & Randall,D.J., pp. 313-350, New York & London Academic Press.

6. Forster,R.P. (1967) Renal transport mechanisms In Proceedings of an International Symposium on Comparative Pharmacology, ed. Cafruny,E.J., pp. 1008-1019, Bethesda,Md. Fed.Am.Soc.Exptl.Biol.

7. Burg,M. & Weller,P. (1969) Iodopyracet transport by isolated perfused flounder proximal renal tubules. Am.J.Physiol. 217, 1053-1056.

8. Braun,E.J. & Dantzler,W.H. (1972) Function of mammalian-type and reptilian-type nephrons in kidney of desert quail. Am.J.Physiol.222,617-29.

9. Skadhaunge,E. & Schmidt-Nielsen,B. (1967) Renal function in domestic fowl. Am.J.Physiol. 212, 793-798.

10. Marshall,E.K.,Jr. (1934) The comparative physiology of the kidney in relation to theories of renal secretion. Physiol.Rev. 14, 133-159.

11. Sperber,I. (1944) Studies on the mammalian kidney. Zool.bidragfran Uppsala, 22, 249-431.

12. Schmidt-Nielsen,B. (1964) Organ systems in adaptation: the excretory system. In Handbook of Physiology, section 4: Adaptation to the Environment, pp.215-243, Washington, Am.Physiol. Soc.

13. Smith,H.W. (1953) From Fish to Philosopher. Boston,Little Brown & Company.

14. Giebisch,G. (1961) Measurements of electrical potential differences on single nephrons of the perfused *Necturus* kidney. J.Gen.Physiol. 44, 659-678.

15. Himmelhoch,S.R. & Karnovsky,M.J. (1961) Oxidative and hydrolytic enzymes in the nephron of *Necturus maculosus*, Histochemical, biochemical and electron microscopical studies. J.Biophys. Biochem.Cyt. 9, 893-908.

16. Richards,A.N. & Walker,A.M. (1937) Methods of collecting fluid from known regions of the renal tubules of amphibia and of perfusing the lumen of a single tubule. Am.J.Physiol. 118, 111-120.

17. Walker,A. & Oliver,J. (1941) Methods for the collection of fluid from single glomeruli and tubules of the mammalian kidney. Am.J.Physiol. 134, 562-579.

18. Walker,A., Bott,P., Oliver,T. & MacDowell,M. (1941) The collection and analysis of fluid from single nephrons of the mammalian kidney. Am.J.Physiol. 134, 580-595.

19. Walker,A. & Hudson,C. (1937) The reabsorption of glucose from the renal tubule in amphibia and the action of phlorhizin upon it. Am. J. Physiol. 118, 130-143.

20. Walker,A., Hudson,C., Findley,T. & Richards,A. N. (1937) The total molecular concentration and NaCl concentration of fluid from different segments of the renal tubule of amphibia. Am.J. Physiol. 118, 121-129.

21. Gottschalk,C.W. (1962-63) Renal tubular function: Lesions from micropuncture. Harvey Lecture Series, 58, 99-123.

22. Gottschalk,C.W. (1961) Micropuncture studies of tubular function in the mammalian kidney. Physiologist 4, 35-55.

23. Giebisch,G. & Windhager,E.E. (1964) Renal tubular transfer of sodium, chloride and potassium. Am.J.Med. 36, 643-669.

24. Bott,P.A. (1962) Micropuncture study of renal excretion of water,K, Na and Cl in *Necturus*. Am.J.Physiol. 203, 662-666.

25. Giebisch,G. (1956) Measurements of pH, chloride and inulin concentrations in proximal tubule fluid of *Necturus*. Am.J.Physiol. 185, 171-174.

26. Lassiter,W.E., Gottschalk,C.W. & Mylle,M.(1961) Micropuncture study of net transtubular movement of water and urea in nondiuretic mammalian kidney. Am.J.Physiol. 200, 1139-1147.

27. Lassiter,W.E., Gottschalk,C.W. & Mylle,M.(1963) Micropuncture study of renal tubular reabsorption of calcium in normal rodents. Am.J.Physiol 204, 771-775.

28. Strickler,J.C., Thompson,D.D., Klose,R.M. & Giebisch,G. (1964) Micropuncture study of inorganic phosphate excretion in the rat. J.Clin. Invest. 43, 1596-1607.

29. Malnic,G., Klose,R.M. & Giebisch,G. (1964) Micropuncture study of renal potassium excretion in the rat. Am.J.Physiol. 206, 674-686.

30. Gottschalk,C.W., Lassiter,W.E. & Mylle,M.(1960) Localization of urine acidification in mammalian kidney. Am.J.Physiol. 198, 581-585.

31. Bergeron,M. & Morel,F. (1969) Amino acid transport in rat renal tubules. Am.J.Physiol. 216,

1139-1149.

32. Rector,F.C. (1971) Renal secretion of hydrogen. In The Kidney,Morphology,Biochemistry, Physiology., Rouiller,C. & Müller,A.F. ed., vol.3, pp. 209-252, New York, Academic Press.

33. Malnic,G. & Giebisch,G. (1972) Mechanism of renal hydrogen ion secretion. Kidney International 1, 280-296.

34. Glabman,S., Klose,R.M. & Giebisch,G. (1963) Micropuncture study of ammonia excretion in the rat. Am.J.Physiol. 205, 127-132.

35. Hayes,C.P. jr., Mayson,J.S., Owen,E.E. & Robinson,R.R. (1964) A micropuncture evaluation of renal ammonia excretion in the rat. Am. J. Physiol. 207, 77-83.

36. Sonnenberg,H., Oelert,H. & Baumann,K. (1965) Proximal tubular reabsorption of some organic acids in the rat kidney in vivo. Arch.ges. Physiol. 286, 171-180.

37. Gottschalk,C.W. & Mylle,M. (1959) Micropuncture study of the mammalian urinary concentrating mechanism: evidence for the counter-current hypothesis. Am.J.Physiol. 196, 927-936.

38. Berliner,R.W. & Bennet,C.M. (1967) Concentration of urine in the mammalian kidney. Am. J. Med. 42, 777-789.

39. Wirz,H. (1956) Der osmotische Druck in den corticalen tubuli der Rattenniere. Helv.physiol. Pharmacol.Acta. 14, 354-362.

40. Bott,P.A. (1952) Renal excretion of creatinine in Necturus. A reinvestigation by direct analysis of glomerular and tubular fluid for creatinine and inulin. Am.J.Physiol. 168, 107-111.

41. Giebisch,G. (1971) Renal potassium excretion. In The Kidney, Morphology, Biochemistry, Physiology. Rouiller,C. & Müller,A.F. ed., vol.III pp. 329-382., New York, Academic Press.

42. Malnic,G., Klose,R.M. & Giebisch,G. (1966) Micropuncture study of distal tubular potassium and sodium transport in rat nephron. Am.J.

Physiol. 211, 529-547.

43. Malnic,G., Klose,R.M. & Giebisch,G. (1966) Microperfusion study of distal tubular potassium and sodium transfer in rat kidney. Am. J. Physiol. 211, 548-559.

44. Malnic,G., Mello-Aires,M. & Giebisch,G.(1971) Potassium transport across renal distal tubules during acid-base disturbances. Am.J.Physiol. 221, 1192-1208.

45. Diezi,J., Michoud,P., Aceves,J. & Giebisch,G. Micropuncture study of electrolyte transport across the papillary collecting duct of the rat. Am.J.Physiol. submitted for publication.

46. Hierholzer,K. (1961) Secretion of potassium and acidification in collecting ducts of mammalian kidney. Am.J.Physiol. 201, 318-324.

47. Jamison,R.L., Burrkert,J. & Lacy,F. (1971) A micropuncture study of collecting tubule function in rats with hereditary diabetes insipidus J. Clin.Invest. 50, 2444-2452.

48. Hierholzer,K., Wiederholt,M., Holzgreve,H., Giebisch,G., Klose,R.M. & Windhager,E.E. (1965) Micropuncture study of renal transtubular concentration gradients of sodium and potassium in adrenalectomized rats. Arch.ges.Physiol. 285, 193-210.

49. Uhlich,E., Baldamus,C.A. & Ullrich,K.J. (1969) Einfluss von Aldosterone auf den Natrium transport in den Sammel-rohren der Saügetieruiere. Arch. ges.Physiol. 308, 111-126.

50. Deyrup,I.J. (1964) Water balance and kideny. In Physiology of the Amphibia, Moore,J.A. ed., pp. 251-328, New York, Academic Press.

51. Giebisch,G., Klose,R.M., Malnic,G., Sullivan, W.J. & Windhager,E.E. (1964) Sodium movement across single perfused proximal tubules of rat kidneys. J.Gen.Physiol. 47, 1175-1194.

52. Windhager,E.E., Whittembury,G., Oken,D.E., Schatzmann,H.J. & Solomon,A.K. (1959) Single proximal tubules of the *Necturus* kidney. III. Dependence of H20 movement on NaCl concentra-

tion. Am.J.Physiol. 197, 313-318.

53. Whittembury,G., Oken,D., Windhager,E.E. & Solomon,A.K. (1959) Single proximal tubules in Necturus kidney. IV. Dependence of H_2O movement on osmotic gradients. Am.J.Physiol. 197, 1121-1127.

54. Ullrich,K.J., Rumrich,G. & Fuchs,G. (1964) Wasserpermeabilität und transtubulärer Wasserfluss corticalen nephron abschnitte bei verschiedenen Diureseustäuden. Arch.ges.Physiol. 280, 99-119.

55. Windhager,E.E., Lewy,J.E. & Spitzer,A. (1969) Intrarenal control of proximal tubular reabsorption of sodium and water. Nephron 6, 247-259.

56. Green,R., Windhager,E.E. & Giebisch,G. The effects of transepithelial oncotic pressure gradients on proximal tubular fluid movement in the rat. Abstr.V.Int.Congr.Nephrol. In press.

57. Montgomery,H., Pierce,J.A. (1937) The site of acidification of the urine within the renal tubule in Amphibia. Am.J.Physiol. 118, 114-152.

58. Walker,A.M. (1940) Ammonia formation in the amphibian kidney. Am.J.Physiol. 131, 187-194.

59. Balagura-Baruch,S. (1971) Renal metabolism and transfer of ammonia. In The Kidney, Morphology,, Biochemistry, Physiology. Rouiller,C. & Müller, A.F. ed., pp.253-327., New York, Academic Press

60. Boulpaep,E.L. & Seely,J.F. (1971) Electrophysiology of proximal and distal tubules in the autoperfused dog kidney. Am.J.Physiol. 221, 1084-1096.

61. Walker,A., Hudson,C.L., Findley,T. jr. & Richards,A.N. (1937) The total molecular concentration and the chloride concentration of fluid from different segments of the renal tubule of amphibia. The site of chloride reabsorption. Am.J.Physiol. 118, 121-129.

62. Boulpaep,E.L. (1971) Electrophysiological properties of the proximal tubule: Importance of

cellular and intercellular transport pathways. In: Electrophysiology of Epithelial Cells, ed. G.Giebisch, pp. 91-112, Stuttgart & New York, F.K. Schattauer Verlag.

63. Kashgarian,M., Stöckle,H., Gottschalk,C.W. & Ullrich,K.J. (1963) Transtubular electrochemical potentials of sodium and chloride in proximal and distal renal tubules of rats during antidiuresis and water diuresis (diabetes insipidus). Arch. ges.Physiol. 277, 89-106.

64. Malnic,G. & Mello-Aires,M. (1970) Microperfusion study of chloride, bicarbonate and sulfate transfer in proximal tubules of rat kidney. Am. J.Physiol. 218, 27-32.

65. Boulpaep,E.L. (1972) Permeability of the proximal tubule of *Necturus* during saline loading. Am.J.Physiol. 222, 517-531.

66. Spring,K. Voltage clamp analysis of fluid transport in *Necturus* proximal tubule. J.Gen. Physiol. In press.

67. de Mello,G.B. & Malnic,G. (1970) Algunas caracteristicas electrofisiologicas de tubulos proximales de rins de rato. Proc.XXII.Congr.Soc. Brasileiro parao Progreso da Ciencia, p.326.

68. Burg,M.B. & Orloff,J.(1970) Electrical potential difference across proximal convoluted tubules. Am.J.Physiol.219, 1714-1716.

69. Kokko,J.P. (1972) Proximal tubule potential difference: dependence on glucose, HCO_3 and amino-acids. Clin.Res. 20, 698.

70. Kokko,J. & Rector,F.C. (1971) Flow dependence of transtubular potential difference in isolated perfused segments of rabbit proximal convoluted tubule. J.Clin.Invest. 40, 2745-2750.

71. Frömter,E., Müller,C.W. & Wick,T. (1971) Permeability properties of the proximal tubular epithelium of the rat kidney studied with electrophysiological methods. In Electrophysiology of Epithelial Cells, pp. 119-146, ed. G.Giebisch, Stuttgart, New York, F.K.Schattauer Verlag.

71a. Keynes,R.D. (1969) From frog skin to sheep rumen: a survey of transport of salts and water across multicellular structures. Quart.Rev. Biophys. 2, 177-281.

72. Hegel,U. & Frömter,E. (1967) Der elektrische Wandwiderstand des proximalen Konvolutes der Rattennieve. Arch.ges.Physiol. 294, 274-290.

73. Giebisch,G. (1968) Some electrical properties of single renal tubule cells. J.Gen.Physiol. 51, 315s-325s.

74. Windhager,E.E., Boulpaep,E.L. & Giebisch,G. (1966) Electrophysiological studies on single nephrons. 3rd Int.Congr.Nephrol.,Washington, vol.I, pp.35-47.

75. Whittembury,G. & Rawlins,F.A. (1971) Evidence for a paracellular pathway for ion flow in the kidney proximal tubule: electron microscopic demonstration of lanthanum precipitate in the tight junction. Arch.ges.Physiol.330, 302-309.

76. Frizzell,R.A. & Schultz,S.G. (1972) Ionic conductances of extracellular shunt pathway in rabbit epithelium: influence of shunt on transmural Na transport and electrical potential differences. J.Gen.Physiol. 59, 318-346.

77. Frömter,E. & Diamond,J. (1972) Route of passive ion permeation in epithelia. Nature New Biology 235, 9-13.

78. Cortney,M.A., Mylle,M., Lassiter,W.E. & Gottschalk,C. (1965) Renal tubular transport of water, solute and PAH in rats loaded with isotonic saline. Am.J.Physiol. 209, 1199-1205.

78a. Tanner,G. & Isenberg,M. (1970) Secretion of p-aminohippurate by rat proximal tubules. Am. J.Physiol. 219, 889-892.

79. Tanner,G. (1967) Micropuncture study of PAH and Diodrost transport in *Necturus* kidney. Am. J. Physiol. 212, 1341-1346.

80. Loewenstein,W.R., Sokolar,S.T., Hogashino,S., Kanno,Y. & Davidson,N. (1965) Intercellular communication: renal, urinary bladder, sensory and salivary gland cells. Science,149 295-298.

81. Long,S. Renal handling of urea in Rana catesbeiana. Am.J.Physiol. In press.

82. Bennet,C.M., Clapp,J.R. & Berliner,R.W. (1967) Micropuncture study of the proximal and distal tubule in the dog. Am.J.Physiol.213,1254-1262.

83. Bennet,C.M., Brenner,B.M. & Berliner,R.W.(1968) Micropuncture study of nephron function in the Rhesus monkey. J.Clin.Invest. 47, 203-216.

84. Colindres,R.E. & Lechene,C. Technical problems associated with collection of distal tubular fluid in the rat. Yale J.Biol. & Med. In press.

85. Wiederholt,M., Sullivan,W.J., Giebisch,G., Curran,P.F. & Solomon,A.K. (1971) Potassium and sodium transport across single distal tubules of Amphiuma. J.Gen.Physiol.57, 495-525.

86. Duarte,C., Chomety,F. & Giebisch,G. (1971) Effect of amiloride, ouabain and furosemide on distal tubular function in the rat. Am. J. Physiol. 221, 632-640.

87. Giebisch,G., Malnic,G. & Curran,P.F. (1971) Proc.Int.Union Physiol.Sc. IX, p.203.

88. Kempton,R.T. (1937) The dimensions of the renal tubules of Necturus maculosus. J.Morphol.61, 51.

89. Giebisch,G., Boulpaep,E.L. & Whittembury,G. (1971) Electrolyte transport in kidney tubule cells. Proc.Roy.Soc.(B)London 262,175-196.

90. Kinter,W. (1959) Renal tubular transport of Diodrast I^{131} and PAH in Necturus: evidence for simultaneous reabsorption and secretion. Am.J. Physiol. 196, 1141-1149.

91. Bentzel,C.J., Anagnostopoulos,T. & Pandit,H. (1970) Necturus kidney: its response to effects of isotonic volume expansion. Am.J.Physiol. 218, 205-213.

92. Tanner,G.A. & Kinter,W.B. (1966) Reabsorption and secretion of p-aminohippurate and diodrast in Necturus kidney. Am.J.Physiol.210,221-231.

Comparative Physiology, eds. L,Bolis, K. Schmidt-Nielsen and S. H. P. Maddrell

SEA-WATER TELEOSTS: PRESENCE OF α AND β ADRENERGIC RECEPTORS IN THE GILL REGULATING SALT EXTRUSION AND WATER PERMEABILITY

P.PIC, N.MAYER-GOSTAN & J.MAETZ
Groupe de Biologie Marine du Département de Biologie du Commissariat à L'Energie Atomique,
Station Zoologique,
Villefranche-sur-Mer, France,

& Station marine d'Endoume,
Marseille, France.

SUMMARY

Epinephrine injected intraperitoneally into <u>Mugil capito</u> adapted to sea water produces within minutes a simultaneous inhibition of the branchial Na and Cl effluxes, and an increase of the diffusional water efflux. Na influx is also reduced. Part of this effect is due to a considerable diminution of the Na ingested by drinking. A loss of weight is observed in the treated fish suggesting an increase of the osmotic water loss by the gill.

Epinephrine produces a depression of the Na extrusion pump as evidenced by the decrease in the branchial Na net flux and by a 80% reduction of the Na/K exchange pump which is responsible for active Na extrusion. The Na and Cl leak across the gill remains unchanged.

The depressing effect of epinephrine on salt extrusion is mediated by α receptors as both tolazoline and phentolamine restore normal Na and Cl excretion. These blocking agents do not affect the water permeability changes induced by epinephrine however.

The stimulating effect of epinephrine on the water permeability is mediated by β receptors as propranolol alleviates the effects of epinephrine.

This blocking agent is not able to restore the effects of epinephrine on salt extrusion.Arguments are presented suggesting that the water permeability changes result from a direct effect of epinephrine on the gill epithelium rather than from hemodynamic effects. The α and β receptors are probably located in different epithelial cells.

Keys and Bateman (11) first showed that in the seawater adapted eel heart-gill preparation epinephrine produces an increase of the chloride concentration of the perfusion medium, while in control preparations the chloride level decreases during perfusion. More recently Pickford, Srivastava and Pang (34) in *Fundulus heteroclitus* and Pic (33) in *Mugil capito* observed that the plasma osmolarity rises after epinephrine administration. In the mullet, this increase is at least in part, due to a rise of Na plasma level.

This effect may result from either an increased water permeability of the gill epithelium hastening water loss down the osmotic gradient or to a reversal of the net salt extrusion activity which occurs across the gill. This reversal may be the consequence either of an increased passive ionic influx which has become bigger than the efflux or of a decreased efflux and its active extrusion component. All these effects would be the consequence of a *direct action* of the cathecholamines on the branchial epithelium suggesting the existence of adrenergic receptors similar to those recently discovered in frog skin epithelium (1,37,43).

There is however, the possibility that the effects of catecholamine on the gill may be the result of an *indirect action via* changes in the branchial haemodynamics. Epinephrine increases blood flow across the gill, while it decreases blood flow in peripheral vascular beds (11,12,32). Recent morphological studies (40,42) suggest that branchial vasodilatation is the result of an augmented lamellar circulation at the expense of the central filamental by-pass. According to Conte (5), the mitochondria-rich chloride' cells which almost certainly are the site of the active salt extrusion activity of the gill are located in the interlamellar region irrigated by the central-core blood supply. Thus vascular actions of catecholamines may well be accompanied by alterations in the salt extrusion efficiency were the blood supply to these cells a limiting factor. Alternatively, the lamellar epithelium may be either more permeable to water than the interlamellar region or the site of the diffusive ionic exchanges and thus increased circulation in this region may favour passive water and ion movements and diminish the efficiency of the active salt excretion mechanism.

The present report describes the effects of catecholamines on sodium and chloride exchanges across the gill. Diffusional water permeability was also measured with the help of isotopic techniques. The branchial osmotic permeability was evaluated indirectly by estimating the drinking rate which balances water loss across the gills and the kidney in sea water teleosts. With the help of pharmacological agents - α or β blockers - the nature of the adrenergic receptors was ascertained. Epinephrine depresses the sodium and chloride extrusion pumps *via* α adrenergic receptors, while the leak pathway remains unchanged. Epinephrine stimulates water permeability *via* β adrenergic receptors.

BIOLOGICAL MATERIAL AND METHODS

Juvenile *Mugil capito*, weighing from 6 to 41g were used. They were caught in the Mediterranean sea along the coast between Toulon and Marseille. The fish were stored in running sea water tanks and fed daily. One week before the experiments, feeding was discontinued and the fish were transferred into closed-circuit tanks at controlled temperature (16 $\pm$ 2°C).

For most of the experiments, the fish were prepared as follows: they were fitted with an indwelling intraperitoneal catheter. Anaesthesia was performed by keeping the fish in an ice-bath. Fish were placed in individual flux-chambers for at least 24 hrs before the flux measurements. Since urinary output plays a negligible role in the ionic and water exchanges (13,27), the urinary papilla was not cannulated.

Na, Cl and H_2O efflux measurements.

In most of the experiments these effluxes were measured simultaneously. ^{24}Na (in an isotonic sodium sulphate solution, about 100 µC per fish), ^{36}Cl (in a 2 molar sodium chloride solution, about 100 µC per fish) and HTO (in the form of a dilution in distilled water, about 100 µC per fish) were injected through the indwelling catheter, the volume of injected fluid being about 250 µl. To obtain homogenous distribution of the isotopes in the fish, injection was performed 30 to 45 min. before flux measurements.

Flux experiments consisted in closing the circuit, the volume of the aquarium water being kept constant (about 300 ml per fish), the temperature being adjusted at 16 $\pm$ 1° C by means of heat-exchangers. The techniques for monitoring the appearance of the β or γ emitters in the external bath have been described elsewhere (15,44). The appearance of the tracers in the external medium was followed by flow counters associated with a digital printout and with a visual display (Intertechnique). Figures 1-4 illustrate typical experiments making use of these techniques. Simultaneously, one ml samples were taken from the bath every 5 min. in order to measure the ^{24}Na conc. in a well counter (Mecaserto) and after decay of the isotope (about 150 hrs), a 250 µl fraction of the sample was added together with 750 µl distilled water to 10 ml Bray solution in order to measure the HTO and ^{36}Cl conc. in a scintillation counter (Intertechnique SL 40). These two isotopes were measured simultaneously taking advantage of their widely different β energies. A small computer associated with the counter permitted the automatic calculation of their respective concentrations taking into account their relative quenching. The radioactivity of the external samples was compared to that of a diluted plasma sample. A blood sample was drawn from the heart at the end of the flux experiment, immediately centrifuged and diluted. Na^+ conc. was measured

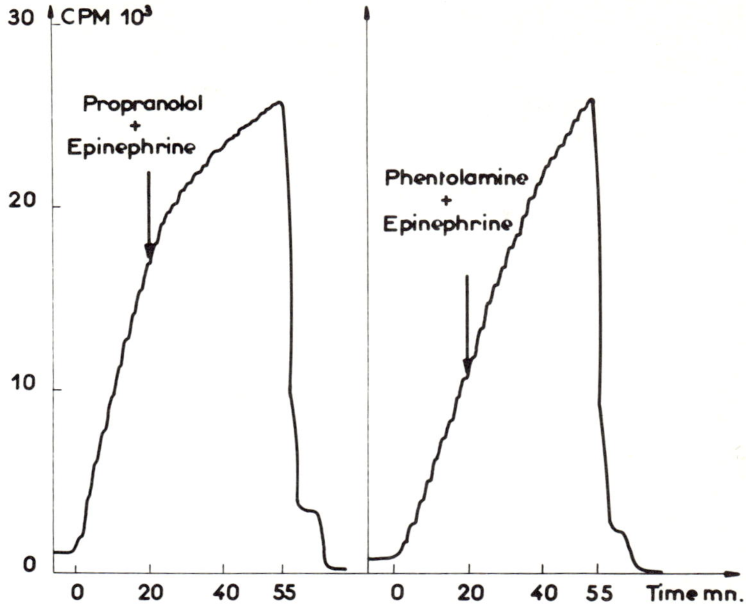

Figure 1 - Recording of the ^{24}Na appearance from the seawater adapted mullet
Ordinate: external ^{24}Na conc. in 10^3 cpm
Abscissa: Time in min. The printout was every 2 min.
At 20 min (arrow): injection of the drugs.
Left: effect of epinephrine (5 µg) with a β blocking agent, propranolol (100 µg)
Right: effect of epinephrine (5 µg) with an α blocking agent, phentolamine (200 µg).
Note the depressing effect of epinephrine on the efflux which is seen despite the addition of the β blockers.
Note the absence of effect of epinephrine with an α blocker.

by flame photometry (Eppendorf) and Cl^- conc. by amperometric titration (Aminco-Cotlove). The specific radioactivity of the various isotopes (in cpm or dpm μEq^{-1} for Na^+ and Cl^- or in dpm ml^{-1} for HTO) was calculated.

The efflux experiments were divided into three periods given in the following tables as periods I,

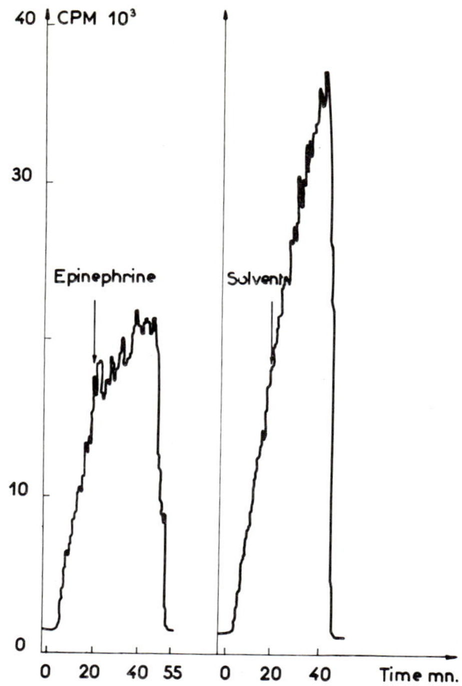

Figure 2 - Recording of the ^{36}Cl appearance from the seawater adapted mullet. Coordinates as in Fig.1 except that printout was every minute and that it is ^{36}Cl appearance which is recorded.
Left : effect of epinephrine (5 µg).
Right: control experiment with solvent injection alone.
Note the depressing effect of epinephrine.

II and III. Period I was a control period of 20 min., at the end of which the drug to be tested or the solvent was quickly injected by means of the indwelling catheter. Period II and III lasted respectively 15 and 20 or 25 min. For each of these periods the plasma specific radioactivity was corrected taking into account the radioactive loss into the external medium and the space of the isotope. For HTO, this space is taken as 70 ml/100g weight (13) for Na and Cl the space was taken as 25 and 20 ml/100g (21, and R.Kirsch, pers.comm.) in the controls and 20 and 15 ml in the periods

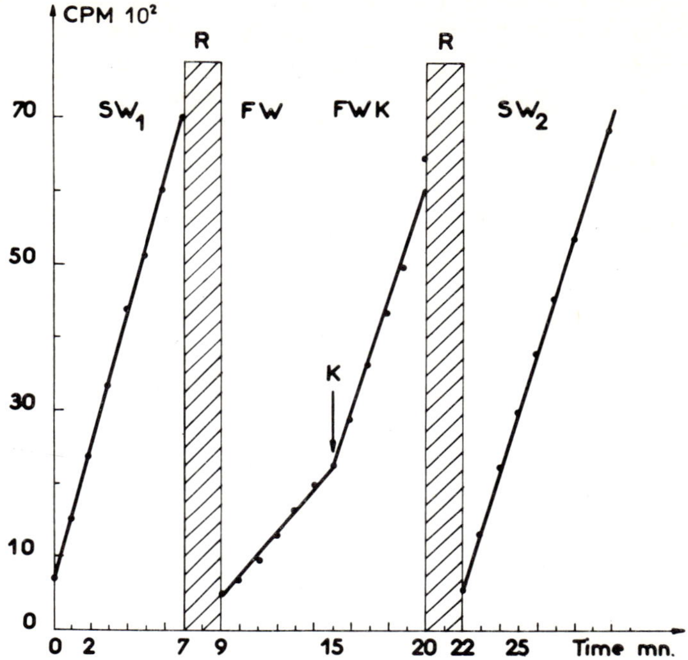

Figure 3 - A typical "K - test" for testing the branchial Na/K exchange across the gill of the mullet (fish treated with epinephrine 20 µg).
Ordinates: Ext ^{24}Na conc.in 10^2 cpm.
Abscissa: time in min.
SW_1, FW, FWK, SW_2 successive test periods in SW (control)
FW, FW with 10mM KCl added at arrow and in SW (return)
R rinsing periods of 2 min. for changing the external medium.
Note change of slope following addition of external K.

following epinephrine injection as during the course of these investigations it was found that epinephrine reduces the Na and Cl space by about 5 ml (see results). The radioactive backflux from the external medium was negligible in the case of ^{24}Na and ^{36}Cl, but not in the case of HTO. For the HTO flux measurements the difference between internal and external specific radioactivities was

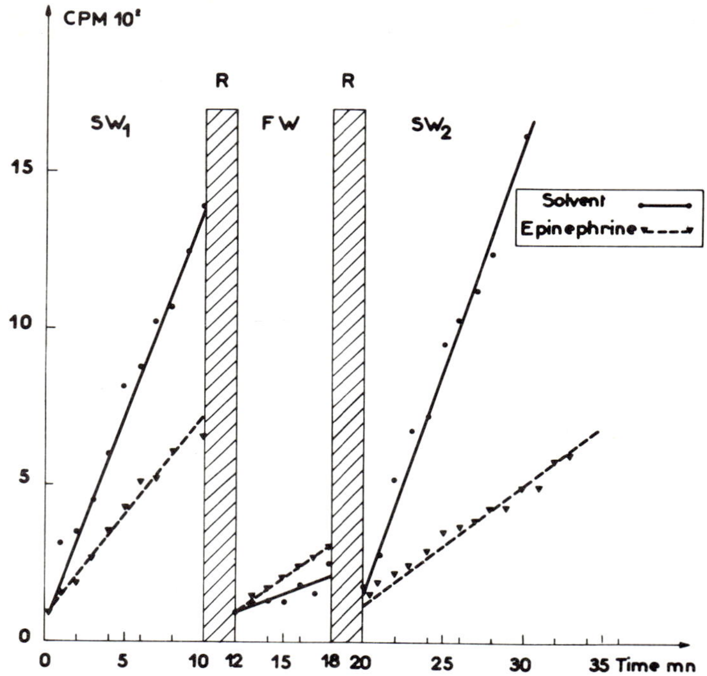

Figure 4 - ^{36}Cl efflux readjustment following transfer from SW to FW. Coordinates as in preceding figure except that it is ^{36}Cl which is recorded.
Circles : control fish
Triangles: fish treated with 5 µg epinephrine
Note efflux reduction upon transfer to FW for ^{36}Cl as for ^{24}Na.
This reduction is less important after epinephrine injection.

used. The correction amounted to about 30% for the period III. The details of the flux calculations have been given elsewhere (3).

Fluxes are expressed in $\mu Eq^{-1}.hr^{-1} 100g^{-1}$ for ions and in ml hr^{-1} $100g^{-1}$ for water. The weight of the fish was taken after the cardiac puncture at the end of the experiment at the same time as the volume of the external medium was measured in a graduated cylinder.

Assessment of the efflux components in the case of Na^+ and Cl^-.

Sodium and chloride efflux may be divided into 3 or 4 components according to recent studies (3, 27).

1) a small Na or Cl extrusion component balancing the intestinal absorption effected by a Na/K exchange pump. This component represents about 5 to 10% of the efflux.

2) a large efflux component dependent upon the external Na or Cl concentration which is considered to be an exchange-diffusion component.

3) A small leak component observed after transfer of the fish into deionized water or fresh water. This component is about 20 to 30% of the total efflux and is independent of external Na or Cl concentration.

4) In the case of Na efflux, it remains possible that part of the large component may be due to simple diffusion (17).

The characteristics of the Na/K exchange pump have been tested using the rapid transfer technique described elsewhere (3) with the following modifications. To avoid stressing the fish, animals were kept in the same chamber throughout the test. The changes of fluids were obtained by the admission of either SW or FW at very high flow rates in the chamber to permit a complete rinse within 2 min. K was added to FW in the form of a molar KCl solution 1 ml/100 ml to avoid transferring the fish from FW to a KCl 10 mMolar solution in FW. Only the relative effluxes observed in SW, FW and FWK were calculated. No blood was taken at the end of the experiment and thus the specific radioactivity of the Na or Cl tracer was not measured.

In the case of the Cl efflux, the test consisted of a SW-FW-SW transfer. The relative effluxes observed in these media were calculated.

Sodium influx and sodium space.

The technique is similar to that given by Bornancin, Cuthbert and Maetz (3) for eels. As the fishes used here were of small size the following modifications of the technique were introduced. At the end of the influx period (1 hr), the fish were quickly rinsed in FW containing MS 222 as an anesthetic. A blood sample was taken by cardiac puncture. The fish was cut into small pieces to fit into polythene tubes to be counted in the well counter and comparated with an aliquot of the external medium. This aliquot permitted the calculation of the external Na specific activity. For the influx calculation, the radioactive backflux from the fish was taken into account (3). A plasma sample was also counted to allow for the calculation of the Na space and for the determination of internal Na specific activity and the radioactive backflux. The flux is given in $\mu Eq.hr^{-1}100g^{-1}$ and the space in ml 100^{-1}.

Drinking rate.

The technique (27) employing colloidal gold ^{198}Au (from Saclay center) was applied. Body weight was taken before and after the one-hour period in labelled water to monitor the weight changes incurred by the catecholamine treatment. The osmotic water flow across the gill was calculated as the difference between the drinking rate and the weight change expressed in $\mu l.100g^{-1}$ assuming a density of one for the fish. As the urine flow is at most 100 $\mu l\ hr^{-1}\ 100g^{-1}$ in fishes of this size (see review by Lahlou, 13), it was neglected.

Drugs.

Epinephrine (Rhône Poulenc) was kept in HCl solution (pH about 3). The dose given was either 5 or 20 µg per fish.

As α-blocking agents, we used either phentolamine chloride (I.C.I.) in a 2 mg/ml solution in distilled water or tolazoline (CIBA) in a 10 mg. ml^{-1} solution. The doses given were respectively 200 µg and 1 mg per fish. The β-blocking agent was propranolol (AVLON) in a 1 mg ml^{-1} solution, the dose injected being 100 µg per fish.

The blocking agents were injected simultaneously with epinephrine by means of the indwelling catheter in the efflux experiments. For the influx and drinking rate experiments, the fish were handled for the injection just before placing the fish in the radioactive sea water.

For the control injection, the solvent was injected in an equivalent volume.

RESULTS

Na^+ and Cl^- effluxes.

The results pertaining to the Na^+ efflux are given in table 1. As may be seen the various treatments concerned fishes of widely different body weights. Correspondingly the fluxes observed during the control period (I) were found to differ significantly ($P < 0.02$) for the statistical comparison between the 28 and 8g groups). The highest fluxes are found in the smallest fish. Thus the effects of the various treatments were compared statistically by the method of 'paired differences' between the values obtained in the control period and those observed in post-treatment periods.

Solvent injection is followed by a small increase in Na efflux, statistically not significant. Epinephrine (20µg per fish) produces a clear-cut reduction of the Na efflux attaining 40% of the initial flux during period III. In a few supplementary experiments not reported in the table, smaller doses of epinephrine (1 to 5 µg per fish) proved as efficient as 20 µg in inhibiting Na efflux. Accordingly the effects of the blocking agents were tested on a 5 µg dose of epinephrine. Both α blocking agents, phentolamine and tolazoline were found to yield similar effects. Thus the results given in table 1 are pooled. These agents proved effective in blocking the depressing effect of epinephrine on the Na efflux. Propranolol, the β blocking agent, on the contrary proved ineffective. Figure 1 shows two recordings illustrating the appearance of ^{24}Na in the external bath. On the left, inhibition of the radioactive efflux is clearly observed within 5 to 10 min.following the injection of epinephrine together with propranolol. On the right, phentolamine was administered together with epine-

TABLE 1

Sodium efflux (μEq.hr^{-1}.100g^{-1})

Effect of epinephrine and of α and β blocking agents

Injection	n	Body weight (g)	Flux periods			Paired differences		
			I	II	III	Δ (II – I)	Δ (III – I)	Δ (III – II)
Solvent	5	28.0 ± 4.0	3927 ± 923	4625 ± 796	5202 ± 693	+ 697 ± 307 n.s	+ 1275 ± 682 n.s	+ 577 ± 415 n.s
Epinephrine*	8	18.2 ± 1.7	5489 ± 725	4287 ± 548	3305 ± 471	–1201 ± 269 $P < 0.01$	–2081 ± 554 $P < 0.01$	–982 ± 341 $0.02 < P < 0.01$
Epineph** + α blocker°	8	17.5 ± 3.1	4399 ± 468	4208 ± 335	4408 ± 466	–191 ± 374 n.s	+ 82 ± 402 n.s	+ 273 ± 343 n.s
Epineph** + β blocker°°	7	8.0 ± 0.6	7767 ± 885	6247 ± 785	5465 ± 380	–1412 ± 665 n.s.	–2301 ± 658 $0.01 < P < 0.02$	–904 ± 649 n.s.

* Dose: 20 μg/fish
** : 5 μg/fish
° Phentolamine 200 μg or Tolazoline 1 mg
°° Propanolol 100 μg

Injection between period I and period II (15 min.)
Period III: 20 min.

TABLE 2
Chloride efflux (μEq.hr^{-1}.100g^{-1})
Effect of epinephrine and of α and β blocking agents

Treatment	n	Body weight (g)	Flux periods			Paired differences		
			I	II	III	Δ (II − I)	Δ (III − I)	Δ (III − II)
Solvent	6	18.2 ± 4.5	2094 ± 567	2691 ± 498	2793 ± 533	+596 ± 132 $P < 0.01$	+701 ± 284 n.s.	−105 ± 228 n.s.
Epinephrine*	7	16.9 ± 3.3	3124 ± 520	1893 ± 433	1199 ± 406	−1231 ± 605 n.s.	−1925 ± 450 $P < 0.01$	−693 ± 316 n.s.
Epineph* + α blocker°	3	19.5 ± 7.5	1455 ± 828	2145 ± 375	2905 ± 694	+690 ± 602 n.s.	+1450 ± 915 n.s.	+760 ± 423 n.s.
Epineph* + β blocker°°	4	7.2 ± 0.7	6073 ± 783	3799 ± 757	2701 ± 579	−2273 ± 303 $P < 0.01$	−3372 ± 375 $P < 0.01$	−1098 ± 378 n.s.

* Dose: 5μg/fish
° Phentolamine 200 μg or Tolazoline 1mg
°° Propranolol 100 μg
Periods as in table 1

phrine. The ^{24}Na appearance rate remains unchanged except for the progressive reduction of the radioactive efflux which is accounted for by the parallel decrease of the internal Na specific activity.

The results of the effects of epinephrine on the Cl^- efflux are collected in table 2. A dose of 5 µg was used throughout. A small increase of the efflux is observed after solvent injection, while epinephrine depresses the Cl^- efflux even more than the Na efflux, the reduction attaining 60% during the period III. For Na^+ as for Cl^-, the α blockers were effective in blocking the adrenergic response, while propranolol remains ineffective. Figure 2 illustrates two recordings of the ^{36}Cl appearance into the external bath. On the right, it may be seen that solvent injection alone has no effect, while on the left, the epinephrine injection produces significant diminution of the ^{36}Cl efflux within a few minutes. In addition, the following points emerge from the comparison of tables 1 and 2. For Cl as for Na, the efflux values seem to be inversely related to the body weight. The Cl efflux is definitely smaller than the Na efflux.

Components of the Na^+ and Cl^- effluxes.

Table 3 collects the data concerning the relative effluxes measured after transfer from SW into FW and following the addition of K to the external bath. Figure 3 illustrates a typical experiment on a mullet injected with epinephrine.

Transfer from SW to FW is accompanied by an instantaneous reduction of the Na efflux by about 80% in control fish and 70% in epinephrine (dose 20 µg) treated animals. The addition of K produces an instantaneous increase of the Na efflux which regains its original rate in both control and treated fish. This is shown in figure 3 by the parallel slopes of the linear ^{24}Na appearance in SW control, SW return and FWK periods. In the same table, absolute values of the fluxes in SW, FW and in FWK are given; they were calculated by multiplying the relative slopes by the flux values given in table 1 (period III) for control and epinephrine treated animals. It may be seen that the Na leak

TABLE 3 - Components of the Na efflux. Effects of epinephrine.

1. Relative ^{24}Na effluxes in SW, FW and in FWK

	n	SW	FW	FWK
Solvent	7	100	22.2 ± 4.0	100.0 ± 14.8
Epineph. (20µg)	7	100	32.1 ± 5.2	87.4 ± 7.0

The mean ^{24}Na appearance rate observed in the control and in the return periods in sea water (SW) was taken as reference 100. The return flux was 46.8 ± 6.2% of the control flux in the solvent injected fish and 67.8 ± 6.7% in the epinephrine treated fish.
FW : fresh water; FWK: fresh water with 10mM KCl.

2. Absolute value of the effluxes taking period III from table 1 as reference.

	Solvent	Epinephrine
Total efflux (from Table 1)	5202	3305
leak (flux after transfer to FW)	1155	1061
FWK flux	5202	2889
Na/K exchange (FWK - FW flux)	4047	1828

Fluxes in $\mu Eq.hr^{-1}.(100g)^{-1}$

observed in FW remains unchanged while the difference between the FWK and FW fluxes differ by at least a factor 2. This difference has been shown to be an index of the Na/K exchange or Na extrusion pump activity (17). The conclusion is thus that epinephrine inhibits the sodium extrusion activity of the gill.

Transfer from SW to FW is accompanied by an instantaneous reduction of the Cl^- efflux by about 75% in control fish. Figure 4 illustrates typical experiments of this kind permitting the comparison of the efflux readjustments in control and in epinephrine treated fish (dose : 5 µg). It may be seen from the figure, and from table 4 which summarizes the results, that the reduction is significantly less in the treated fish attaining only 50%. The absolute values of the Cl effluxes in SW and in FW are given in the same table, using period III of table 2 as the reference values for the SW period.

TABLE 4 - Components of the Cl efflux. Effects of epinephrine

1. Relative ^{36}Cl effluxes in SW and in FW

	n	SW	FW
Solvent	6	100	24.2 ± 4.8
Epineph. (5µg)	5	100	50.3 ± 7.1

2. Absolute value of the fluxes taking period III in table 2 as reference.

	Solvent	Epinephrine
Total efflux	2793	1199
Leak (FW flux)	676	603
Exchange-diffusion component	2117	596

It may be seen that the leak flux observed in FW remains unchanged while the exchange-diffusion component (difference between the SW and FW fluxes) is decreased by at least a factor of 3.

Sodium influx and sodium space.

Table 5 compares total Na influx in control

TABLE 5 - Sodium influx: branchial and oral components. Sodium Space. Effects of epinephrine.

Treatment	n	body weight	Influx Total	Influx oral (x)	Influx branchial	space
Solvent	6	12.5 ± 1.4	4709±303	844	3865	27.7 ± 4.9
Epineph. (20µg)	6	12.7 ± 0.7	3394±240	186	3208	17.3 ± 1.1
Statistical sign. of dif.			$P < 0.01$			n.s.

(x) calculated from the drinking rate multiplied by the Na conc. of SW (550 mM)

Fluxes in $\mu Eq.h^{-1}.(100g)^{-1}$

and epinephrine treated fish, both groups of fish being in the same weight range. It may be seen that Na influx is reduced by about 30% by epinephrine administration (dose 20 µg). The Na space is also reduced, although not significantly because of the wide scatter of the individual values in control fish.

In the same table, the rate of Na ingestion has been given, calculated by multiplying the drinking rate given in the following table by the Na concentration of SW (550mM). The difference between total Na efflux and Na ingestion rate gives a measure of the branchial Na influx, which is found to be reduced by about 20% following epinephrine treatment.

Comparing the values for Na efflux and for Na influx, it appears that both are reduced by epinephrine. The sodium efflux is reduced to a bigger extent however. Thus the difference between branchial Na efflux and influx decreases, confirming the inhibitory effect of the drug on the net extrusion activity of the gill.

Diffusional and osmotic water permeability of the gill.

Table 6 gives the results of the drinking rate measurements. In epinephrine (20 µg) treated fish, the drinking rate is considerably reduced. Body weight changes measured simultaneously are also reported in the same table, expressed in $\mu l\ hr^{-1}$ $100g^{-1}$. Epinephrine induces weight loss. The calculated, net osmotic flow across the gill has been calculated, neglecting urine flow. An increased loss by 65% is seen after epinephrine treatment. This increase suggests an augmented osmotic water permeability of the gill which is probably even more important, as the osmotic gradient is certainly smaller in epinephrine treated fish whose plasma osmolarity increases as shown by Pic (33).

The effect of epinephrine on the branchial water permeability is confirmed by the results concerning tritiated water efflux reported in table 7.*)

During the control period the water efflux, like the ionic effluxes, is inversely related to

*) see page 324

TABLE 6 - Drinking rate and weight change. Effects of epinephrine.

Treatment	n	Body weight	Drinking rate ($\mu l.hr^{-1}100g^{-1}$)	Weight change (in mg $100g^{-1}$)	Osmotic net flow° ($\mu l.hr^{-1}.100g^{-1}$)
Solvent	6	9.4 ± 2.4	1533 ± 345	-200 ± 680°	-1733
Epineph.(20µg)	6	8.6 ± 1.6	337 ± 84 (xx)	-2500 ± 480 (x)	-2837

Paired differences between initial and final weight °: n.s.

(x): $P < 0.01$

Difference between means of the control group and epinephrine treated group: (xx) $P < 0.01$

° neglecting urine flow

body weight. Solvent injection is followed by a slight non-significant rise. Epinephrine (20 µg) produces a 70 to 75% stimulation of the diffusional water efflux and this is highly significant. Lower doses (1 - 5µg) were also found to be effective. The α blocking agents phentolamine and tolazoline remain without effect on this stimulation, the relative doses being the same as those previously utilised in the study of the ionic effluxes. That the observed stimulation is due to β adrenergic receptors is shown by the results concerning the simultaneous injection of epinephrine (5µg) and propranolol (100 µg). Propranolol completely abolishes the stimulating effect of epinephrine.

In conclusion, α receptors are concerned with active salt extrusion control by means of an inhibitory effect. β receptors seem to be of importance in the control of the passive water permeability by means of a stimulatory action.

DISCUSSION

1. Salient features of the ionic and water exchanges in *Mugil capito.*

The mullet used in this study is characterized by high rates of ionic exchange, amounting to about 80 and 120% of the exchangeable Cl^- and Na^+ per hr. Such a higher turnover rate for Na has already been reported for the related species *Mugil cephalus* (24,25). Moreover, the Na exchange is faster than the Cl exchange and a similar difference has been reported for *Platichthys flesus* (24, 25) and for *Pholis gunnellus* (6).

Transfer into FW is followed by an instantaneous reduction of the Na and Cl efflux similar to the 'Na or Cl - free effect' described earlier for the flounder (24,25). Here the reduction in Na efflux is 78% compared to the 87% reduction found in *Mugil cephalus* (24,25).

Addition of K to the external medium restores the Na efflux observed in SW. In the flounder, on the contrary, only about 50% of the SW efflux is recovered by K addition (16). In this respect, the mullet resembles *Anguilla* (26). This would mean that for both mullet and eel all the Na-free effect results from an exchange-diffusion mechanism and that the competition between external Na and K for the Na/K exchange carrier (16-17,26) is probably responsible for the Na/Na exchange.

The diffusional water flux is also very fast, at least 100% higher than those reported previously (27) for the seawater adapted *Platichthys*,*Anguilla* or *Serranus*. The animals studied were of much smaller size which accounts for the difference.The fluxes in *Tilapia* which were of the same size as the mullet studied here are very similar to those given in Table 7 (36). From tables 1,2 and 7, an inverse relationship between body weight and ionic or water effluxes is evident. Such a relationship had previously been reported for water flux of *Tilapia mossambica* (36) and for various other teleosts (7).

Finally, the drinking rate was also found to be very high in the mullets discussed in table 6. (see ref. 27, for comparison). The rate is however similar to those reported for *Fundulus heteroclitus* (35) and *Tilapia* (36) of the same body weight.

The Na ingested by the fish corresponds to less than 20% of the total Na influx as shown in table 5. This observation confirms the earlier conclusion of Motais and Maetz (28) about the relative importance of the gut and gill in the ionic exchanges in the flounder and the sea perch.

2. Mechanism of the osmoregulatory effects of epinephrine.

The results reported above give us a comprehensive picture of the osmoregulatory disturbance produced by the epinephrine. Both the increase in the branchial water permeability and the depression of the active Na and Cl extrusion pumps act to increase the internal osmolarity (cf.33). The 2.5% water loss observed during the hour following epinephrine injection is however insufficient to account for the 20% increase of the osmolarity described by Pic (33) in a previous paper. Thus,the drop in active salt efflux across the gill constitutes the major component of the osmotic perturbance produced by epinephrine.

The decreased drinking rate following epinephrine injection results from the presence of adrenergic receptors mediating contraction of the smooth musculature of the stomach in fish (4,30,31). This effect certainly accentuates the weight loss incurred by the fish. In a series of complementary experiments, it was found that the rate of water ingestion decreases further after 20 µg epinephrine injection; after 2 hrs it is 143 ± 53 µl and after 4 to 5 hrs it is 171 ± 65 µl hr^{-1} $100g^{-1}$.

The osmoregulatory perturbance produced by epinephrine is of physiological importance as it has been shown that stress is accompanied by an important increase of the circulating epinephrine (23, 29). This is one of the reasons why rather high doses of epinephrine were used in the present study, the second reason being the intraperitoneal route chosen for injecting the drug. Stress produces various physiological effects which may be explained in the light of our results. Evans (7) observed in the trout that stress increases the diffusional water permeability. A similar effect has been reported for the goldfish (14). Laboratory diuresis observed after handling of freshwater fish also reflects the increased osmotic permeability of the gill (8). A body weight increase has also been reported in the rainbow trout as an effect of forced swimming (43).All these results suggest that the water permeability increases after endogenous catecholamine release or epinephrine administration in fresh water as in sea water adapted teleosts. Indeed, such an increase has been observed in the FW adapted mullet too. In a series of 6 fish, an augmentation of the water efflux from 64 ± 20.5 ml hr^{-1} $100g^{-1}$ (period I) to 111.5 ± 15.6 (period II) and to 163.5 ± 21.8 (period III) was observed after injection of 20 µg epinephrine (Pic, Mayer-Gostan and Maetz, in preparation).

In seawater adapted fish, stress has been associated with weight loss and increased plasma osmolarity (34). Disturbance of the ionic balance in response to handling have been studied in our laboratory on _Fundulus heteroclitus_ (19) _Anguilla anguilla_ (20) and _Tilapia mossambica_ (Dharmamba, Bern, Mayer-Gostan and Maetz, in preparation). In all these fish, an _increase_ of the Na turnover rate has been observed, however, 24 to 48 hrs after the initial shock. This is in apparent contradiction to the <u>decrease</u> observed after epinephrine injection. It may be suggested that what is observed is a delayed compensatory response to the increased plasma osmolarity incurred immediately after the stress. Such a compensatory feed-back stimulation of the Na efflux has been experimentally demonstrated in response to salt-loading in the eel (22).

The effect of epinephrine on the salt extrusion pump remains to be discussed. In the case of Na, the inhibition of the pump is illustrated by the 55% decrease of the Na/K exchange fluxes (see table 4). The Na/Na exchange is probably reduced to more or less the same extent as shown by the diminution of Na influx (see table 5). The exchange-diffusion concept suggests that Na influx and efflux are tightly coupled. There are, however, exaples of hormonal effects which demonstrate that Na/K and Na/Na exchanges may be uncoupled, for example the observations of Dr. R. Langford in our laboratory concerning the effects of insulin (25 IU given in 3 days) on the seawater adapted eel. Apparently insulin blocks completely the Na/K exchange, while the Na/Na exchange remains unchanged. The inhibition of the pump is confirmed by an increase of the internal Na level by 25%. Independence between Na/K and Na/Na exchanges is seen also after Na SCN injection in the eel. As shown recently by Dr. F.H. Epstein in our laboratory this inhibitor reduces drastically Cl efflux and leaves the Na efflux unchanged. The Na pump is however, inhibited in parallel to the Cl pump, as shown by the 70% reduction of the Na efflux observed in FWK. The parallel depression of the Na and Cl pump activities is also indicated by the progressive increase of the Na and Cl plasma levels following SCN administration (Epstein and Maetz, in preparation). In addition, the experiments on SCN inhibition demonstrated that in the eel at least, the Cl pump is tightly linked to the exchange diffusion mechanism because the 'Cl-free effect' was abolished after the SCN injection. In the mullet, epinephrine produces a parallel depression of both the Cl pump illustrated by an increase of the internal Cl level (Pic, in preparation) and the Cl exchange-diffusion fluxes.

In contrast to the significant decrease observed in the active effluxes of Na and Cl, it may be noted that the passive leak-components of the Na and Cl effluxes remain unchanged. The latter evidence is at variance with the reports of Kirschner (12) and of Randall. Baumgarten and Malyusz (38) who found an increased rate of ^{22}Na loss across the gills in the isolated head of eel or trout adapted to fresh water. An enhancement of the sodium uptake by the isolated gill of the trout under the influence of epinephrine has also been reported (41). In the freshwater adapted mullet, however, preliminary observations on the sodium exchange *in vivo* showed no significant effect of epinephrine on both influx and efflux. More work is necessary as our results were obtained on fish without catheterization of the urinary papilla.

3. The presence of α and β adrenergic receptors in the gill.

The results obtained with the α and β blocking agents demonstrate clearly that the rates of Na and Cl transport are regulated by α receptors, while the water permeability is controlled by β receptors. The former mediate an inhibition of active ion transport while the latter correspond to an enhancement of water permeability.

In recent experiments (18,39) on the perfused eel gill preparation, we described the vascular effects of catecholamines. An increased perfusion flow at constant hydrostatic pressure was observed confirming previous observations (see introduction). This hemodynamic effect was found to be blocked by propranolol showing that β receptors were involved. Phentolamine remained without effect. Similar results were recently reported for the perfused carp gill (2). In the conger eel, however, α and β receptors were found to be involved in the control of blood flow. The former mediate an increased vascular resistance with norepinephrine being the more effective hormone, whereas the latter, as in the carp, mediates a decreased vascular resistance with epinephrine being the more active in this case. These observations on branchial hemodynamics have been interpreted in the framework of the model proposed by Steen and Kruysse (42) who have suggested that epinephrine favours lamellar flow probably at the expense of the blood flow passing through the central core of the filament. Blood flow could be a limiting factor in HTO diffusion as a result of unstirred layers in the lamellar blood compartment, the specific activity of the lamellar blood being less than that of the blood circulating in the filament. The stimulating effect of epinephrine could result from an increased lamellar blood flow with a concomitant increase of the specific activity of the lamellar blood and thus an increased HTO efflux. No increase of water permeability would be involved in such a mode of action of epinephrine. The β adrenergic receptor would be branchial musculature rather than the epithelium itself.

This hypothesis cannot be accepted without reservations, however, as epinephrine also stimulates the osmotic water loss across the gill (see table 6). While diffusional water fluxes involve an important fraction, up to 30%, of the cardiac output (27), the osmotic net flux of water involves about 30 times less. It is doubtful whether in control fish restriction of the blood flow in the lamellar region could be important enough to produce a significant decrease of the osmotic gradient across the epithelium because the oxygen uptake by the gill would be simultaneously impaired. Finally the effects of small doses of epinephrine on the isolated frog skin epithelium (37) which show an increased osmotic permeability also as a result of β adrenergic stimulation, suggest that frog skin and teleostean gill epithelia are similar in this respect.

The depressing effect on salt transport on the contrary cannot be the result of a diversion of the blood flow from the central core to the lamellae because it is blocked by α and not by β blocking agents. Furthermore, preliminary histological examination of the repartition of the chloride cells in the branchial epithelium of the mullet performed by A.Masoni show definitely that the chloride cells are to be found along the whole lamellar surface on the afferent side. That the chloride cells are the site of active salt transport has been demonstrated beyond any doubt by the recent observations of Kamiya (10). By the use of fractionation techniques, the chloride cells were isolated and shown to contain most of the Na-K dependent ATP-ase activity. Thus the chloride cells almost certainly contain the α adrenergic receptors which regulate the active ionic transport. The fact that epinephrine stimulates simultaneously at the same dose the water permeability via β adrenergic receptors suggests that the β receptors are located elsewhere in the gill. The situation with respect to frog skin epithelium is more complicated because low doses of epinephrine increase active sodium transport and passive water permeability, both effects being mediated by β adrenergic receptors. High doses of epinephrine, on the contrary, produce an inhibition of the active sodium transport and water permeability by the agency of α adrenergic receptors (37).

The results obtained by Rajerison, Montegut, Jard and Morel (37) strongly suggest that the α receptors are concerned with an inhibition of the adenylcyclase-system which has been characterized for example in frog skin (see the review by Jard, 9). The β receptors on the contrary may affect a stimulation of the same enzymatic system. Our results could be interpreted in the same way providing adenylcyclase activity can be demonstrated in the teleostean gill. This hypothesis is currently being investigated in our laboratory.

REFERENCES

1. Bastide,F. & Jard,S. (1968) Actions de la noradrénaline et de l'ocytocine sur le transport actif de sodium et la perméabilité à l'eau de la peu de grenouille. Rôle du 3',5'-AMP cyclique. Biochim.Biophys.Acta 150, 113-123.

2. Belaud,A., Peyraud-Waitzenegger,M. & Peyraud,C. (1971) Etude comparée des réactions vasoosmotrices des branchies perfusées de deux téléostéens: la Carpe et le Congre. C.R.Soc.Biol.165, 1114-1118.

3. Bornancin,M., Cuthbert,A.W. & Maetz,J. (1972) The effects of calcium on branchial sodium fluxes in the sea-water adapted eel, Anguilla anguilla L. J.Physiol. (Lond) 222, 487-496.

4. Burnstock,G. (1958) The effect of drugs on spontaneous mobility and on response to stimulation of the extrinsic nerves of the gut of a teleostean fish. Br.J.Pharmac.Chemother. 13, 216-226.

5. Conte,F.P. (1969) Salt secretion. In Fish Physiology, vol.1, pp. 241-292, ed. Hoar,W.S. & Randall,D.J., New York:Academic Press.

6. Evans,D.H. (1969) Sodium chloride and water balance of the intertidal teleost, Pholis gunnellus. J.exp.Biol. 50, 179-190.

7. Evans,D.H. (1969) Studies on the permeability of selected marine, freshwater and euryhaline teleosts. J.exp.Biol. 50, 698-703.

8. Holmes,R.M. (1961) Kidney function in migrating salmonids. Rep.Challenger Soc. n.13, 23.

9. Jard,S. (1971) Le mécanisme d'action de l'hormone antidiuretique. J.Physiol. (Paris) 63, 99-146.

10. Kamiya,M. (1972) Sodium-potassium-activated adenosine triphosphatase in isolated chloride cells from eel gills. Comp.Biochem. Physiol. (in the press).

11. Keys,A.B. & Bateman,J.B. (1932) Branchial responses to adrenalin and to pitressin in the eel. J.Physiol. (Lond.) 75, 327-336.

12. Kirschner,L.B. (1969) Ventral aortic pressure and sodium fluxes in perfused eel gills. Am. J.Physiol. 217, 596-604.

13. Lahlou,B. (1970) La fonction rénale des téléostéens et son rôle dans l'osmorégulation. Bull.Inf.Sci.Tech.-C.E.A. 144, 17-52.

14. Lahlou,B. & Giordan,A. (1970) Le contrôle hormonal des échanges et de la balance de l'eau chez le Téléostéen d'eau douce Carassius auratus intact et hypophysectomisé. Gen.Comp.

Endocrinol. 14, 491-509.

15. Maetz,J. (1964) Recherches sur la perméabilité aux électrolytes de la branchie des poissons et sa régulation endocrinienne. Bull.Inf.Sci. Tech. C.E.A. 86, 11-70.

16. Maetz,J. (1969) Seawater Teleosts: Evidence for a sodium-potassium exchange in the branchial sodium excreting pump. Science 166, 613-615.

17. Maetz,J. (1971) Fish gills: mechanisms of salt transfer in freshwater and in seawater. Phil. Trans.Roy.Soc.Lond.,B. 262, 209-249.

18. Maetz,J. & Rankin,J.C. (1969) Quelques aspects du rôle biologique des hormones neurohypophysaires chez les poissons. Colloques Internationaux du C.N.R.S. Paris 177, 45-55.

19. Maetz,J., Sawyer,W.H., Pickford,G.E. & Mayer,N. (1967) Evolution de la balance minérale du sodium chez *Fundulus heteroclitus* au cours du transfert d'eau de mer en eau douce. Effets de l'hypophysectomie et de la prolactine. Gen. Comp.Endocrinol. 8, 163-176.

20. Mayer,N. & Maetz,J. (1967) Axe hypophyso-interrénalien et osmorégulation chez l'anguille en eau de mer. Etude de l'animal intact avec notes sur les perturbations de l'équilibre minéral produites par l'effet de "choc". C.R.Acad.Sci. Paris 264, 1632-1635.

21. Mayer,N. & Nibelle,J. (1969) Sodium space in freshwater and seawater eels. Comp.Biochem. Physiol. 31, 589-597.

22. Mayer,N. & Nibelle,J. (1970) Kinetics of the mineral balance in the eel *Anguilla anguilla* in relation to external salinity changes and intravascular saline infusions. Comp.Biochem. Physiol. 35, 553-566.

23. Mazeaud,M. (1964) Influence de divers facteurs sur l'adrénalinémie et la noradrénalinémie de la Carpe. C.R.Soc.Biol. 158, 2018-2021.

24. Motais,R. (1967) Les mécanismes d'échanges ioniques branchiaux chez les téléostéens. Ann.

Inst.Océanog.(Monaco) 45, 1-83.

25. Motais,R., Garcia-Romeu,F. & Maetz,J. (1966) Exchange-diffusion and euryhalinity in Teleosts J.Gen.Physiol. 50, 391-422.

26. Motais,R. & Isaia,J. (1972) Evidence of an effect of ouabain on the branchial sodium-excreting pump of marine teleosts: interaction between the inhibitor and external Na and K. J. exp.Biol. 57 (in the press).

27. Motais,R., Isaia,J., Rankin,J.C. & Maetz,J. (1969) Adaptive changes of the water permeability of the teleostean gill epithelium in relation to external salinity. J.exp.Biol.51, 529-546.

28. Motais,R. & Maetz,J. (1965) Comparaison des échanges de sodium chez un téléostéen euryhalin (le Flet) et sténohalin (le Serran). Importance relative du tube digestif et de la branchie dans ces échanges. C.r.hebd.Séanc.Acad.Sci. Paris 261, 532-535.

29. Nakano,T. & Tomlinson,N. (1967) Catecholamine and carbohydrate concentrations in rainbow trout (Salmo gairdnerii) in relation to physical disturbance. J.Fish.Res.Bd.Canada 24, 1701-1715.

30. Nilsson,S. & Fänge,R. (1967) Adrenergic receptors in the swimbladder and gut of a teleost (Anguilla anguilla). Comp.Biochem.Physiol. 23 661-664.

31. Nilsson,S. & Fänge,R. (1969) Adrenergic and cholinergic vagal effects on the stomach of a teleost (Gadus morhua). Comp.Biochem.Physiol. 30, 691-694.

32. Östlund,E. & Fänge R. (1962) Vasodilatation by adrenaline and noradrenaline and the effects of some others substances on perfused fish gills. Comp.Biochem.Physiol. 5 307-309.

33. Pic,P. (1972) Hypernatrémie consécutive à la administration d'adrénaline chez Mugil capito (Téléostéen Mugilides) adapté à l'eau de mer. C.R. Sé.Soc.Biol. 166, 131-136.

34. Pickford,G.E., Srivastava,A.K., Slicher,A.M. & Pang,P.K.T. (1971) The stress response in the abundance of circulating leucocytes in the killfish, Fundulus heteroclitus. II. The role of catecholamines. J.exp.Zool. 177, 97-108.

35. Potts,W.T.W. & Evans,D.H. (1967) Sodium and chloride balance in the killfish Fundulus heteroclitus. Biol.Bull.Wood's Hole 133, 411-425.

36. Potts,W.T.W., Foster,M.A., Rudy,P.P. & Parry Howells,G. (1967) Sodium and water balance in the cichlid teleost, Tilapia mossambica. J. exp.Biol. 47, 461-470.

37. Rajerison,R.M., Montegut,M., Jard,S. & Morel,F. (1972) The isolated frog skin epithelium: presence of α and β adrenergic receptors regulating active sodium transport and water permeability. Pflügers Arch. 332, 313-331.

38. Randall,D.J., Baumgarten,D. & Malyusz,M.(1972) The relationship between gas and ion transfer across the gills of fishes. Comp.Biochem. Physiol. 41A, 629-637.

39. Rankin,J.C. & Maetz,J. (1971) A perfused teleostean gill preparation: vascular actions of neurohypophysial hormones and catecholamine. J.Endocr. 51, 621-635.

40. Richards,B.D. & Fromm,P.O. (1969) Patterns of blood flow through filaments and lamellae of isolated-perfused rainbow trout (Salmo gairdneri) gills. Comp.Biochem.Physiol. 29, 1063-1070.

41. Richards,B.D. & Fromm,P.O. (1970) Sodium uptake by isolated-perfused gills of rainbow trout (Salmo gairdnerii). Comp.Biochem.Physiol. 33, 303-310.

42. Steen,J.B. & Kruysse,A. (1964) The respiratory function of teleostean gills. Comp.Biochem. Physiol. 12, 127-142.

43. Stevens,E.Don (1968) The effect of exercise on the distribution of blood to various organs in rainbow trout. Comp.Biochem.Physiol.25, 615-

625.

44. Tanguy,R. (1970) Ensembles électroniques destinés à l'étude de l'osmorégulation d'animaux aquatiques. Bull.Inf.Sci.Techn.-CEA 144, 11-15.

45. Watlington,C.O. (1969) α adrenergic inhibition of Na^+ transport: the interaction of vasopressin and 3'-5'-AMP. Biochim.Biophys.Acta 193, 394-402.

TABLE 7

Water efflux (ml.hr^{-1}.100g^{-1})

Effect of epinephrine and of α and β blocking agents

Treatment	n	Body weight (g)	Flux periods			Paired differences		
			I	II	III	Δ (II – I)	Δ (III – I)	Δ (III – II)
Solvent	6	25.2 ± 4.3	24.4 ± 3.9	28.3 ± 2.6	29.7 ± 3.0	+3.9 ± 3.3 n.s.	+5.3 ± 2.6 n.s.	+1.4 ± 2.2 n.s.
Epinephrine*	8	18.3 ± 1.9	34.7 ± 5.5	60.5 ± 7.1	57.1 ± 6.2	+25.8 ± 3.1 $P < 0.001$	+22.3 ± 5.6 $P < 0.001$	–3.4 ± 3.0 n.s.
Epineph** + α blocker°	8	17.5 ± 3.1	38.6 ± 3.9	63.1 ± 4.4	68.8 ± 6.8	+24.5 ± 5.6 $P < 0.01$	+30.2 ± 6.0 $P < 0.01$	+5.7 ± 9.4 n.s.
Epineph** + β blocker°°	7	8.0 ± 0.6	52.3 ± 6.3	52.9 ± 5.0	50.2 ± 7.8	+0.6 ± 5.3 n.s.	–2.1 ± 3.5 n.s.	–2.7 ± 6.8 n.s.

* Dose 20 μg/fish
** 5 μg/fish
° Phentolamine 200 μg or Tolazoline 1 mg
°° Propanolol 100 μg
Periods as in table 1.

Comparative Physiology, eds. L. Bolis, K. Schmidt-Nielsen and S. H. P. Maddrell

ON THE AUTONOMIC NERVOUS CONTROL OF ORGANS IN TELEOSTEAN FISHES

S. NILSSON
Department of Zoophysiology
University of Göteborg, Sweden

The common picture of the autonomic nervous system obtained from mammals involves a sympathetic and a parasympathetic part. The sympathetic nerves leave the central nervous system at the thoraco-lumbar level, and the parasympathetic fibres originate from the cranial and sacral portions of the CNS. The two systems are in several cases antagonistic on the organ they innervate, e.g. the heart is accelerated by sympathetic nervous activity but inhibited by parasympathetic activity.

The parasympathetic ganglia are situated close to the effector organ, while the cell bodies of the sympathetic post-ganglionic fibres are found in the paravertebral sympathetic ganglion chain or in collateral ganglia. The sympathetic chain receives preganglionic fibres from the spinal cord via the white rami communicantes. Some of the postganglionic fibres do not continue independently to the effector organ, but re-enter and run with the spinal nerves via the grey rami communicantes.

In lower vertebrates, however, the situation is less clear. Parasympathetic fibres in teleosts are found in the oculomotor and vagus nerves,while in elasmobranchs facial and glossopharyngeal fibres also exists (4,16,24,25). No sacral parasympathetic outflow occurs in these groups (4).

Sympathetic paravertebral ganglia are found in elasmobranchs, but they are not connected to form a distinct sympathetic chain (4,16,25). In teleosts, however, there are well developed sympathetic ganglionic chains which make connections with the cranial nerves, sending sympathetic fibres to run in them (4,16,24). Adrenergic fibres, which have been demonstrated in the vagus, may be of sympathetic origin (9,18,21). A splanchnic nerve runs from the two first sympathetic ganglia on the right side (23,24) and carries sympathetic cholinergic (3) as well as adrenergic fibres (19) to the viscera. Excitatory sympathetic fibres to the urinary bladder were demonstrated by Young (26).

Double antagonistic innervation of teleost viscera was not demonstrated until recently, when Gannon and Burnstock (10) provided some evidence for the existence of cholinergic inhibitory as well as adrenergic excitatory fibres to the trout heart.

In the Department of Zoophysiology, University of Göteborg, Sweden, studies on the autonomic innervation of various organs in teleost fish have been going on for some years. Among other things, we have investigated the mode of innervation of the uro-genital system, and parts of the vascular system, and some of these results will be briefly outlined here. All experiments described here are made with the cod, _Gadus morhua._

VASCULAR SYSTEM

Experiments with isolated-perfused vascular beds have been carried out in the gas gland (19) and spleen (6), and an adrenergic vasoconstrictor innervation is thought to be responsible for the regulation of the vessels in these organs. Preliminary results have also been obtained concerning the autonomic nervous regulation of the vascular resistance of the gill vessels. The latter studies are presented below.

The gill arches receive autonomic innervation from the branchial branches of the vagus nerve, and sympathetic fibres to the gills are also thought to run in these. In our experiments the afferent and efferent arteries of one isolated gill arch were catheterized, and the gill vessels were perfused at a constant flow rate, the perfusion pressure being recorded. When the vagus branch was stimulated electrically, an increase in perfusion pressure was recorded, reflecting an increased vascular resistan ce (Fig. 1). The nature of this innervation has not been studied so far, nor has the pattern of blood flow through the gill filaments.

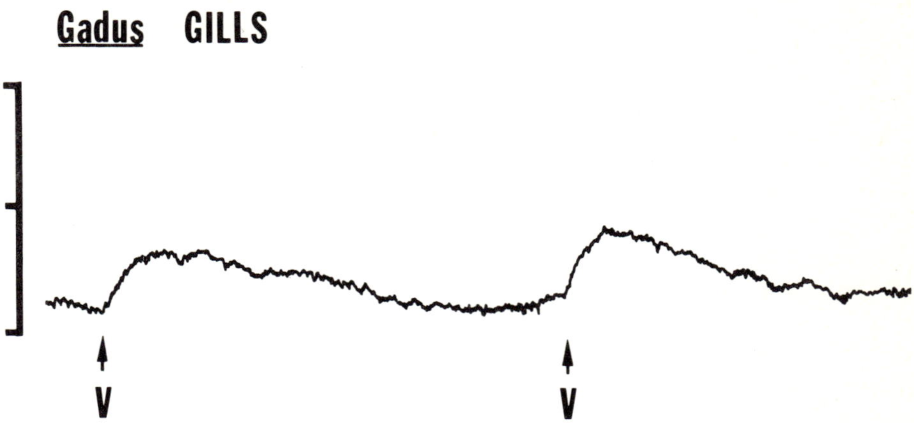

Figure 1 - *Gadus morhua*. One isolated gill arch perfused with constant flow rate, the perfusion pressure being recorded. A rise in perfusion pressure reflecting an increased vascular resistance can be seen after stimulation of the branchial branches of the vagus (v).

In experiments with free-swimming, un-anesthetized cods, the blood pressure was recorded pre- and post-branchially (11,12). Injection of the beta-receptor blocking agent propranolol caused an elevation of the pre-branchial and a decrease in the post-branchial blood pressure. This may be explained by assuming that propranolol decreases the influence of an adrenergic vasomotor innervation or of circulating catecholamines which normally act by decreasing vascular resistance (11, 12). Decreased vascular resistance in the branchial vessels after adrenaline treatment has been demonstrated (27) and may be due to a beta-adrenergic effect (22).

It may perhaps be concluded from these highly preliminary data that the gill vessels in the cod are regulated by vasoconstrictor fibres, the nature of which is not yet known, and by a beta-adrenergic vasodilator mechanism, either nervous or due to circulating catecholamines.

URINARY BLADDER

The urinary bladder of the cod consists of the fused ureters forming a single sac. This is divided into two lobes, one on each side of the rectum (Fig. 2). The vesicular nerve originates from the sympathetic chains, and runs along the ureters, branching over the bladder. After anastomoses and plexus formations, part of the nerve fibres running in the vesicular nerve are directed to the gonads (Fig. 2).

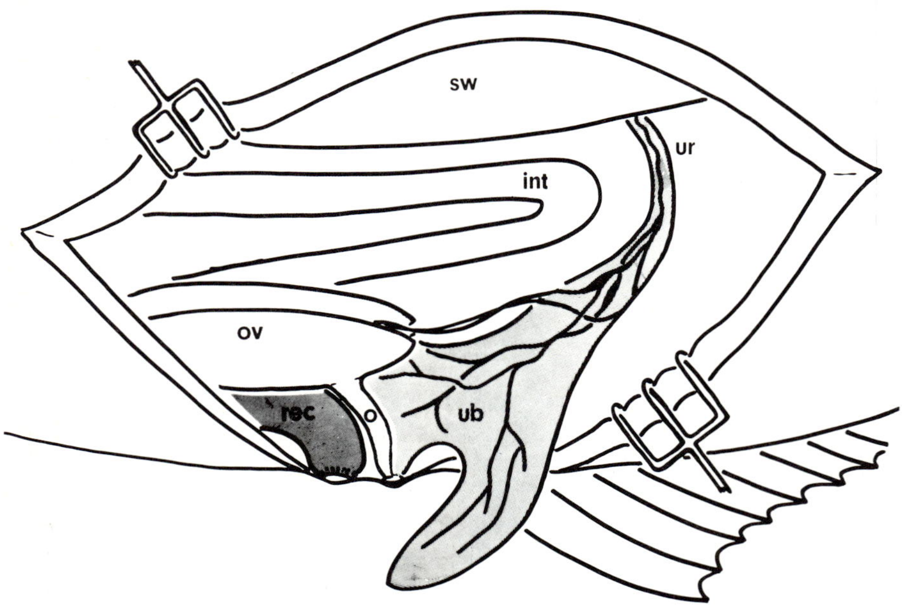

Figure 2 - Gadus morhua. The body wall has been opened laterally, exposing the urinary bladder and gonads. Legend: int, intestine; o, oviduct; ov, ovary; rec, rectum; sw, swimbladder; ur, ureters, ub, urinary bladder.

Contraction of the urinary bladder and gonads can be produced by electrical stimulation of the sympathetic chains *in situ* or the vesicular nerve *in vitro*. The contractions produced in this way are rapidly abolished if atropine is present in the saline bathing the organs. In the urinary bladder, a small relaxation is seen during each stimulation after atropinization, and this effect is blocked by adrenergic beta receptor blocking agent propranolol. Contractions can be produced by acetylcholine and relaxation by noradrenaline or adrenaline in both the urinary bladder and gonads, and these effects are abolished by atropine and propranolol respectively. Contractions in the urinary bladder produced by electrical stimulation are diminished or abolished by the ganglionic blocking agent hexamethonium, if the vesicular nerve is stimulated far from the lobes of the bladder, while no effect or a very small effect is seen with this drug if the vesicular nerve is stimulated close to the lobes.

The results obtained from the experiments suggest that an excitatory cholinergic innervation originating from the sympathetic chains is responsible for the contractions seen during nerve stimulation. Adrenergic fibres following the same path from the sympathetic chains cause the relaxation occurring when the effect of the excitatory fibres is blocked by atropine. Cholinergic synapses along the vesicular nerve are suggested by the fact that hexamethonium blocks the effect of nerve stimulation when this is carried out at some distance from the lobes of the bladder (17).

Further evidence for the existence of an adrenergic innervation in the urinary bladder has been obtained by the Falck-Hillarp histochemical method (20). Varicose fibres in the bladder wall intermingling with smooth muscle bundles have been detected, and fluorescent fibres have also been observed surrounding small blood vessels (20). The catecholamine of the fluorescent fibres appears to be noradrenaline.

The existence of excitatory sympathetic fibres to the urinary bladder was demonstrated in *Uranoscopus* by Young (26). The nature of the transmitter however, was not clear but may have been acetylcholine. Non-cholinergic excitatory fibres to the urinary bladder have been suggested in a number of vertebrates (4), but the results from the cod indicate that the excitatory fibres are purely cholinergic (17). In lower vertebrates, cholinergic excitatory fibres have also been demonstrated in the urinary bladder of amphibians (1,2,5,15) and reptiles (4).

Adrenergic fibres in the urinary bladder have also previously been demonstrated in amphibians (13,15) and reptiles (4,14) but are inhibitory only in reptiles and higher vertebrates. The results from cod indicate the presence of inhibitory adrenergic fibres (17,20), and thus the urinary bladder of this fish receives a double antagonistic innervation.

CONSLUSIONS

The existence of inhibitory innervation in the cod urinary bladder does not agree with the view that this kind of innervation is a late evolutionary development (4). Furthermore, the stomach of the cod is regulated in part by vagal excitatory fibres which are possibly cholinergic (21). It has been suggested that the primitive vagus supply to the stomach is non-adrenergic inhibitory (4,7,8) and that the excitatory response seen after vagal stimulation in fishes is due to a rebound contraction. In the cod the contraction seen during electrical stimulation is primary, starting immediately after the start of the stimulation (21). It cannot, however, be excluded that the cholinergic excitatory fibres in the cod vagus are sympathetic,coming from the sympathetic chains (4,9,24).

Vasoconstriction in the cod appears to take place mainly by noradrenergic fibres (6,11,12,19), although effects of circulating catecholamines cannot be excluded. Acetylcholine has only a very weak effect,if at all,on the vessels of the perfused gas gland (19) or spleen (6) of the cod. Adrenergic fibres have been demonstrated by fluorescent histochemistry in arteries of the cod (20), and such fibres may be of greater importance in regulating blood pressure in the cod than in other species, which have a mainly cholinergic vasoconstrictor innervation, which has been considered to be dominant in fishes (4).

It is therefore concluded that from an "autonomic nervous" point of view, the cod is a rather highly developed animal, which in some respects may resemble higher vertebrates. This strongly indicates that much more work is necessary on different species of teleostean fishes, to obtain a true picture of this interesting group of animals.

REFERENCES

1. Bell,C. (1967) A histochemical study of the esterases in the bladder of the toad (*Bufo marinus*). Comp.Biochem.Physiol. *21*, 91-98.
2. Bell,C. & Burnstock,G. (1965) Cholinesterases in the bladder of the toad (*Bufo marinus*). Biochem.Pharmacol. *14*, 78-79.
3. Burnstock,G. (1958) The effect of drugs on spontaneous motility and on the response to stimulation of extrinsic nerves of the gut of a teleostean fish. Brit.J.Pharmacol.Chemother. *13*, 216-226.
4. Burnstock,G. & Wood,M.E. (1967) Innervation of the urinary bladder of the sleepy lizard (*Trachysaurus rugosus*). II. Physiology and Pharmacology. Comp.Biochem.Physiol. *20*,675-690.
5. Burnstock,G., O'Shea,J. & Wood,M. (1963) Comparative physiology of the vertebrate autonomic system. I. Innervation of the urinary bladder of the toad (*Bufo marinus*). J.Exp.Biol. *40*,403-420.
6. Börjesson,M. & Nilsson,S. (1971) Unpublished observations.
7. Campbell,G. (1970) The autonomic nervous system. In *Fish Physiology* vol.IV, pp.109-132, ed. Hoar W.S. & Randall,D.J., New York & London Academic Press.
8. Campbell,G. & Burnstock,G. (1968) The comparative physiology of gastro-intestinal motility,

In Handbook of Physiology, Sect.6, vol.IX, pp. 2213-2266, ed. Code,C.F., Washington American Physiological Society.

9. Fänge,R. (1953) The mechanism of gas transport in the euphysoclist swimbladder. Acta Physiol. Scand. 30, suppl. 110, 1-333.

10. Gannon,B. & Burnstock,G.(1969) Excitatory adrenergic innervation of the fish heart. Comp. Biochem.Physiol. 29, 765-773.

11. Helgason,S.S. & Nilsson,S. (1972) Drug effects on blood pressure and heart rate in a free-swimming fish. Acta physiol.scand. Abstract in press.

12. Helgason,S.S. & Nilsson,S., Drug effects on pre- and post-branchial blood pressure and heart rate in a free-swimming marine teleost, *Gadus morhua*. Manuscript in preparation.

13. McLean,J.R. & Burnstock,G. (1966) Histochemical localization of catecholamines in the urinary bladder of the toad (*Bufo marinus*). J.Histochem. Cytochem. 14, 538-548.

14. McLean,J.R. & Burnstock,G. (1967) Innervation of the urinary bladder of the sleepy lizard (*Trachysaurus rugosus*). I. Fluorescent histochemical localization of the catecholamines. Comp. Biochem.Physiol. 20, 667-673.

15. McLean,J.R., Bell,C. & Burnstock,G. (1967) Histochemical and pharmacological studies on the innervation of the urinary bladder of the frog (*Rana temporaria*). Comp.Biochem.Physiol. 21, 383-392.

16. Nicol,J.A.C. (1952) The autonomic nervous system in lower chordates. Biol.Rev. (Cambridge) 27, 1-49.

17. Nilsson,S.(1970) Excitatory and inhibitory innervation of the urinary bladder and gonads of a teleost, *Gadus morhua*. Comp.gen.Pharmac. 1, 23-28.

18. Nilsson,S. (1971) Adrenergic innervation and drug responses of the oval sphincter in the swimbladder of the cod (*Gadus morhua*). Acta

physiol.scand. 83, 446-453.

19. Nilsson,S.- Autonomic vasomotor innervation in the gas gland of the swimbladder of a teleost (Gadus morhua). Comp.gen.Pharmac. - In press.

20. Nilsson,S.- Fluorescent histochemistry of the catecholamines in the urinary bladder of a teleost (Gadus morhua). Comp.gen.Pharmac. - In press.

21. Nilsson,S. & Fänge,R. (1969) Adrenergic and cholinergic vagal effects on the stomach of a teleost (Gadus morhua). Comp.Biochem.Physiol. 30, 691-694.

22. Rankin,J.C. & Maetz,J.(1971) A perfused teleostean gill preparation: vascular actions of neurohypophysial hormones and catecholamines. J.Endocr. 51, 621-635.

23. Stannius,H. (1849) Das peripherische Nervensystem der Fische. Rostock, Stiller.

24. Young,J.Z. (1931) On the autonomic nervous system of a teleostean fish Uranoscopus scaber. Quart.J.Microscop.Sci. 74, 491-535.

25. Young,J.Z. (1933) The autonomic nervous system of selachians. Quart.J.Microscop.Sci. 75,571-624.

26. Young,J.Z. (1936) The innervation and reactions to drugs of the viscera of teleostean fish. Proc.Roy.Soc. Ser B. Biol.Sci. 120, 303-318.

27. Östlung,E. & Fänge,R. (1962) Vasodilation by adrenaline and noradrenaline and the effects of some other substances on perfused fish gills. Comp.Biochem.Physiol. 5, 307-309.

Comparative Physiology, eds. L. Bolis, K. Schmidt-Nielsen and S. H. P. Maddrell

RECTAL UPTAKE OF WATER IN INSECTS

J. NOBLE-NESBITT
School of Biological Sciences,
University of East Anglia,
University Plain,
Norwich, England.

INTRODUCTION

The rectum is the drying plant of the insect. One of its main functions is the extraction of water from the gut contents before they are voided to the exterior through the anus, as Wigglesworth pointed out in 1932 in his study of the function of the rectal glands (26). The urine, produced by the Malpighian tubules and mingling with the midgut contents as both enter the hindgut, is concentrated as it moves down the gut, the final, all-important stage in this drying process being carried out within the rectum. Here, the faeces are dried down, often to a fine, hydroscopic powder or hygroscopic pellet. These remarks apply especially to those terrestrial insects which in their natural environments find themselves in situations of water stress where conservation of water assumes great importance. Such are the insects which inhabit particularly dry areas,including stored products, deserts and the arid conditions of boiler houses, where free water is often unavailable. These facts have been appreciated for some considerable time. To this aspect of the function of the rectum must now be added the absorption of water vapour from subsaturated atmospheres, an involvement first demonstrated in the firebrat (*Thermobia domestica*) and the mealworm larva (*Tenebrio molitor*) (10,11).

In recent years, the insect rectum has come under close scrutiny, and much progress has been made in the determination of its ultrastructure,and in the relation of this to its physiological properties. Much of the impetus for the ultrastructural investigations derived from the elegant experimental work of Phillips on the locust (14-17) and blowfly (14,18), which first provided clear, unequivocal evidence of the transport of water from the rectal lumen to the haemocoele, without concomitant net transfer of solute. These studies have allowed a more precise analysis of the possible mechanisms for this water transport. Two excellent recent reviews have put together much of this work (8,21). It is my intention here to outline the significant aspects of these studies, and to deal in some detail with the relevant aspects of my own work with *Thermobia* and *Tenebrio*.

The Rectal Systems

One fact which these studies have clearly revealed is that there is no single structural and physiological rectal system applicable to all insects. At least three main systems can be recognised: first, the rectum having rectal papillae,pads or glands (the papillate rectum); secondly, the rectum with associated cryptonephric tubules (the crytonephric rectum), which in common with the papillate rectum has a subepithelial, or perirectal, space; and thirdly, the rectum with chambers (anal sacs) in its posterior region. I will consider examples of these systems in turn.

The Papillate Rectum

I include in this section such types as exemplified by the blowfly, *Calliphora* (2,3,5,6) the cockroach, *Periplaneta* (13,24,25), and the locust, *Schistocerca* (19), all of which have been closely studied in recent years by the authors indicated.

The ultrastructural arrangement of the blowfly's rectal papillae is shown in Figure 1. The modified epithelial cells of the cortex are closely joined at the apical (luminal) and basal extremities of their lateral plasma membranes by tight junctions and septate desmosomes which serve to iso

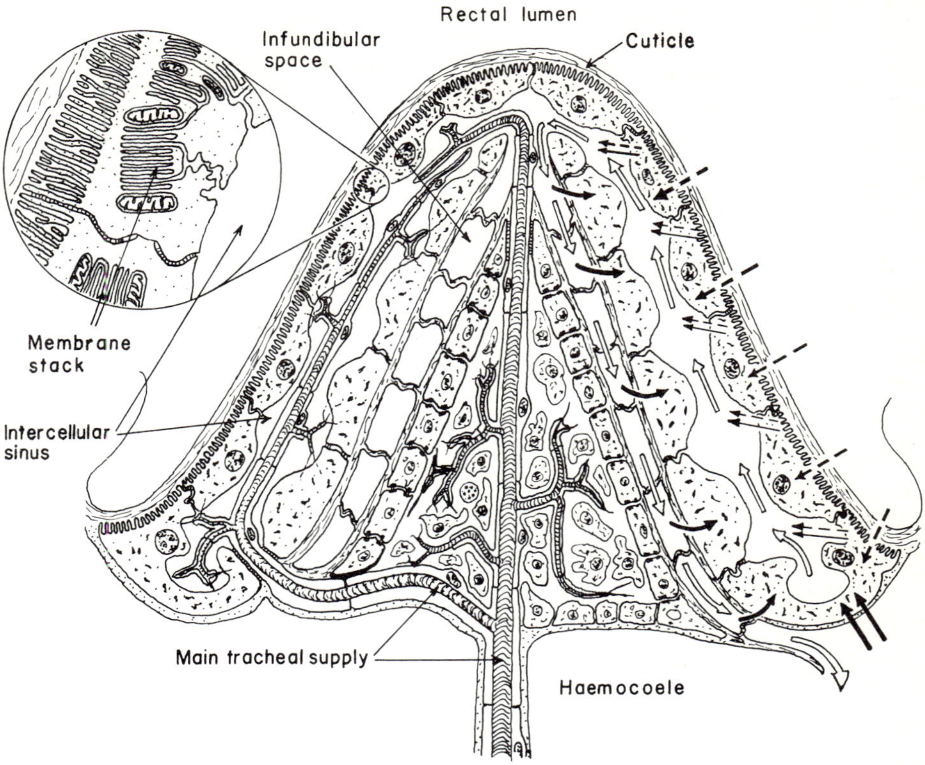

Figure 1 - Diagrammatic representation of the ultrastructural organisation of the rectal papilla of *Calliphora*, showing the postulated ion movements linked to the absorption of water from the lumen. Solid continuous arrows represent active movements of solute; solid broken arrows represent passive movements of water; and open continuous arrows represent the flow of fluid along the intercellular spaces (based on 6 and 8).

late the rectal lumen from the system of sinuses within and below the epithelium.

Over the rest of their length, the lateral plasma membranes are more widely spaced, forming channels which open into the sinuses. The tracheal supply penetrates into these sinuses, which are confluent throughout the epithelium, and run into the system of subepithelial spaces between medulla and cortex, and thence through a non-return flap-valve arrangement into the haemocoele. The lateral plasma membranes are further thrown into pleated stacks, with mitochondria in close association. A magnesium-activated ATPase is located in the stacks on the cytoplasmic surface of the membranes. The stacks are ideally suited for active ion transport into the inter-cellular spaces creating high osmotic gradients, with subsequent transport of water into the spaces causing a mass flow of fluid into the sinuses and so into the haemocoele. In common with most rectal epithelial cells, the apical plasma membrane is thrown into plicate folds, and is invested with a coating of particles on its cytoplasmic aspect.

The rectal pads of the cockroach have a basically similar arrangement, as shown in Figure 2. Again, channels formed between the epithelial cells in the rectal pads drain into a subepithelial (perirectal) space, which has limited points of connection into the haemocoele. In the rectal glands of the locust, a very similar arrangement is again found.

The working of these systems implies the production of a hyperosmolar fluid in the intercellular channels, and analysis of this fluid in the cockroach has shown that this occurs (25). During periods of rapid uptake, dilation of these spaces may be expected, and this has been observed in the cockroach by the same workers.

The operation of the mechanism for water transfer as suggested in these instances requires that ions should be made available continuously for recruitment at the luminal face of the specialised cells, or at least in the immediate vicinity of the pleated lateral plasma membranes. As pointed out earlier, Phillips has shown that water transport occurs in the absence of net solute transport from lumen to haemocoele. More recently, Irvine and Phillips (7), using selective metabolic inhibitors and ouabain, have demonstrated the continued transport of water in the absence of net transport of ions, and even in the face of a reversed net transport of ions.

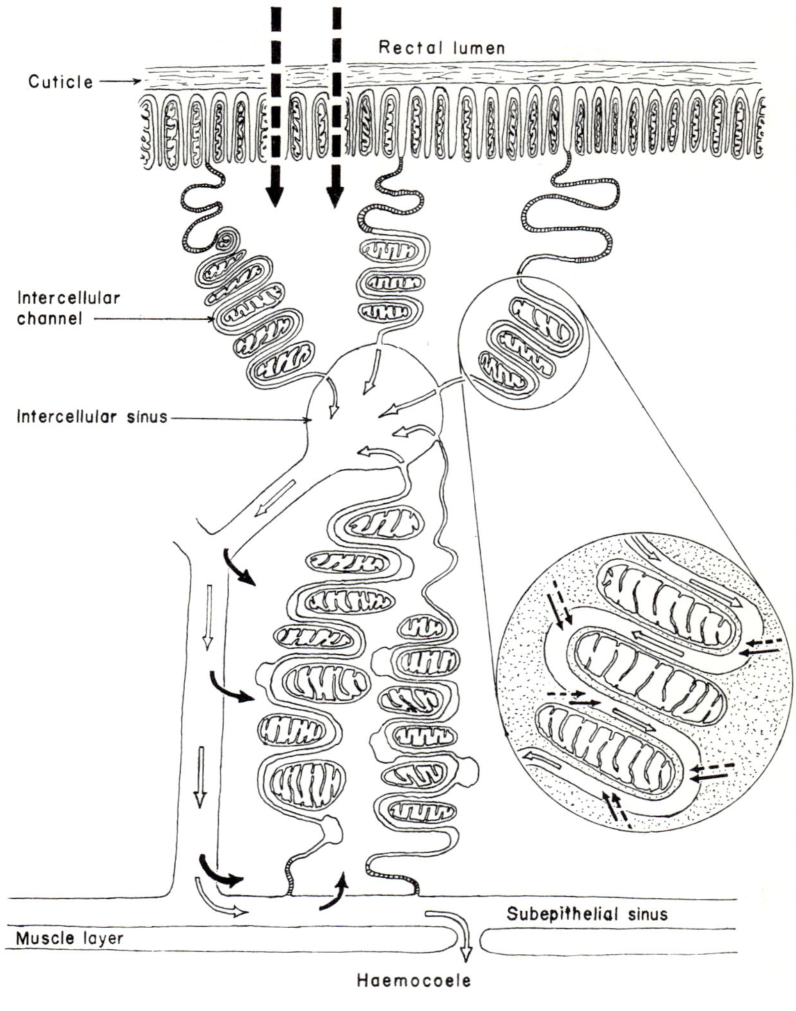

<u>Figure 2</u> - Diagrammatic representation of the ultrastructural organisation and suggested functioning of the rectal pad of <u>Periplaneta</u>; a scheme which would apply generally to <u>Calliphora</u> and <u>Schistocerca</u> also. Solid continuous arrows represent active movements of solute; solid broken arrows represent passive movements of water; and open continuous arrows represent the flow of fluid along the intercellular spaces (based on 13 and 8).

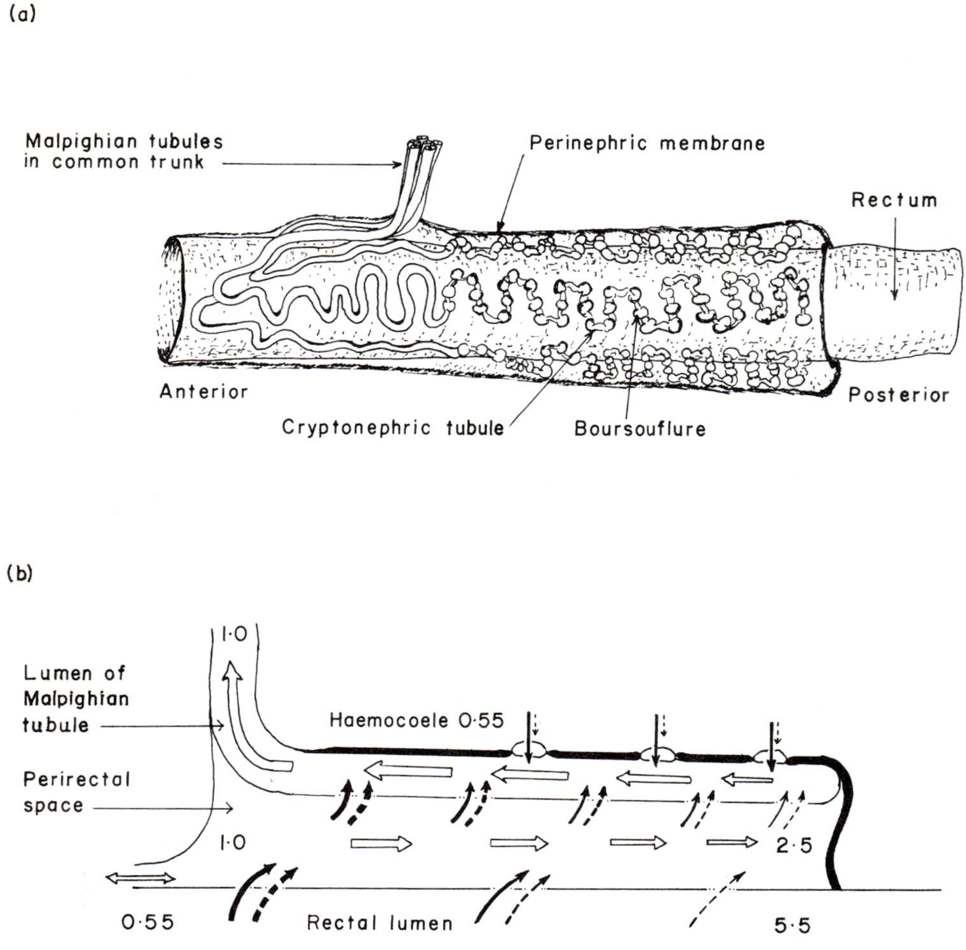

Figure 3 - The rectal complex of *Tenebrio.* (a) Diagram showing the cryptonephridial arrangement of the Malpighian tubules round the posterior part of the rectum. (b) Model proposed for the mechanism involved in the transfer of water from the rectal lumen, showing the functionally important compartments with the osmolarities of their contents. Solid continuous arrows represent active movements of solute; solid broken arrows represent passive movements of water; and open continuous arrows represent the flows of fluid in the compartments (based on 4).

This does not rule out the possibility of local recycling within the epithelium, as pointed out by Stobbart (23), and Irvine and Phillips favour this explanation for their results. Wall and Oschman recognised this need for epithelial recycling of ions and incorporated it into the scheme which they proposed for the cockroach. The planar nature of the membranes lining the channels at the point of re-entry would facilitate the transport of ions without linked water movements in iso-osmotic proportions.

The Cryptonephric Rectum

The rectal epithelium of the mealworm larva, Tenebrio, is relatively unspecialised, with none of the elaborations of the lateral plasma membranes as seen in the papillate rectum. The main elaborations in this system occur in the distal ends of the Malpighian tubules which are hidden within a membranous bag surrounding the rectum. The arrangement as described by Ramsay (20), and Grimstone, Mullinger and Ramsay (4) is shown in Figures 3 and 4. Gut fluid entering the rectum is isosmolar to the haemolymph (0.55 osmol/l). The fluid leaving the cryptonephric tubules at the anterior end, iso-osmolar to the perirectal fluid at that point, is hyperosmolar to the haemolymph (1.0 osmol/l), as is perirectal fluid escaping into the haemolymph. At the posterior end, the osmolarity of the perirectal fluid is even higher (2.5 osmol/l). At this point, the rectal contents have become markedly hyper-osmolar (5.5 osmol/l), and are contained in air spaces with humidities well below saturation. Two concentrating systems in series are apparently in operation. Active transport of ions from the haemolymph to the lumen of the cryptonephric tubule via the leptophragmata, without water following in iso-osmotic proportions, could account for the hyper-osmolarity of the tubule contents required to bring about transfer of water from the perirectal space, though Ramsay (21) doubts the capacity of the cells involved to perform the task assigned to them. Diffusion of ions from the tubule into the perirectal space would account for the gradient of osmo-

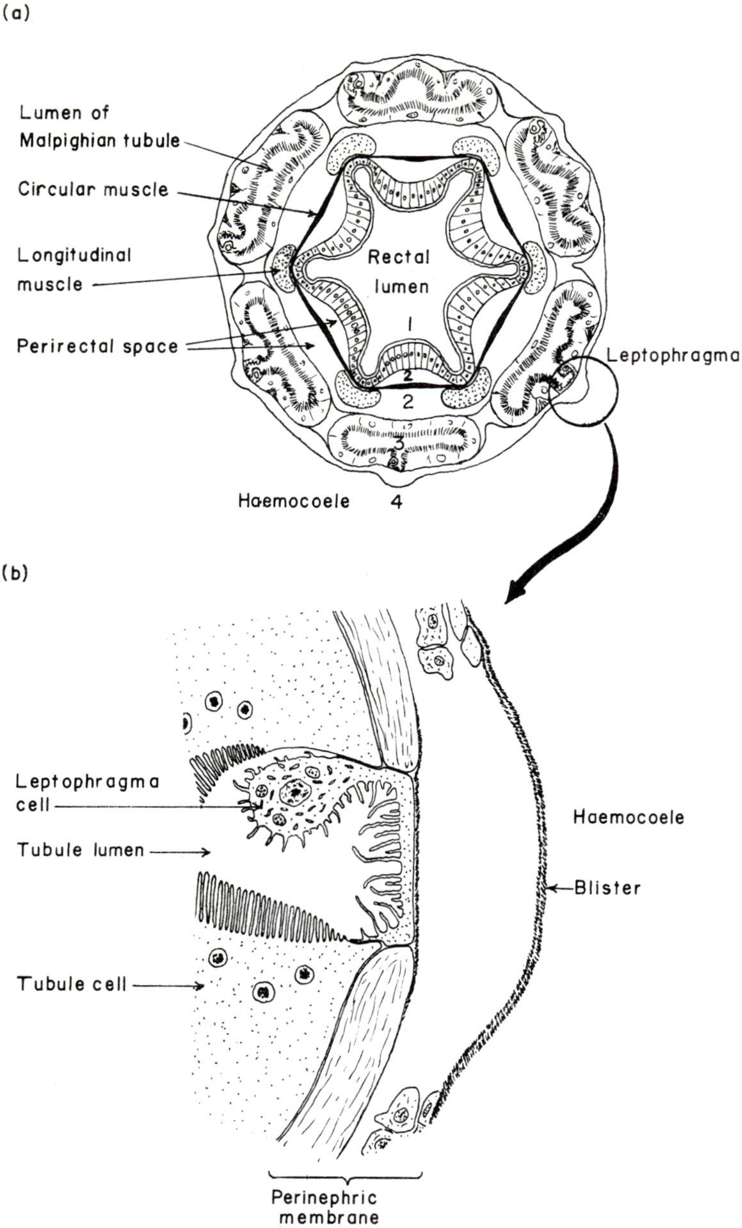

Figure 4 - The rectal complex of Tenebrio.
(a) Diagrammatic transverse section of the cryptonephridial system.
(b) Detail of the ultrastructural arrangement of a leptophragma (based on 20 and 4 and 8).

larity along its length, and for transfer of water from the anterior end of the rectum. The large osmotic gradient between rectal contents and perirectal fluid at the posterior end reflects a transport of water by the rectal epithelium by some method at present unknown. We shall have cause to return to this intriguing problem, in relation to the absorption of water vapour from subsaturated atmospheres.

Rectum with Anal Chambers (Sacs)

In the firebrat, *Thermobia*, a specialised rectal epithelium with no clear sub-epithelial (perirectal) space or elaborations of the lateral plasma membranes, is found in chambers in the posterior, anal region of the rectum (10-12). Its involvement in the uptake of water from subsatured atmospheres was demonstrated in experiments utilising molten wax to block the anus which abolished uptake. This conclusion was corroborated by similar results in *Tenebrio*; in this case uptake could recommence following the early removal of the wax cap over the anus (Table 1).

Table 1 - Percentage of treated *Tenebrio* showing uptake before and after removal of anal capping, for varying periods of capping. The figures in parentheses indicate the percentage of insects showing both evident secondary blockage effects and failure of uptake, probably as a result of this secondary blockage; note the increase with the length of time the capping remains in position.

% Showing Uptake		Day of Removal of Block
Before Removal of Block	After Removal of Block	
0	80 (20)	6
0	20 (60)	10
0	0 (100)	15

The reversible nature of this effect has now been confirmed in further experiments on Thermobia using a preparation in which it is possible to expose the anterior and posterior parts of the insect's body separately to atmospheres of different humidity.Full uptake by previously desiccated insects occurred only when the posterior end was exposed to an atmosphere of high humidity. In insects failing to take up water when only the anterior parts of the body were exposed to high humidity,uptake subsequently occurred when the posterior end was then exposed to high humidity (Figure 5).

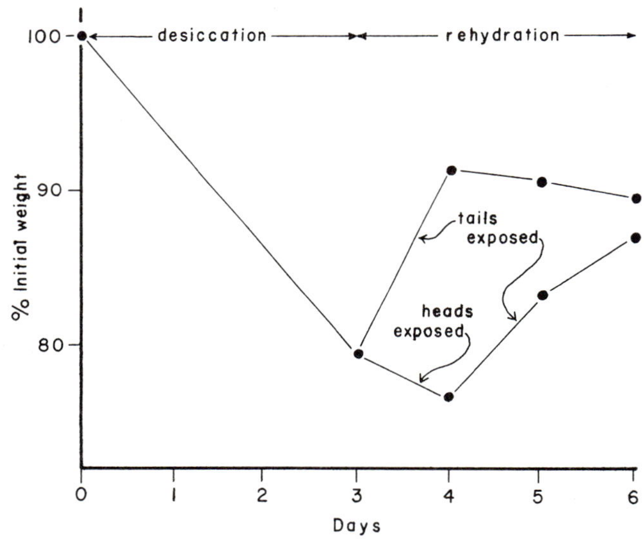

Figure 5 - Effects of exposing head end or tail end alone to high humidity on subsequent rehydration of previously desiccated Thermobia. Experiments conducted at 37°C; desiccation over granular calcium chloride; rehydration in 83% R.H. Mean values for the weights of the insects are plotted and are expressed as per cent initial weight against time in days. The lines joining the mean values indicate the time-course of weight change.

In their recent study of the sensory structures of the thysanuran Ctenolepisma lineata pilera, Slifer and Sekon (22) mention that this insect is not seen to drink water from the supply provided in the culture dishes, but is often seen orientated with its tail near to the water source. This interesting behavioural observation suggests again that water vapour uptake may occur through the anus and rectum.

In Thermobia, the anal chambers of the posterior rectum open to the exterior through a roughly triangular opening closed by three valves, one dorsal and two ventro-lateral. When observed in desiccating conditions, the aperture is seen to be firmly closed by the valves. When the insect is subsequently placed in hydrating conditions, incipient movements of the valves are soon followed by larger movements, culminating in cycles of wide-opening and closing. These brief cycles are regularly and rapidly repeated. In well hydrated insects the cycles are less regular and less frequent. It is clear that these pulsating movements, regularly and rapidly repeated, cause a continuous replenishment of the air in contact with the lining of the anal chambers. This tidal ventilation will aid diffusion between the atmosphere and the rectal epithelium. Calculations based on the estimated volume of the rectal chambers show that the maximum recorded rate of uptake required the replenishment of the air in the chambers on average every 1-2 seconds. This agrees closely with the duration of the cycles of opening and closing which occur approximately every 1-2 seconds.

Investigations of the ultrastructure of the rectum of Thermobia have indicated the specialised nature to be expected of an epithelium invested with the physiological properties of the magnitude exhibited in this case (Figure 6). The apical (luminal) face of the simple epithelium lining the anal chambers of the posterior rectum is invested with cuticle. The epithelium is thrown into folds, forming the chambers. The main feature of significance is the deeply infolded nature of the apical border of the constituent cells, associated with elongated rod-like mitochondria. The lateral and basal surfaces have simple plasma membranes. Tight

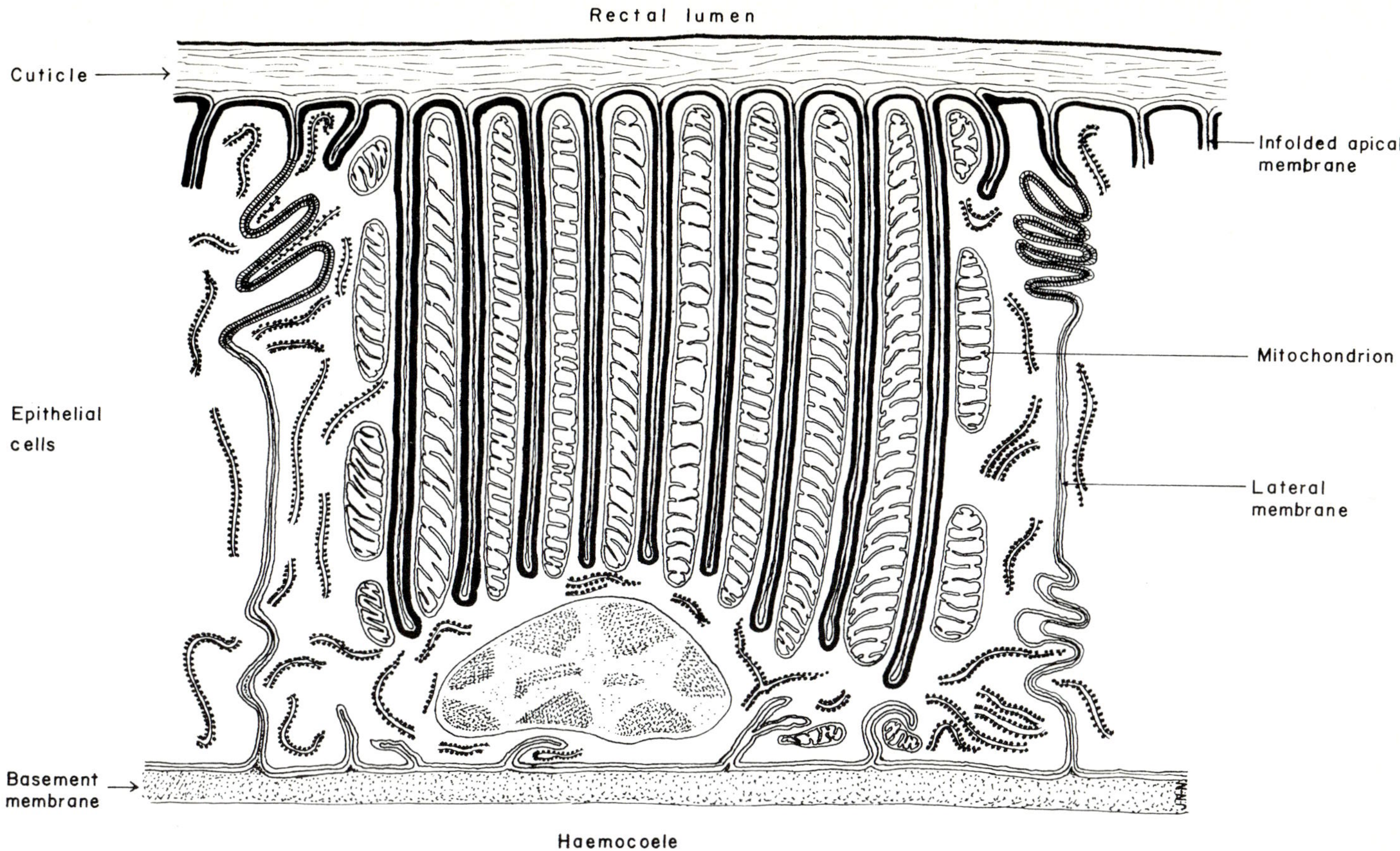

Figure 6 - Diagrammatic representation of the ultrastructural organisation of a rectal epithelial cell in the anal chambers (sacs) of *Thermobia*.

junctions, with septate desmosomes, occur near the apex between the lateral plasma membranes of adjacent cells. In common with rectal epithelial cells of other insects, a 15 nm-wide particulate coating is found on the inner, cytoplasmic side of the apical plasma membrane. This coating is in very close association with the outer membrane of the adjoining mitochondrion. In normal, hydrated insects there is no gap between neighbouring outer plasma membrane surfaces in the infoldings, except for occasional small ones, such as the base of the infolding (Figure 7)

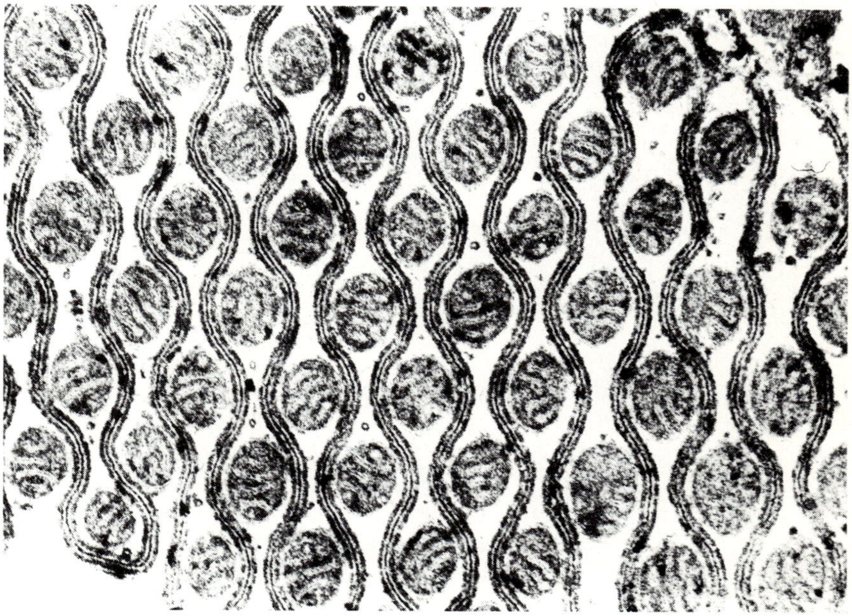

Figure 7 - Electron micrograph of a section through the apical region of an epithelial cell from the anal chambers (sacs) of *Thermobia* rectum, parallel with the luminal surface, showing the regular, hexagonal array of mitochondria, the closely-opposed apical plasma membranes in the infolds, and the particulate coating on the cytoplasmic surface of the apical plasma membrane in close association with the outer mitochondrial membranes. Normal condition.

Distinct extracellular spaces occur in the infolds in desiccated insects,however (Figure 8).

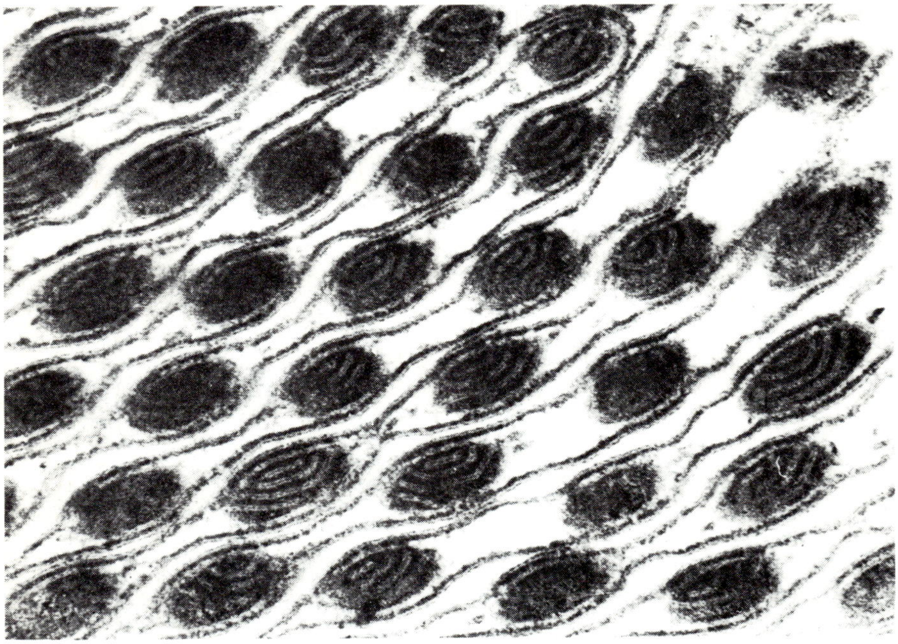

Figure 8 - Electron micrograph of a section through the apical region of an epithelial cell from the anal chambers (sacs) of *Thermobia* rectum, showing the condition found in desiccated insects,thought to be associated with osmotic swelling during fixation. Note the widely-opposed apical plasma membranes in the infolds, forming wide channels.

These spaces appear to result from swelling during fixation, and to indicate areas where water is extracted from the aqueous fixative before complete fixation has occurred. The osmotic nature of this swelling was confirmed by increasing the osmolarity of the fixative, which eliminated the extracellular spaces. This suggests that these areas have a high osmotic activity.

It would seem that the high density of mitochondria and their large surface area in close asso ciation with a large apical cell surface are important features of the mechanism for the uptake of water. The deep infoldings suggest that the gradient of water potential from the apical (luminal) side (low) to the basal (haemolymph) side (high) of the epithelium may be set up by differential activity along the channel-like infoldings.

The involvement of this specialised epithelium in uptake is further substantiated by results of ultrastructural investigations carried out during the moulting cycle. Uptake is arrested during the moult. The epithelium during this period undergoes a complete reorganisation, and the apical membrane-mitochondria complex disappears, to be reformed again as the moult reaches its conclusion.

Mechanisms of Water Transport

The transfer of water from rectal lumen to haemocoele could be accomplished either by a primary transport of water itself, or as has generally been thought in the past to be more reasonable, by a secondary transport depending upon a primary transport of solute. In the former case, a gradient of water potential would be expected in the direction of rectal lumen (low) to haemocoele (high), maintained by a transport of water in the direction rectal lumen to haemocoele. Certainly the measurements on *Tenebrio* indicate the presence of this gradient from rectal lumen (5.5 osmol/l) to perirectal space (1.0 - 2.5 osmol/l) to haemocoele (0.55 osmol/l) (4). Again, in the cockroach, this gradient from rectal lumen to perirectal space is found (24). However, as I have earlier mentioned, the epithelial channels in this case of the cockroach are hyperosmolar with respect to the rectal lumen, and secondary transport is a very definite possibility in the papillate rectum.

When we turn to consider the removal of water vapour from the atmosphere, we are struck by the enormity of the task. As illustrations, the mealworm is capable of absorbing water from atmosphere of 88% relative humidity and higher (9) and it brings its rectal air spaces down to relative humidities as low as 75% (20), whilst the firebrat can absorb water vapour when the relative humidity is as low as 45% (1). A solution of potassium chloride would be concentrated to saturation point by *Tenebrio* and to dryness by *Thermobia*. Indeed, so would be sodium chloride solution, by *Thermobia* at least. Active transport and recycling of these inorganic ions clearly does not account for the observed transport of water in these instances. Large organic molecules of the type found in the perirectal fluid of the mealworm (4) may be more adequately equipped for the job (8), but even with them there remains the problem of releasing the water entrained by them and making it available to the internal tissues, and of presenting them alternately to the rectal lumen and to the haemocoele. A valve-like arrangement, perhaps involving the cuticle, could be invoked. But there is another possibility which I have already alluded to: that a standing gradient is set up in the narrow channels of the infolded apical membrane, and maintained by the activity of the dense array of mitochondria. A primary water pump could be the answer in this case, but this is speculation, and the notion requires more experimental study.

ACKNOWLEDGEMENT

I am grateful to the Agricultural Research Council for a grant in support of this work.

REFERENCES

1. Beament,J.W.L., Noble-Nesbitt,J. & Watson,J.A.L. (1964) The waterproofing mechanism of arthropods. III. Cuticular permeability in the firebrat, *Thermobia domestica* (Packard). J. exp. Biol. 41, 323-330.

2. Berridge,M.J. & Gupta,B.L. (1967) Fine-structural changes in relation to ion and water transport in the rectal papillae of the blowfly, *Calliphora*. J.Cell.Sci. 2, 89-112.

3. Berridge,M.J. & Gupta,B.L. (1968) Fine-structural localisation of adenosine triphosphatase in the rectum of *Calliphora*. J.Cell Sci. 3,17-32.

4. Grimstone,A.V., Mullinger,A.M. & Ramsay,J.A. (1968) Further studies on the rectal complex of the mealworm *Tenebrio molitor*, L. (Coleoptera, Tenebrionidae). Phil.Trans.Roy.Soc.Lond. B 253, 343-382.

5. Gupta,B.L. & Berridge,M.J. (1966a) A coat of repeating subunits on the cytoplasmic surface of the plasma membrane in the rectal papillae of the blowfly, *Calliphora erythrocephala* (Meig.), studied *in situ* by electron microscopy. J.Cell Biol. 29, 376-382.

6. Gupta,B.L. & Berridge,M.J. (1966b) Fine structural organization of the rectum in the blowfly, *Calliphora erythrocephala* (Meig.), with special reference to connective tissue, trachea and neurosecretory innervation in the rectal papillae. J.Morph. 120, 23-81.

7. Irvine,H.B. & Phillips,J.E. (1971) Effects of respiratory inhibitors and ouabain on water transport by isolated locust rectum. J.Insect Physiol. 17, 381-393.

8. Maddrell,S.H.P. (1971) The mechanisms of insect excretory systems. In Advances in Insect Physiology, ed. Beament,J.W.L. et al., vol.8, pp. 199-331, London & New York, Academic Press.

9. Mellanby,K. (1932) The effect of atmospheric humidity on the metabolism of the fasting mealworm (*Tenebrio molitor* L., Coleoptera). Proc. Roy.Soc.Lond. B 111, 376-390.

10. Noble-Nesbitt,J. (1970a) Water uptake from subsaturated atmospheres: its site in insects. Nature Lond. 225, 753-754.

11. Noble-Nesbitt,J. (1970b) Water balance in the firebrat, *Thermobia domestica* (Packard).Exchanges of water with the atmosphere. J.exp.Biol. 52, 193-200.

12. Noirot,Ch. & Noirot-Timothee,C. (1971) Ultrastructure du proctodeum chez le Thysanoure

Lepismodes inquilinus Newman (= Thermobia domestica Packard) II. Le sac anal. J.Ultrastruct. Res. 37, 335-350.

13. Oschman,J.L. & Wall,B.J. (1969).The structure of the rectal pads of Periplaneta americana L. with regard to fluid transport. J.Morph. 127, 475-510.

14. Phillips,J.E. (1961) Rectal absorption of water and salts in the locust and blowfly. Ph.D. Thesis, University of Cambridge.

15. Phillips,J.E. (1964a) Rectal absorption in the desert locust, Schistocerca gregaria Forskal I.Water. J.exp.Biol. 41, 15-38.

16. Phillips,J.E. (1964b) Rectal absorption in the desert locust, Schistocerca gregaria Forskal II. Sodium, potassium and chloride. J.exp.Biol. 41, 39-67.

17. Phillips,J.E. (1964c) Rectal absorption in the desert locust, Schistocerca gregaria Forskal III. The nature of the excretory process. J. exp.Biol. 41, 68-80.

18. Phillips,J.E. (1969) Osmotic regulation and rectal absorption in the blowfly, Calliphora erythrocephala. Can.J. Zool. 47, 851-863.

19. Phillips,J.E. (1970) Apparent transport of water by insect excretory systems. Am.Zool. 10, 413-436.

20. Ramsay,J.A. (1964) The rectal complex of the mealworm Tenebrio molitor L. (Coleoptera, Tenebrionidae). Phil.Trans. Roy.Soc. Lond. B 248, 279-315.

21. Ramsay,J.A. (1971) Insect Rectum. Phil.Trans. Roy Soc.Lond. B 262, 251-260.

22. Slifer,E.N. & Sekon,S.S. (1970) Sense organs of a thysanuran, Ctenolepisma lineata pilifera, with special reference to those on the antennal flagellum (Thysanura, Lepismatidae). J.Morph. 132, 1-26.

23. Stobbart,R.H. (1968) Ion movement and water transport in the rectum of the locust, Schisto-

cerca gregaria. J. Insect Physiol. 14, 269-275.

24. Wall,B.J. & Oschman,J.L. (1970) Water and solute uptake by rectal pads of *Periplaneta americana*. Am.J.Physiol. 218, 1208-1215.

25. Wall,B.J., Oschman,J.L. & Schmidt-Nielsen,B. (1970) Fluid transport: concentration of the intracellular compartment. Science, N.Y. 167, 1497-1498.

26. Wigglesworth,V.B. (1932) On the function of the rectal glands in insects. Q.Jl.Microsc.Sci. 75, 131-150.

Comparative Physiology, eds. L. Bolis, K. Schmidt-Nielsen and S. H. P. Maddrell

AMINO ACID METABOLISM AND CELL VOLUME REGULATION

E. SCHOFFENIELS
Department of Biochemistry,
University of Liège,
Liège, Belgium.

Since the amount of free amino acids is higher in tissues of marine animals than in tissues of freshwater ones, it may well be that free amino acids play an important role in the osmoregulation of the intracellular fluid. It is thus of interest to compare the situation of a euryhaline marine invertebrate in a concentrated and in a diluted medium. This was first accomplished by Florkin and co-workers with the Chinese crab (7,3). In *Eriocheir sinensis* adapted to freshwater,the inorganic constituents of the muscles represent about 40% of the osmotically active constituents. The rest of the osmotic pressure is due to small organic molecules, of which the determined amino acids constitute more than 50%. When animals are adapted to seawater, an increase is observed in all the osmotically active constituents. The total concentration of the free amino acids is approximately doubled. This variation of the amino acid component can only depend on an active modification since there is only a slight modification of the tissue hydration (55, 3). Therefore Florkin (11,12) has proposed that one ought "to consider the variation of the amino acid component resulting from a change in the medium concentration, as exerting an intracellular osmotic regulation".

Many studies in recent years have dealt with the free amino acid content of euryhaline invertebrates in relation to the environment. These observations show that the tissue level of free amino acids varies with environmental salinity. This phenomenon appears to occur systematically in euryhaline aquatic invertebrates since it has been re-

TABLE 1

Amino acid concentration and variations of concentration in muscles of several crustaceans adapted to media of various salinities[a]

	Eriocheir sinensis[b]			*Carcinus maenas*[c]			*Leander serratus*[d]			*Leander squilla*[d]			*Astacus astacus*[e]			*Callinectes sapidus*[f]		
	Fresh-water	Sea-water	Varia-tion %	Sea-water	50% Sea-water	Varia-tion %	Sea-water	30% Sea-water	Varia-tion %	Sea-water	30% Sea-water	Varia-tion %	Fresh-water	50% Sea-water	Varia-tion %	Sea-water	50% Sea-water	Varia-tion %
Alanine	1.39	3.37	142.5	2.05	0.96	−53.2	1.69	0.26	−84.6	0.79	0.63	−20.3	1.32	2.11	59.8	2.05	1.24	−40.0
Arginine	2.99	4.13	107.5	3.62	3.37	− 6.9	2.58	2.07	−19.8	2.58	2.47	− 4.3	3.70	4.42	19.5	7.44	5.24	−29.6
Aspartic acid	0.29	0.86	196.6	0.39	0.27	−30.8	0.18	0.05	−72.3	0	0.08	–	0.80	0.95	18.8	0.42	0.17	−59.5
Glutamic acid	0.84	2.11	151.2	3.60	1.71	−52.2	0.32	0.11	−65.6	0	0	–	3.43	5.05	47.2	0.55	0.19	−65.5
Glycine	4.64	8.00	72.4	10.07	7.13	−29.2	11.2	9.5	−15.2	8.0	5.7	−28.8	2.09	2.43	16.3	19.75	15.50	−21.5
Histidine	g	g	g	0.01	0.004	−60.0	0	0	–	0	0	–	0.06	0.16	166.7	0.13	0.04	−69.2
Isoleucine	0.08	0.24	200.0	0.17	0.03	−82.4	0.21	tr	–	tr	tr	–	0.17	0.34	100.0	0.1	0.09	−10.0
Leucine	0.14	0.40	185.7	0.26	0.05	−80.8	0.34	tr	–	tr	tr	–	0.21	0.40	90.5	0.19	0.19	0
Lysine[g]	1.16	1.38	18.9	0.19	0.10	−47.4	0.16	0.05	−68.7	0.09	0.07	−22.2	0.56	0.71	26.8	0.19	0.005	−97.4
Phenylalanine	0	tr	–	0.05	0.012	−76.0	tr	0	–	tr	0	–	0.09	0.06	−33.4	0.23	0.038	−83.5
Proline	0.77	3.50	354.5	9.84	1.83	−81.4	2.17	0.64	−70.5	2.95	2.26	−23.4	0.78	2.57	229.5	4.04	2.68	−33.7
Serine	0.21	0.47	123.8	–	–	–	0.55	0.12	−78.2	0.50	0.35	−30.0	–	–	–	2.86	0.28	−90.2
Taurine	1.67	2.06	23.3	–	–	–	2.56	2.32	− 9.4	2.40	2.24	− 6.7	–	–	–	3.79	2.05	−45.9
Threonine	0.36	1.14	216.7	0.33	0	–	0.21	tr	–	0	tr	–	0.27	0.56	107.4	–	–	–
Tyrosine	0	tr	–	0.03	0.006	−80.0	tr	0	–	tr	0	–	0.12	0.10	−16.7	–	–	–
Valine	0	0.50	–	0.33	0.046	−86.1	0.37	tr	–	tr	tr	–	0.11	0.46	318.2	0.29	0.30	3.4
Total	14.54	28.16	93.7	30.9	15.5	−49.8	22.5	15.12	−32.8	17.31	13.80	−20.3	13.71	20.32	48.2	42.03	28.013	−22.0

a Values are given in μmules/100 mg wet weight, tr indicates trace; dashes indicate that values were not determined. b Bricteux-Gregoire *et al.* (1962). c Duchâteu *et al.* (1959) d Jeuniaux *et al.* (1961). e Duchâteau and Florkin (1961). f Gilles, R. unpublished. g Histidine is measured together with lysine.

TABLE II

FREE AMINO ACIDS IN VARIOUS TISSUES OF <u>CALLINECTES</u> ACCLIMATIZED TO SEA-WATER OR TO TWICE-DILUTED SEA-WATER (AFTER GILLES, UNPUBLISHED)

	Hepatopancreas		Gill		Nerve		Muscle		Blood	
	SW	SW/2	SW	SW/2	SW	SW/2	SW	SW/2	SW	SW/2
Taurine	55.10	54.64	-	-	15.52	12.01	49.06	22.22	.14	.22
Aspartic acid	Tr	Tr	1.29	1.21	226.72•	134.40	5.52	2.94	.02	.04
Threonine	2.59	1.98	1.01	Tr	-	-	-	-	-	-
Serine	4.90	4.23	2.74	2.68	2.77	3.85	3.70	5.02	.17	.27
Glutamic acid	9.81	11.26	7.12	11.09	27.22	20.47	7.19	2.31	.05	.07
Proline	31.39	14.90	15.86	12.64	37.61	12.62	52.31	46.27	.65	.91
Glycine	41.54	28.81	16.64	11.04	9.85	14.32	255.89	198.85	1.17	.77
Alanine	19.62	11.92	17.36	17.36	22.59	14.25	26.58	21.33	.37	.25
Valine	5.77	4.63	.96	.77	1.55	.72	3.75	1.22	.03	.02
Methionine	2.27	1.98	.57	.25	.48	Tr	4.17	0.21	.01	Tr
Isoleucine	3.17	2.64	.57	.54	.70	.32	1.25	0.50	.01	.01
Leucine	10.67	8.94	.72	.69	1.06	.45	2.42	0.94	.02	.01
Tyrosine	5.77	5.29	1.34	.56	.77	.32	1.41	-	.01	.01
Phenylalanine	6.05	5.29	.59	.33	.54	-	3.00	-	.02	.02
Total	198.75	156.51	66.77	59.16	347.38	213.73	416.25	301.81	2.67	2.60

Concentrations are given in mM/Kg tissular H_2O; Tr : traces; - : not determined

• : Aspartic acid + undetermined ninhydrin positive compound.

corded in all the phyla studied so far (54).

It is interesting to note that the euryhalinity of several species depends only on an intracellular isosmotic regulatory mechanism. As a matter of fact, Asterias rubens does not show any extracellular anisosmotic regulation (2),neither do Arenicola marina (45, 1) or Mytilus edulis (40). Thus in all the cases of euryhaline invertebrates so far studied, isosmotic regulation is found,while anisosmotic regulation is not always present. We may therefore consider the isosmotic regulation as the more primitive mechanism, to which, in several species, the anisosmotic regulation adds a new range of possibilities.

The question now arises: which amino acids take part in the isosmotic regulation process in Arthropoda? Table I gives values for the amino acid level in the muscle of several euryhaline species in their normal environment as well as the percentage of variations for each individual amino acid when animals are adapted to media of different salinities. From these results, it can be concluded that the adaptation to a new medium affects the level of most of the amino acids determined, but mainly alanine, arginine, aspartic acid, glutamic acid, glycine, and proline. Proline which is, with glycine, the most concentrated amino acid, undergoes large variations.

The amino acid content of other tissues is also affected by osmotic stress. Table II shows the results obtained with Callinectes sapidus. It is interesting to note that with the exception of gills and blood, amino acids are used to regulate the osmotic pressure.

Thus, with the exception of arginine, the contribution of the amino acid pool to the isoosmotic regulation is mainly due to the so-called nonessential amino acids. As to the origin of these amino acids, the problem is to know if they originate extracellularly or if they are formed within the cell.

Schoffeniels (47) has shown that the total amino nitrogen level of isolated nerves of Eriocheir sinensis is higher for nerves incubated in a saline corresponding to the blood concentration of a seawater animal than for nerves placed in a saline corresponding to the blood concentration of an Eriocheir adapted to freshwater. This observation favors the idea of an intracellular origin for the amino acids involved in the cell volume regulation process in Eriocheir.

More recently, Gilles and Schoffeniels (23) have confirmed and extended this finding. The results obtained suggest that the mechanism responsible for the regulation of the amino acid level during osmotic stress is not under hormonal control and involves at least two processes (Table III,IV).

Table III - Amino acids, ammonia and protein contents of nerves isolated from a fresh-water adapted Eriocheir and placed in a sea-water saline (After Gilles and Schoffeniels (23)).

	Eriocheir 1		*Eriocheir* 2			
	Nerves in fresh-water saline (a)	Nerves in sea-water saline (b)	Nerves in fresh-water saline (c)	Nerves in sea-water saline (d)	Difference (b) − (a)	Difference (d) − (c)
Taurine	0·976	1·712	1·151	1·520	0·736	0·369
Aspartic acid	3·114	5·802	2·987	4·081	2·688	1·094
Threonine	0·066	0·137	0·092	0·145	0·071	0·053
Serine	0·228	0·312	0·288	0·299	0·084	0·011
Glutamic acid	0·681	1·151	0·685	1·013	0·470	0·328
Proline	0·495	1·316	0·461	1·060	0·821	0·599
Glycine	0·367	0·627	0·419	0·497	0·260	0·078
Alanine	0·883	1·955	1·034	1·973	1·072	0·939
Cystine	0·050	0·057	Tr	0·112	0·007	—
Valine	—	0·099	0·075	0·097	—	0·022
Methionine	—	0·137	0·044	0·099	—	0·055
Isoleucine	—	0·032	0·050	0·074	—	0·024
Leucine	—	0·051	0·078	0·094	—	0·016
Tyrosine	—	0·022	0·028	0·041	—	0·013
Phenylalanine	—	0·015	0·029	0·039	—	0·010
Lysine	—	0·033	0·085	0·095	—	0·010
Histidine	—	0·029	0·028	0·034	—	0·006
Arginine	—	0·375	0·696	0·816	—	0·120
Ammonia	—	0·048	2·978	0·159		−2·819
Proteins	17·01	17·55	22·09	21·44	0·54	−0·65

Amino acids and ammonia levels are expressed in μM/100 mg wet wt. Protein content is given as amino-N in μM of alanine/100 mg wet wt.

Tr: traces; —: not determined.

The concentration of some amino acids such as tyrosine, phenylalanine, leucine, isoleucine, or valine appears to be regulated by the permeability of the nerve membrane. As a matter of fact, the intracellular modification of the concentration of these amino acids is accompanied by a reverse modification of the concentration of about the same

Table IV - Same experiments as in Table III. Analysis of the incubating medium.

	Eriocheir 1		*Eriocheir* 2			
	Fresh-water saline (a)	Sea-water saline (b)	Fresh-water saline (c)	Sea-water saline (d)	Difference (b) − (a)	Difference (d) − (c)
Taurine	0·382	0·410	0·089	0·108	0·028	0·019
Aspartic acid	2·322	2·844	1·665	1·882	0·522(+)	0·217(+)
Threonine	0·066	0·064	Tr	0·040	−0·002	—
Serine	0·170	0·180	0·088	0·104	0·010	0·016
Glutamic acid	0·491	0·496	0·461	0·481	0·005	0·020
Proline	0·291	0·411	0·146	0·403	0·120(+)	0·257(+)
Glycine	0·260	0·278	0·279	0·293	0·018	0·014
Alanine	0·462	0·478	0·486	0·481	0·016	−0·005
Cystine	Tr	Tr	Tr	Tr	Tr	Tr
Valine	—	0·053	0·049	0·020	—	0·029(+)
Methionine	—	Tr	Tr	Tr	Tr	Tr
Isoleucine	—	0·028	0·080	0·065	—	−0·015(+)
Leucine	—	0·044	0·100	0·087	—	−0·013(+)
Tyrosine	—	0·020	0·041	0·023	—	−0·018(+)
Phenylalanine	—	0·023	0·047	0·034	—	−0·013(+)
Ammonia	—	1·444	2·228	1·549		−0·679(+)

Amino acids and ammonia levels are expressed in μM/100 mg wet wt. of nerve incubated.
Tr: traces; —: not determined.
(+): indicates that the differences are significant.

importance in the incubating medium. The concentration of the so-called nonessential amino acids is regulated by a modification of their metabolism since the variation of the amino acid level in the nerve tissues is not followed by a concomitant variation in the medium. Moreover, the mechanism involved could result from the activity of a synthesizing process and not from a modification of the steady-state between amino acids and proteins,since the content of nerve proteins does not vary significantly during adaptation. Such a conclusion is also supported by experiments on protein patterns during adaptation of Eriocheir (14,50,13).

This may not be the case with all species of the Crustacea and Siebers (57) has advanced some arguments favoring the idea that in conditions of hyperosmotic stress, the increased concentration of amino acids could well be explained by the hydrolysis of some proteins. He also found an increase in the concentration of neutral sugars and organic acids of low molecular weight.

The fact that in our experiments ammonia

undergoes a significant variation in nerve tissue as well as in the medium (Tables III-IV) also agrees with our interpretation since a net synthesis of amino acids from their keto precursors would require ammonia. Concerning the control of proline concentration two mechanisms seem to be involved. This amino acid, although synthesized in the isolated nerve during its adaptation from a freshwater saline to a seawater medium, is nevertheless not catabolized but rather released from the nerve during the reverse adaptation.

As shown by the results of Tables V and VI the loss of proline into the incubating medium is greater than the decrease of this amino acid in the nerve. On the other hand, there is a decrease in the concentration of glutamate and alanine in the medium. If the extra proline loss into the medium could be accounted for by an extra synthesis, using as precursors glutamate and alanine, one could find a direct correlation between the decrease in the levels of glutamate and alanine and the increased loss of proline.

Table V - Amino acids, ammonia and protein contents of nerves isolated from a sea-water adapted *Eriocheir* and placed in a fresh-water saline (after Gilles and Schoffeniels (23)).

	Eriocheir 1		*Eriocheir* 2			
	Nerves in sea-water saline (a)	Nerves in fresh water saline (b)	Nerves in sea-water saline (c)	Nerves in fresh-water saline (d)	Difference (b) – (a)	Difference (d) – (c)
Taurine	0·950	0·720	1·296	0·991	–0·230	–0·305
Aspartic acid	3·804	1·856	4·354	1·888	–1·948	–2·466
Threonine	—	0·088	0·193	0·162	—	–0·031
Serine	0·268	0·245	0·615	0·572	–0·020	–0·043
Glutamic acid	0·818	0·777	0·559	0·484	–0·041	–0·075
Proline	2·394	0·693	3·274	1·279	–1·701	–1·995
Glycine	0·446	0·400	0·706	0·597	–0·046	–0·109
Alanine	2·314	1·212	2·239	1·487	–1·102	–0·752
Cystine	Tr	Tr	Tr	Tr	Tr	Tr
Valine	0·057	0·043	0·107	0·096	–0·014	–0·011
Methionine	0·035	0·021	0·019	Tr	–0·014	—
Isoleucine	0·030	0·026	0·093	0·056	–0·004	–0·037
Leucine	0·045	0·035	0·084	0·060	–0·010	–0·024
Tyrosine	0·026	0·020	0·056	0·040	–0·006	–0·016
Phenylalanine	0·029	0·026	0·056	0·045	–0·003	–0·011
Lysine	0·050	0·047	0·073	0·078	–0·003	+0·005
Histidine	0·034	0·047	0·094	0·091	+0·013	–0·003
Arginine	0·691	0·669	0·558	0·267	–0·022	–0·291
Ammonia	1·191	0·304	1·841	0·404	–0·887	–1·437
Proteins	15·45	15·96	27·05	25·59	0·41	–1·46

Amino acids and ammonia levels are given in μM/100 mg wet wt. Protein content is expressed as amino-N in μM of alanine/100 mg wet wt.

Tr: traces; —: not determined.

Table VI - Same experiments as in Table V. Analysis of the incubating medium.

	Eriocheir 1		*Eriocheir* 2			
	Sea-water saline	Fresh-water saline	Sea-water saline	Fresh-water saline	Difference (b) – (a)	Difference (d) – (c)
Taurine	0·100	0·084	0·130	0·153	–0·016	+0·023
Aspartic acid	1·660	1·623	2·046	2·058	–0·037	+0·012
Threonine	0·030	0·012	0·042	0·018	–0·018(+)	–0·024(+)
Serine	0·105	0·118	0·163	0·154	0·013	–0·009
Glutamic acid	0·548	0·288	0·448	0·264	–0·260(+)	–0·184(+)
Proline	1·484	3·507	1·460	3·561	2·023(+)	2·101(+)
Glycine	0·257	0·253	0·202	0·212	–0·004	+0·010
Alanine	1·372	1·309	1·116	1·048	–0·063	–0·068
Cystine	Tr	Tr	Tr	Tr	Tr	Tr
Valine	0·037	0·052	0·054	0·071	+0·015(+)	+0·017(+)
Methionine	Tr	Tr	Tr	Tr	Tr	Tr
Isoleucine	0·031	0·038	0·029	0·056	0·007	0·027(+)
Leucine	0·043	0·055	0·038	0·067	0·012(+)	0·029(+)
Tyrosine	0·027	0·032	0·029	0·042	0·005	0·013(+)
Phenylalanine	0·021	0·034	0·035	0·051	0·013(+)	0·016(+)
Ammonia	1·382	2·803	1·611	3·202	1·421	1·591

Amino acids and ammonia levels are given in μM/100 mg wet wt. of nerve incubated.
Tr: traces; —: not determined.
(+): indicates that the difference is significant.

That this is indeed the case is demonstrated in Table VII. This is certainly a reasonable assumption since glutamate is a precursor of proline.

It is probable that the mechanism described in isolated *Eriocheir* nerves is also at play in the intact animal. As a matter of fact Vincent-Marique and Gilles (62) have shown that during adaptation of *Eriocheir* from seawater to freshwater, the decrease in nonessential amino acids in muscle is accompanied by a decrease in their amount in the blood, thus suggesting an intracellular degradation process or loss by excretion. The fact that in the intact *Eriocheir* modification of the amino acid level is paralleled by concomitant modification in nitrogen excretion (26,14) is in agreement with this hypothesis. As far as proline is concerned, the important decrease in muscle concentration is accompanied by a similar increase in the blood (Figure 1). This can be interpreted as the consequence of an extrusion of proline from the tissues during adaptation of *Eriocheir* from seawater to freshwater (62); a conclusion that raises the question of the further degradation of proline.

Table VII

Balance sheet for proline, glutamate and alanine in experiments described in Tables III-VI

Difference between loss in sea-water and fresh-water salines	Decrease in nerve concentration		Difference between loss in sea-water and fresh-water salines		
Proline	Proline	Algebraic sum	Glu	Ala	Algebraic sum
1	2	1 + 2	3	4	3 + 4
+ 2·023	− 1·701	0·322	− 0·260	− 0·063	− 0·323
+ 2·201	− 1·995	0·206	− 0·184	− 0·068	− 0·252

Explanations see text.

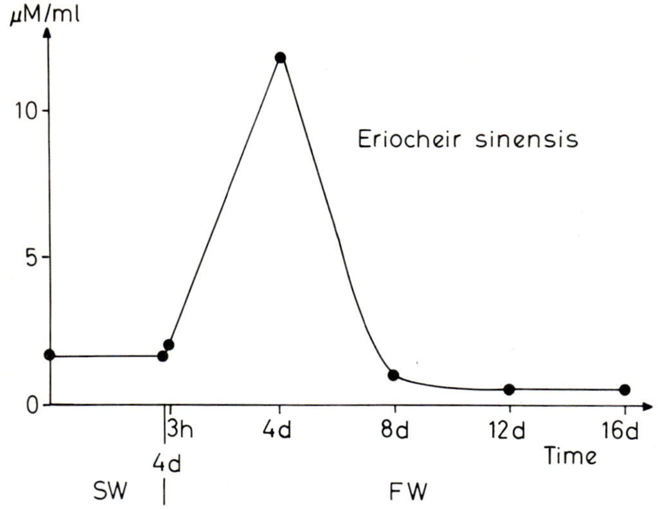

Figure 1 - Modification of the blood proline level of a single *Eriocheir sinensis* as a function of the time of adaptation from seawater to freshwater. Results taken from Vincent-Marique and Gilles (62).

It has already been demonstrated that during adaptation of *Eriocheir* to freshwater, the modification of the ammonia level in the medium accounts for 73% to 94% of the total modification of the nitrogen excretion (26). It seems therefore unlikely that proline can be excreted by Eriocheir.

Vincent-Marique and Gilles (62) have demonstrated the existence of "proline oxidase" activity in the gill preparation of *Eriocheir sinensis*. It is worth noting that an appreciable enzymic activity can only be found in the gills and not in the other tissues examined (Table VIII). Moreover, one can also see that the "proline oxidase" activity is much greater in the three posterior pairs of gills than in the three anterior ones. It is only the three posterior gills which show active transport of sodium (31, 28). Thus, the degradation of proline in these gills may perhaps be related to this phenomenon. It is interesting to note that proline can be a readily available source of energy in insects (4,5). Further experiments are required to shed more light on this interesting problem.

The mechanisms leading to the modification

Table VIII - Proline oxidase activity from various tissues of *Eriocheir sinensis*[a].

Hepatopancreas	Muscle	Nerve	Gills: Anterior	Gills: Posterior
900	--	300	500	2900
--	--	--	600	2500
100	--	300	200	2200
--	--	--	300	1900
--	--	--	800	3000
--	--	--	700	2300

[a]Activities are given as a variation of the optical density in 8 min/g per wet weight tissue. The modification of the optical density is due to the reduction of 2,6-dichlorophenolindophenol by proline in the presence of the enzymic preparations.

in the cell permeability to amino acids are far from being elucidated. The interest of research workers has been mainly centered on the ability of euryhaline invertebrates to modify their amino acid metabolism with respect to the modification of the osmotic pressure of the surroundings. We shall therefore try to investigate 1) the part played by the different pathways leading to amino acid synthesis in the establishment of the amino acid pattern during the adaptation process, and 2) the primary cause for the modification of the activity of these different pathways during adaptation to media having different salinities.

If, in the incubating medium of isolated nerves from *Eriocheir sinensis*, an increase in osmotic pressure is achieved by addition of sucrose to the diluted medium, nerves become electrically inactive within a few hours and the intracellular amino nitrogen level considerably decreases instead of undergoing the increase which occurs when the rise in osmotic pressure is obtained by addition of seawater (47). This experiment clearly demonstrates that it is not the osmotic pressure *per se* which is responsible for the regulation of amino acid levels during adaptation to concentrated media.

It is also evident from the experiments performed on isolated nerves (47, 23) that the isosmotic regulation process is not under hormonal control since the isolated tissue is able to regulate its amino acid level with respect to the osmotic pressure of the incubating medium. An argument favoring this view may also be found in the results of Duchâteau-Bosson and Florkin (9) who showed that removal of the eyestalk gland does not alter the isosmotic regulation power of *Eriocheir sinensis.*

It can therefore be suggested that it is the modification of the ionic concentration to which the cell is submitted during adaptation which controls the mechanism responsible for the modification of the amino acid level (47). In connection with this hypothesis, Gilles and Schoffeniels (17, 21) have demonstrated that some substances, the actions of which are classically explained by a modification of the intracellular ionic composition, have an effect on the synthesis of amino acids.

To explain such a type of control, two major possibilities exist. Either the ionic composition of the incubating medium acts through an intermediary substance on the amino acid metabolism, or some ionic species control directly the activity of key enzymes involved in the amino acid metabolism.

In the first interpretation, one would be dealing with the kind of mechanism generally used to explain hormonal action, thus postulating the production of an intermediary effector such as 3',5'-AMP. So far no attempt has been made to bring experimental evidence to bear on such an interpretation.

As to the second possibility, we have been looking for enzymes involved directly or indirectly in the nitrogen metabolism, the activity of which would be affected by the ionic composition of the incubating medium. Our studies were performed on the following enzyme systems.

1. Those involved directly in the amino acid metabolism such as glutamate dehydrogenase (EC 1.4.1.2), aspartate aminotransferase (EC 2.6.1.1), alanine transferase (EC 2.6.1.2), serine hydrolase

(EC 4.2.1.13), aspartate decarboxylase (EC 4.1.1. 11) or glyoxylate reductase (EC 1.1.1.26).

2. Those involved in the supply of the keto precursors required for amino acid synthesis, such as isocitrate dehydrogenase (EC 1.1.1.42), malate dehydrogenase (EC 1.1.1.37 and EC 1.1.1.40), malate hydrolyase (EC 4.2.1.2), and oxalacetate decarboxylase (EC 4.1.1.3).

3. Those involved in the fate of reducing equivalents in the intracellular medium such as lactic dehydrogenase (EC 1.1.1.27) or 3-glycerophosphate dehydrogenase (EC 1.1.1.8).

These experiments have been performed mainly in my laboratory on tissue preparations from stenohaline and euryhaline species. They have been summarized in recent monographs (50, 13) as well as in review articles (19, 16, 51,54). I shall therefore only briefly summarize the main conclusions.

Table IX reports some of the data obtained.It can be seen that at concentrations of NaCl rather close to those assumed to be found in crustacean cells, there is no difference in the effect of NaCl on the enzymic systems implicated in the metabolism of amino acids extracted from the euryhaline crayfish or from the stenohaline lobster. On the contrary, interesting differences are observed when considering the effect of NaCl on lactic dehydrogenase and 3-glycerophosphate dehydrogenase. These enzymes are generally considered to be the most important among those controlling the extramitochondrial ratio NADH/NAD (lactic dehydrogenase) as well as the transfer of reducing equivalent originating in the cytoplasm to the mitochondrial respiratory chain (3-glycerophosphate dehydrogenase) (32).

Also the isozymic pattern of lactic dehydrogenase is modified under the influence of an osmotic stress, at least in species that can only be adapted *progressively* to media of various concentrations. *Eriocheir,* a perfect euryhaline species that can be subjected abruptly either to seawater or to freshwater has only one isozyme of lactic dehydrogenase in the various organs we have studied (heart, nerve and muscle of the walking legs, gills, claw and thoracic muscle) (52).It appears therefore that a biochemical difference between

TABLE IX

EFFECT OF NaCl ON THE ACTIVITY OF VARIOUS ENZYMES EXTRACTED FROM MUSCLE OF *Homarus vulgaris* OR *Astacus fluviatilis*

Enzyme	Animal	NaCl concentration (mM)[a] 0	50	100	200	300	400	600	Reference
Glutamate dehydrogenase (EC 1.4.1.2)	Lobster	100	130	160	180	—	250	160	Gilles and Schoffeniels (1970)
	Crayfish	100	150	230	270	—	310	220	
Aspartate aminotransferase (EC 2.6.1.1)	Lobster	100	95	108	100	—	95	—	Gilles (1969)
	Crayfish	100	100	100	95	—	102		
Serine hydro-lyase (EC 4.2.1.13)	Lobster	100	74	36	12	—	—	—	Gilles (1969)
	Crayfish	100	63	38	3	—	—	—	
Aspartate decarboxylase (EC 4.1.1.11)	Lobster	100	—	—	99	—	104	—	Gilles and Schoffeniels (1966)
	Crayfish	100	—	—	95	—	86	—	
3-Glycerophosphate dehydrogenase (EC 1.1.1.8)	Lobster	100	—	145	117	93	50	33	Schoffeniels (1968)
	Crayfish	100	—	366	315	217	195	116	
Lactic dehydrogenase (EC 1.1.1.27) Pyruvate (5×10^{-4} M)	Lobster	100	—	58	12	5.1	4.1	5.1	Schoffeniels (1968)
	Crayfish	100	94	66	47	34	28	—	
Pyruvate (10^{-2} M)	Lobster	100	125	125	125	125	125	—	Schoffeniels (1968)
	Crayfish	100	100	100	78	78	68	—	
Malate dehydrogenase (E.C 1.1.1.37)	Lobster	100	115	113	105	—	78	—	Gilles (1969)
	Crayfish	100	103	112	83	—	—	—	
Malate hydro-lyase (EC 4.2.1.2)	Lobster	100	142	117	70	—	—	—	Gilles (1969)
	Crayfish	100	74	57	30	—	—	—	
Isocitrate dehydrogenase (EC 1.1.1.42)	Lobster	100	83	75	52	—	25	—	Gilles (1969)
	Crayfish	100	87	73	63	—	50	29	
Malate dehydrogenase (EC 1.1.1.40)	Lobster	100	80	69	59	—	48	16	Gilles (1969)
	Crayfish	100	68	58	51	—	43	31	
Oxalacetate decarboxylase (EC 4.1.1.3)	Lobster	100	98	95	89	—	85	80	Gilles and Schoffeniels (1966)
	Crayfish	100	96	86	72	—	65	58	
Glyoxylate reductase (EC 1.1.1.26)	Lobster	100	—	42	24	15	10	6	Schoffeniels (1968)
	Crayfish	100	—	30	15	7	6	3	

[a] Results are given as percentage of the control activity taken as 100. The activity is given:

(1) By the variation in optical density. At 340 mμ due to NADH oxidation for glutamate dehydrogenase, 3-glycerophosphate dehydrogenase, lactic dehydrogenase, malate dehydrogenase (EC 1.1.1.37), and glyoxylate reductase; at 340 mμ due to NADP reduction for malate dehydrogenase (EC 1.1.1.40); at 280 mμ due to the disappearance of oxalacetate in the case of aspartate aminotransferase; and at 240 mμ due to the decrease in fumarate concentration in the case of malate hydro-lyase.

(2) By the quantity of pyruvate appearing after a 20-min incubating period in the case of serine hydro-lyase.

(3) By the amount of CO_2 produced after a 30-min incubating period in the case of aspartate or oxalacetate decarboxylase.

See also Schoffeniels and Gilles (1963) Schoffeniels (1964, 1965) and Florkin and Schoffeniels (1969).

euryhaline and stenohaline species must be looked for in the control of fate of the reducing equivalents.

Moreover, the results presented in Table IX show an effect of the ionic composition on the activity of different enzymes directly involved in the synthesis and in the degradation of amino acids and particularly on glutamate dehydrogenase. It is apparent, when considering the standard oxido-reduction potentials of the system E_O,(NADH/NAD)= - 0.32 volt, E_O' (glutamate/2-oxoglutarate) = - 0.108 volt, that the enzyme favors the reductive amination of 2-oxoglutarate. This fact indicates that this enzyme must play an important part in the nitrogen metabolism by controlling the entry of ammonia into the amino acid pool. The importance of glutamate dehydrogenase in controlling the synthesis of glutamate is recognized by many authors (30, 59, 37,38). It is of interest to note that this enzyme is activated by an increase in ionic concentration of the incubating medium. On the contrary, under the same experimental conditions, the activities of aspartate aminotransferase and alanine aminotransferase (25) are not affected,and the activity of serine hydro-lyase is inhibited (Table IX). It can therefore be tentatively concluded that an increase in ionic concentration induces an increase in glutamate synthesis. Under conditions of increased ionic concentration,serine hydro-lyase is inhibited thus leading to a reduced deaminating activity. In turn, the increased amount of glutamate would induce an increase in the amount of the other amino acids, the activity of the aminotransferase being unaffected (Figure 2). This concept is in agreement with observations showing an increase in the free amino acid pool and a decrease in the nitrogen excretion (35, 26, 14) during adaptation of euryhaline species to concentrated media.

If the synthesis of glutamate is dependent on the availability of 2-oxoglutarate and NH_3, it is also dependent on the availability of reducing equivalents. It is therefore worth noting that under conditions of ionic concentrations inducing an increase in the activity of glutamate dehydrogenase, 3-glycerophosphate dehydrogenase is inhibited (Table IX).

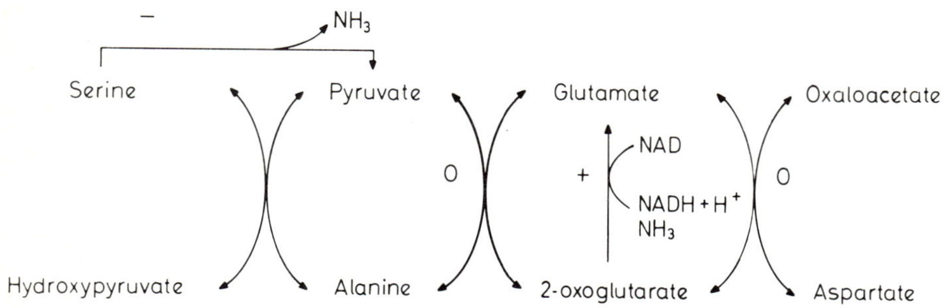

Figure 2 - Effect of NaCl on the activity of enzymes implicated in the metabolism of amino acids. +, indicates an increase in the enzymic activity, -, indicates a decrease in the enzymic activity; O, no effect on the enzymic activity. From Gilles (16).

Therefore the cell metabolism is geared toward anaerobic conditions which prevent some of the reducing equivalents from entering the respiratory chain (Figure 3). Furthermore, under the same

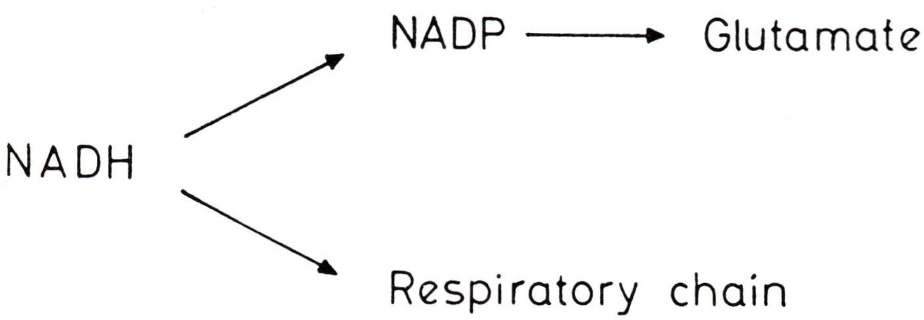

Figure 3 - Competition between glutamate synthesis and the respiratory chain. After Schoffeniels (51).

conditions of ionic concentration, lactic dehydrogenase activity is also inhibited thus making the reducing equivalents available for other purposes (e.g., glutamate synthesis). If such a scheme can be applied to crustacean amino acid metabolism,one may consider that a competition must exist between the various metabolic sequences needing reducing equivalents. As a result, one should find some kind of interaction between the respiratory chain and the synthesis of glutamate (Figure 3). This interpretation finds experimental support in the results obtained by Gilles and Schoffeniels (18) on the isolated nerve chain of the crayfish. They demonstrate that addition of NH_4Cl to the incubating medium increases the intracellular pool of free amino acids (see also (21)) and decreases the oxygen consumption. This is at variance with the results obtained with the lobster nerve chain in which the oxygen consumption is unaffected. Such a difference between the euryhaline crayfish and the stenohaline lobster seems to indicate once more that a biochemical difference between euryhaline and stenohaline species must be looked for in the control of the fate of reducing equivalents.

Another observation providing additional support to the above hypothesis is described in Tables X and XI. The CO_2 production and oxygen consumption of isolated nerves of <u>Callinectes sapidus</u> have been measured under osmotic stress using labelled amino acids as substrate. It can be seen that the Q_{O_2} and Q_{CO_2} increase during the hypoosmotic stress, the reverse being true during the hyperosmotic stress.

Another problem of interest, brought about by the above observations, is that of the origin of the keto acids used in amino acid synthesis. Classically, the glycolysis and the Krebs cycle should provide keto precursors such as pyruvate, oxaloacetate and α-ketoglutarate. Using various labelled substrates, e.g. $^{14}CO_2$, glucose-U-^{14}C, glucose-1-^{14}C, glucose-6-^{14}C, pyruvate-1-^{14}C, pyruvate-2-^{14}C, D-arabinose-u-^{14}C, it has been possible to establish the relative importance of the various metabolic sequences involved in the carbohydrate metabolism. Thus the demonstration that pyruvate-1-^{14}C leads to the production of labelled glu-

Table X

Effect of the hypoosmotic stress on the CO_2 production, the oxygen consumption and the $^{14}CO_2$ production of isolated axons of <u>Callinectes sapidus</u> incubated with various amino acids uniformly labelled. Amino acid concentration: 20 mM; Radioactivity 2 uC/ml SW = 1100 mOsm/L; SW/2 = 850 mOsm/L. (After Gilles, unpublished)

Amino acids	$^{14}CO_2$ CPM/h/mg.d.w.		QO_2 ul/h/mg.d.w.		QCO_2 ul/h/mg.d.w.		CO_2 specific activity CPM/µM	
	SW	SW/2	SW	SW/2	SW	SW/2	SW	SW/2
Glutamate	85.91	113.08	5.56	6.38	4.16	4.86	462.56	521.25
Arginine	36.31	44.81	2.16	2.42	1.62	1.91	501.98	525.50
Leucine	17.76	21.13	2.77	3.46	2.09	2.62	190.17	180.54
Alanine	371.11	595.96	4.88	10.38	3.86	7.96	2152.60	1677.70
Aspartate	175.40	219.35	4.08	5.62	3.00	4.30	1234.91	1142.62
Serine	31.76	51.18	3.60	5.35	2.68	4.10	265.44	279.55

Table XI

Effect of hyperosmotic stress on the CO_2 production, the oxygen consumption and the $^{14}CO_2$ production of isolated axons of <u>Callinectes sapidus</u> incubated with various labelled amino acids. Conditions as in Table X.

Amino acids	$^{14}CO_2$ CPM/h/mg.d.w.		QO_2 µl/h/mg.d.w.		QCO_2 µl/h/mg.d.w.		CO_2 specific activity CPM/µM	
	SW/2	SW	SW/2	SW	SW/2	SW	SW/2	SW
Glutamate	114.03	75.32	2.65	1.92	2.05	1.43	1245.88	1179.80
Arginine	32.67	23.87	2.35	1.60	1.84	1.20	397.60	445.53
Leucine	17.50	14.34	2.53	1.71	2.04	1.30	192.19	247.07
Alanine	192.94	140.02	2.89	2.09	2.17	1.53	1991.58	2050.04
Aspartate	223.74	179.81	4.06	3.34	2.92	2.58	1716.28	1561.05
Serine	127.61	84.31	3.34	1.98	2.59	1.49	1103.65	1267.39

tamate and aspartate is indicative of the fact that they must be synthesized following a pathway that does not involve a decarboxylation of pyruvate but rather a carboxylation. This is a well known method in vertebrates where it has been shown that the carboxylation of pyruvate and phosphoenolpyruvate contributes appreciably to the synthesis of dicarboxylic amino acids.

As shown by the results of Gilles and Schoffeniels (20) this is also true in the case of lobster, a stenohaline crustacean. Also, by comparing the specific radioactivity of alanine obtained by using pyruvate-1 or 2-^{14}C, it was necessary to assume the existence of a sequence involving the decarboxylation of C_1 in pyruvate (17).This interpretation has found experimental support in the observation showing the existence of an oxaloacetate and aspartate decarboxylase in Crustacea (19). Taurine is labelled in the presence of $^{14}CO_2$, thus suggesting the existence of a still unknown metabolic sequence that could well be a biochemical characteristic of the Crustacea. Finally the fact that taurine, ethionine sulfoxide, glutamate and aspartate are labelled when D-arabinose-u-^{14}C is used as substrate seems to indicate that 2-ketoglutarate could be produced from the glucuronate pathway via the formation of arabonate (Figure 4). If this interpretation is correct the specific radioactivity of glutamate should be higher than that of aspartate. That it is the case is shown by the results of Table XII.

Malate dehydrogenase with oxaloacetate decarboxylase constitutes the main pathways of entrance of pyruvate in the Krebs cycle by carboxylation. The importance of these pathways in the supply of oxidizable material to the tricarboxylic acid cycle is demonstrated by the important rate of $^{14}CO_2$ incorporation into glutamate and aspartate which have intermediates of the cycle as keto precursors (6,20).

The fact that both malate dehydrogenase and oxaloacetate decarboxylase are inhibited by an increased ionic concentration does not necessarily mean that the synthesis of malate and oxaloacetate from pyruvate is inhibited. It must indeed be borne in mind that the enzyme activity is tested by

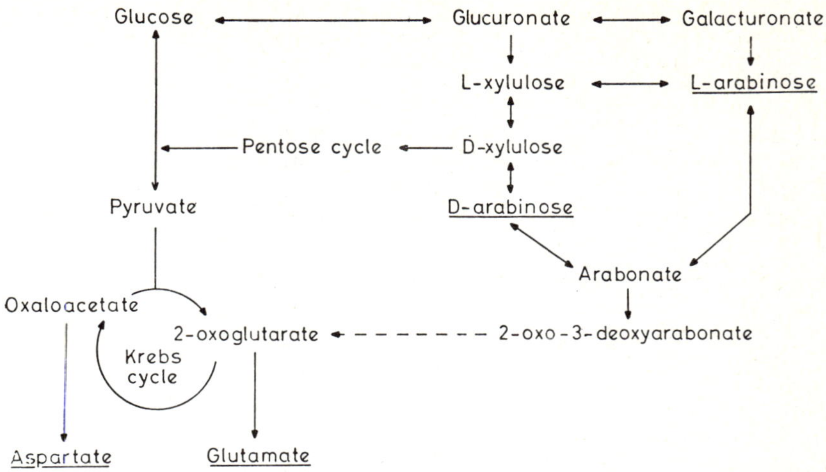

Figure 4 - Anaplerotic sequence of synthesis of 2-oxoglutarate. (After Gilles and Schoffeniels (22)).

Table XII - Specific activity of dicarboxylic amino acids and taurine labelled from D-arabinose-$-u^{14}C$. (After Gilles and Schoffeniels (22)).

Amino acids	
Taurine	18.956
Aspartate	3.041
Glutamate	9.797
Taurine	35.847
Aspartate	2.248
Glutamate	26.822
Taurine	20.715
Aspartate	237
Glutamate	8.098
Taurine	12.924
Aspartate	3.638
Glutamate	43.789

The specific activity is given in counts per µM.

measuring the formation of pyruvate from either malate or oxaloacetate.If our results show an inhibitory effect of NaCl, we do not have any information as to the effect of these ions on the reverse reactions that could well be unaffected, inhibited, or even stimulated.

It is worth noting that the activity of both isocitrate dehydrogenase and malate hydro-lyase of the euryhaline crayfish is inhibited by an increased ionic concentration. It can therefore be concluded that the Krebs cycle is inhibited in the euryhaline crayfish when the extracellular ionic concentration rises. This conclusion is in agreement with the inhibition of the TCA cycle observed when incubating different tissues in the presence of high NaCl concentrations or under conditions inducing a modification of the intracellular ionic content (60,61,20,21, 25). This conclusion is also in agreement with the fact that when the euryhaline crayfish is transferred from a diluted medium to a more concentrated one, an important decrease in the oxygen consumption is recorded (44, 56, 39; see also Tables X and XI).

The production of $^{14}CO_2$ by the crayfish nerve cord is much more important than that by the lobster nerve cord whatever the substrate used (Table XIII; more specifically, the decarboxylation of

Table XIII - Production of $^{14}CO_2$ by lobster and the crayfish nerve cord incubated with different substrates.(After Gilles and Schoffeniels (24)).

Species		Pyr-1-C^{14}	Glu-U-C^{14}	Glu-1-C^{14}	Glu-6-C^{14}
Lobster		3211·62	17·42	5·84	2·68
		3264·94	16·96	9·59	2·24
		3635·36	22·60	8·60	1·08
		2893·52	12·38	9·33	1·20
		3275·80	16·04	9·02	1·58
	Mean	3256·25	17·08	8·47	1·75
Crayfish		34,233·66	376·28	62·08	8·28
		29,720·90	322·95	32·26	12·56
		31,140·60	247·20	47·66	10·48
		32,406·80	564·59	55·09	9·08
		28,836·80	228·00	41·64	10·18
	Mean	31,267·75	347·80	47·74	10·11

The results are expressed in counts/min per mg wet wt. and per hr of incubation. Substrate concentration, 25 mM; radioactivity, 2 μc/ml saline.

C-1 from pyruvate is ten times slower in the lobster than in the crayfish. This could mean that the turnover of the Krebs cycle intermediates is very slow in the lobster. This fact could be related to the low oxygen consumption of lobster tissues (43, 27, 18). The fact that the production of $^{14}CO_2$ from glucose-1-^{14}C is about five times more important than that obtained from glucose-6-^{14}C cannot be interpreted if we consider that the Krebs cycle is the main pathway of decarboxylation. As it is well known, the carbons 1 and 6 of glucose are found in position 3 on pyruvate, thus making their origin on pyruvate undistinguishable. If on the other hand, part of the glucose follows the hexosemonophosphate shunt, the result is easily explained. MacWhinnie and Corkill (34) and Puyear _et al._ (41) working with crayfish have observed the same difference in CO_2 production. They have interpreted their results as indicating the existence of the pentose cycle in the crayfish.

Our results demonstrate that the biochemical events which occur when a euryhaline species such as _Astacus_ or _Carcinus_ is transferred from a diluted to a more concentrated medium (i.e., increase of the free amino acid pool, decrease of nitrogen excretion and oxygen consumption) can be explained, at least partially, by the effect the ionic concentration exerts on the activity of enzymic systems related to the amino acid metabolism. Such an explanation cannot however be generalized as such since the decrease in the oxygen consumption is not observed for all euryhaline Crustacea. As pointed out by Kinne (29), four types of responses can be observed: 1) oxygen consumption is higher in lower salinities than in higher salinities; 2) oxygen consumption is high in both low and high salinities; 3) oxygen consumption is low in both low and high salinities; and 4) oxygen consumption is unaffected by salinity.

It is however important to realize that this conclusion refers to the situation of the animal when it is adapted and not during adaptation.

It must be noted here that _Astacus_ and _Carcinus_ belong to category 1) which includes the majority of the osmoregulating Arthropoda. However, the most efficient regulators such as _Eriocheir sinen-_

sis, Artemia salina, or Palaemonetes vulgaris respire at the same rate in a wide range of salinities (33,15,56). It is worth noting that during osmotic adaptation of Eriocheir, the nitrogen excretion which, as we have seen, can be considered to be a reflection of the amino acid metabolism, shows only a transitory alteration (14). On the contrary, in a species showing an increased respiratory rate in diluted media such as Carcinus, it seems that a definite alteration in nitrogen excretion occurs under osmotic stress (35). It thus appears that the situation when dealing with osmotic regulation is far more complex than previously imagined and that for studying many aspects of this problem separate consideration of individual species will often have to be made. In the literature on Arthropoda metabolism there is little information concerning the possibility of using for glutamate synthesis the reducing equivalents implicated in the classic energy metabolism sequences. On the contrary, the important work of Slater and coworkers has revealed many interesting facts concerning the utilization of the Krebs cycle in the reductive amination of 2-oxoglutarate in mammals. These authors have produced evidence suggesting how reducing equivalents could be produced from intermediates of oxidative phosphorylation, which points to the importance, for amino acid synthesis, of energy-rich compounds that can drive the respiratory chain between succinate and NAD (58, 59).

Figure 5 summarizes this interpretation. In this scheme a constant amount of energy-rich intermediate is produced. Its utilization between ATP or glutamate synthesis is well balanced so that the oxygen consumption is not affected by the pathway that is followed. In the case of a decrease in glutamate synthesis, as happens when the animal is in freshwater, more of the energy-rich intermediate could be used for ATP synthesis that would,in turn, serve for the active transport of sodium from the dilute medium. If such correlations are correct, they would implicate direct relations between the amino acid metabolism and energy metabolism. These considerations seem to indicate that the differences between stenohaline and euryhaline species as well as those between the various cate-

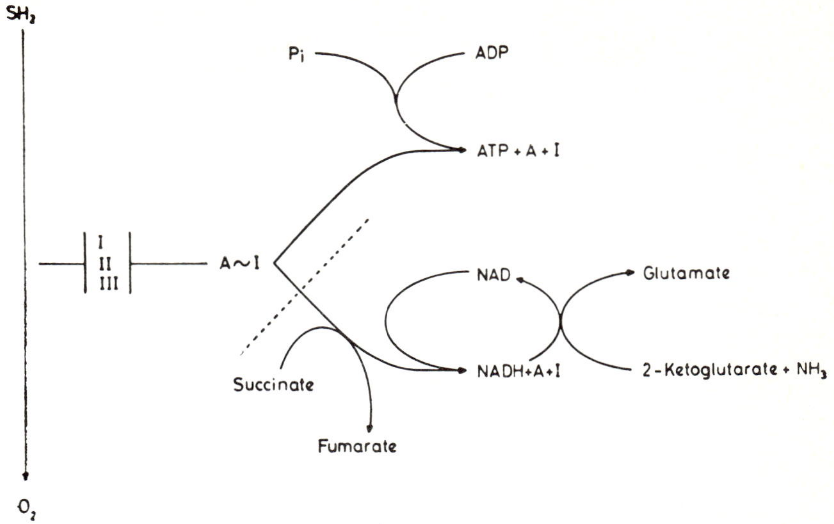

Figure 5 - Glutamate synthesis as uncoupler of oxidative phosphorylations. After Schoffeniels (51).

gories of osmoregulators must be searched for in the way the reducing equivalents are geared toward amino acid or energy metabolism (Figs.3 and 5).

Recently, Gilles, using a split-beam spectrophotometer has been able to show that the components of the respiratory chain and more specifically NAD get more reduced when a muscle of *Callinectes* is submitted to a hyposmotic stress; this is in agreement with the above interpretation.

The proposed scheme must be considered as being tentative, since more information on the intracellular localization of the enzyme involved and on the ionic composition prevailing locally is needed before a more complete picture can be produced. The relationships suggested above, if they explain satisfactorily some of the observations, are by no means considered to be exclusive of other regulatory mechanisms. For example, Schlieper (44; see also 42) has suggested that the changes in respiration rate which occur during adaptation are a consequence of changes in the hydration of the tissues,which may affect the activity of some enzymic systems. A change of intracellular osmotic pressure will cause a swelling of the tissue mitochondria thus altering the activity of the mechano-enzyme complexes. Schlieper (46) cites an example: the close correlation between the volume of *Nereis* and its respiratory rate after transfer to diluted media. However, no direct experimental evidence has yet been given in support of this hypothesis.

Though the changes in hydration of the tissues of a euryhaline species undergoing an osmotic stress are slight, as already observed by Florkin and his colleagues (3), they do exist but they cannot account for the change in the concentrations of the various osmotic effectors. The regulation of the cell volume during a hyposmotic stress seems to be faster than in the reverse situation. This seems to be true, not only when dealing with isolated tissues such as nerves, but also when working with the intact animal. The fact that proline is released from the cell interior provides an easy explanation. As we have already pointed out above, during a hyperosmotic stress, amino acids have to be synthesized, a process that may require some time before being completed. One could also argue that in our *in vitro* experiments with isolated nerves, the ratio of the volumes of extracellular space/intracellular space is very different from the *in vivo* situation.Though in Crustacea we do not have precise data as to the relative volume of the blood, with respect to that of the intracellular space (available data are 8% of body weight for *Maia*, 17% for *Homarus* and 38% for *Eriocheir* as far as the blood volume is concerned)[1], it is clear that in our experiments in which we use 100 mg of nerve in an extracellular volume of 5 ml, we are very far from the situation the animal encounters in nature. As pointed out by Olmstead (36) the greater the extracellular volume, the greater are the effects of an osmotic stress. In the case of *Eriocheir* adapted abruptly to seawater there is a shrinkage of the muscle fibers,amounting to up to 8% of the initial weight,that lasts about 10 days.

Thus, even in the intact animals, the change

(1 The extracellular space is thus greater in the euryhaline species. It should be useful to check the validity of such a correlation by studying systematically a large number of species.

in the osmotic pressure of the blood, as well as the volume of the extracellular space with respect to that of the intracellular space, leads to a small change in the hydration of the cell that takes several days to be controlled adequately through the regulation of the ionic and amino acid metabolism.

To conclude, euryhalinity cannot be unequivocally defined as resulting solely from the control of enzyme systems by inorganic ions. Obviously, it results from the association of a number of adaptations - some morphological, some biochemical. Euryhalinity implies the possibility of regulating adequately both the osmotic pressure of the blood and of the cells. As far as the regulation of the osmotic pressure of the blood is concerned, gills and the excretory organs are certainly of prime importance. It appears that an efficient control of the amino acid and energy metabolism is necessary for the regulation of the osmotic pressure of the cell.

REFERENCES

1. Beadle,L.C. (1937) Adaptation to changes of salinity in the polychaetes. I.Control of body volume and of body fluid concentration in *Nereis diversicolor*. J.exp.Biol. 14, 56-70.
2. Binyon,J. (1961) Salinity tolerance and permeability to water of the starfish *Asterias rubens* L. J.mar.Biol.Assoc. U.K. 41, 161-174.
3. Bricteux-Gregoire,S., Duchâteau-Bosson,Gh., Jeuniaux,C. & Florkin,M. (1962) Constituants osmotiquement actifs des muscles du crabe chinois *Eriocheir sinensis*, adapté à l'eau douce ou à l'eau de mer. Arch.internat.Physiol. Biochim. 70, 273-286.
4. Bursell,E. (1963) Aspects of the metabolism of amino acids in the tsetse fly, *Glossina* (diptera). J.Ins.Physiol. 9, 439-452.
5. Bursell,E. (1966) Aspects of the flight metabolism of tsetse flies (*Glossina*). Comp.Biochem. Physiol. 19, 809-818.
6. Cheng,S.C. & Mela,P. (1966) CO_2 fixation in the

nervous system. II. Environmental effects on CO_2 fixation in lobster nerve. J.Neurochem.13, 281-287.

7. Duchâteau,Gh. & Florkin,M. (1955) Concentration du milieu extérieur et état stationnaire du pool des acides aminés non protéiques des muscles d'Eriocheir sinensis, Milne Edwards. Arch.internat.Physiol.Biochim. 63, 249-251.

8. Duchâteau-Bosson,Gh. & Florkin,M. (1961) Change in intracellular concentration of free amino acids as a factor of euryhalinity in the crayfish Astacus astacus. Comp.Biochem.Physiol.3, 245-249.

9. Duchâteau-Bosson,Gh., & Florkin,M. (1962) Régulation isosmotique intracellulaire chez Eriocheir sinensis après ablation des pédoncules oculaires. Arch.internat.Physiol.Biochim. 70, 393-396.

10. Duchâteau,Gh., M.Florkin, & Jeuniaux,C. (1959) Composante aminoacide des tissus, chez les Crustacés. I.Composante amino-acide des muscles de Carcinus maenas L. lors du passage de l'eau de mer à l'eau saumâtre et au cours de la mue. Arch.internat.Physiol.Biochim. 67, 489-500.

11. Florkin,M.(1956) Vergleichende Betrachtung des stationären Zustandes der nicht-eiwissgebundenen Aminosäuren der Tiere. In Vergleichend biochemischen Fragen, 6. Colloquium der Gesellschaft für physiologische Chemie.pp.62-99, Berlin, Springer-Verlag.

12. Florkin,M. (1962) La régulation isosmotique intracellulaire chez les invertébrés marins euryhaline. Bull.Cl.Sci.Acad.Roy.Belgique 48, 687-694.

13. Florkin,M. & Schoffeniels,E.(1969) Molecular Approaches to Ecology. New York, Academic Press.

14. Florkin,M., Duchâteau-Bosson,Gh.,Jeuniaux,C. & Schoffeniels,E. (1964) Sur le mécanisme de la régulation de la concentration intracellulaire en acides aminés libres chez Eriocheir sinensis au cours de l'adaptation osmotique.

Arch.Internat.Physiol.Biochim. 72, 892-906.

15. Gilchrist,B.M. (1956) The oxygen consumption of *Artemia salina* (L) in different salinities. Hydrobiologica 8, 54-63.

16. Gilles,R. (1969) Effect of various salts on the activity of enzymes implicated in amino-acid metabolism. Arch.internat.Physiol.Biochim. 77, 441-464.

17. Gilles,R. & Schoffeniels,E. (1964) La synthèse des acides aminés de la chaîne nerveuse ventrale du homard. Biochim.Biophys.Acta 82,518-524.

18. Gilles,R. & Schoffeniels,E. (1965) Consommation d'oxygène et synthèse des acides aminés. Arch.internat.Physiol.Biochim. 73,144-145.

19. Gilles,R. & Schoffeniels,E. (1966) Décarboxylation des acides aspartique et oxaloacétique chez le homard et l'écrevisse. Bull.Soc.Chim. Biol. 48, 397-417.

20. Gilles,R. & Schoffeniels,E. (1968) Fixation de $^{14}CO_2$ par les acides aminés de la chaîne nerveuse ventrale du crustacé *Homarus vulgaris* M. Edw. Arch.internat.Physiol.Biochim. 76, 441-451.

21. Gilles,R. & Schoffeniels,E. (1968) Influence du NH_4Cl et du sulfate de vératrine sur la synthèse et le pool des acides aminés au niveau de la chaîne nerveuse ventrale de deux crustacés (*Homarus vulgaris* M.Edw. et *Astacus fluviatilis* F.). Arch.internat.Physiol.Biochim. 76, 452-464.

22. Gilles,R. & Schoffeniels,E. (1969) Metabolism of arabinose in the ventral nerve cord of the lobster *Homarus vulgaris* (M.Edw.). Comp.Biochemistry Physiol. 28, 1145-1152.

23. Gilles,R. & Schoffeniels,E. (1969) Isosmotic regulation in isolated surviving nerves of *Eriocheir sinensis* Milne Edwards. Comp. Biochem. Physiol. 31, 927-939.

24. Gilles,R. & Schoffeniels,E. (1969) Metabolic fate of glucose and pyruvate in the nerve cord of *Homarus vulgaris* (M.Edw.) and *Astacus flu-*

viatilis (F.). Comp.Biochem.Physiol. 28, 417-423.

25. Huggins,A.K. & Munday,K.A. (1968) Crustacean metabolism. In Advances in Comparative Physiology and Biochemistry, vol.3, pp.271-378, New York, Academic Press.

26. Jeuniaux,C. & Florkin,M. (1961) Modification de l'excrétion azotée du crabe chinois au cours de l'adaptation osmotique.Arch.internat. Physiol.Biochim. 69, 385-386.

27. Kermack,W.O., Lees,H. & Wood,J.D. (1954) Enzymes of lobster tissues. Biochem.J. 57, xxii.

28. King,E.N. & Schoffeniels,E. (1969) "In vitro" preparation of crab gill for use in ion transport studies. Arch.internat.Physiol.Biochim. 77, 105-111.

29. Kinne,O.(1964) The effect of temperature and salinity on marine and brackish water animals. II. Salinity and temperature-salinity combination. Oceanograph.Marine Biol. Ann.Rev. 2, 281-339.

30. Klingenberg,M. & Pette,D. (1962) Proportions of mitochondrial enzymes and pyridine nucleotides. Biochem.Biophys.Res.Commun. 7, 430-432.

31. Koch,H.J. (1954) Cholinesterase and active transport of sodium chloride through the isolated gills of the crab Eriocheir sinensis (M. Edw.) In Recent development in cell physiology, ed. Kitching,J.A., pp.15-27, New York,Academic press.

32. Lehninger,A. (1964) The Mitochondrion, New York, W.A.Benjamin.

33. McFarland,W.N. & Pickens,P.E.(1965) The effects of season temperature and salinity on standard and active oxygen consumption of the grass-shrimp Palaemonetes vulgaris (Say). Can.J.Zool. 43, 571-585.

34. McWhinnie,M.A. & Corkill,A.J. (1964) The hexosemonophosphate pathway and its variation in the intermolt cycle in crayfish. Comp.Biochem.

Physiol. 12, 81-93.

35. Needham,A.E. (1957) Factors affecting nitrogen-excretion in Carcinides maenas (Pennant). Physiologie comparata et Oecologia 4, 209-239.

36. Olmstead,E. (1963) Importance of the extra-cellular-intracellular water volume ratio in mammalian cell water exchange. Life Sci. 2, 745-750.

37. Papa,S., Tager,J.M., Francavilla,A. & Quaglia-riello,E. (1969) NAD(P)-linked oxidoreductions and the nicotinamide nucleotide specificity of glutamate dehydrogenase in rat-liver mitochondria. Biochim.Biophys.Acta 172, 20-29.

38. Papa,S., Tager,J.M., Guerrieri,F. & Quagliariello,E. (1969) Effect of monovalent cations on oxidative phosphorylation in submitochondrial particles. Biochim.Biophys.Acta 172, 184-186.

39. Peters,H. (1935) Über den Einfluss des Salzgehaltes in Aussenmedium auf den Bau und die Funktion des Exkretions organes dekapoder Crustacean Potamobius fluviatilis, Homarus vulgaris. Z.Morphol.Oekol.Tiere 30, 355-381.

40. Potts,W.T.W. (1958) The inorganic and amino acid composition of some lamellibranch muscles. J.exp.Biol. 35, 749-764.

41. Puyear,R.L., Wang,C.H. & Pritchard,A.W. (1965) Catabolic pathways of carbohydrate in the intermolt crayfish, Pacifastacus leninsculus. Comp.Biochem.Physiol. 14, 145-153.

42. Remane,A. & Schlieper,C. (1958) Die Biologie des Brackwassers. In Die Binnengewässer, ed. Thienemann,A., vol.XXII, Stuttgart,Schweizerbartsche Verlags Buchhandlung.

43. Scheer,B.T., Schwabe C.W. & Scheer,M.A.R. (1950-52) Tissue oxidation in Crustacea. III. On the hormonal regulation of metabolism in Crustacea. Physiol.Comp.Oecol. 2, 327-338.

44. Schlieper,C. (1929) Über die Einwirkung niederer Salzkonzentrationen auf marine Organismen. Z.Vergleich.Physiol. 9, 478-514.

45. Schlieper,C. (1929) Neue Versuche über die osmoregulation wasserlebenden Tiere. Sitzber. Ges.Befoerder.Ges.Naturw. Marburg 64, 143-156.

46. Schlieper,C. (1958) Sur l'adaptation des invertébrés marins à l'eau de mer diluée. Vie et Milieu 9, 139-152.

47. Schoffeniels,E. (1960) Origine des acides aminés intervenant dans la régulation de la pression osmotique intracellulaire de Eriocheir sinensis Milne Edwards. Arch.internat.Physiol. Biochim. 68, 696-698.

48. Schoffeniels,E. (1964) Effect of inorganic ions on the activity of L-glutamic acid dehydrogenase. Life Sci. 3, 845-850.

49. Schoffeniels,E. (1965) L-glutamic acid dehydrogenase activity in the gills of Palinurus vulgaris Latr. Arch.internat.Physiol.Biochim. 73, 73-80.

50. Schoffeniels,E. (1967) Cellular Aspect of Membrane Permeability. Oxford, Pergamon Press.

51. Schoffeniels,E. (1968) The control of intracellular hydrogen transport by inorganic ions. Arch. internat.Physiol.Biochim. 76,319-343.

52. Schoffeniels,E.(1971) Adaptation at the molecular scale. In Biochemical Evolution and the Origin of Life, ed.Schoffeniels,E., vol.2, pp. 314-335, Amsterdam, North Holland Publ.Co.

53. Schoffeniels,E. & Gilles,R. (1963) Effect of cations on the activity of L-glutamic acid dehydrogenase. Life Sci. 2, 834-839.

54. Schoffeniels,E. & Gilles,R. (1970) Osmoregulation in aquatic arthropods. In Chemical Zoology, ed.Florkin,M. & Scheer,B.T., vol.V, pp.255-285, New York,Academic Press.

55. Scholles,W. (1933) Über die Mineralregulation wasserlebender Evertebraten. Z.Vergleich. Physiol. 19, 522-554.

56. Schwabe,E. (1933) Über die Osmoregulation verschiedener Krebse (Malacostracen).Z.Vergleich. Physiol. 19, 183-236.

57. Siebers,D. (1972) Mechanismen der intrazellulären isosmotischen Regulation der Aminosäurekonzentration bei dem Flusskrebs Orconectes limosus. Z.Vergleich.Physiol. 76, 97-114.

58. Slater,E.C. & Tager,J.M. (1963) Synthesis of glutamate from α-oxoglutarate and ammonia in rat-liver mitochondria. Biochim.Biophys.Acta 77, 276-300.

59. Tager,J.M. & Slater,E.C. (1963) Synthesis of glutamate from α-oxoglutarate and ammonia in rat-liver mitochondria. I.Comparison of different hydrogen donors. Biochim.Biophys.Acta 77, 227-245.

60. Tustanoff,E.R., Stewart,H.B. (1965) The effect of neutral salts on particle preparations from rat liver. I.The effect of neutral salt on pyruvate oxidation. Can.J.Biochem. 43, 341-357.

61. Tustanoff,E.R. & Stewart,H.B. (1965) The effect of neutral salts on particle preparations from rat liver. II.Effects of neutral salts on tricarboxylic acid cycle reactions. Can.J.Biochem. 43, 359-372.

62. Vincent-Marique,C. & Gilles,R. (1970) Changes in the amino-acid concentration in blood and muscle of Eriocheir sinensis during hypoosmotic stress. Life Sci. 9, 509-512.

Comparative Physiology, eds. L. Bolis, K. Schmidt-Nielsen and S. H. P. Maddrell

SOME ASPECTS OF TRYPTOPHAN METABOLISM IN A LOCUST (*Schistocerca gregaria* Forsk)[x]

G.COLOMBO, S.PINAMONTI & A.PETRIS
Istituto di Zoologia,
Università di Ferrara,
Ferrara, Italy.

In Insects tryptophan metabolism via kynurenine to ommochromes follows the path outlined in the scheme shown below (Figure 1).

This metabolic chain has been widely studied because of the fact that in *Drosophila melanogaster* and in other Insects, several of its steps were, long ago, demonstrated to be under gene control (cfr. Ziegler,(18)).

In our laboratory this metabolic chain was investigated with a different approach, more precisely with the aim of determining the role of the different organs and of finding out their physiological control. For this purpose we used a locust, *Schistocerca gregaria* Forsk, which unfortunately has few mutants (only an albino strain was available), but it is comparatively large, making the dissection and the collection of a relatively large amount of material rather easy. Moreover we have breeding facilities and therefore animals of desired stages of development are available.

There were two main experimental approaches: 1) the determination of tryptophan metabolites and of their distribution in the different body parts;

---

x The researches were supported by contracts of the Italian "Consiglio Nazionale delle Ricerche". The results are due to the work of the Authors with the collaboration of the research workers and students, G.D'Orazio, M.Miliani, G.Chiarelli-Alvisi, O.Bellagamba, F.Ottaviani and M. Pinzetta.

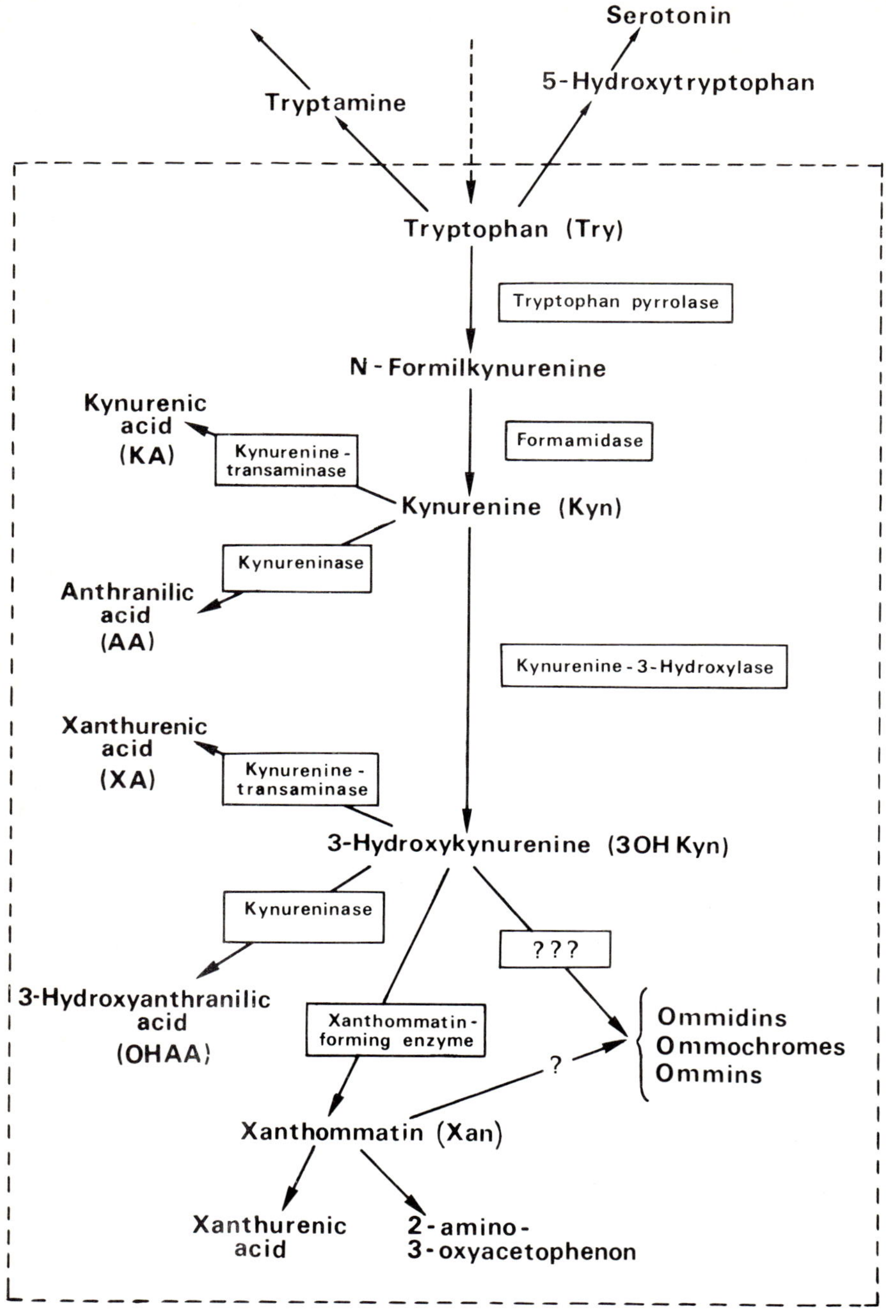

Figure 1 - Scheme of tryptophan metabolism via kynurenine to ommochromes.

2) the identification of the enzymatic systems involved in this pathway in the different organs and in the subcellular fractions.

The analytical methods for the metabolites' determination were those used by Coppini *et al.* (6), Butenandt *et al.* (1,2,3); Linzen (10), slightly modified in order to adapt them to our material.

For the enzymatic tests the methods used for mammalian systems were generally followed, modified according to the characteristics of the used tissues and of the tested enzymes.

Most of the data reported here have already been published, while others await publication (4,5,11,12,13,14,15,16,17).

Metabolites distribution

A summary of results of metabolite determinations is shown in table 1. They show that a small amount of kynurenine is stored in the newly laid eggs. Its formation by the embryos occurs late in development and later on the other metabolites appear. Enzymatic determination on tryptophan pyrrolase showed that this enzyme is present in small quantity at the early stages of development, probably produced by the mother and stored in the eggs, and only in later stages is it synthesized by the embryo. In conclusion the metabolic chain develops in the embryos along with tissue differentiation.

In the adults, as well as in nymphs according to our data not reported here, tryptophan, absorbed by the intestinal tract, is transformed into kynurenine only in the fat body. Through the hemolymph it is distributed to the other body parts, and in the eyes and epidermis, kynurenine is transformed into 3OH-kynurenine which is the source compound of the ommochromes. 3OH-kynurenine was never found in the blood, therefore it is likely that is used where it is formed. Xanthommatin and other ommochromes are synthesized only in the eyes and in the epidermal cells.

Some other points are worth emphasizing. No anthranilic and hydroxyanthranilic acids were ever detected in the hemolymph and tissues. Kynurenic acids and xanthurenic acids are present in blood; the former was also found in the fat body, the latter being found in the tegument and in the eyes.

In the malpighian tubes, the main excretion organs, almost all the metabolites, except anthranilic and hydroxyanthranilic acids, were found; and this is not easily understandable, unless one assumes an accumulation of compounds which are present in hemolymph in too low concentration to be detected with our methods. Moreover the degradation of some ommochromes cannot be excluded.

Enzymatic data

The enzymatic investigations clarify some of the above results but introduce other problems. A summary of our results is reported in Table 2.

Tryptophan pyrrolase was found only in the fat body cells; kynurenin-3-hydroxylase and the xanthommatin-forming enzyme only in the epidermis and in the ommatidia. In no organs was kynureninase detected. These data are in good agreement with results on the metabolite distribution.

Kynurenine transaminase was found only in fat body. This enzyme shows a poor specificity: *in vitro* xanthurenic acid was also formed by this enzyme, but 3OH-kynurenine was never found in the fat body cells and in few stages in the hemolymph. Xanthurenic acid is present in the eyes and in the epidermal cells where the transaminase was not detected. Xanthurenic acid can be formed by degradation of xanthommatin and of other ommochromes.

The metabolic pathway in the locust

From all the above data, a descriptive model of the tryptophan metabolic system in the locust was drawn, and is reported in Figure 2.

The proposed model is the first necessary approach to a physiological study. In this model the central endocrinological system is also indicated, by assuming that the system which strictly controls the growth, sexual maturation and egg production in the locust (7) also affects the tryptophan metabolism. We have not yet obtained conclu-

Distribution of tryptophan metabolites in different body parts

The values are the ranges of the metabolites content in adult females and males of different ages, found in several observations

($\mu g/mg$ of proteic N_2)

	Kyn.	KA	3OH-Kyn.	XA.	Xanth.	Ommi-dine	Ommo-chromes
<u>EMBRYOS</u>							
- Before blastokynesis	0.5 [1]	–	–	–	–	–	–
- After blastokynesis	1-1.5 [1]	0.1-0.5 [1]	2.5 [1]	–	traces	–	–
- At the hatching	1-1.5 [1]	3 [1]	20-25 [1]	0.2-1.2	+ +	+ +	+ +
<u>ADULTS ♀♀ and ♂♂</u>							
- Intestinal tracts	–	–	–	–	–	–	–
- Fat body	0.04-0.85	0.08-1.1	–	0.04-0.10	–	–	–
- Emolymph	1.0 - 12	4.0 - 12.0	–	0.04 - 1.0	?	?	?
- Malpighian tubes	0.04 - 1.10	0.04 - 2.85	0.04 - 1.30 [2]	0.04-0.75	?	–	?
- Epiderm	0.04	0.04 - 5.10	0.51 - 8.80	0.04-3.50	60-450	–	+ +
- Eyes	0.9 - 9.0 [3]	0.9 [3]	7.3 - 16.8	0,75 [3]	200-300	+ +	–

(1) $\mu g/g$ of wet weight

(2) only in females

(3) in females and males only few days after the ecdysis

Enzymatic activities of different body parts

Nymphs and adults ♀♀ and ♂♂ (1)	Tryptophan-pyrrolase system $\frac{\mu g \text{ of Kyn.}}{mg\ N_2 \cdot hr}$	Kynurenine-transaminase $\frac{\mu g \text{ of KA}}{mg\ N_2 \cdot hr}$	Kynureninase $\frac{\mu g \text{ of AA}}{mg\ N_2 \cdot hr}$	3OH-Kynurenin-hydroxylase $\frac{\mu g \text{ of 3OH-Kyn.}}{mg\ N_2 \cdot hr}$	Xanthommatin-forming enzymatic system $\frac{\mu g \text{ of Xan.}}{mg\ N_2 \cdot hr}$
FAT BODY	7.0	1.0	–	–	–
EYES	–	–	–	38	110
TEGUMENTS	–	–	–	13	350 (3)
MALPIGHIAN TUBES	–	–	–	–	–
HEMOLYMPH	0.20 (2)	–	–	–	–

(1) Enzymatic source: tissue homogenate after centrifugation at 40,000 x g

(2) μg of kyn/hr/ml of hemolymph only in egg laiding females

(3) Determined on mithochondrial fraction

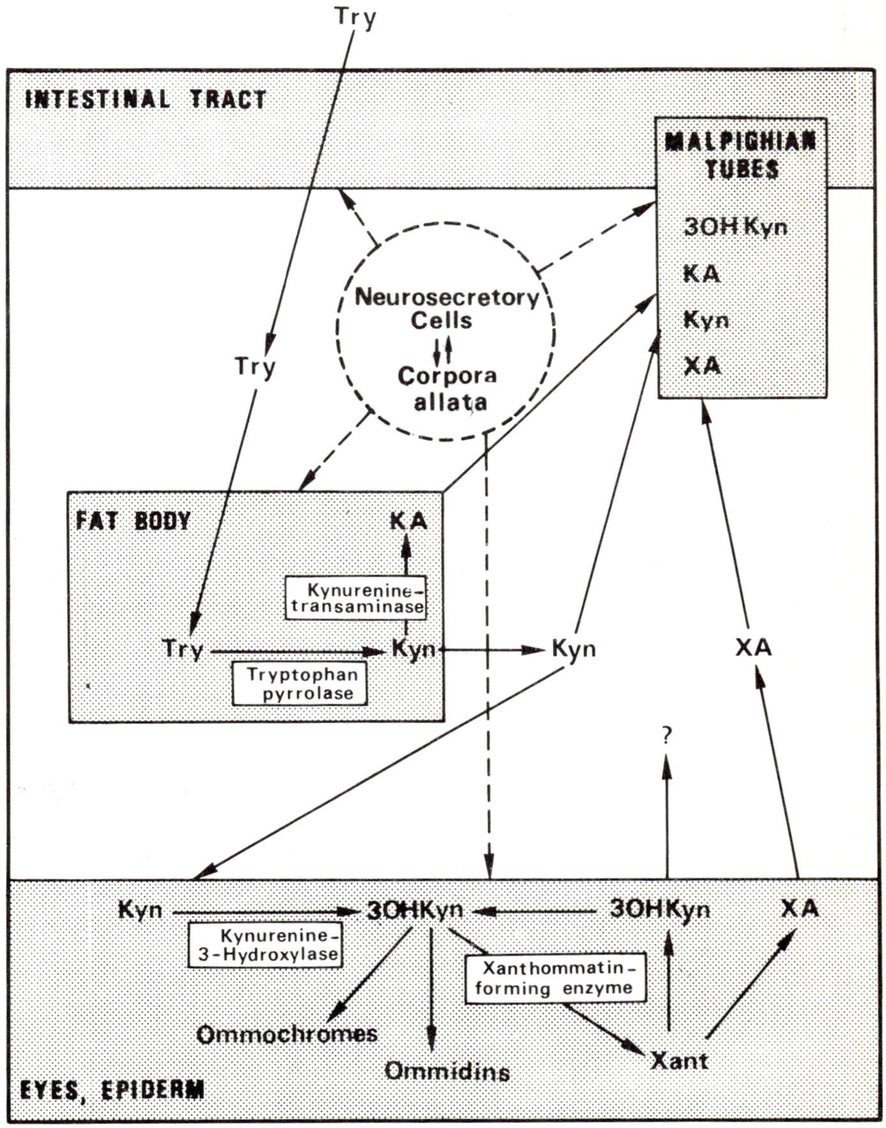

Figure 2 - Model describing the tryptophan pathway to ommochromes in *S. gregaria.*

sive experimental results on a direct endocrinological control of the tryptophan metabolic chain, but most of our descriptive biochemical observations are related to the stages of development and growth, therefore, some general relationships are suggested. We have carefully investigated the changes in pigments and in some enzymes of adult females and males in relation to sexual maturation and aging. Striking differences between sexes and among ages were evident; tryptophan pyrrolase of fat body is high in old egg-laying females while it decreases in old males.

Ommochromes of the teguments also decrease in sexually mature males when they change from pink-brown to yellow pigmentation. In allatectomized males this change in epidermal pigments does not occur and the metabolites and the enzymatic activities remain at high level at an age when the control and the sham-operated animals show a sharp modification.

Experiments on tryptophan-pyrrolase

Possible hormonal regulation of this metabolic activity does not seem to us to be properly studied in locusts by experiments of this kind, because it is not possible to tell which organs are directly affected by the endocrine system. There is evidence (8, 9) of a pattern of feeding of the adult locusts, therefore the enzymatic activities of fat body can be controlled also by the hemolymph content and by the concentrations of the available substrates.

This problem was approached by other experiments *in vivo* and *in vitro*; no definite conclusion was drawn but the results (not yet published) throw some light on the methodology to be followed.

Several experiments on starvation, on feeding and injection of tryptophan did not produce significant changes of tryptophan pyrrolase activities of fat body.

We approached the problem with experiments *in vitro* on kynurenine formation by fat body, taken out from adult females and males of different ages and incubated in locust physiological saline in the presence and absence of tryptophan, and with and without adding glucose of the incubation medium.

Some results are reported in the following tables.

Table 3 - Kynurenine released in incubation medium by fat body of males and females of various ages. Fat body was incubated for 3 hours at 37°C in Hoyle's saline supplemented with $1 \cdot 10^{-3}$M tryptophan in the presence (+) or absence (-) of 1% glucose. Kynurenine was determined spectrophotometrically with Bratton and Marshall colorimetric method on an aliquot (2 ml) of incubation medium.
Hoyle's solution has the following composition:
NaCl 701 mg; KCl 75,5 mg; $CaCl_2 \cdot 2H_2O$ 30 mg; $MgSO_4 \cdot 7H_2O$ 99 mg; $NaH_2PO_4 \cdot H_2O$ 55 mg; $Na_2HPO_4 \cdot 2H_2O$ 107 mg;
Penicillin 100 mg;Streptomycin 50 mg; bidistilled water 100 ml.

Days after the last moult	µg Kynurenine/h/mg total N_2 of fat body			
	Males		Females	
	-glucose	+glucose	-glucose	+glucose
0	1.75	1.45	3.40	1.40
10	9.70	4.10	8.10	3.30
15			9.89	3.82
20			13.06	2.96

We would prefer not to discuss the cellular mechanisms, because experiments on this point are not yet completed; we would like, however, to emphasize other aspects from a physiological point of view.

First, the stage of the development and the sex of the insect play a fundamental role in the reaction of the fat body cells *in vitro*; clearly

Table 4 - Tryptophan pyrrolase activity of homogenated fat body of males and females at various ages. Fat body was incubated for 12 hours at 37°C in Hoyle's saline with (+) or without (-) tryptophan $1 \cdot 10^{-3}$ M, and then immediately homogenized by sonication in KCl 0.14 M and centrifuged for 1 h at 40.000 x g. The clear supernatant solution was used as enzymatic source. The enzyme activity is expressed as µg of kynurenine formed per hour per mg of N_2.
Control = tryptophan pyrrolase activity of fat body homogenated soon after dissection. The standard reaction mixture was: 0.3 ml of clear supernatant, 0.1 ml of $2 \cdot 10^{-3}$M ascorbic acid; 0.8 ml of $2 \cdot 10^{-3}$M L-tryptophan; 0.8 ml Na phosphate buffer (0.1 M; pH 7.35).

Days after the last moult	MALES			FEMALES		
	Control	-Trypt.	+Trypt.	Control	-Trypt.	+Trypt.
0	3.2	6.9	11.4	7.5	11.9	16.1
7-8	2.3	5.3	13.6	2.0	3.6	13.8
18	-	-	-	1.53	7.1	31.5

Table 5 - Tryptophan pyrrolase activity (TPO) of fat body extract and of the incubation medium used as enzyme source. Fat body was incubated for three hours in Hoyle's saline containing $1 \cdot 10^{-3}$M tryptophan, in the presence (+) or absence (-) of glucose, homogenised by sonication, centrifuged at 40.000 x g; the clear supernatant was used as enzyme source. An aliquot of the incubation mixture was concentrated on Sephadex G-25 (final volume one third of the initial one) and used in enzymatic assay. The activity is expressed as µg of kynurenine per hour per mg of total N_2 of fat body. Control = Tryptophan pyrrolase activity of fat body homogenated soon after dissection. The enzymatic extract was prepared as indicated in the legend of Table 2.

	Control	+ GLUCOSE TPO extracted from fat body	+ GLUCOSE TPO released in incubation mixture	- GLUCOSE TPO extracted from fat body	- GLUCOSE TPO released in incubation mixture
10 days adult males	3.06	6.94	13.09	17.71	16.56
10 days adult females	4.69	8.23	7.41	20.39	7.84

Table 6 - Tryptophan metabolites determined by ion exchange resin and paper chromatography; μM of metabolite (K = kynurenine, KA = kynurenic acid) released in incubation medium per hour per mg of total N_2 by fat body of 10-12 days-old adults incubated in the presence (+) or absence (-) of glucose.

	- GLUCOSE		+ GLUCOSE	
	K	KA	K	KA
Males	0.034	0.020	0.003	0.0016
Females	0.015	0.003	0.002	0.012

the effects of the control by the endocrine system last for a long time and determine the further reactions.

Second, the reaction of the fat body to the composition of the incubating medium (and *in vivo* to the hemolymph composition) is immediate.

The model, previously proposed, is therefore not only descriptive but also shows causal relationships; the plan of our further researches is to give a numerical value to each relationship between the overall rates in the system and to the rates of each subsystem.

REFERENCES

1. Butenandt,A., Biekert,E. & Linzen,B. (1958) Isolierung und Charakterisierung von Omminen. Z.Physiol. Hoppe Seyler, 312, 227-236.

2. Butenandt,A., Biekert,E., Kubler,H. & Linzen,B. (1960) Zur Verbreitung der Ommatine im Tierreich Neue Methoden zu ihrer Identifizierung und quantitativen Bestimmung. Z.Physiol.Hoppe Seyler 319, 238-256.

3. Butenandt,A., Schiedt,U., Biekert,E. & Korman,P. (1954) Isolierung von Xanthommatin und Rhodommatin un Ommatin C aus den Schlupfsekreten von

Vanessa urticae. Liebigs Ann.Chem. 586, 217-228.

4. Colombo,G. & Pinamonti,S. (1965) Metaboliti del Triptofano via chinurenina in *Schistocerca gregaria* Forsk durante lo sviluppo embrionale. Rend.Classe Scienze Fisiche,Matematiche e Naturali 38, 728-731.

5. Colombo,G., Facchini,L. & Pinamonti,S. (1966) Gli acidi nucleici nel corso dello sviluppo embrionale di *Schistocerca gregaria* Forsk (Orthoptera) ed andamento dell'attività triptofan-pirrolasica. Atti del XIV Convegno del G.E.I. Acta Medica Romana 4, 38-42.

6. Coppini,D., Benassi,C.A. & Montorsi,M. (1953) Quantitative determination of tryptophan metabolites (via kynurenine) biological fluids. Clinical Chem. 5, 391-401.

7. Highman,K.C. & Hill,L. (1969) *The comparative endocrinology of the invertebrates.* London, Edward Arnold Ltd.

8. Hill,L., Luntz,A.J. & Stelle,P.A. (1968). The relationships between somatic growth, ovarian growth, and feeding in the adult desert locust. J.Insect Physiol. 14, 1-20.

9. Hill,L. & Goldsworthy,G.J. (1968) Growth, feeding activity and the utilization of reserves in larvae of *Locusta*. J.Insect Physiol. 14, 1085-1098.

10. Linzen,B. (1966) Ommidine ein neuer Typus von Ommochromen aus Orthopteren. Zeitschrift für Naturforschung 21b, 1038-1047.

11. Pinamonti,S., D'Orazio,G. & Veronese,F.M.(1964) Ricerche sul metabolismo del triptofano via chinurenina in *Schistocerca gregaria* Forsk (Ortotteri). Boll.Zool. 33, 567-571.

12. Pinamonti,S. & Petris,A. (1966) Studies on tryptophan pyrrolase of *Schistocerca gregaria* Forsk (Orthoptera, Acrididae). Comp.Biochem.Physiol. 17, 1079-1087.

13. Pinamonti,S., Petris,A. & Colombo,G. (1966) Nucleic acids, proteins, and Tryptophan pyrro-

lase activity of the fat body of Schistocerca gregaria Forsk (Orthoptera) during ovarian maturation. J.Insect Physiol. 12, 1403-1410.

14. Pinamonti,S., Petris,A. & Miliani,M. (1970) Studies on the kynurenine transaminase of Schistocerca gregaria Forsk (Orthoptera Acrididae). Comp.Biochem.Physiol. 37, 311-320.

15. Pinamonti,S., Petris,A. & Ottaviani,F. (1971) Studio sulla chinurenin-3-idrossilasi in Schistocerca gregaria Forsk (Orthoptera, Acrididae). Atti della Acc. delle Scienze di Ferrara 47, 1-12.

16. Pinamonti,S., Chiarelli-Alvisi,G. & Colombo,G. The ommochromes of the eyes and the integument of albino and pigmented locusts Schistocerca gregaria Forsk (Orthoptera) (submitted for publication).

17. Pinamonti,S., Chiarelli-Alvisi,G. & Colombo,G. The xanthommatin forming enzyme system of Schistocerca gregaria Forsk (Orthoptera, Acrididae) (submitted for publication).

18. Ziegler,I. (1961) Genetic aspects of ommochrome and pterine pigments. Advanc.Genet. 10, 349-403.

Comparative Physiology, eds. L. Bolis, K. Schmidt-Nielsen and S. H. P. Maddrell

INTESTINAL ABSORPTION AND SECRETION IN VERTEBRATES

V. CAPRARO & G. ESPOSITO
Istituto di Fisiologia Generale,
Università di Milano,
Milano, Italy.

The most important function of the intestinal mucosa is obviously the absorption of organic food materials, water and mineral ions. In addition to this function, an intestinal secretion concerning enzymes, mucopolysaccharides, water, hormones and mineral ions is also present.

The present survey deals only with some features of epithelia and particularly with the intestinal epithelium, i.e. cell polarity, absorption and secretion of ions, water, organic food materials and some differences between the *in vivo* and *in vitro* conditions.

1. Cell polarity

The intestinal epithelium is a chemical engine which promotes a net flux of matter from one side to the other of the barrier, perpendicularly to the surface of the epithelium. That is the reason why the epithelium shows a typical morphological and chemical polarity, e.g. a sensitivity to ouabain of the membrane facing the serosal side and a sensitivity to phlorizin of the luminal membrane (brush border) of the absorbing cell. For the same reason, the physical characteristics of the luminal and serosal membranes of the epithelium cells differ too.

If we determine in the rat small intestine *in vitro* the concentration profile of an uncharged, chemically inert and passively diffusing substance, such as acetamide, in a condition close to the steady state, we observe that the concentration gradient across the serosal side facing the membrane is much higher than that across the brush border (Figure 1). This asymmetric behaviour may be

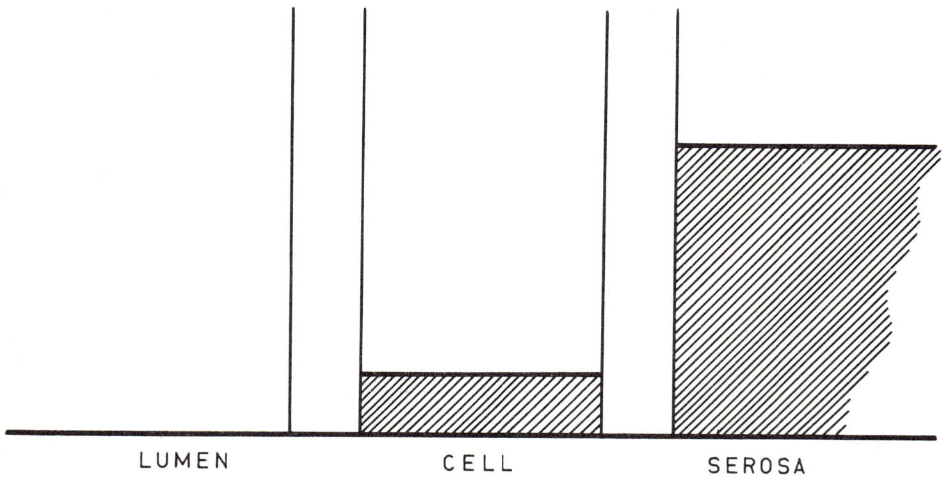

Figure 1 - Concentration profile of acetamide, a passively diffusing substance, in the three compartments of an *in vitro* preparation (everted and cannulated jejunum of rat), in a nearly steady state condition. Concentration difference between serosal and cellular compartments in more than four times the concentration difference between cellular and luminal compartments.

due to a difference in permeable area and/or to a difference in permeability per unit area (18). A further asymmetry is also represented by the ohmic resistance of the two membranes (Table 1). These observations were made on the absorbing cells of rabbit gall-bladder (12).

2. Absorption of mineral ions

The isolated intestine incubated *in vitro* is able to absorb sodium against an electrochemical gradient (Table 2). If the mucosal and serosal perfusing solutions are at the same sodium concentration not far from the concentration of the perfusing fluids.

Table 1 - Electrical resistances of rabbit gall bladder mucosa. R_w, transmural total resistance, R_e, resistance of the epithelial layer, R_m, resistance of the mucosal pole of a single epithelial cell, R_s, resistance of the serosal pole of a single epithelial cell. Mean values ± S.E. of the mean and number of utilized gall bladders, in parentheses, are reported.

ERRATUM

COMPARATIVE PHYSIOLOGY:

LOCOMOTION, RESPIRATION, TRANSPORT AND BLOOD.

Editors: L. Bolis, K. Schmidt-Nielsen, S. H. P. Maddrell

p. 401, 5th and 6th line of legend to Table 1 should read "…. pole of the epithelial cell ….".

	(8)

But, at least in fishes, there are examples of different behaviour. In the intestine of the marine teleost Cottus scorpius (33,34) there is a flow of a solution much more concentrated (1.6. Equiv) than the perfusing Ringer phosphate solution (0.2 Equiv). On the contrary, the solution transferred by the intestine of the fresh water teleost, Carassius auratus (52), has a sodium content which is lower (0.04 Equiv.) than the perfusing Ringer-phosphate solution.

In all classes of Vertebrates the transfer of sodium is an active process; therefore, it depends on metabolism (34,54,10,9,39,43,2,22,7,41). This conclusion is based on the discrepancy between the observed sodium flux ratio and the ratio calculated from the flux equation:

$$\frac{\phi_{1-2}}{\phi_{2-1}} = \frac{C_1}{C_2} e^{\frac{-F\Delta E}{RT}}$$

where ϕ_{1-2} is the mucosa-serosa unidirectional flux and ϕ_{2-1} is the unidirectional flux in the opposite direction, C_1 and C_2 are the concentrations in the two compartments, F is the Faraday, R the gas constant, T the absolute temperature and ΔE is the transmural potential difference.

When the potential difference (ΔE) is close to zero and C_1 and C_2 are identical, the two fluxes through the intestine are not the same, but there is a net flux from 1 to 2.

In fishes (Table 2), the transmural potential difference (ΔE) is close to zero; in other vertebrates (Table 2) the condition is experimentally achieved by short circuiting the intestine. The transmural potential difference is very low in fishes (34), of the order of a few millivolts only in the small intestine of all classes of vertebrates. In these cases, a sodium chloride pump is presumably present; sodium as well as chloride ion is actively transported (43,14,5). Other anions, such as bicarbonate have been thought to cross actively the intestinal barrier (human ileum and jejunum _in vivo_ (53), rat jejunum _in vitro_ (20)). A direct demonstration of a chloride pump is based, as previously said, on the discrepancy between the experimental data and those predicted by the flux equation (34,43,14).

There is also an indirect proof of a chloride pump, namely the observation that the Q_{10} of the unidirectional chloride flux from mucosa to serosa is higher than the Q_{10} of the unidirectional chloride flux from serosa to mucosa (5).

The potential difference across the large intestine is higher (40-50 mV) in all classes of vertebrates (Table 2). In such cases the chloride ion is probably driven by the electric field due to the active flux of sodium ion; but in the large intestine of Greek Tortoise it seems that there is an active chloride absorption (26).

Sodium pool of the intestinal epithelial layer as well as transepithelial active sodium flux seem to be correlated to the sodium concentration

Table 2 - Intestinal absorption of water and mineral ions in vertebrates. Transport data, when possible, are referred to 1 cm^2 and 1 hour in "in vitro" perfusion (except for *Anguilla*). Numbers in parentheses are the reference numbers.

Species	Water net transport	Transep potential	Na net flux from mucosa to serosa	Cl net flux from mucosa to serosa
Cottus scorpius (33,34)	8.1 $\mu l\ cm^{-2}\ h^{-1}$	0.6 mV	12.9 $\mu Eq\ cm^{-2}\ h^{-1}$	nearly the same
Opsanus tau (1)	25.7 $\mu l\ g^{-1}\ h^{-1}$			
Anguilla anguilla (50)	104 $\mu l\ cm^{-2}\ h^{-1}$		32 $\mu Eq\ cm^{-2}\ h^{-1}$	42 $\mu Eq\ cm^{-2}\ h^{-1}$
Carassius auratus (52)	2.5 $\mu l\ cm^{-1}\ h^{-1}$		0.2 $\mu Eq\ cm^{-1}\ h^{-1}$	
Ameiurus nebulosus (15)		2.5 mV		
Rana catesbeiana large intestine (10)		45 mV	2.25 $\mu Eq\ cm^{-2}\ h^{-1}$	1 $\mu Eq\ cm^{-2}\ h^{-1}$
Rana catesbeiana small intestine (43)		3 mV	1.48 $\mu Eq\ cm^{-2}\ h^{-1}$	1.17 $\mu Eq\ cm^{-2}\ h^{-1}$
Bufo marinus large intestine (9)			0.67 $\mu Eq\ cm^{-2}\ h^{-1}$	
Bufo arenarum large intestine (39)		40-50 mV	1.94 $\mu Eq\ cm^{-2}\ h^{-1}$	
Testudo hermanni small intestine (2)		2-4 mV	0.73 $\mu Eq\ cm^{-2}\ h^{-1}$	
Testudo hermanni large intestine (2,32)		20-50 mV	0.67 1.58 $\mu Eq\ cm^{-2}\ h^{-1}$	
Albino rat small intestine (7)	60 $\mu l\ cm^{-2}\ h^{-1}$	5-10 mV	8 $\mu Eq\ cm^{-2}\ h^{-1}$	
Albino rat large intestine	163 $\mu l\ g^{-1}\ h^{-1}$ / 10 (41)	12-15 mV (17)		

in the mucosal fluid by a saturation curve (4) as it happens in the epithelium of a urinary bladder of the toad *Bufo marinus* (24).

3. Water absorption

Water transport between two isosmotic media or against an osmotic gradient seems to depend on metabolism and to follow an active solute transport (14,8).

In the everted small intestine of albino rat incubated in a Ringer-bicarbonate solution added to with glucose, the net fluid movement from mucosa to serosa ranges between 80-120 µl $cm^{-2}h^{-1}$; with a maximum rate in the middle part of the small intestine (3). The rate of water transfer decreases when an opposite osmotic gradient is artificially imposed and it nearly stops when this gradient is above 100 mOsm/Kg (39,30). Anyway, the freezing point of the transported fluid in mammals at least, is always only 2-3% lower than that of the luminal perfusing fluid (46).

In order to explain the relationship between water and solute transport, a model of two in series membranes have been put forward by Curran (14). This model has been widely supported by many experimental and morphological observations (16,36). The intercellular space can be considered a space delimited by the in series membranes, i.e. the plasma cell membrane facing the intercellular compartment (first membrane) and the subepithelial layers (second membrane). The intercellular space becomes hypertonic because of an active transport of solutes (sodium salts, organic food materials) across the first membrane. Water moves in the same direction, driven by the osmotic force, and enlarges the intercellular space; during this process water comes mainly through the first membrane because it has a higher reflexion coefficient. The enlargement of the intercellular space against a mechanical resistance produces an hydrostatic pressure which in turn causes a net flow of water mainly through the second membrane (Figure 2). As a matter of fact, the second membrane has a higher hydraulic permeability coefficient. Therefore, a net flow of a nearly isotonic fluid crosses the whole

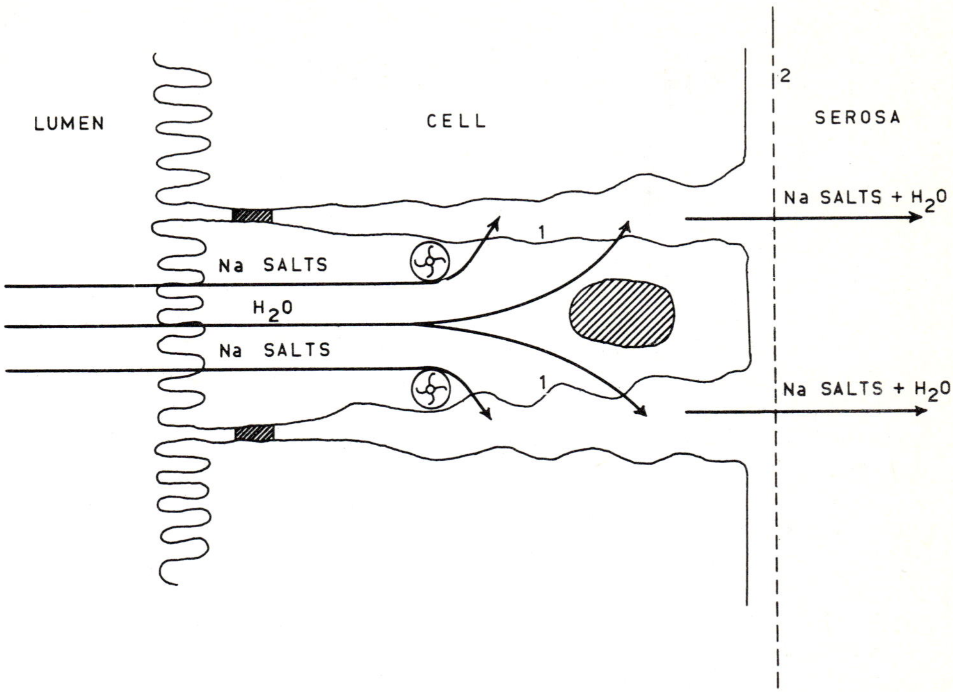

Figure 2 - Schema of the model of two in series membranes applied to the intestinal epithelium. The first membrane (1) is the plasma cell membrane facing the intercellular compartment. The subepithelial layers are the second membrane (2).

barrier.

4. Secretion of inorganic ions

Particularly in the small intestine, monovalent ions are not subjected only to an absorption process. In the isolated small intestine of the Greek Tortoise, there is a direct demonstration of an active secretion of chloride ion into the lumen (26). In the same intestinal tract there is indirect evidence that sodium ion is actively secreted as well (2,29). It could be supposed that in the jejunum the mucosa to serosa influx of chloride ion, as well as the serosa to mucosa outflux of the same ion, are, in part at least, active processes, but the outflux is greater than the influx (29). The two processes of absorption and secretion presumably take place in different and antagonistically polarized epithelial cells. As mentioned above, the direct demonstration is based on the discrepancy between the experimental and the calculated fluxes ratio of chloride ion. On the other hand the indirect evidence is based on the sensitivity of one or both fluxes to the inhibition by dinitrophenol and ouabain and to a stimulation activity by glucose.

5. Absorption of organic food materials

The absorption of monosaccharides, aminoacids and fats takes place in the small intestine. This phenomenon has been widely studied in mammals and, to a smaller extent, in other classes of vertebrates. As to this absorption, three characteristics are common to all classes of vertebrates:

1) the rate of influx of a solute from the lumen into the epithelium, as a function of its concentration in the lumen, follows the Michaelis-Menten equation; therefore, the absorption is presumably a chemically mediated process:

$$\dot{n} = AD \frac{[C_r]}{\Delta_x} \frac{[C_s]}{K_m + [C_s]}$$

where $\dot{n}$ is the solute influx, A is the area of the membrane, $[C_r]$ is the concentration of the carrier, Δ_x is the thickness of the membrane, $[C_s]$ is the concentration of the solute, D is the diffusion coefficient of the complex carrier + transported substance within the membrane and K_m is the Michaelis constant.

2) The "in vitro" absorption is usually followed by an uphill accumulation of the solute in the absorbing cells and in the serosal compartment (active transport).

3) The absorption is dependent on the ionic composition of the luminal solution and particularly on the presence of sodium.

As to points 1) and 2) we remember the observations on sugar absorption in Amphibia (42) and in Mammalia (rat, hamster and rabbit) (51,6,22,11, 30) and the observations on amino acid absorption in Reptilia (27) and in Mammalia (21,55,35).

The Michaelis-Menten equation predicts "per se" only a saturation kinetics but not an accumulation against a concentration gradient. The accumulation is possible if the influx follows kinetics different from that of the outflux. This difference can be explained if we take into account point 3), i.e. the sodium dependence of the absorption rate. The entry of sugars and amino acids into the cell and the exit in the opposite direction are very sensitive to the corresponding ionic environment.

Since the well known experiments of Riklis and Quastel (44) on the transmural transfer of sugars in the guinea pig small intestine incubated "in vitro" in a Ringer-bicarbonate solution, a lot of work has been performed on this topic.

The accumulation of D-glucose and other actively transported sugars in the perfusing fluid, as well as of amino-acids, are sodium dependent processes in fishes (37), in Amphibia (13,45) and in Mammalia (6,49,25).

Conversely, the presence of glucose or other actively transported sugars strongly enhances the short circuit current, i.e. the sodium active transport.

The rapidity of the substrate effect suggests an action other than the one depending on the metabolic energy. This hypothesis is supported by the fact that the same stimulation is performed also by non metabolized sugars (48). Furthermore, the uptake of sodium by the isolated rabbit ileum is increased by the presence of 3-O-methyl-glucose and, at every sodium concentration tested there is a 1:1 ratio between sugar uptake and sugar dependent Na uptake (30).

Based on these data is Crane's hypothesis (11) that the carrier in the brush border forms a ternary complex with the transported substance as well as with the monovalent cations present in the surrounding fluid.

The predominant binding with sodium in the mucosal fluid positively affects the affinity of the carrier to the transported molecule and, on the other hand, the predominant binding with potassium in the intracellular fluid decreases this affinity.

Such an asymmetry could allow an accumulation of a substance in the cell compartment and a secondary accumulation in the serosal space. The lowering of the electrochemical activity of crossing cations could be the driving energy responsible for the uphill movement of organic food materials (gradient hypothesis).

6. "In vivo" absorption

Recent experiments performed "in vivo" in human and rat small intestine (47, 23) seem to give less importance to the sodium dependent uptake and transmural transport of glucose. In these experiments, however, the amount of sodium in the intestinal lumen is never nil. Furthermore the "in vivo" condition seems to be, to some extent, very different from the "in vitro" (19). First of all, the "in vivo" extrusion of sugars and sodium from the cell into the serosal compartment is higher than in the isolated intestine (Table 3).

Table 3 - Comparison between "in vitro" and "in vivo" net transport of sodium and D-glucose in the rat jejunum. Data are referred to 1 g dry tissue weight and 1 hour. Mean values $\pm$ S.E. of the mean and number of experiments in parentheses are reported. Perfusing fluids are in both cases Krebs-Henseleit-bicarbonate solution with addition of D-glucose 13.9mM. The experimental temperature of "in vitro" experiments is 38°C. Values concerning the isolated everted sac are calculated from data reported by Capraro, Bianchi & Lippe (Arch.Sci.Biol. Bologna, 47,238,1963)(wet weight:dry weight = 6.7). D-glucose data reported in the same paper have been obtained by sugar reducing methods, so that an over estimation has to be taken into account.

Net transport from mucosa to serosa or blood	Everted sac of rat jejunum	Rat jejunum perfused «in vivo»	
Sodium μequiv. $g^{-1}.h^{-1}$	924 $\pm$ 80 (10)	1618 $\pm$ 262 (7)	$P < 0.05$
D-glucose μmoles. $g^{-1}.h^{-1}$	142 $\pm$ 29 (10)	261 $\pm$ 51 (7)	$P < 0.1$

Sugars and sodium do not accumulate within the cell as they do in the isolated intestine (Figures 3 and 4). Sugars so far tested (D-glucose and 3-0-methyl-glucose are always present in the cell at a lower concentration than in the lumen and blood.

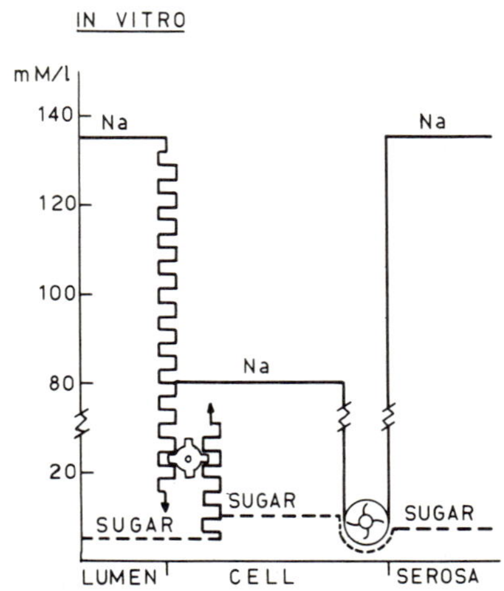

Figure 3 - Concentration profile of sodium and sugar in the three compartments of an "in vitro" preparation (everted cannulated jejunum of rat). Metabolically dependent sodium and sugar extrusion processes are located at the serosal pole of the absorbing cell whereas a linkage between cations and sugar (gradient hypothesis) is located at the mucosal pole of the cell; in this condition the intracellular sodium concentration is higher than "in vivo" (see fig.4) and sugar accumulates within the cell (Data from: A.Faelli, G.Esposito and V. Capraro, Arch.Sc.Biol. Bologna, 1966, 50, 234; Bronk J.R. & Parsons,D.S., J.Physiol. 1965, 179, 323 and Flores,P. & Schedl,H.P., Am.J.Physiol., 1968, 214, 725).

Therefore, the extrusion of sugars from the inside of the epithelial cell towards the blood seems to be an active process due to a sugar pump. The rate of sugar entry from the lumen into the cell depends on the favourable concentration gradient across the brush border and presumably it is also enhanced by the ionic asymmetry between the two sides of the

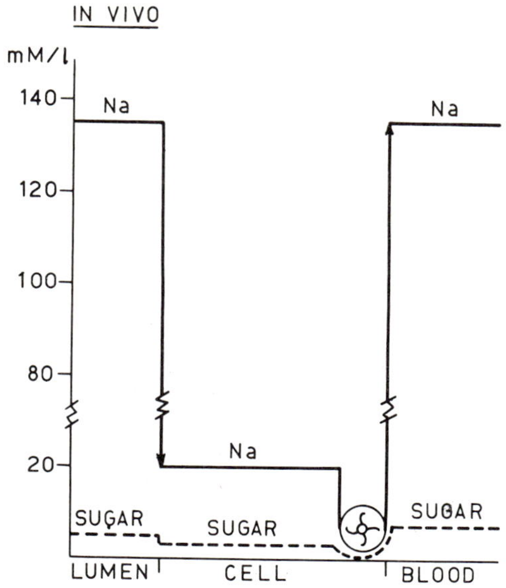

Figure 4 - Concentration profile of sodium and sugar in the three compartments of an "in vivo" preparation of rat jejunum. Metabolically dependent sodium and sugar pumps are located at the serosal pole of the absorbing cell. In this condition also the sugar pump is working against the sugar concentration gradient. Both sugar and sodium enter the cell downhill from the lumen. The linkage between cation and sugar movement (gradient hypothesis) enhances the rate of sugar entry.

brush border itself. On the contrary, the rate of "in vitro" sugar entry is not counterbalanced by the rate of extrusion so that they accumulate within the cell, according to the gradient hypothesis.

REFERENCES

1. Aull,F. (1966) Absorption of fluid from isolated intestine of the toad fish Opsanus tau. Comp.Biochem.Physiol. 17, 867-870.

2. Baillien,M. & Schoffeniels,E. (1961) Origin of bioelectrical potentials of the intestinal epithelium of the Greek tortoise. Biochim.Biophys.

Acta 53, 537-548.

3. Barry,B.A., Matthews,J. & Smyth,D.H. (1961) Transfer of glucose and fluid by different parts of the small intestine of the rat. J.Physiol. 157, 279-288.

4. Bianchi,A. & Vecli,A. (1965) Effetti della sostituzione del cloruro di sodio della soluzione perfondente sul trasporto attivo di sodio da parte della cute isolata di *Rana esculenta*. Boll.Soc.It.Biol.Sper. 41, 1590-1593 (Fig.2).

5. Bianchi,A.Giordana,B. & Repetto,F. (1973) Chloride transport in rabbit isolated jejunum and ileum. Arch.It.Fisiol. - in press.

6. Bihler,I. & Crane,R.K. (1962) Studies on the mechanism of intestinal absorption of sugars. V. The influence of several cations and anions on the active transport of sugars, *in vitro*,by various preparations of hamster small intestine. Biochim.Biophys.Acta 59, 78-93.

7. Capraro,V. Some aspects of the functional polarity of epithelium. In: *Radioisotope in der Gastroenterologie*, *Dritte Jahrestagung* (see here references of previous papers). Shottauer Verlag, Stuttgart.

8. Clarkson,T.W. & Rothstein,A. (1960) Transport of monovalent cations by the isolated small intestine of the rat. Am.J.Physiol. 199, 898-906.

9. Cofré,G. & Crabbé,J. (1967) Active sodium transport by the colon of *Bufo marinus*, stimulation by aldosterone and antidiuretic hormone. J.Physiol. 188, 177-190.

10. Cooperstein,H. & Hogben,C.A.M. (1959) Ionic transfer across the isolated frog large intestine. J.Gen.Physiol. 42, 461-473.

11. Crane,R.K. (1965) Na^+-dependent transport in the intestine and other animal tissues. Fed. Proc. 24, 1000-1006.

12. Cremaschi,D., Henin,S. & Ferroni,A. (1972) Electrical potential profile across the rabbit gall bladder. Atti dell'Accademia Nazionale

dei Lincei. Classe Sc.Fis.Mat e Nat. 52,563-568.

13. Csaky,T.Z. & Thale,M. (1960) Effect of ionic environment on intestinal sugar transport. J.Physiol. 151, 59-65.

14. Curran,P.F. (1960) Na, Cl and water transport by rat ileum in vitro. J.Gen.Physiol. 43,1137-1148.

15. Datta,S. & Savage,N.B. (1968) Investigation of the electrical potential on the isolated midgut of the freshwater teleost Ameiurus nebulosus. Experientia 24, 572-574.

16. Diamond,J.M., Tormey,J. (1966) Role of long extracellular channels in fluid transport across epithelia. Nature 210, 817-820.

17. Edmonds,C.J. & Morriott,J. (1964) Electrical potential and short-circuit current of an in vitro preparation of rat colon mucosa. J. Physiol. 194, 479-494.

18. Esposito,G., Faelli,A. & Capraro,V. (1970) Effect of sodium on passive permeability of non-electrolytes through the intestinal wall. In Permeability and Function of Biological Membranes, ed. L.Bolis, Katchalsky,A., Keynes, R.D., Loewenstein,W.R. & Pethica,B.A. pp.74-85, Amsterdam, North Holland Publishing Company.

19. Esposito,G., Faelli,A. & Capraro,V. (1972) "In vivo" intestinal absorption of sugars and electrolytes. Experientia 28, 1182-1183.

20. Faelli,A., Esposito,G. (1971) Bicarbonates and transintestinal potential differences of the jejunum rat intestine incubated "in vitro". 1st Europ.Biophys.Congr., ed.Broda,E. & Locker, A., E VIII/29 pp. 317-321, Wien H.Springer-Lederer.

21. Finch,L.R. & Hird,F.J.R. (1960) The uptake of aminoacids by isolated segments of rat intestine. 1. A survey of factors affecting the measurement of uptake. Biochim.Biophys.Acta 43, 268-277.

22. Fisher,R.B. & Parsons,D.S. (1953) Glucose move-

ments across the wall of the rat small intestine. J.Physiol. 119, 210-223.

23. Förster,H.(1972) In: Na-linked transport of organic solutes. ed.Heinz,E., p. 134, Springer-Verlag.

24. Frazier,H.G., Dempsey,E.F. & Leaf,A.(1962) Movement of sodium across the mucosal surface of the isolated toad bladder and its modification by vasopressin. J.Gen.Physiol. 45,529-543.

25. Fuisz,R.E., Schultz,S.G. & Curran,P.F.(1966) Effect of sodium on L-alanine flux across the mucosal border of intestine. Biochim.Biophys. Acta 112, 593-595.

26. Gilles-Baillien,M. & Schoffeniels,E. (1967) Fluxes of inorganic ions across the isolated intestinal epithelium of the Greek tortoise. Arch.int.Physiol. et Biochem. 75, 754-762.

27. Gilles-Baillien,M. (1970) Permeability characteristics of the intestinal epithelium and hibernation in Testudo Hermanni hermanni (Gmelin) Arch.Int.Physiol. et Biochem. 78, 327-338.

28. Giordana,B., Bianchi,A. & Lippe,C. (1969) Trasporto di Na dalla sierosa al lume attraverso il tenue isolato di Testudo graeca. Effetto del 2.4-dinitrofenolo e della ouabaina. Rend. Ist.Lombardo, classe B. 103, 115-119.

29. Giordana,B., Bianchi,A., Lippe,C. (1970) Chloride transport across the isolated intestinal mucosa of Testudo Graeca. Comp.Biochem.Physiology 36, 395-401.

30. Goldner,A.M., Schultz,S.G. & Curran,P.F.(1969) Sodium and sugar fluxes across the mucosal border of rabbit ileum. J.Gen.Physiol. 53, 362-383.

31. Green,K., Seshadri,B. & Matty,A.J. (1962) Independence of transfer of solutes and solvent across the rat ileum. Nature 196,1322-1323.

32. Hajjar,H.J. (1971) Ion transport in proximal and distal halves of turtle colon. Comp. Biochem.Physiol. 40A, 39-44.

33. House,C.R. & Green,K. (1963) Sodium and water transport across isolated intestine of a marine teleost. Nature 199, 1293-1294.

34. House,C.R. & Green,K. (1965) Ion and water transport in isolated intestine of the marine teleost, Cottus scorpius. J.Exp.Biol. 42,177-189.

35. Jervis,E.L. & Smyth,D.H. (1959) The effect of concentrations of aminoacids on their rate of absorption from the intestine. J.Physiol. 149, 433-441.

36. Kaye,G.J., Weeler,H.O., Whitlock,R.T. & Lane, N. (1966) Fluid transport in rabbit gallbladder. A combined physiological and electronic microscopy study. J. Cell Biol. 30 ,237-268.

37. Kitchin,S.E. & Morris,D. (1971) The effect of acclimatation temperature on aminoacid transport in the goldfish intestine. Comp.Biochem. Physiol. 40A, 431-443.

38. Levin,R.J. (1966) Transmural potentials across the small and large intestine of the bullfrog, Rana catesbeiana. Proc.Soc.Exp.Biol.Med. 121, 1033-1038.

39. Lew,V.L. (1970) Short circuit current and ionic fluxes in the isolated colonic mucosa of Bufo arenarum. J.Physiol. 206, 509-528.

40. Parsons,D.S. & Wingate,D.L. (1958) Fluid movements across wall of rat small intestine "in vitro". Biochim.Biophys.Acta 30, 666-667.

41. Parsons,D.S. & Paterson,C.R. (1960) Movements of fluid and glucose in an everted sac preparation of rat colonic mucosa. Biochim.Biophys. Acta 41, 173-175.

42. Parsons,D.S. & Prichard,J.S. (1968) Disaccharide absorption by amphibian small intestine "in vitro". J.Physiol. 199, 137-150.

43. Quay,I.F. & Armstrong W. (1969) Sodium and chloride transport by isolated bullfrog small intestine. Am.J.Physiol. 217, 694-702.

44. Riklis,E. & Quastel,J.H. (1958) Effects of cations on sugar absorption by isolated survi-

ving guinea pig intestine. Comp.Biochem. Physiol. 36, 347-361.

45. Robinson,J.W.L. (1970) Comparative aspects of the response of the intestine to its ionic environment. Comp.Biochem.Physiol. 34,641-655.

46. Rossi,S. & Capraro,V. (1961) Net transport of water and solutes by an "in vitro" rat intestinal preparation. Experientia 17, 543-544.

47. Saltzman,D.A., Rector,F.C. & Fordtran,J.S. (1972) The role of intraluminal sodium in glucose absorption "in vivo". J.Clin.Invest. 51, 876-885.

48. Schultz,S.G. & Zalusky,R. (1963) The interaction between active sodium transport and active sugar transport in the isolated rabbit ileum. Biochim.Biophys.Acta 71, 503-505.

49. Schultz,S.G., Fuisz,R.E. & Curran,P.F. (1966) Amino acid and sugar transport in rabbit ileum. J.Gen.Physiol. 49, 849-866.

50. Skadhauge,E. & Maetz,J.(1967) Etude "in vivo" de l'absorption intestinale d'eau et d'electrolytes chez Anguilla anguilla adapté à des milieux de salinités diverses. C.R. Acad.Sc.Paris 265, 347-350.

51. Smyth,D.H. & Taylor,C.B. (1957) Transfer of water and solutes by an "in vitro" intestinal preparation. J.Physiol. 136, 632-648.

52. Smith,M.W. (1964) The "in vitro" absorption of water and solutes from the intestine of goldfish, Carassius auratus. J.Physiol.175,38-49.

53. Turnberg,L.A., Fordtran,J.S., Carter,N.W. & Rector,F.C. (1970) Mechanism of bicarbonate absorption and its relationship to sodium transport in the human jejunum. J.Clin.Invest.49, 548-556.

54. Ussing,H.H. & Andersen - Cited by Cofré and Crabbé (see 9).

55. Wiseman,G. (1953) Absorption of aminoacids using an "in vitro" technique. J.Physiol. 120, 63-72.

Comparative Physiology, eds. L. Bolis, K. Schmidt-Nielsen and S. H. P. Maddrell

CARBONIC ANHYDRASES OF THE INTESTINAL TRACT

D.S. PARSONS
Department of Biochemistry,
University of Oxford,
Oxford, England.

1. BACKGROUND

Acid base changes during salt absorption. Our interest in carbonic anhydrase derives from experiments on salt absorption in the small intestine and colon of small mammals (Parsons, 40,41). These experiments provided evidence that the metabolism of carbon dioxide is closely related to salt absorption by the intestine. Solutions of sodium chloride, isotonic with normal extracellular fluid and buffered in the physiological range of pH with $NaHCO_3$ and CO_2 are absorbed approximately isotonically. However, rates of absorption of sodium and of chloride differ, and changes in the acid base composition of the intestinal contents are observed. In the rat, differences between the relative rates of absorption of sodium and of chloride are found in different regions of the intestinal tract, and these differences are related to the fact that bicarbonate ions rapidly disappear from jejunal contents but are secreted into the contents of the ileum and colon during the absorption of simple Ringer-bicarbonate solutions. During the absorption of such solutions from the colon and ileum, the bicarbonate content of the luminal contents may rise to values as high as 70 mM, whereas during absorption from jejunal segments, the concentration may be reduced to less than 10 mM in man, rat and dog (41). During the absorption of such solutions from the jejunum, the absorbate apparently contains a mixture of sodium bicarbonate and sodium chloride, whereas in the ileum and colon the absorbate contains no bicarbonate and there is a net entry of bicarbonate ions into the lumen (40,42,54). Evidently these bicarbonate ions may enter the lumen either in exchange for an anion, e.g. chloride, or accompanied by a cation such as sodium or potassium. The secretion of bicarbonate ions into rat ileum appears to require the presence of chloride ions in the intestinal lumen (Hubel, 24). The upper small intestine of the hamster, rabbit and guinea pig appears to resemble rat ileum and colon in that bicarbonate is secreted into the luminal contents of this region (See 45, 61).

I originally supposed that, at least in the jejunum of the rat, some of the sodium was absorbed in exchange for hydrogen ions which reacted with the bicarbonate present in the intestinal contents, releasing CO_2. This would mean that the bicarbonate was 'absorbed' as free CO_2. Supporting evidence for this was found in the values of pCO_2 found in the intestinal contents at different locations along the intestinal tract (40,54) (Table 1).

TABLE 1

	Equilibrium concentration of total CO_2 mM/L	Final pCO_2 mmHg $\pm$s.e.of mean (n)	
	dog	dog	rat
Jejunum	9.2	46.3$\pm$0.7 (57)	59.1$\pm$3.6 (9)
Ileum	77.4	36.1$\pm$0.7 (61)	51.0$\pm$3.6 (9)
Colon	72.9	29.8$\pm$0.5 (105)	46.5$\pm$2.1 (11)

Equilibrium concentrations of total CO_2 observed in contents of dog intestine absorbing saline-bicarbonate solution together with calculated partial pressure of CO_2 during absorption of saline-bicarbonate solution by dog and rat intestine. (Data of Parsons (40) and of Swallow and Code (54)).

Interactions between CO_2 and salt absorption. Further evidence of interrelations between carbon dioxide and salt absorption is that CO_2 and bicarbonate ions stimulate sodium and water absorption by the human small intestine *in vivo* (21,51) and also in rat jejunum in vitro (see 41). This effect of carbon dioxide on luminal contents appears to be absent in the ileum and colon of the rat (42).

In addition, it has been found that administration of the carbonic anhydrase inhibitor, acetazolamide, has a marked inhibitory effect on salt absorption from rat small intestine in vivo (40).

All these observations suggest that there may be interactions between CO_2 metabolism and the transport of salt across the intestinal mucosa; indeed it does not require much imagination to construct hypotheses involving movements interrelated between chloride and sodium on the one hand and bicarbonate ions and hydrogen ions derived from CO_2 and water on the other.

It is for these reasons that it seemed worthwhile to look for the presence of the enzyme carbonic anhydrase in the intestinal tract.

2. CARBONIC ANHYDRASE

Carbonic anhydrase (EC 4.2.1.1) catalyses the hydration of carbon dioxide to form carbonic acid; it can be called carbon dioxide hydratase. It is a curious feature of the enzyme that when the "backward" reaction is catalysed, i.e.the dehydration of carbonic acid,the substrates are, in fact, bicarbonate ions and protons (17). The enzyme also catalyses the hydrolysis of some esters (52,57) and the dehydration of aldehydes (44).

Meldrum and Roughton (36,37,38) separated carbonic anhydrase from erythrocytes, the enzyme acting as a catalyst for the hydration of carbon dioxide and for the dehydration of carbonic acid. It is to these authors that we owe the name of the enzyme. Horace Davenport was the first to demonstrate the presence of carbonic anhydrase in tissues, and described its presence in the stomach (Davenport,13) (Davenport & Fisher,15) and in the kidney (16; see also 14). Carbonic anhydrase appears to exist in two distinct major forms. (For a recent survey see 19). One type, which we call 'high activity' carbonic anhydrase (HACA) is sometimes known as carbonic anhydrase C, or carbonic anhydrase II. This variety possesses a much higher carbon dioxide hydratase specific activity than the other type, the 'low activity' carbonic anhydrase (LACA), carbonic anhydrase B, or carbonic anhydrase I. Both major sorts of enzyme possess molecular weights of about 30,000 and each is formed of a single polypeptide chain to which is attached one zinc atom. The two sorts of enzyme show certain characteristic differences in amino acid composition and in some physicochemical properties (9,18).

3. PROPERTIES OF THE TWO VARIETIES OF CARBONIC ANHYDRASE

Amino acid composition. - The amino acid composition of guinea pig 'high activity' and 'low activity' varieties of the carbonic anhydrase differ in quite a characteristic way. It appears to be a general rule that the 'low activity' variety of the enzyme in man, monkey, horse and guinea pig contains about one-and-a-half times as many serine residues as the corresponding 'high activity' variety of the enzyme (9). In addition, the number of residues of the basic amino acids lysine and histidine are greater in the molecules of the 'high activity' variety of the enzyme. Such differences in composition are consistent with the view that the 'high activity' and 'low activity' carbon anhydrase varieties arose through a gene duplication (see 9). The fact that, as will be shown below, the varieties of the enzyme are distributed very differently in the gastrointestinal tract, together with the observation that in ox,a variety absent in erythrocytes is present in the rumen epithelium, also implies the existence of independently-controlled genetic loci for the two major isoenzymes. When the amino-acid composition of the two sorts of enzyme from human erythrocytes are compared the similarities appear to be sufficiently significant to indicate that both varieties sprang originally from a common precursor (22). The amino acid composition of the two kinds of guinea pig enzyme is sufficiently similar to suggest that the same may be true in this species.

Influence of ions. - In the presence of chloride ions, the activity of both varieties of carbonic anhydrase is inhibited. At pH 7.2, 0°C, the sensitivity of 'high activity' enzyme of guinea pig tissues to this inhibition is considerably less than that of the'low activity' variety; for example, under these conditions the concentration of chloride necessary to produce 50% inhibition of the 'low activity' enzyme was about 14 mM, while that for the 'high activity' variety was about 400 mM. Organic monovalent ions such as acetate, pyruvate and butyrate have also been found to inhibit carbonic anhydrase. It seems that these ions do not show the same preferential inhibition of the 'low activity' variety of the guinea pig enzyme (12).

4. CARBONIC ANHYDRASES OF BLOOD

Before considering the nature and distribution of carbonic anhydrase in the tissues of the gastrointestinal tract, something must be said of the varieties of the enzyme in animal red cells. Although many species, including man, possess both varieties of the enzyme in erythrocytes, in some species only the 'high activity' is present (Table 2).

TABLE 2

Species with erythrocytes containing only high specific activity enzyme	Author
Ox	(7,18)
Dog	(5)
Sheep	(55)
Dolphin	(49)
Chicken	M.J. Carter (unpublished observations)
Shark	(34)
European moose	(29)

Absence of low specific activity variety of carbonic anhydrase from blood of certain species.

There is evidently a large excess of total activity of carbonic anhydrase in erythrocytes; recent estimates suggest that this excess is approximately 500 times (33). However, Professor Sir Hans Krebs (26) has pointed out that excess activities of this order are not exceptional for enzymes catalysing the reactions which are non-rate-limiting; also estimates of total carbonic anhydrase activity *in vivo* vary according to whether allowance is made for the inhibitory effects on HACA of the Cl ions present in the erythrocytes.

No obvious explanation is forthcoming as to why there are present in the red cells of some species two varieties of an enzyme, both forms of which seem to perform an apparently straightforward role (but see below). According to Edsall (18) both varieties of the human red cell enzyme contribute substantially to the overall carbonic anhydrase activity. When measured at 0°C, the kinetic properties of the varieties of carbonic anhydrase found in rat and guinea pig (12,35) seem to suggest that the contribution of the 'low activity' enzyme to total activity is small in erythrocytes of these species. However, at 38°C the relative contributions of the two enzymes could be rather different. The fact that the 'low activity' isoenzyme is absent from the red cells of some mammalian species (Table 2), and may occasionally be absent from human erythrocytes (47), leads to the conclusion that whatever function it has in other tissues, the 'low activity' variety of carbonic anhydrase is not essential for the occurrence of adequate rates of exchange of carbon dioxide in erythrocytes *in vivo*.

5. CARBONIC ANHYDRASE IN STOMACH

Monogastric animals. - Davenport & Fisher (15) showed that the carbonic anhydrase activity of gastric mucosa is of the same order of magnitude as that of the blood. Davenport later related the amount of enzyme in the tissue to the number of parietal cells present, but his data indicate that the enzyme is also present in non-parietal cells (13). In the acid-secreting part of the guinea pig stomach the carbonic anhydrase activity (Table

3) is due to the presence of an homogeneous variety of carbonic anhydrase which was found to consist almost entirely of the same 'high activity' variety as that present in the erythrocytes. The 'low activity' variety appears to be absent. The 'high activity' type of enzyme is present at a concentration of more than 0.5 mg enzyme per 100 mg gastric mucosal protein (Figure 1).

TABLE 3

	Total carbonic anhydrase activity units
Whole blood	240 ± 8 (6)
Mucosa of:-	
Acid-secreting portion of stomach	830 ± 37 (6)
Gall bladder	340 ± 44 (5)
Jejunum	23 ± 3 (6)
Ileum	33 ± 3 (6)
Caecum	280 ± 24 (6)
Proximal colon	560 ± 25 (6)

Carbonic anhydrase activity of some guinea pig tissues. Data of Carter & Parsons (11), q.v. for details of assay. Units are μM CO_2 hydrated min^{-1} mg^{-1} protein. Values are mean ± s.e. of mean (number of observations).

Adult ruminants. - Carbonic anhydrase is present in the epithelial cells of the mucosa of ox rumen; the apparent specificity of this tissue being about one-third of that of whole blood. The activity of the rumen epithelium is due to the presence of a single type of enzyme. This enzyme has been extracted and purified; from both the kinetic data and from the study of its amino acid composition the type of enzyme is clearly closely related to the 'low activity' variety found in other species (7). It will be recalled that, unlike some other mammalian species, the erythrocytes of

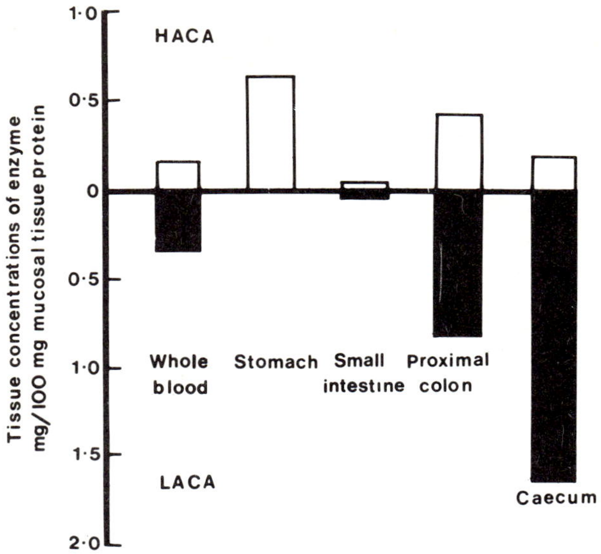

Figure 1 - Estimated concentrations as mg of enzyme protein/100 mg mucosal tissue protein of the 'high specific activity' (HACA) and 'low specific activity' (LACA) varieties of carbonic anhydrase in blood and gastrointestinal tract of guinea pig.
From Carter & Parsons (11).

ox blood contain only one variety of carbonic anhydrase and this is of the 'high activity' type. It therefore follows that the enzyme present in the ox rumen epithelium is quite distinct from the type present in the blood. The concentration of this enzyme in the rumen mucosal tissue is high, being of the order of one mg of enzyme protein per 100 mg mucosal epithelial cell protein (7).

6. CARBONIC ANHYDRASE IN SMALL INTESTINE

Presence in small intestinal mucosa. - The carbonic anhydrase activity of rat small intestinal mucosa was first studied by Davenport and Fisher (15) who showed that in parts of the small intestine the activity was much less than that of the gastric mucosa. More recently, these results have received qualitative confirmation for the dog and rat (27,33), guinea pig, hamster and rat (8,11) (Table 3 and Fig.1).

Carter and Parsons (8,11) have presented evidence in support of the view that the small amount of carbonic anhydrase activity found in the mucosa of the small intestine of the guinea pig, hamster and rat was not due to contamination of the samples with blood. They concluded that, although the exact location of the carbonic anhydrase activity and its relation to transport processes in the mucosal cells of the small intestine was unknown, the slight amount of activity which is indubitably present suggests that the enzyme might be restricted to one specific cell type which occurs relatively infrequently in the whole mucosal cell population. Whatever may be the biochemical basis of the ionic exchanges which constitute the processes of salt transport in the small intestine, they do not seem to depend upon the presence of carbonic anhydrase. The phenomena of net salt absorption and bicarbonate secretion appear to be very similar in the ileum and colon; it is therefore possible that the carbonic anhydrase found in the caecum and large intestine (see below) has functions other than those concerned with salt transport.

Effects of carbonic anhydrase inhibitors. The finding that there are only relatively small amounts of carbonic anhydrase in the small intestinal mucosa, yet that an inhibition of salt absorption is found in animals that have been treated with carbonic anhydrase inhibitors, may well be an example of an important principle. This is the fact that a specific inhibitor of an enzyme may also have effects on other, quite different, biochemical systems. Nevertheless, the affinity of the inhibitor for the differing systems upon which it acts may be different, and hence the different dose may produce different effects. In the experiments reported by Parsons (40) rather large doses of acetazolamide were used. In this connection it is of interest that Hogben (23) has pointed out that inhibitors of carbonic anhydrase have a more marked effect on the chloride secretion than on the proton secretion by gastric mucosa. He concluded that the inhibitors may react with some anion transporting system unrelated to carbonic anhydrase.

7. CARBONIC ANHYDRASE IN CAECUM

The mucosa of the caecum of the guinea pig has as much total carbonic anhydrase as does whole blood (Table 3). The enzymes in caecum have been isolated and it is found that the total activity is due to the presence of a mixture of both 'high activity' and 'low activity' varieties of carbonic anhydrase. The 'low activity' was present at about ten times the concentration of the 'high activity' variety (Figure 1). From observations made on the pure enzymes, isolated from the mucosa, data on kinetic properties and on amino acid compositions were obtained. On this basis it was found that the 'low activity' variety of the enzyme obtained from the caecal mucosa is indistinguishable from the corresponding variety of the enzyme present in the erythrocytes. A striking feature of the caecal enzyme from the guinea pig is the large concentration in the tissue. No less than between 1.5 - 2.0 mg of total carbonic anhydrase enzyme protein per 100 mg of tissue protein are found in the caecal mucosa of adult guinea pigs.

8. CARBONIC ANHYDRASE IN COLON

Kuriaki and Magee (27) described the presence of carbonic anhydrase activity in the mucosae of the proximal colon of dogs and also rats. The total carbonic anhydrase activity was estimated to be similar to that of the gastric mucosa. Maren (33) reported the presence of carbonic anhydrase in rat colon, but his data indicate that the specific activity of the colonic enzymes is less than that of blood. In the guinea pig, the carbonic anhydrase activity of the proximal colonic mucosa is about twice the enzymic activity of guinea pig whole blood (Table 3). The enzyme from guinea pig colon has been purified and is found to consist of a mixture of the 'high activity' and 'low activity' varieties in proportions approximately the same as that found in the erythrocytes (9,10,11) (see also

Fig.1). From data obtained on purified enzymes derived from the blood and mucosa it is clear that colonic mucosal enzymes are identical in composition with the corresponding enzymes present in the blood. There appears to be some evidence that the mucosa of the distal colon of the guinea pig possesses only about one twentieth of the carbonic anhydrase activity of the proximal colonic mucosa. Nevertheless, the quantity of the 'low activity' variety of the enzyme present in the distal colon is quite appreciable (11).

9. POSSIBLE PHYSIOLOGICAL ROLE OF CARBONIC ANHYDRASE IN INTESTINAL TRACT

The origin of the studies on the distribution and properties of the carbonic anhydrase enzymes in the mucosal epithelium in the intestinal tract which I have described stems from a consideration of an earlier hypothesis. This hypothesis was invoked to explain the facts that sodium and chloride are absorbed to some extent independently from saline solutions in the intestinal tract and that associated with the absorption there were concomitant movements of CO_2. It was suggested that some of the sodium was absorbed in exchange for protons and some of the chloride by an independent process involving an exchange with bicarbonate. In order to explain the different acid base changes observed in the luminal contents at different locations in the small intestine, the hypothesis also requires that the sodium/proton exchange system is more active in the upper small intestine and the chloride/bicarbonate exchange system more active in ileum and colon. It is now evident that if the postulated ionic exchanges do occur, carbonic anhydrase does not appear to be implicated in them because, in spite of the fact that acetazolamide inhibits salt transport in the small intestine and colon, very little of the enzymes is found in either the jejunum or ileum of the species studied.

Distribution of enzymes. - We may summarise the distribution of the varieties of carbonic anhydrase in the intestinal tract as follows. The 'high activity' variety of the enzyme is always found in blood, being present in high concentrations in all erythrocytes. It is also found in gastric mucosa and in colon and caecum. It appears that when the various enzymic activities of the 'high activity' variety of the carbonic anhydrase are compared between different species, the relative hydrolytic esterase and carbon dioxide hydratase activities are similar. In contrast, the relative hydrolytic and hydratase activities of the 'low activity' variety vary much more widely between different species (58).

The 'low activity' variety of the isoenzyme is found in some, but not all, erythrocytes (Table 1).

As far as I know no comparative study has been made of the presence of this variety of the enzyme in the two species of sheep, i.e. those containing 'high' and those containing 'low' concentrations of potassium in the erythrocytes. The erythrocytes of cattle and of dogs possess a high sodium content. Incidentally, so do those of cats, but in this case no information is available on the nature of the blood carbonic anhydrase. Although the data are incomplete, it is tempting to associate the absence of the 'low activity' variety of the enzyme with the presence of a relatively low potassium content in the erythrocytes.

Role of gastrointestinal tract. - The 'low activity' variety of the enzyme is not present in guinea pig gastric mucosa, but is present in very large amounts in caecum and rumen epithelium and also particularly in the proximal colon. The characteristic distribution of the 'low activity' variety of the enzyme in the epithelia enclosing the fermentation chambers of the intestine, such as rumen, caecum and colon, may be related to a transport and metabolism of the products of fermentation (Figure 2). A possible connection between carbonic anhydrase and the absorption of volatile fatty acids from the rumen has been suggested (Aafjes, 1) and I have already presented data which indicate that fatty acid anions do influence carbonic anhydrase.

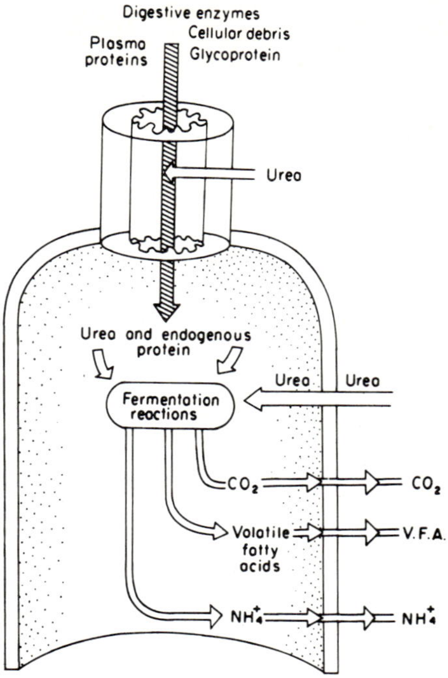

Figure 2 - The caecum as a fermentation vat. Endogenous protein and urea, and exogenous undigested material, e.g. cellulose, entering mainly through the ileocaecal valve, are converted by microbiological action to NH_3, CO_2 and volatile fatty acids VFA (e.g. propionate and butyrate) which are absorbed. An intriguing question is whether a similar system exists in marine animals, which must have a very different intestinal flora from that of terrestrial animals. Marine mammals possess relatively a very short colon and a small caecum.

Ammonia - Another product formed during microbial action in fermentation chambers within the intestinal canal is ammonia. In non-ruminant animals by far the largest and most constant source of ammonia in the blood is the large intestine (20, 39, 50). This ammonia may be of nutritive value (see, for example, 46, 53). It is formed by the breakdown of proteins and of urea, and there is evidence of a special carbon dioxide - bicarbonate-dependent mechanism existing for its absorption. There appears to be no effect of ammonia on the activity of either variety of carbonic anhydrase (12,48). Nevertheless, I would like to point out that there may be a close relationship between carbonic anhydrase and ammonia in several tissues.

In the urinogenital tract of humans,mammals and birds, movements of ammonia may be related to the presence of carbonic anhydrase. Carbonic anhydrase is known to be present in kidney (16) and it has been known for a long time that there are positive correlations between the titratable acidity and the secretion of ammonia in the urine of carnivorous animals (60). Claude Bernard (4) pointed out that carnivorous animals excrete a more acid urine than herbivorous animals, and it is known that the kidneys of herbivorous animals secrete little ammonia (3). It now appears that the kidneys of herbivorous animals contain much less carbonic anhydrase activity (11,59). Carbonic anhydrase has also been implicated in the excretion of ammonium ions by fish gills (31). In the shell-gland of birds carbonic anhydrase and ammonia have been implicated in the complex process of shell formation (6). In the rabbit a rise in the carbonic anhydrase activity of the endometrium during late pregnancy (30) may be related to the appearance of large quantities of ammonia in the uterine vein at that time (39).

<u>Carbon dioxide of fermentation.</u> - The classical role for carbonic anhydrase is to promote the hydration of carbon dioxide and the dehydration of carbonic acid although, as pointed out earlier, it appears that the real substrate for the latter reaction is in fact bicarbonate ion and hydrogen ion. It is therefore possible that the presence of the variety of the carbonic anhydrase which possesses a 'low specific activity' may be related to the fact that the gas in the contents of the gastrointestinal tract of many animals contains a very high proportion of CO_2, often up to 70% by volume in the case of rumen and also in the human colon (Table 4). The CO_2 is derived from decarboxylation of organic acids and also from the hydro-

TABLE 4

Organ	Species	CO_2 content of gas (per cent CO_2)	Equilibrium concentration, 38°C 0.2M NaCl (mM/Kg H_2O)	Author
Rumen	ox,goat,lamb,cow	45 - 70	10.7 - 16.7	(25,56)
Caecum	human	12.5	3.0	(32)
Colon	human	43 - 70	10.2 - 16.7	(56)
		8.5 - 21.1	2.0 - 5.0	(43)
Rectum	human	42.9	10.2	(32)
		36.4	8.7	(56)

CO_2 content of intestinal gases. Also shown is the equilibrium concentration of CO_2, dissolved at the appropriate partial pressure, in 0.2M NaCl solution at 38°C. Se also Levitt & Bond (28).

lysis of urea. (See, for example, Annison & Lewis, 2). Equilibrated at 38°C in saline solution these gas mixtures will yield up to 16 mM/L of CO_2 in solution (Table 4). The presence of the enzyme in the epithelium lining fermentation chambers such as the rumen and caecum may then be related to the metabolism of carbon dioxide produced within these chambers by fermentation reactions. It would be interesting to attempt to correlate the development of the fermentation process in the lumen and the appearance of the characteristic enzyme in the mucosal epithelium of young animals. In particular, a study of the forestomach of young ruminants (e.g. calves) and the caecum of young herbivorous animals (e.g. rabbits) would be well worthwhile.

The experiments I have described may be used to exemplify two general principles. In the first place, you can only discover the enzyme you have first of all thought of; in other words you have to think of a substrate, say X, and test it on the preparation you are interested in to see if it is transformed. If it is, then you call it X-ase. But the fact that there is an X-ase in a tissue does not mean that the physiological role of the enzyme *in vivo* is related to the metabolism of X or indeed that X is an important metabolite for that tissue. The enzyme may have a completely different role, by reacting with another completely different substrate, which no-one has yet thought of and tested. The second principle, of course, is that most beautiful hypotheses are destroyed by ugly facts.

SUMMARY

Acid base changes occur during absorption of isotonic salt solutions from the small and large intestine. Thus, bicarbonate ions are secreted into the contents of the duodenum and jejunum of the rabbit and guinea pig. In rat, dog and man bicarbonate disappears from the contents of the upper small intestine, possibly as CO_2 following H-ion entry in exchange for Na. Salt absorption is inhibited by inhibitors of carbonic anhydrase and, in the jejunum, stimulated by CO_2. Carbonic anhydrase exists in two distinct forms, one of high specific activity (HACA) and one of low specific activity (LACA). Both varieties are present in the blood of many species including primates, but LACA is absent from that of cattle, sheep,dog, chicken, European moose and shark. In the small intestinal mucosa of rat, chicken, and guinea pig very little carbonic anhydrase is present, yet large quantities are present in the colonic and caecal mucosae of rat and guinea pig. In the colonic mucosa of the guinea pig both HACA and LACA are present in concentrations exceeding that of blood. In the caecum the enzyme is present in very high concentrations (1.5 mg enzyme per 100 mg mucosal cell protein) mostly as LACA. In guinea pig gastric mucosa carbonic anhydrase is present almost entirely as HACA (0.5 mg per 100 mg mucosal cell protein). In the contents of the caecum and colon of all mammalian species fermentation of endogenous urea, protein and other materials yields CO_2, NH_3 and volatile fatty acids (VFA). We find that the mucosa of cattle rumen contains large quantities of carbonic anhydrase of LACA type although this variety is entirely absent from the blood of the animal. The characteristic changes which accompany salt absorption in the small intestine therefore do not depend upon carbonic anhydrase. The presence of large amounts of LACA in epithelia lining the walls of fermentation organs such as the rumen and caecum suggests that this variety of the enzyme has a role in the mucosal transport or metabolism of the products of fermentation in the intestinal lumen.

ACKNOWLEDGEMENTS

Many of the experiments I have described have been undertaken in my laboratory by various young workers whom it has been my privilege to have as colleagues. Among these Dr.Garth Powis, Dr.Gordon Sladen, Miss Christine Wills and particularly Dr. Martin Carter have been responsible for much of the work described here. I have benefited greatly from their lively and critical discussion of the ideas derived from their work. I am grateful to the Medical Research Council for financial support.

REFERENCES

1. Aafjies,J.H. (1967) Carbonic anhydrase in the wall of the forestomachs of cows. Br.Vet.J. 123, 252-255.

2. Annison,E.F. & Lewis,D. (1959) Metabolism in the Rumen, London, Methuen, ch.6.

3. Beccari,L. (1917) Cited by Dukes,H.H. in The Physiology of Farm Animals, New York,Comstock, 1947.

4. Bernard,C. (1856) An introduction to the study of experimental medicine. Transl. by H.C.Greene New York, Schuman, 1927, p.152.

5. Byvoet,P. & Gotti,A. (1967) Isolation and properties of carbonic anhydrase from dog kidney and erythrocytes. Mol.Pharmacol. 3,142-152.

6. Campbell,J.W. & Speeg,K.V. (1969) Ammonia and biological deposition of calcium carbonate. Nature, Lond. 224, 725-726.

7. Carter,M.J. (1971) The carbonic anhydrase in the rumen epithelial tissue of the ox. Biochim. Biophys.Acta 235, 222-236.

8. Carter,M.J. & Parsons,D.S. (1968) Carbonic anhydrase activity of mucosa of small intestine and colon. Nature,Lond. 219, 176-177.

9. Carter,M.J. & Parsons,D.S. (1970) The purification and properties of carbonic anhydrases from guinea pig erythrocytes and mucosae of the gastrointestinal tract. Biochem.J.120,797-808.

10. Carter,M.J. & Parsons,D.S. (1970) The carbonic anhydrases of some guinea pig tissues. Biochim. Biophys.Acta 206, 190-192.

11. Carter,M.J. & Parsons,D.S. (1971) The isoenzymes of carbonic anhydrase: Tissue,subcellular distribution and functional significance, with particular reference to the intestinal tract. J.Physiol. 215, 71-94.

12. Carter,M.J. & Parsons,D.S. (1972) The isoenzymes of carbonic anhydrase: Kinetic properties with particular reference to the functions in

the intestinal tract. J.Physiol. 220,465-478.

13. Davenport,H.W. (1939) Gastric carbonic anhydrase. J.Physiol. 97, 32-43.

14. Davenport,H.W. (1946) Carbonic anhydrase in tissues other than blood. Physiol.Rev.26,560-573.

15. Davenport,H.W. & Fisher,R.B. (1938) Carbonic anhydrase in the gastrointestinal mucosa. J.Physiol. 94, 16P.

16. Davenport,H.W. & Wilhelmi,A.E. (1941) Renal carbonic anhydrase. Proc.Soc.exp.Biol.Med.48, 53-56.

17. DeVoe,H. & Kistiakowsky,G.B. (1961) The enzyme kinetics of carbonic anhydrase from bovine and human erythrocytes. J.Am.Chem.Soc. 83,274-280.

18. Edsall,J.T. (1968) Multiple molecular forms of carbonic anhydrase in erythrocytes. Ann.N.Y. Acad.Sci. 151, 41-63.

19. Edsall,J.T. (1969) Carbonic Anhydrase,in CO_2: Chemical Biochemical and Physiological Aspects, pp.87-177, ed. Forster,R.E., Edsall,J.T.,Otis A.B. & Roughton,F.J.W., Washington D.C. National Aeronautics and Space Administration.

20. Folin,O. & Denis,W. (1912) The origin and significance of the ammonia in the portal blood. J.Biol.Chem. 11, 161-167.

21. Fordtran,J.S., Rector,F.C. jr. & Carter,N.W. (1968) The mechanisms of sodium absorption in the human small intestine. J.Clin.Invest. 47, 884-900.

22. Henderson,L.E., Henriksson,D., Nyman,P.O. & Strid,L. (1972) On the primary structure of the B and C isoenzymes of human erythrocyte carbonic anhydrase. In Oxygen Affinity of Haemoglobin and Red Cell Acid Base Status, pp.341-352,ed.Rørth,M. & Astrup,P.,N.York,Acad;Press.

23. Hogben,C.A.M. (1967) The chloride effect of carbonic anhydrase inhibitors. Mol.Pharmacol. 3, 318-326.

24. Hubel,K.A. (1967) Bicarbonate secretion in rat

ileum and its dependence on intraluminal chloride. Am.J.Physiol. 213, 1409-1413.

25. Kleiber,M., Cole,H.H. & Mead,S.W. (1943) Bloat in cattle and composition of rumen gases. J.Dairy Sci. 26, 929-933.

26. Krebs,H.A. (1963) Renal gluconeogenesis. Adv. in Enzyme Regulation 1, 385-400.

27. Kuriaki,K. & Magee,D.F. (1964) On the carbonic anhydrase activity of the alimentary canal and pancreas. Life Sci. 3, 1377-1382.

28. Levitt,M.D. & Bond,J.H. (1970) Volume, composition and source of intestinal gas.Gastroenterology 59, 921-929.

29. Lindskog,S. (1972) In Oxygen Affinity of Hemoglobin and Red Cell Acid Base Status (Alfred Benzon Symposium IV, Copenhagen,1971) p.409, ed. Rørth,M. & Astrup,P., Copenhagen,Munksgaard New York, Academic Press.

30. Lutwak-Mann,C. (1955) Carbonic anhydrase in the female reproductive tract. Occurrence,distribution and hormonal dependence. J.Endocrinol. 13, 26-38.

31. Maetz,J. & Garcia Romeu,F. (1964) The mechanism of sodium and chloride uptake by the gills of a freshwater fish, Carassius auratus. II. Evidence for NH^+/Na^+ and HCO_3^-/Cl^- exchanges. J.Gen.Physiol. 47, 1209-1227.

32. Magendie,F. (1833) Précis Elémentaire de Physiologie, pp.89 et seq., 3rd Edn. vol.2,Paris, Méquignon-Marvis Père et Fils.

33. Maren,T.H. (1967) Carbonic anhydrase:chemistry, physiology and inhibition. Physiol.Rev.47, 595-781.

34. Maynard,J.R. & Coleman,J.E. (1971) Elasmobranch carbonic anhydrase. Purification and properties of the enzyme from two species of shark. J.Biol.Chem. 246, 4455-4464.

35. McIntosh,J.E.A. (1969) Carbonic anhydrase isoenzymes in the erythrocytes and dorsolateral prostate of the rat. Biochem.J. 114, 463-476.

36. Meldrum,N.M. & Roughton,F.J.W. (1932) The CO_2 catalyst present in blood. J.Physiol. 75,3P.

37. Meldrum,N.M. & Roughton,F.J.W. (1932) Some properties of carbonic anhydrase, the CO_2 enzyme present in blood. J.Physiol. 75, 15P.

38. Meldrum,N.M. & Roughton,F.J.W. (1933) Carbonic anhydrase. Use, preparation and properties. J.Physiol. 80, 113-142.

39. Parnas,J.K. & Klisiecki,A. (1926) Über den Ammoniakgehalt und die Ammoniakbildung im Blute. Biochem.Z. 173, 224-248.

40. Parsons,D.S. (1956) The absorption of bicarbonate-saline solutions by the small intestine and colon of the white rat. Quart.J.exp.Physiol. 41, 410-420.

41. Parsons,D.S. (1971) Salt Transport. J.Clin. Path. 24, suppl.5, 90-98.

42. Parsons,D.S. & Powis,G. (1971) Some properties of a preparation of rat colon perfused *in vitro* through the vascular bed. J.Physiol. 217, 641-663.

43. Pembrey,M.S. (1898) Chemistry of respiration. In *Textbook of Physiology*, vol.I, pp.692-781, ed. Schäfer,E.A., Edinburgh,Pentland.

44. Pocker,Y. & Meany,J.E. (1965) The catalytic versatility of erythrocyte carbonic anhydrase. I. Kinetic studies of enzyme-catalysed hydration of acetaldehyde. Biochemistry 4,2535-2541.

45. Powell,D.W., Malawer,S.J. & Plotkin,G.R. (1968) Secretion of electrolytes and water by the guinea pig small intestine *in vivo*. Am.J. Physiol. 215, 1226-1233.

46. Richards,P.A., Metcalfe-Gibson,A., Ward,E.E., Wrong,O. & Houghton,B.J. (1967) Utilisation of ammonia nitrogen for protein synthesis in man, and the effect of protein restriction and uraemia. Lancet 11, 845-849.

47. Rieder,R.F. & Weatherall,D.J. (1964) A variation in the electrophoretic pattern of human erythrocyte carbonic anhydrase. Nature 203, 1364-1366.

48. Roughton, F.J.W. & Booth, V.H. (1946) The effect of substrate concentration, pH and other factors upon the activity of carbonic anhydrase. Biochem.J. 40, 319-330.

49. Shimizu, C. & Matsuura, F. (1962) Some physicochemical properties of crystalline carbonic anhydrase from erythrocytes of ox, blue-white dolphin and yellowfin tuna. Bull.Jap.soc.Sci. Fisheries 28, 924-929.

50. Silen, W., Harper, H.A., Mawdsley, D.L., Weirich, W. L. (1955) Effect of antibacterial agents on ammonia production within the intestine. Proc. Soc.exp.Biol.Med. 88, 138-140.

51. Sladen, G.E. & Dawson, A.M. (1968) Effect of bicarbonate on sodium absorption by the human jejunum. Nature Lond. 218, 267-268.

52. Schneider, F. & Lieflander, M. (1963) Über die Reaktion von Carbonat-hydrolase mit p-nitrophenylacetat. Z.phys.Chemie 334, 279-282P.

53. Snyderman, S.E. (1967) Urea as a source of unessential nitrogen for the human. In Urea as a Protein Supplement, pp.441-453, ed. Briggs, M.H., Oxford, Pergamon Press.

54. Swallow, J.H. & Code, C.F. (1967) Intestinal transmucosal fluxes of bicarbonate. Am.J.Physiol. 212, 717-723.

55. Tanis, R.J., Tashian, R.E., & Yu, Y-S.L. (1970) Properties of carbonic anhydrase isoenzymes isolated from porcine erythrocytes. J.Biol.Chem. 245, 6003-6009.

56. Tappeiner, H. (1898) Quoted by Pembrey, M.S., Chemistry of respiration, in Textbook of Physiology, vol.1, Schäfer, E.A., ed., Edinburgh, Pentland. p.729.

57. Tashian, R.E., Douglas, O.P., & Yu, Y-S. (1964) Esterase and hydrase activity of carbonic anhydrase-I from primate erythrocytes. Biochem. Biophys.Res.Commun. 14, 256-261.

58. Tashian, R.E., Tanis, R.J., & Ferrel, R.E. (1972) Comparative aspects of the primary structure and activities of mammalian carbonic anhydrase.

In Oxygen Affinity of Hemoglobin and Red Cell Acid Base Status (Alfred Benzon Symposium IV, Copenhagen,1971), pp.353-362, ed.Rørth,M. & Astrup,P., Copenhagen, Munksgaard; New York, Academic Press.

59. van Goor,H. (1948) Carbonic anhydrase, its properties, distribution and significance for carbon dioxide transport. Enzymologia 13,73-164.

60. Walter,F. (1877) Untersuchungen über die Wirkung der Säuren auf den thierischen Organismus. Arch.exp.Path.Pharmak. 7, 148-182.

61. Wilson,T.H., Kazyak,L. (1957) Acid-base changes across the wall of hamster and rat intestine. Biochim.Biophys.Acta 24, 124-132.

Comparative Physiology, eds. L. Bolis, K. Schmidt-Nielsen and S. H. P. Maddrell

TWO CARRIERS FOR FREE MONOSACCHARIDES OF THE GLUCOSE TYPE IN HAMSTER SMALL INTESTINE

P.HONEGGER
Eidg.Technische Hochschule Zürich,
Laboratorium für Biochemie,
Zürich, Switzerland.

The existence of two different transport systems for glucose was first suggested in 1966 by Newey, Sanford and Smyth (4) who found an uranyl-sensitive and an uranyl-insensitive glucose transport system in rat small intestine. Meeuwisse (3) found in persons suffering from glucose-galactose-malabsorption a small but definite absorption of glucose. Recently, Debnam and Levin (2) observed in rat small intestine *in vitro* and *in vivo* a nutritionally induced differential change in the apparent K_m of the transfer of several hexoses such as glucose, galactose, α-methyl glucoside and 3-0-methyl glucose (3-0-MG) and suggested the presence of separate carriers or transfer mechanisms for each hexose.

Since these indirect data suggested either the existence of two transport systems for free glucose in the enterocytes, or separate carriers for each hexose, we have attempted to obtain additional data by kinetic experiments using hamster small intestine. Our method (5), in which pieces of everted small intestine are mounted in frames, allows exposure of the mucosal side alone to the medium with only a small correction for the extracellular space. Nevertheless we did correct for the extracellular space using different markers such as ^{3}H- and ^{14}C-mannitol, inulin or 2-deoxy-D-galactose.

The medium-cell unidirectional flux of each monosaccharide was first studied over a wide concentration range and the results are shown in Eadie-Hofstee-Augustinsson plots (Figures 1 to 4). It is apparent that the data fall on one sin-

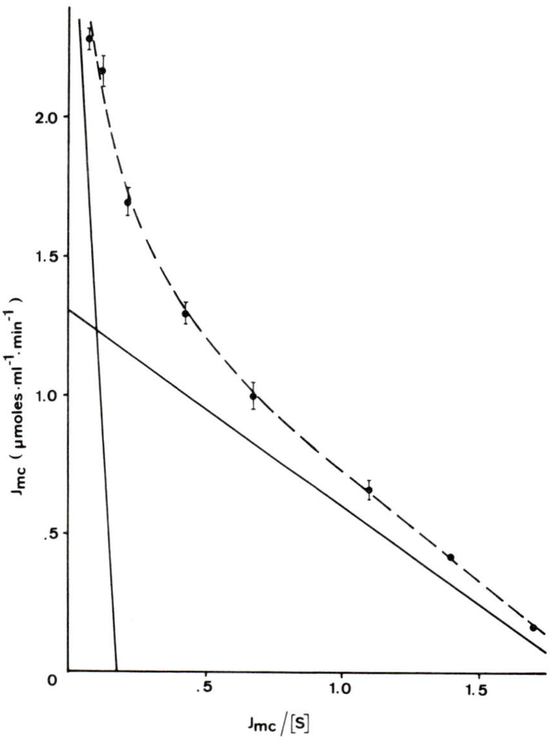

Figure 1 - Unidirectional flux of D-glucose from the medium into hamster small intestine (J_{mc}) as a function of glucose concentration (Eadie plot). Concentration range: 0.1 to 60 mM; 2-min incubations at 37° C in modified Krebs-Henseleit buffer, pH 7.2, containing 93 mM Na^+. ● ,average of 16 experiments; the bars indicate the S.E.M. Full lines, theoretical lines computed (6) by iteration for each of the two carriers; dashed line, calculated line for the sum of the two carriers.

gle straight line in the cases of 3-0-MG and 6-deoxy-D-glucose (6-DG), whereas those for glucose and galactose can be described satisfactorily only by the sum of two Michaelian functions. Thus, the

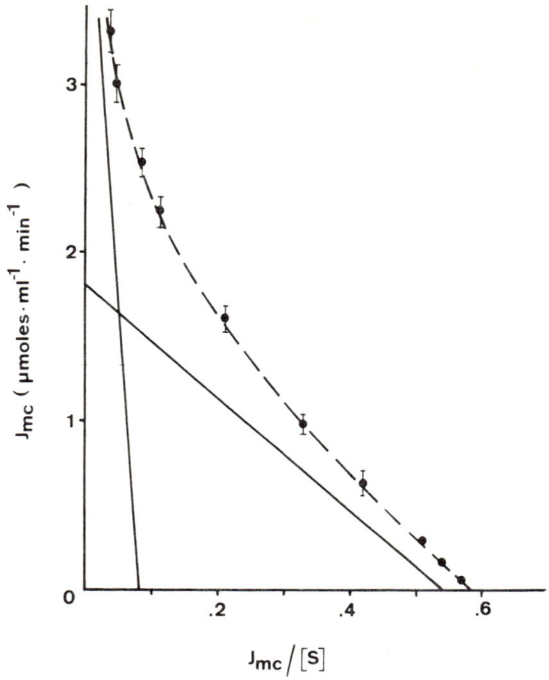

Figure 2 - Unidirectional flux of D-galactose from the medium into hamster small intestine (J_{mc}) as a function of galactose concentration (Eadie plot). Concentration range: 0.1 to 100 mM ● , average of 8 experiments. For further details see legend to Fig.1.

apparent $K_{m1,2}$ and $V_{1,2}$ values were computed (6) and listed in Table 1.

Table 1 - Kinetic parameters of the two transport systems for glucalogues in the small intestine of adult hamster (calculated from the data in figs.1 to 4).

Monosaccharide (conc. range mM)		K_{m1} (mM)	K_{m2}	V_1 (μMole·ml^{-1}·min^{-1})	V_2
D-gluc	(0.1-60)	0.7	15	1.3	1.5
D-gal	(0.1-100)	3.5	50	1.9	2.2
6-DG	(0.1-30)	2.8		2.0	
3-O-MG	(0.3-100)	28		3.5	

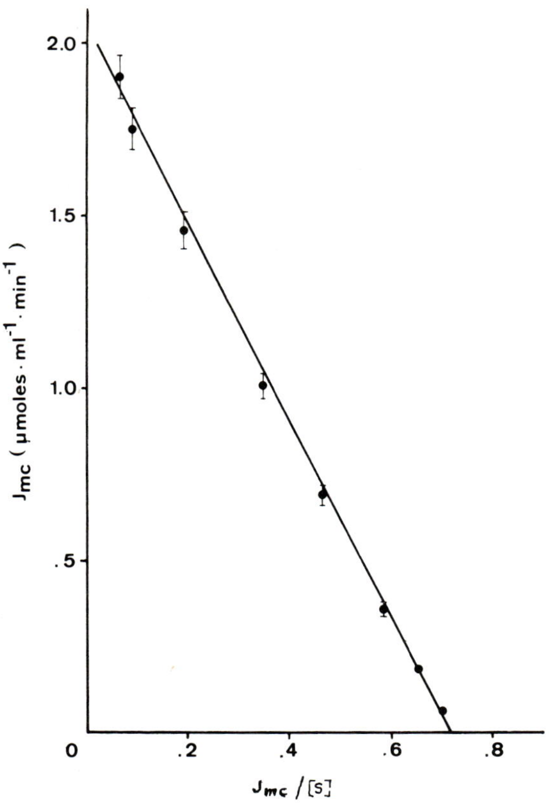

Figure 3 - Unidirectional flux of 6-deoxy-D-glucose from the medium into hamster small intestine (J_{mc}) as a function of 6-DG concentration (Eadie plot). Concentration range: 0.1 to 30mM. ● , average of 6 experiments. The experimental points fall on one single regression line. Conditions as in Fig. 1.

In order to check the validity of the different transport parameters for each monosaccharide, experiments on mutual inhibition among these monosaccharides were performed. The results obtained with glucose or galactose are better described by assuming the presence of two carriers rather than of a single carrier. Thus it seemed reasonable to conclude that hamster small intestine has two transport systems for free monosaccharides at the mucosal surface: one for all glucalogues tested

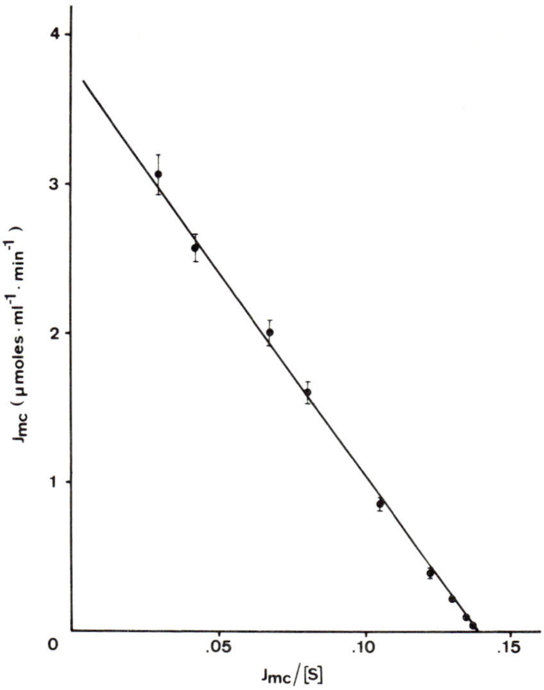

Figure 4 - Unidirectional flux of 3-O-methyl-D-glucose from the medium into hamster small intestine (J_{mc}) as a function of 3-O-MG concentration (Eadie plot). Concentration range: 0.3 to 100 mM. ●, average of 8 experiments. Conditions as described in Fig.1. The experimental points fall on one single regression line.

and another, having larger K_m values, but essentially specific for glucose and galactose.

In further experiments the inhibition of glucalogue uptake by arbutin (β -hydroquinone-glycopyranoside) was examined. This substance is described in the literature as a good substrate of the common glucalogue carrier with a K_m of about 2-3 mM (1). In previous experiments with relatively low substrate concentrations it was found that 3-O-MG and 6-DG were inhibited competitively with a K_i of about 2 mM, but glucose and galactose were inhibited significantly less than expected on the assumption of a single carrier system. Therefore, the uptake of glucalogues over a broad concentration range was examined in the presence and absence of 30 mM arbutin (Figures 5 to 7). Under these conditions arbutin apparently inhibits only the glucalogue carrier having broad specificity.

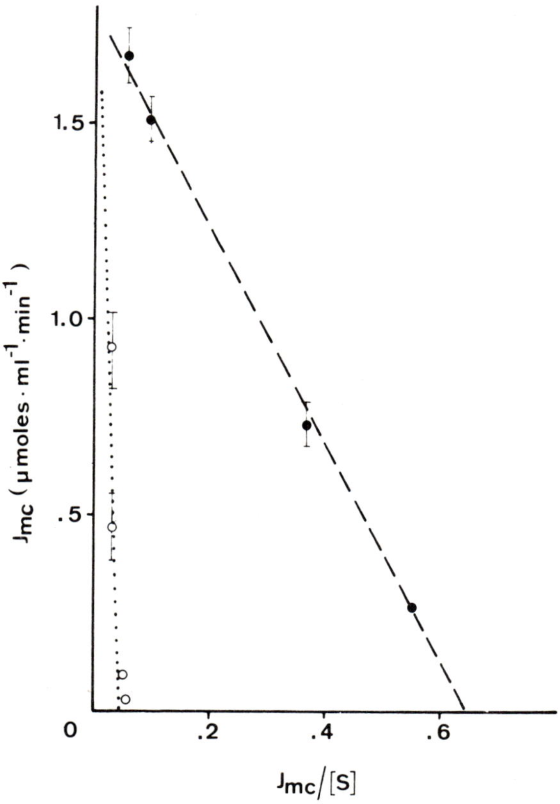

Figure 5 - Unidirectional flux of 6-deoxy-D-glucose from the medium into hamster small intestine (J_{mc}) in the presence (o) or in the absence (•) of arbutin (30 mM). Average of 4 experiments. The dotted line was calculated from the assumption that arbutin inhibits 6-DG transport with a K_i = 2 mM. Conditions as described in Fig. 1.

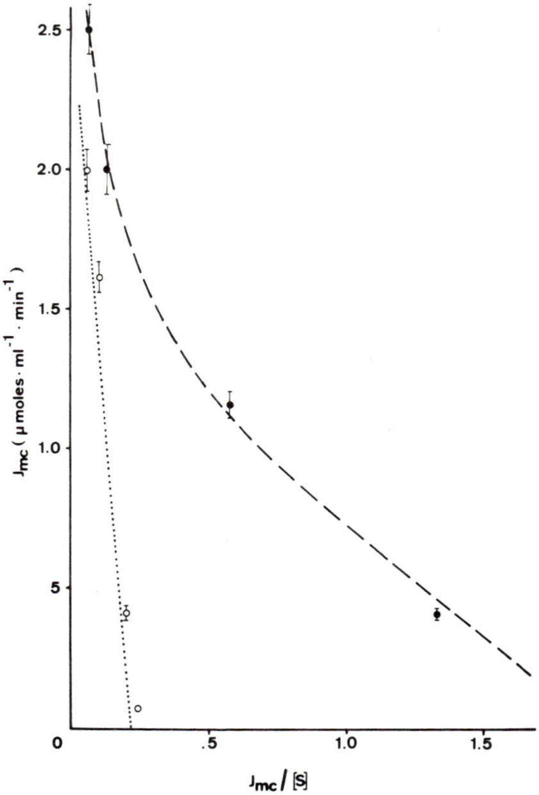

Figure 6 - Unidirectional flux of D-glucose from the medium into hamster small intestine (J_{mc}) in the presence (o) or in the absence (●) of arbutin (30 mM). Average of 5 experiments. The dashed line is the same as in Fig.1; the dotted line was calculated from the assumption that arbutin inhibits the less specific, low-K_m, carrier alone, with a K_i = 2 mM. Other conditions, as in Fig.1.

Similar experiments with phlorizin indicated that both 3-O-MG and 6-DG were competitively inhibited with a K_i of $1.7 \cdot 10^{-6}$M. By contrast inhibition of glucose and galactose was incompatible with a single carrier system. Assuming K_{i1} for phlorizin of $1.7 \cdot 10^{-6}$M as calculated with 3-O-MG and 6-DG, the apparent K_{i2} values for phlorizin with

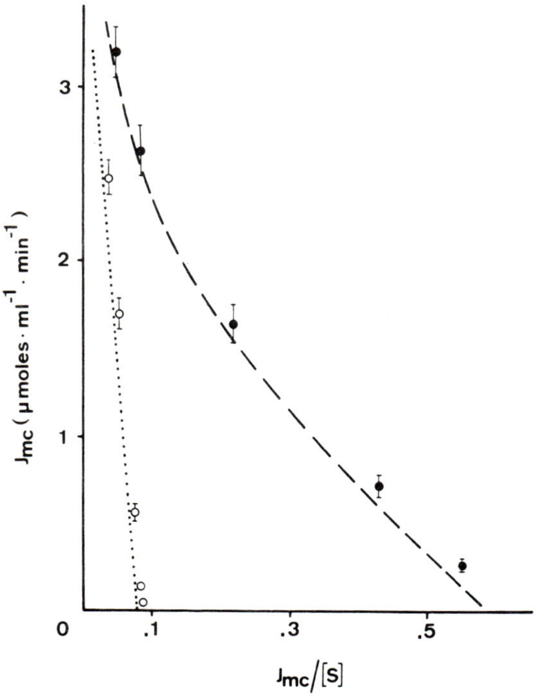

Figure 7 - Unidirectional flux of D-galactose from the medium into hamster small intestine (J_{mc}) in the presence (o) or in the absence (●) of arbutin (30 mM). Average of 4 experiments. For further details see legend to Fig.5.

galactose was found to be $2 \cdot 10^{-5}$M and with glucose $1.5 \cdot 10^{-4}$M. Both glucalogue carriers are Na^+ dependent.

In the experiments reported above, adult animals were used. We also studied glucalogue uptake in baby hamsters and again found the same glucalogue carriers with apparent $K_{m1,2}$ values similar to the adult animals but with a larger V_1/V_2 ratio. This finding indicates an age dependence of the second glucalogue carrier and gives further support to the existence of two different glucalogue carriers.

In summary, we can say that the results of

experiments on the uptake of glucose and galactose over a large concentration range, the inhibition of their uptake by arbutin and phlorizin, and their age dependence, all lead to the conclusion that at least 2 different transport systems, with overlapping specificity, exist for free glucalogues in hamster small intestine.

REFERENCES

1. Alvarado,F. & Crane,R.K. (1964) Studies on the mechanism of intestinal absorption of sugars. VII. Phenylglucoside transport and its possible relationship to phlorizin inhibition of the active transport of sugars by the small intestine. Biochim.Biophys.Acta 93, 116-135.

2. Debnam,E.S. & Levin,R.J. (1971) Evidence for separate hexose carriers in the small intestine uncovered by fasting and glucose alimentation. J.Physiol. 218, 38-39P.

3. Meeuwisse,G.W. (1970) Glucose-galactose malabsorption. An inborn error of carrier-mediated transport. Thesis, Lund, Sweden.

4. Newey,H., Sanford,P.A. & Smyth,D.H. (1966) The effect of uranyl nitrate on intestinal transfer of hexoses. J.Physiol. 186, 493-502.

5. Semenza,G. (1969) Studies on intestinal sucrase and sugar transport. VII. A method for measuring intestinal uptake. The absorption of the anomeric forms of some monosaccharides. Biochim. Biophys.Acta 173, 104-112.

6. Spears,G., Sneyd,J.G.T. & Loten,E.G. (1971) A method for deriving kinetic constants for two enzymes acting on the same substrate. Biochem. J. 125, 1149-1151.

Comparative Physiology, eds. L. Bolis, K. Schmidt-Nielsen and S. H. P. Maddrell

COMPARATIVE ASPECTS OF THE EFFECT OF PHENOLS ON SUGAR TRANSPORT

F. ALVARADO
Department of Physiology & Biophysics,
University of Puerto Rico Medical School,
San Juan, Puerto Rico.

Phlorizin is a well-known, fully competitive (type 1a[x]) inhibitor of the mechanism of sugar active transport in the small intestine, which implies that transport inhibition by this glucoside is both instantaneous and reversible (2,3).

Such actions of phlorizin have been found in all mammals studied, in birds (e.g.,the chicken) and in some fish (see 4); similar actions also seem to occur in the mammalian kidney proximal tubule (5), and in the absorptive tegument of some worms (see 6). All these consist of monocellular epithelia with apical brush-borders having rather similar sugar transport mechanisms, susceptibility to phlorizin inhibition being just one among several apparently common properties.

The polyphenol phloretin, the aglycone of phlorizin, is also an inhibitor of sugar transport, but its mechanism of action is not well understood. In the red blood cell, phloretin appears to be a fully competitive inhibitor (type 1a), whereas its glucoside, phlorizin, is essentially inert (7).Here the red cell would appear to behave in a diametrically different manner from the intestine and other epithelia mentioned above.

In the small intestine, considerable controversy still exists regarding the mechanism of action of phloretin. According to Alvarado, phloretin is a type b, i.e.,allosteric (or allotopic) inhibitor of both sugar and amino acid transport in

---

x The nomenclature of Dixon and Webb will be used throughout (1).

hamster small intestine (2,3), although this worker did not originally try to define whether the inhibition was Ib or IIb. Both mechanisms would be compatible with the interpretation, proposed by Alvarado, that phloretin inhibits sugar transport indirectly by binding to a site different from, but contiguous to, the transport site (2).

Sahagian, using guinea pig small intestine, found that both phlorizin and phloretin were fully competitive (type 1a) inhibitors of sugar transport, phloretin having about one-fifth the affinity of phlorizin for the sugar transport mechanism or carrier (8). This work, however, is open to the double criticism that the kinetic study was not done using initial transport velocities (as we shall see, the time factor is very important with regard to the action of phloretin), and no reference was made to the purity of the compounds used. Most commercial phloretin preparations are heavily contaminated by phlorizin.

Diedrich (9) showed that the effect of phloretin on sugar transport in hamster intestine is noncompetitive (type II), but he made no distinction whether the inhibition was of the *a* or *b* type. More recently, Colombo and Semenza (10) observed that, as the phloretin concentration increases, sugar transport in hamster intestine is fully inhibited (the transport velocity tends to zero, see Figure 3 in ref. 10). Under these conditions, therefore, phloretin inhibition is likely to be fully non-competitive, i.e., type IIa.

In the present communication I present evidence that the inhibition of sugar transport by phloretin in guinea pig intestine changes with time, first being an allotopic, partially non-competitive (type IIb) inhibition, as described by Alvarado (2, 3); and later becoming what appears to be a fully non-competitive (type IIa) inhibition as in the experiments of Colombo and Semenza (10) with hamster intestine.

MATERIALS AND METHODS

Transport was determined *in vitro* by the "tissue accumulation method", as previously described (see 11). In the experiment shown in Figure 5, a modification of this method was used that unequivocally allows determination only of those fluxes that occur through the brush border membrane of the intestinal epithelial cells. Rather than using "rings" with open (cut) ends, empty everted sacs were used which, after incubation, were blotted with filter paper, separated from their ligatures and digested in KOH as described (11). Both procedures (substrate accumulation into either rings or sacs) gave essentially identical results (Alvarado, unpublished). All experiments were carried out using unfasted (250-400g) guinea pigs.

Incubations were in phosphate buffer,pH 7.2., saturated with 100% O_2, with a salt composition similar to that of the Krebs and Henseleit buffer and a Na/K ratio of 25.5 (12). Results are expressed as velocities (= μ moles of substrate accumulated per ml tissue water per minute); or as the final tissue /medium (T/M) concentration ratios, where a ratio higher than 1 would signify uphill transport. When required, the results were corrected for the extracellular space, using C^{14} D-mannitol as the model non-penetrating analog. The water content of the tissues was determined to be about 80%.

The compounds used were chromatographically pure. Phloretin and phlorizin were separated from each other by large-scale chromatography in LH-20 Sephadex, using 60% ethanol as the solvent and a SR-25/45 column from Pharmacia (Alvarado, unpublished).

RESULTS AND DISCUSSION

The initial velocity of transport of the glucose analog, β-methylglucoside was first determined in the presence of variable concentrations of phloretin, for 5 minutes, without prior incubation of the tissues in the inhibitor. Under these conditions (see Figure 1), the inhibition was not complete but reached a plateau at about 50%. This indicates the inhibition to be of the *b* type (i.e., allotopic) as shown previously for sugar (2) and amino acid (3) transport in hamster small intestine.

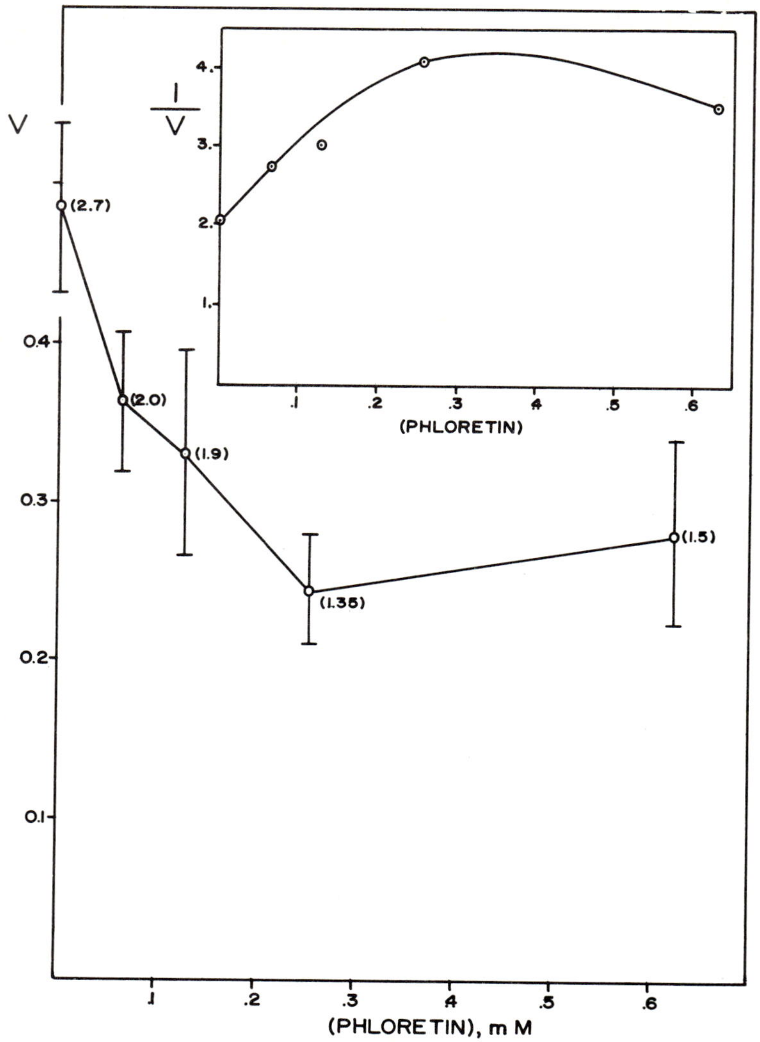

Figure 1 - The effect of phloretin on the active transport of methyl-β-D-glucopyranoside in guinea pig small intestine. Rings of everted intestine were incubated at 37°C for 5 min.in 5 ml phosphate buffer containing 1 mM C^{14}-methyl-glucoside and the indicated concentrations of phloretin. Results were corrected for the extracellular space. Figures in parenthesis indicate T/M ratios, also corrected. Each point is the mean of 4 determinations ± S.D. In the insert,the same results are plotted according to Dixon (see ref.1,pp. 328-330).

Since, when using a single phloretin concentration and variable substrate, again for 5 minutes without preincubation, the inhibition is non-competitive (there is a decrease in the Vmax, with no apparent change in the Km, see Figure 2), the con-

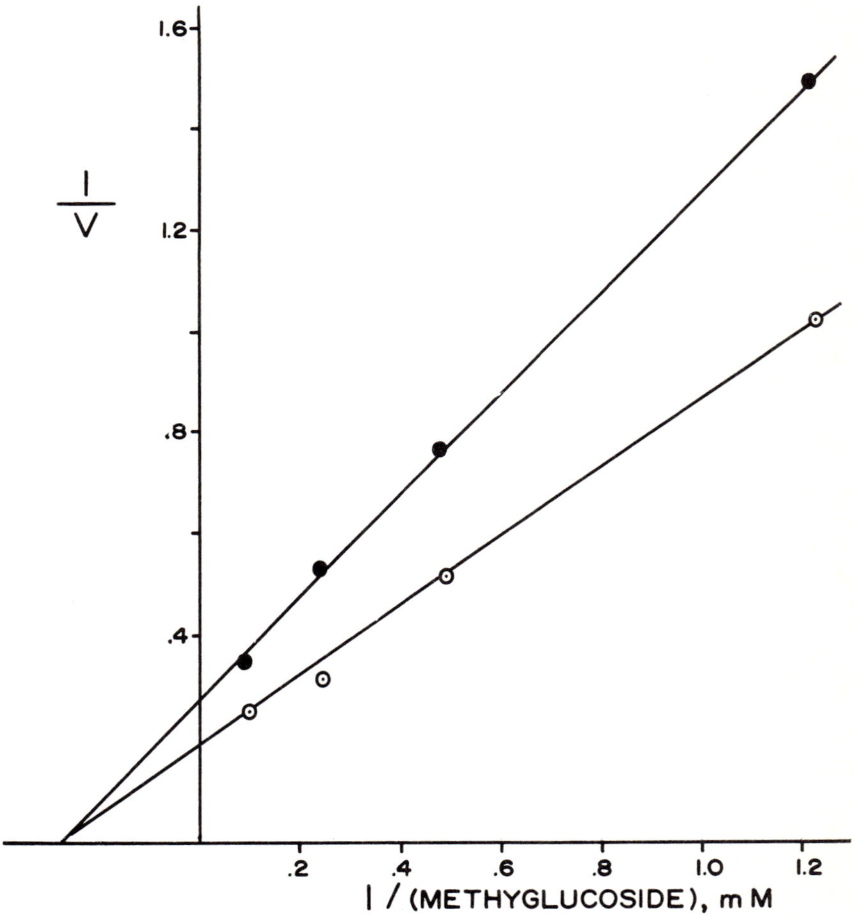

Figure 2 - Non-competitive effect of phloretin on methyl-α-glucopyranoside transport into guinea pig intestinal rings. Incubations were for 5 min. in the presence (•) or in the absence (o) of a 0.32 mM concentration of phloretin, and concentrations of substrate ranging from 0.84 to 10.5 mM. Velocities were corrected for the extracellular space. Each point is the mean of 8 determinations.

clusion seems warranted that phloretin is a partially non-competitive (type IIb) inhibitor of sugar transport in guinea pig as well as in hamster small intestine (13).

However, when similar experiments using longer incubation periods are carried out (we used 40 minutes; Sahagian (8) used 90 minutes and Colombo and Semenza (10) used 15 minute incubations, preceded by a 30-minute preincubation period in the presence of phloretin), the effect of phloretin still appears to be non-competitive (type II) according to a Lineweaver-Burk analysis (result not shown). But, if studied at constant substrate and increasing inhibitor concentrations, as shown in Figure 3, the velocity of transport does not reach a plateau. On the contrary, the transport velocity appears to tend to zero as the phloretin concentration increases, and the experimental points adequately fit a theoretical curve drawn on the assumption that phloretin acts in this case as a fully non-competitive inhibitor (type IIa). In the experiment shown, 95% inhibition of methylglucoside transport would have been attained at 2.47 mM phloretin concentration, well above the solubility limit of this phenol.

These results thus appear to be compatible with the interpretation that phloretin binds to the membrane surface to a site different from, but contiguous to, the sugar-binding site responsible for sugar recognition in the sugar transport mechanism (this would explain inhibition kinetics of the IIb type at short times of contact of the inhibitor with the membrane, i.e., 5 min.). With time, however, phloretin may interact irreversibly with other components of the transport machinery (and this may include interaction with the sugar-binding site itself), thus giving rise to an inhibition of the IIa type. It is important to recognize that the inhibition by phloretin at these long incubations would appear to affect the transport machinery proper; that is, the carrier or that part of the process involved in downhill transport across the membrane, as indicated by the fact (Figure 3) that the transport velocity in the presence of phloretin tends to zero. Operationally, this means that no sugar entry other than that into the extracellular space would occur at saturating phloretin concentrations. If phloretin were acting indirectly, e.g., by inhibiting the energy supplies necessary for uphill or accumulative transport, the kinetics of the inhibition would be different, namely, sugar entry up to a tissue/medium concentration ratio of 1 would take place.

The study of Colombo and Semenza (10) with phloretin indicates that there is "mutual competition" (between phlorizin and phloretin) "irrespective of the kinetic type of the individual inhibitors".

These observations, although suggesting a "common step" in the mechanism of action of the two inhibitors, would not distinguish between two possible mechanisms of action of phloretin, namely: (i) phloretin is an allotopic inhibitor of sugar transport (2,3); and (ii) phloretin generated from phlorizin by the action of a specific hydrolase, would be the species responsible for the fully competitive effect of phlorizin (14).

The observations presented in this paper appear to settle the question in favor of hypothesis (i), in confirmation of other reasoning marshalled before (3), and other evidence which indicates absence of phlorizin hydrolase activity in kidney brush borders (Diedrich, personal communication).

According to hypothesis (i), phlorizin is a fully competitive inhibitor of sugar active transport because its β-glucopyranoside moiety directs the molecule to interact reversibly with the sugar-binding site of the carrier. But the presence of an aglycone (phloretin) moiety in phlorizin leads the inhibitor molecule further to interact with an adjacent "phenol site", which results in some thousandfold increase in the apparent affinity of the glucoside for the sugar carrier (2). As a corollary of the hypothesis, it should be expected that free phloretin, by competing with the aglycone of phlorizin for the phenol-binding site, should be able to counteract the inhibitory action of phlorizin in a concentration-dependent manner. Preliminary experiments carried out to test this prediction failed, however, presumably because of the low solubility of phloretin in water, which did not permit the achievement of phloretin/phlorizin concentration ratios higher than about 70, insufficient to set

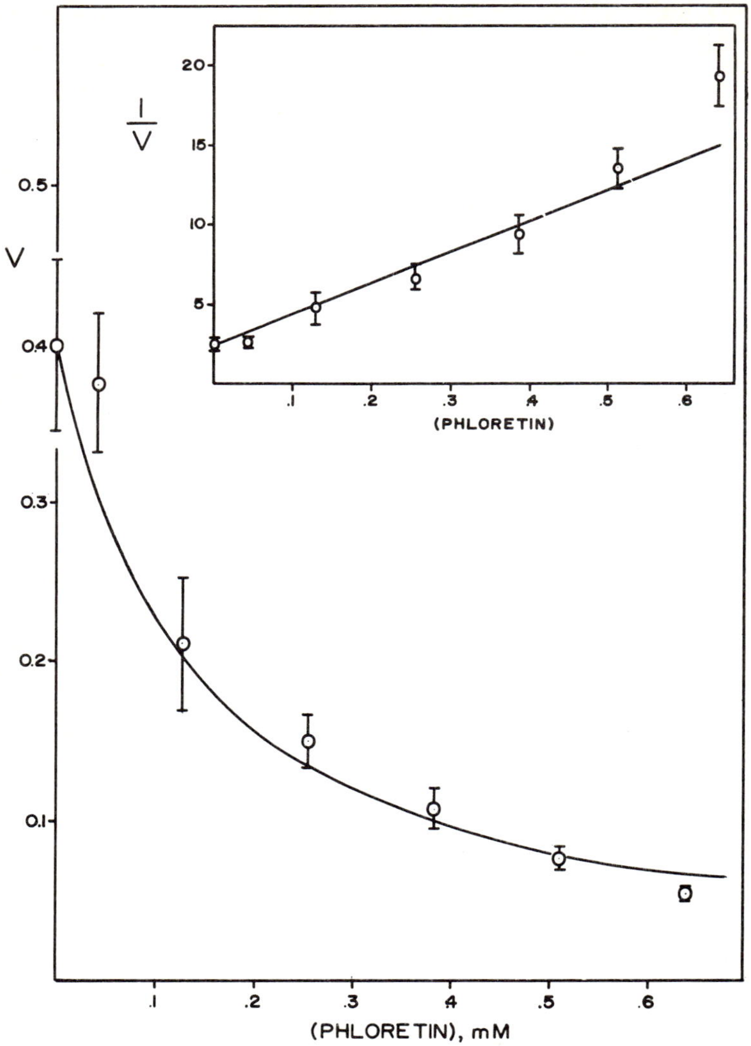

Figure 3 - The effect of phloretin on methyl-β-glucoside transport in guinea pig small intestine. Experiment identical to that in Figure 1, except that the tissues were incubated for 40 min. at 37°C, prior to the determination of β-glucoside transport, this for 5 min. The curves shown were calculated on the assumption that the inhibition is fully non-competitive (type IIa), using the following constants: Km (β-glucoside) = 1.75 mM; Vmax = 1.13 μmoles per minute; average (S) = 0.96 mM; and Ki (phloretin) = 0.13 mM.

the balance of affinity in favor of the free phenol (2).

The problem has now been approached again by using a water-soluble phenol, saligenin, which may in principle be considered a very simplified model of phloretin (see Figure 4). In experiments not

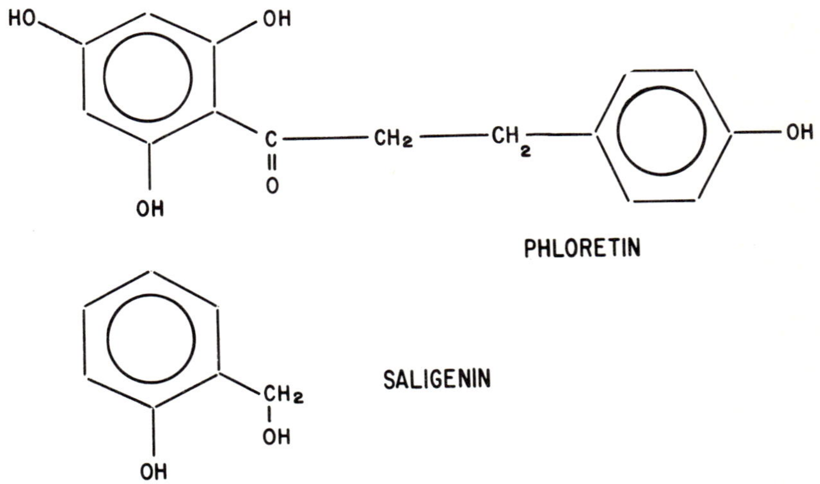

Figure 4

shown here, saligenin was previously found to be an inhibitor of sugar active transport in hamster and guinea pig intestine, although rather high concentrations are needed to obtain the inhibition (e.g. for a 50% inhibition of methyl-α-glucoside transport, 90 mM saligenin is required). Although there is evidence that saligenin has an effect on metabolism, reversibly inhibiting oxygen consumption in guinea pig intestine (Alvarado, unpublished), the prediction was made that it should also bind to the (membrane) phloretin site because of its being a phenol. Even if the affinity of saligenin for this site were low, it would have the advantage over phloretin of being very soluble in water, and thus 90 mM saligenin concentrations could be used (as in the experiment of Figure 5), in contrast with phloretin concentrations 0.64 mM, which are the highest that can be achieved (see above).

The fact that saligenin inhibits metabolism,

thus indirectly inhibiting sugar uphill transport, was made use of as follows. It is well known that intestinal sugar transport can be operationally resolved into two components: (i) a carrier-mediated energy-independent entry component at which level phlorizin inhibition takes place; and (ii) an energy-dependent component of accumulation against the concentration gradient (15). As shown in Figure 5, with the use of either 90 mM saligenin or 0.1% sodium azide, the accumulation component is blocked and substrate (α-methylglucoside) entry occurs down the gradient, fast at first and later more slowly, approaching a T/M ratio of 1, i.e. equilibration. To insure that the accumulation step had been fully blocked, the experiment of Figure 5 was carried out at 25° C. Under these conditions, no significant difference was observed in the rate of sugar entry in the presence of either saligenin or azide; and thus a single control curve (without added phlorizin) is drawn in the figure.

If, in the presence of azide, 0.1 mM phlorizin is also added (lower curve, Figure 5), the entry step of sugar transport is fully blocked, as would have been expected, the resulting curve being practically undistinguishable from that of a non-penetrating analog such as mannitol (not shown).

But, in contrast with this result, in the presence of 90 mM saligenin, 0.1 mM phlorizin is no longer able fully to block the entry of substrate (middle curve,Figure 5), as if, in confirmation of the theoretical expectations mentioned above, saligenin binds to the phloretin site, thereby interfering with the binding of the aglycone moiety of phlorizin to this site. This would cause a significant fall in the apparent affinity of phlorizin for the sugar carrier (2), the result being that the entry stage of sugar transport would not be fully blocked at the phlorizin concentration used.

Although more work is in progress to confirm and extend these observations, the experiment just described would appear to constitute the first direct demonstration of the existence of a phenol site in the membrane at which level a free phenol (saligenin) and the phloretin component of phlorizin compete.

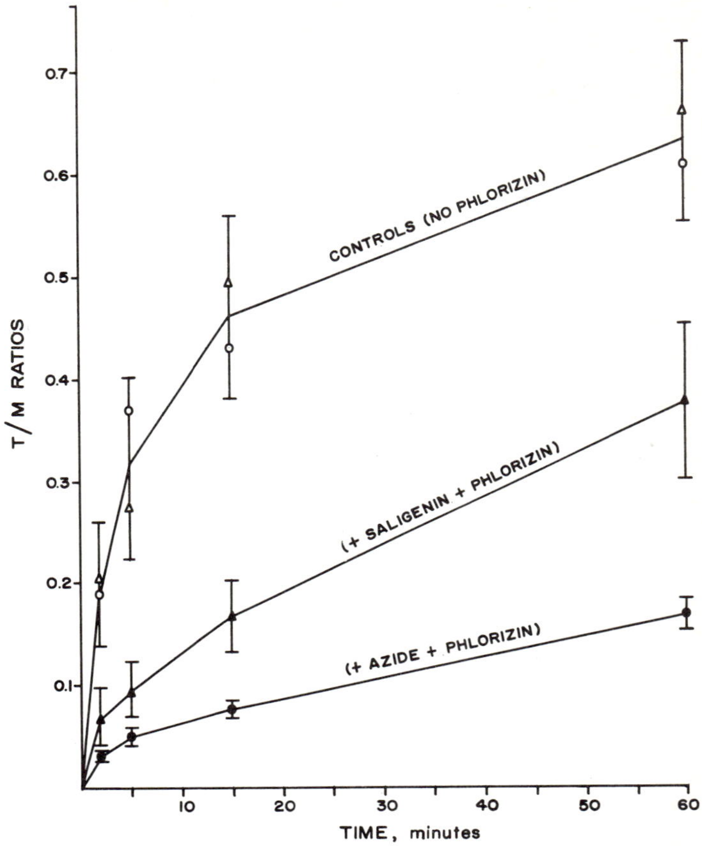

Figure 5 - The effect of phlorizin on the equilibrating (downhill) transport of α-methylglucoside in the presence of either azide or saligenin. All incubations were at 25°C in 5 ml buffer containing 1 mM C^{14} substrate and effectors as follows. One series (circles) contained 0.1% sodium azide, either alone (O) or in the presence (●) of 0.1mM phlorizin. Another series (triangles) contained 90mM saligenin (1.5 ml 0.3M saligenin in 3.5 ml phosphate buffer) either alone (Δ) or in the presence (▲) of 0.1mM phlorizin. The two control series (without phlorizin, O,Δ) gave results statistically undistinguishable, thus allowing calculation of a single control curve using the pooled data (the mean of each independent group is nevertheless shown). Each point represents the mean of either 8 (O,●) or 12 (Δ,▲) determinations ± S.D. The results shown were corrected for the extracellular space.

ACKNOWLEDGEMENTS

Aided by NIH grants AM-13118; 5 S01 RR 05419 and 5D08 ME 00048; and by Research Career Development Award 5-KO-AM 42383 from the National Institute of Arthritis and Metabolic Diseases, U.S.P.H.S. The skilful technical assistance of Mrs.Beatriz Robillard is greatly appreciated.

REFERENCES

1. Dixon,M. & Webb,E.C. (1964) Enzymes, pp.315-330, New York, Academic Press.
2. Alvarado,F. (1967) Hypothesis for the interaction of phlorizin and phloretin with membrane carriers of sugars. Biochim.Biophys.Acta 135, 483-495.
3. Alvarado,F. (1970) Effect of phloretin and phlorizin on sugar and amino acid transport systems in small intestine. FEBS Symposium 20, 131-139.
4. Alvarado,F. & Monreal,J. (1967) Na^+-dependent active transport of phenylglucosides in the chicken small intestine. Comp.Biochem.Physiol. 20, 471-488.
5. Diedrich,D.F. (1966) Glucose transport carrier in dog kidney; its concentration and turnover number. Am.J.Physiol. 211, 581-587.
6. Laurie,J.S. (1971) Carbohydrate absorption by *Gyrocotyle fimbriata* and *Gyrocotyle parvispinosa* (Platyhelmintes). Exptl.Parasitol. 29, 375-385.
7. LeFevre,P.G. (1961) Sugar transport in red blood cell: structure-activity relationships in substrates and antagonists. Pharmacol.Revs. 13, 39-70.
8. Sahagian,B.M.(1965) Active glucose uptake by strips of guinea pig intestine; competitive inhibition by phlorizin and phloretin. Can.J. Biochem. 43, 851-858.
9. Diedrich,D.F. (1966) Competitive inhibition of

intestinal glucose transport by phlorizin analogs. Arch.Biochem.Biophys. 117, 248-256.

10. Colombo,V.E. & Semenza,G. (1972) An example of mutual competition between transport inhibitors of different kinetic type: the inhibition of intestinal transport of glucalogues by phloretin and phlorizin. Biochim.Biophys.Acta 288,145-152.

11. Robinson,J.W.L. & Alvarado,F. (1971) Interaction between the sugar and aminoacid transport systems at the small intestinal brush-border:a comparative study. Pflügers Arch. 326,48-75.

12. Alvarado,F. (1972) Sodium activation of intestinal sugar and amino acid transport: A general or an individual effect? In Na-linked transport of Organic Solutes, pp.147-153, Berlin,Springer-Verlag.

13. Alvarado,F. (1971) Effect of phloretin on sugar active transport in guinea pig small intestine. Biophys.J. 11, 83a.

14. Diedrich,D.F. (1968) Is phloretin the sugar transport inhibitor in intestine? Arch.Biochem. Biophys. 127, 803-812.

15. Alvarado,F. (1966) D-xylose active transport in the hamster small intestine. Biochim. Biophys.Acta 112, 292-306.

Comparative Physiology, eds. L. Bolis, K. Schmidt-Nielsen and S. H. P. Maddrell

INHIBITION OF INTESTINAL SUGAR AND AMINO ACID TRANSPORT BY N-BUTYL-BIGUANIDE.

A.L. LUISIER & W.L. ROBINSON
Département de Chirurgie Expérimentale,
Hôpital Cantonal Universitaire,
Lausanne, Switzerland.

INTRODUCTION

It has been suggested that inhibition of intestinal transport systems might be one of the factors relevant to the anti-diabetic action of substituted biguanides. Animal studies have shown that sugar transport is inhibited both *in vivo* (1,3) and *in vitro* (2,6) by these drugs. The present study was undertaken to investigate whether n-butyl-biguanide has a direct or an indirect action on the transport of sugars and amino-acids in the small intestine *in vitro.*

METHODS

Sections of ileum were removed from anaesthetized rats or guinea pigs and cut into rings of 20-40 mg as described previously (8). The tissues were incubated at 37°C for either one minute or one hour in a solution of ^{14}C-labelled L-phenylalanine or β -methyl-D-glucoside in Krebs bicarbonate buffer to determine respectively the initial velocity of uptake of the substrate or its distribution ratio at equilibrium. After the incubation, the tissues were rinsed, blotted, and dissolved in 0.1 ml 30% KOH prior to counting in a liquid scintillation counter (8). The uptake was then calculated by comparison with the specific activity of the incubation medium, determined under identical conditions.

The action of n-butyl-biguanide was studied by adding it to the incubation medium or to a preincubation medium and comparing its influence with that of ethacrynic acid or ouabain.

The effect of n-butyl-biguanide on Na^+-K^+-ATPase activity in microsomal suspensions was also investigated, the method of preparation and means of determination being identical to those used previously (8).

RESULTS

The sensitivity to inhibition by n-butyl-biguanide of the uptake of sugars or amino-acids by guinea-pig intestinal rings is presented in Figure 1. The two curves are very similar, they are not asymptotic to unity - that is to say, there is a small residual component of non-electrolyte transport that is refractory to biguanide - and the inhibitor concentration provoking a half-maximal inhibition is very similar in the two cases (1.45 mM for sugar transport and 1.25 mM for amino-acid transport). This result suggests strongly that biguanides affect a process which is common to both transport systems.

The studies shown in Figure 2 compare the effects of various inhibitors on the uptake of β-methyl-glucoside by guinea-pig intestinal slices during short and long incubations. Whereas phloridzin, a well known inhibitor of the sugar transport mechanism, exerts its effect during both short and long incubations, ouabain and biguanide are only capable of inhibiting the equilibrium uptake of the substrate.It thus appears that biguanide does not act directly on the carrier mechanism for sugar or amino-acid transport. One possible explanation for the findings in these two figures is that biguanides cause the breakdown of the sodium gradient across the intestinal cell membrane upon which the active uptake of sugars and amino-acids depends. This could occur either by a direct action of the drug on a sodium-pumping mechanism, or by interference with the cellular energetic metabolism responsible for the functioning of the sodium pump.

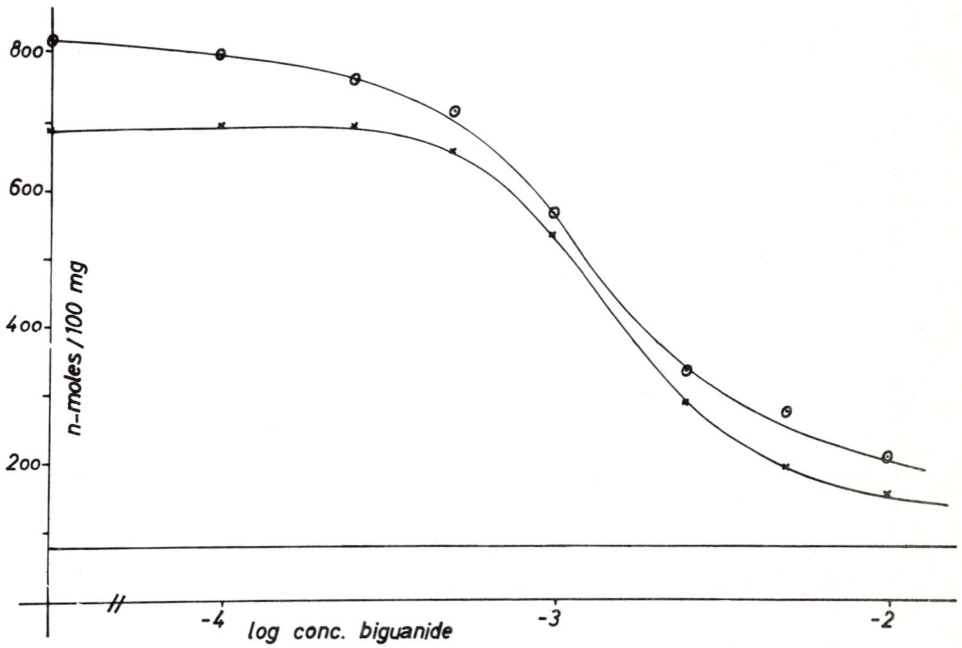

Figure 1 - Inhibition of transport of methyl-glucoside (crosses) and phenyl-alanine (circles) by n-butyl-biguanide in guinea-pig intestinal rings. Tissues incubated for one hour in 1 mM substrate in the presence of different concentrations of the inhibitor. Each curve is the mean of four different experiments. The horizontal line represents the position of equilibrium. Probit analysis (9) shows that the most likely value of the asymptote for the sugar line is 148, and hence a semi-maximal inhibition occurs at a concentration of 1.45 mM biguanide, whereas the most likely asymptote for the amino-acid line is 172, and a semi-maximal inhibition occurs at 1.25 mM biguanide.

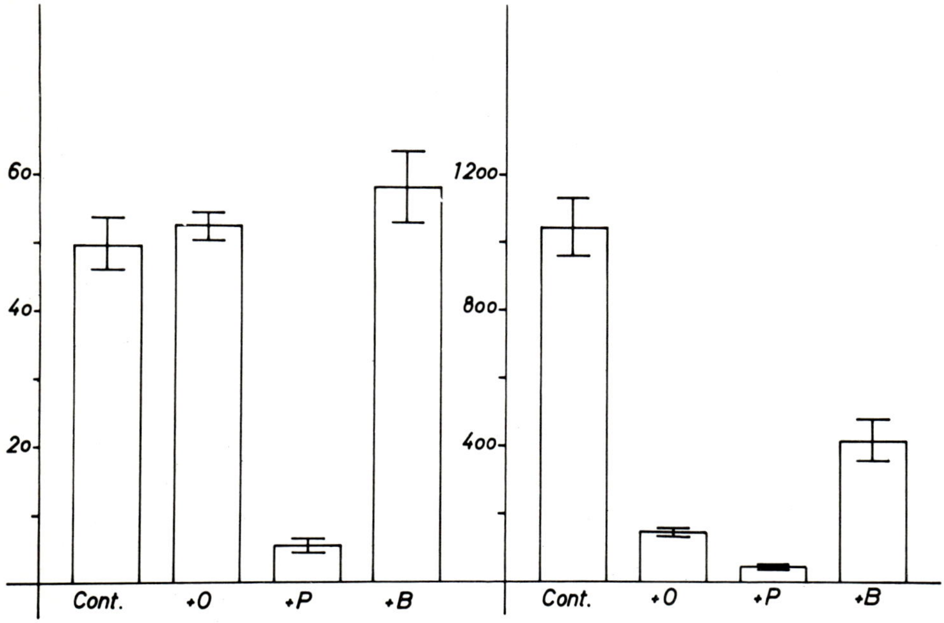

Figure 2 - Inhibition of the uptake of β-methyl-glucoside (1 mM) in guinea pig intestinal rings during incubations of one minute (left) and one hour (right). The inhibitors were 0.5 mM ouabain, 1 mM phloridzin or 3 mM n-butyl-biguanide. Uptake in each case is expressed in n-moles/100 mg tissue (net weight). The results are the means of four experiments, expressed ± SEM.

This hypothesis is upheld by the experiments presented in Table 1. If intestinal rings are pre-incubated in the presence of biguanides for 30 minutes, then the initial rate of uptake of phenylalanine during a subsequent incubation is inhibited, though the drug has no effect on this parameter when present solely in the incubation medium. A control series is included where methionine, a competitive inhibitor of phenylalanine uptake was added to the medium of the second incubation, and

TABLE 1 - Effect of a preincubation in the presence of n-butyl-biguanide, ouabain or ethacrynic acid on the subsequent initial uptake of amino-acid by guinea pig and rat intestine.

Preincubation	Incubation (in presence of amino acid)	Uptake (n-moles/100 mg wet weight) after 1 minute by: Guinea pig intestine	Rat intestine
Buffer	Buffer	39.0	40.5
0.5 mM ouabain	Buffer	12.7	34.2
3 mM n-butyl biguanide	Buffer	21.3	12.2
1 mM ethacrynic acid	Buffer	42.0	13.1
Buffer	0.5 mM ouabain	36.5	36.9
Buffer	3 mM n-butyl-biguanide	43.0	36.7
Buffer	1 mM ethacrynic acid	40.4	36.0
Buffer	5 mM methionine	12.0	17.1
	$D_{0.05}$	5.2	7.7
	$D_{0.01}$	6.9	10.4
	$D_{0.001}$	9.2	13.8

Tissues were preincubated for 30 minutes in Krebs bicarbonate buffer with or without drugs, and were subsequently incubated for one minute in 1 mM L-phenylalanine in the same buffer, with or without drugs. The initial velocity of uptake is only inhibited significantly by methionine, unless the tissues are preincubated in the drugs, which then have species-specific effects. Statistical evaluation by two-way analysis of variance (9), from which is extracted the value of the least significant difference at various probability levels, shown as $D_{0.05}$, etc. in the Table. Results are means by six experiments with guinea-pigs and five experiments with rats.

a strong inhibition of the initial rate of entry of phenylalanine is observed. In these experiments, the effects of biguanide are compared with those of ouabain and ethacrynic acid, both in the rat and the guinea-pig intestine. In the latter, preincubation of intestinal rings with biguanide or ouabain leads to an inhibition of subsequent amino-acid uptake; ethacrynic acid is without effect - it appears that this species is insensitive to this drug. The rat intestine differs in being sensitive to ethacrynic acid, but not to ouabain (5,8), though like biguanide, ethacrynic acid affects the initial uptake only after a preincubation, but has no direct influence on the uptake mechanism. This incidentally corrects a misinterpretation in an earlier publication (10), where it was suggested on the basis of experiments *in vivo* that ethacrynic acid might interfere with the formation of the carrier-substrate complex at the luminal border of the rat intestinal cell - which is refuted by present results.

Table 2 - Effects of ouabain, ethacrynic acid and n-butyl-biguanide and mixtures of the drugs on phenylalanine accumulation by guinea-pig intestinal rings.

Addition	Distribution ratio
None	22.22 ± 3.07
1mM Ouabain	2.88 ± 0.16
10mM n-butyl-biguanide	5.03 ± 0.31
1mM ethacrynic acid	20.51 ± 2.02
1mM ouabain + 10mM n-butyl-biguanide	3.05 ± 0.12
1mM ouabain + 1mM ethacrynic acid	2.53 ± 0.10
1mM ethacrynic acid + 10mM n-butyl-biguanide	4.68 + 0.30

Guinea pig intestinal ring incubated for 1h in 0.1 mM L-phenylalanine. The concentration in the incubation medium was determined after the incubation, for the calculation of the distribution ratio. The uptake was calculated per ml of tissue water (on the basis of 80% tissue water), but corrections for the extracellular space were not applied. The results are means of six experiments and are expressed ± SEM.

Previous studies (11) have shown that in kidney slices, the effects of ouabain and biguanide are additive, a result that has been interpreted as indicating that biguanides, like ethacrynic acid (9), act on a different sodium-pumping mechanism from the ouabain-sensitive pump. On the other hand, in guinea pig intestinal rings, the effects of ouabain and biguanide are not additive, as shown by the results in Table 2. Again ethacrynic acid has no action in this tissues. There is a small but significant component of active phenylalanine uptake that is insensitive to ouabain, the reason for which is unclear, but biguanide is unable to reduce this uptake further. The extremely low substrate concentration and high inhibitor concentrations were employed specifically to exaggerate these results - even under these conditions, no detectable additivity between ouabain and biguanides could be revealed.

Finally the effect of n-butyl-biguanide on a microsomal Na^+-K^+-ATPase from guinea pig intestinal mucosa was examined but no inhibition of this enzyme was observed (Table 3).

Table 3 - Lack of inhibition of ATPase activities of guinea pig mucosa by n-butyl-biguanide.

Inhibitor	Na^+-K^+-ATPase	Mg^+ATPase
None	108 ± 14	85 ± 4
3mM n-butyl-biguanide	97 ± 8	78 ± 6
0.75mM n-butyl-biguanide	105 ± 17	83 ± 6

Total ATPase and residual ATPase (measured in presence of ouabain) determined in microsomal suspensions prepared from guinea pig intestinal mucosa (8). The results are expressed in n-moles of phosphate liberated per minute per mg of protein, and are the means (± SEM) of four determinations. No differences are significant.

DISCUSSION

The results presented in this study demonstrate that n-butyl-biguanide inhibits intestinal sugar or amino-acid transport strongly, but that it does not act on the mechanism responsible for the translocation of these substrates across the luminal membrane of the intestinal epithelial cell. Its mode of action is analogous to that of ouabain in the guinea pig, or ethacrynic acid in the rat, though since it has no effect on intestinal Na^{+}-K^{+}-ATPase, it is unlikely to have the same locus of action as ouabain.

In renal proximal tubular cells, the effects of ouabain and n-butyl-biguanide on sugar and amino acid transport are additive, both inhibitors only partially inhibiting transport when present alone, and it has been suggested that n-butyl-biguanide acts on an ethacrynic-acid-sensitive sodium pump in this tissue (11). Much evidence has been obtained in favour of two separate sodium-pumping mechanisms in the renal tubular cell (13), although an unambiguous demonstration of a similar dichotomy in the epithelial cell of the intestine is not available. The lack of additivity of the effects of ouabain and biguanide in the present experiments,which appears to represent a fundamental difference between the kidney and the intestine, indicates that an analogous inhibition at the level of a hypothetical sodium pump is unlikely. The results could possibly be explained in terms of an action of biguanides on the production of cellular ATP, necessary for the maintenance of the sodium gradient across the cell membrane on which sugar and amino-acid transport depends. Indeed other workers have demonstrated a reduction in intestinal ATP levels after administration of biguanides (7), and yet others have described a mitochondrial locus of action of these drugs (4,12). Such an explanation, if confirmed, would be entirely consistent with the findings in the intestine presented in this study, though at variance with the renal results obtained in our laboratory (11).

SUMMARY

The uptake of L-phenylalanine or β-methyl-D-glucoside by guinea pig intestinal rings is inhibited by n-butyl-biguanide during long incubations, but not during short incubations. The dose-response curve for the inhibition of amino acid transport is almost identical to that of sugar transport, a semi-maximal inhibition occurring at 1.25 mM n-butyl-biguanide in the former case and at 1.45 mM in the latter. Preincubation of intestinal rings in n-butyl-biguanide leads to an inhibition of the initial rate of entry of amino acid into the tissue during a subsequent short incubation. The action of n-butyl-biguanide is analogous to that of ouabain in the guinea pig and of ethacrynic acid in the rat. Although a small component of active transport of phenylalanine in the guinea pig intestine is not inhibited by ouabain nor by biguanide, the effects of the two drugs are not additive, as has been found in the kidney. n-butyl-biguanide has no action on microsomal Na^+-K^+-ATPase. The results suggest that n-butyl-biguanide abolishes the sodium gradient across the membrane of the epithelial cell on which active sugar and amino acid transport depends.

ACKNOWLEDGEMENTS

This study was performed with the aid of grants from the Fonds National Suisse and from F.Hoffmann-La Roche & Co AG. We are grateful to Mlles Christine Brandt and Ursula Pässler for their efficient technical assistance.

REFERENCES

1. Birò,L., Bànyàsz,T., Kovacs,M.B., & Bajor,M. (1961) Die Wirkung des Phenyläthylbiguanids auf die Glucoseresorption. Klin.Wschr. 39,760-762.

2. Caspary,W.F. & Creutzfeld,W. (1971) Analysis of the inhibitory effect of biguanides on glucose absorption: Inhibition of active sugar transport. Diabetologia 7, 379-385.

3. Czyzyk,A., Tawecki,J., Sadowski,J., Ponikowska, I. & Szcepanik,Z. (1968) Effects of biguanides on intestinal absorption of glucose. Diabetes

17, 492-498.

4. Davidoff,F. (1971) Effects of guanidine derivatives on mitochondrial function. III. The mechanism of phenethylbiguanide accumulation and its relationship to in vitro respiratory inhibition. J.Biol.Chem. 246, 4017-4027.

5. Dettmer,D., Müller,F. & Kuhfahl,E. (1967) Zum Problem der Kopplung zwischen aktivem Natrium- und Monosaccharidtransport in der Dünndarmschleimhaut. Acta biol.med.Germ. 18,555-561.

6. Lorch,E. (1971) Inhibition of intestinal absorption and improvement of oral glucose tolerance by biguanide in the normal and in the streptozotocin-diabetic rat. Diabetologia 7, 195-203.

7. Riecken,E.O., Bloch,R., Menge,H.,Schaarschmidt, W.D., Göttesbüren,H. & Goebell,H. (1971) Biochemische, histochemische,histologische und funktionelle Untersuchungen zur Phenforminwirkung auf die Dünndarmschleimhaut bei Ratte und Mensch. Verh.dtsch.Ges.inn.Med. 77, 532-533.

8. Robinson,J.W.L. (1970) The difference in sensitivity to cardiac steroids of (Na^{+}+K^{+})-stimulated ATPase and amino acid transport in the intestinal mucosa of the rat and other species. J.Physiol.(London) 206, 41-60.

9. Robinson,J.W.L. (1972) The inhibition of glycine and β-methyl-glucoside transport in dog kidney cortex slices by ouabain and ethacrynic acid: contribution to the understanding of sodium-pumping mechanisms. Comp.gen.Pharmac. 3, 145-159.

10. Robinson,J.W.L. & Förster,H. (1968) Effect of ethacrynic acid and furosemide on the transport systems of the rat intestine. Naturwissenschaften 55, 232-233.

11. Robinson,J.W.L., & Luisier,A.-L. (1972) Die Wirkung von n-Butyl-Biguanid auf den Dünndarm- und Nierentransport von Zuckern und Aminosäuren. Naunyn-Schmiedebergs Arch.Pharmac.274, R93.

12. Schäfer,G. (1963)Oligomycinähnlicher Effekt von Biguaniden. Biochem.Z. 339, 46-52.

13. Whittembury,G. & Proverbio,F. (1970) Two modes of Na extrusion in cells of guinea pig kidney cortex slices. Pflügers Arch. - Europ.J. Physiol. 316, 1-25.

Comparative Physiology, eds. L. Bolis, K. Schmidt-Nielsen and S. H. P. Maddrell

PROPERTIES OF FISH HEMOGLOBINS: THE HEMOGLOBIN SYSTEM OF TROUT (*Salmo irideus*).

M.BRUNORI, B.GIARDINA, J.BONAVENTURA[(o)], D.BARRA and E.ANTONINI

Laboratory of Molecular Biology,
Institute of Biochemistry,
University of Camerino (Italy) and
C.N.R. Centre of Molecular Biology,
Institutes of Biochemistry and Chemistry,
University of Rome (Italy)

INTRODUCTION

Hemoglobins have evolved to satisfy the oxygen demands of organisms living under very different conditions. This need to fulfill multiple physiological requirements appears particularly obvious in hemoglobins from teleost fish, whose blood must deliver oxygen to a hydrostatic organ (the swim bladder), besides satisfying the normal respiratory demands.

In an attempt to understand the molecular basis of differences in functional properties displayed by hemoglobins from various organisms, we have started extensive work on fish hemoglobins. This paper reports a summary of significant structural and functional features of hemoglobin components from trout blood (*Salmo irideus*). Such studies appear to offer a "rationale" for the multiplicity of hemoglobin components observed in the blood of trout and other teleost fish.

General Features of Allosteric Interactions in Hemoglobin

Modern ideas about allosteric enzymes were

(o) Present address: Duke University, Marine Laboratory, Beaufort, North Carolina, USA.

largely developed from studies on the structure and function of hemoglobin, and this continues to play a significant role in the understanding of the properties of regulatory proteins in general (18). Thus it is known that, for an allosteric enzyme, a plot of reaction velocity against substrate concentration yields a sigmoidal curve which has the same meaning as the sigmoidal O_2 saturation curve of hemoglobin (1). This type of saturation function results from the existence of homotropic interaction effects (heme-heme interactions) by which the reactivity of a given site in the macromolecule is affected by the presence of a ligand to another site in the same molecule. Such a cooperative binding curve is particularly advantageous in hemoglobin, because release of oxygen at the level of the tissues is performed over a limited range of ligand concentration, thereby increasing effectively the total amount of O_2 delivered by the blood under a given set of conditions.

The ligand binding curve of hemoglobin is often represented graphically by a plot of log $(\bar{Y}/1-\bar{Y})$ versus log (X) (Hill plot)(1,19)(Fig.1). In accordance with the Hill equation:

$$\frac{\bar{Y}}{1-\bar{Y}} = K\ (X)^n \qquad (1)$$

such a plot will give a straight line with a slope equal to n, the apparent order of reaction. Although the equation is devoid of physical significance, the use of this type of plot is advantageous because it allows an easy comparison between different dissociation curves and gives an immediate description of the character of the binding process. Thus $n = 1$ implies that, for equivalent sites, no interaction effects are present and corresponds to a hyperbolic dissociation curve; $n > 1$ corresponds to a sigmoidal curve and points to the presence of positive heme-heme interactions; a value of $n < 1$ may indicate either presence of non-equivalent sites (intramolecular heterogeneity) or negative site-site interactions.

The Hill plot may be used as a quantitative measure of the heme-heme interactions and is related to the overall free energy of interaction (ΔF_i) as follows (19):

$$\Delta F_i = \frac{RT}{\bar{Y}(1-\bar{Y})}\left(1 - \frac{1}{n}\right) \qquad (2)$$

ΔF_i is a measure of the favorable effect of saturating one site upon ligand binding to other sites in the macromolecule. The data in a Hill plot tend, at both ends, towards asymptotes of unitary slope; the minimum free energy of interaction per site realized in saturating the protein may be obtained from the perpendicular distance between the two asymptotes multiplied by RT $\sqrt{2}$ (19). Such a situation for sheep hemoglobin is represented in Figure 1.

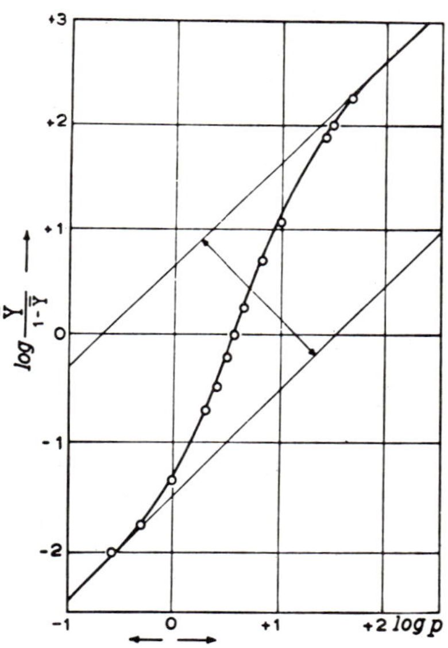

Figure 1 - Hill plot of the oxygen equilibrium of sheep hemoglobin at pH = 9.1 and 19°C. Total free energy of interaction given by this plot is 3.0 Kcal per site. (From Wyman, ref.19).

The interactions between sites which bind different types of ligand, commonly referred to as heterotropic interactions, are exemplified in hemoglobin by the Bohr effect. This is the effect of pH on the oxygen affinity of the molecule (1), and involves a linkage between oxygen binding sites (hemes) and protein binding sites (aminoacid side chains of the protein). Besides protons there are several other ligands (such as CO_2 and organic phosphates) which affect the oxygen dissociation curve of hemoglobin, and their effect, like that of all third components, is formally analogous to that of protons (Bohr effect).

The dependence of the oxygen affinity of hemoglobin on the partial pressure of carbon dioxide is of the utmost physiological importance in view of the double role played *in vivo* by the protein, i.e. O_2 and CO_2 transport between environment and tissues (1,4). The decrease in oxygen affinity brought about by CO_2 is now known to be due partly to a change in pH and partly to a direct binding of CO_2 to the hemoglobin molecule (1,16). Organic phosphates (DPG, ATP), which are normally present in the red blood cells, have a pronounced effect on the oxygen equilibrium curve of hemoglobin, which is shifted towards lower affinities in the presence of the phosphates. This implies that binding of organic phosphates occurs preferentially to deoxygenated hemoglobin (1,2).

It is clear that the simultaneous presence of both *homotropic* and *heterotropic* interaction effects in hemoglobin is of great advantange in enhancing the release of oxygen at the level of the tissues, and more generally in coupling the O_2 transport to metabolic processes in the erythrocyte. Great physiological significance has to be attributed to the action of organic phosphates because their concentration in the erythrocyte depends on complex metabolic relationships and may control the respiratory function of the hemoglobin molecule as well as its adaptation to various physiological and pathological conditions.

It is generally accepted that in hemoglobin conformational changes associated with ligand binding represent the physical basis of functional interactions. Interest in this aspect of the problem has been very vivid because conformational changes associated with binding of substrates appear to be primary events in the regulatory phenomena of enzyme function.

It was shown long ago that crystals of oxy and deoxy hemoglobin are not isomorphous since deoxyhemoglobin crystals will break if oxygen is allowed to diffuse into the system (8). The X ray crystal analysis of Perutz and coworkers has unequivocally demonstrated the ligand linked conformational changes which occur in hemoglobin. A number of physico-chemical measurements in solution have also shown that the two physiologically relevant derivatives differ in conformation, and the kinetics of some of the conformational transitions have been estimated (1).

Ligand-linked conformational transitions in hemoglobin may involve different parts of the molecule and even occur independently from one another. At a phenomenological level, the conformational changes involved in homotropic and heterotropic interactions may appear dissociable. Intra-chain conformational transitions responsible for heterotropic interaction are now well documented and explain a number of chemical data on the Bohr effect. On the other hand, changes in conformation relevant to homotropic interactions should involve other chains besides that where the ligand is bound, and therefore they should be inter-chain as well as intra-chain.

Observations of the crystals of oxy- and deoxy-hemoglobin and comparison with functional properties on modified hemoglobins has yielded significant structural information. On this basis Perutz (12) has proposed a stereochemical model for the allosteric properties of hemoglobin.

The General Purpose of Comparative Work on Hemoglobin

Mammalian hemoglobins have been given by far the greatest attention, and a large number of structural and functional studies have revealed many aspects of the mechanism involved in their function (1). Extension of this type of investigation to hemoglobins from other species, whose knowledge is still quite fragmentary, is highly desirable since it may lead to a verification of what is already known for mammalian hemoglobins. Moreover it may yield information on control mechanisms which have developed to satisfy special requirements.

The concept that all hemoglobin polypeptide chains are arranged similarly and are all derived from a common ancestral form is firmly supported by great similarities between vertebrate and invertebrate hemoglobins. Fish hemoglobins are consequently expected to share some of their fundamental structural and functional characteristics with mammalian hemoglobins, but they also may exhibit some unique features in function and primary structure. In general, fish (as well as reptiles and amphibians) have a greater multiplicity of hemoglobin components as compared with mammals (13); this may reflect some important physiological factors related to the advantages that such a multiplicity may confer for a prompt adaptation to changes in environmental conditions.

With these points in mind we have started an extensive investigation of the hemoglobin from fish, using the trout "*Salmo irideus*" as experimental material. A summary of the status of the problem is given below.

Hemoglobin System from Trout (*Salmo irideus*)

It has been reported in the previous papers (3,6) that in the blood from the trout "*Salmo irideus*" there are four main hemoglobin components which have been named as Hb trout I through Hb trout IV according to the increase in anionic mobility. The major components Hb trout I and IV (smaller and larger electrophoretic mobility) account for about 20% and 65% of the total pigment. It should be noted that among the various individuals, no change is observed in the proportion of the main components. Figure 2 shows the electrophoresis of the hemolyzate from 11 different individuals of different age from 8 months (about 15 cm long and 50 g ave.) to 3 years (about 50 cm long and 1500 g ave.). This experiment shows that a) the various components are present in the same proportion in different individuals, thereby excluding polymorphism as the reason for multiplicity; b) there is no large variation in the proportion of the various components with age over this age range.

Isolation of the various components, performed by chromatography on a DEAE Sephadex A-50 Column using a pH gradient from 9 (Tris - HCl 0.1M)

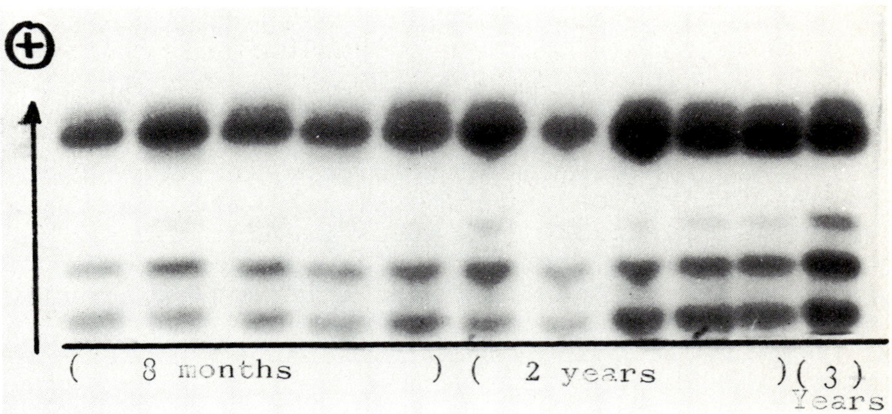

Figure 2 - Starch-gel electrophoresis (at pH 9.6) of trout hemoglobin from 11 different individuals of different age: 8 months (1-5), 2 years (6-10), 3 years (11).

to 6 (KH_2PO_4 0.1M), allows a characterization of functional and physico-chemical properties of the various components in homogeneous form, and especially of Hb trout I and IV. The two other components (Hb trout II and III) have been obtained,up to now, only in smaller amounts and therefore only preliminary information on their properties is available.

Some physico-chemical properties. - The overall molecular architecture of Hb trout I and IV is similar to that of mammalian hemoglobins. Thus it was observed by starch gel electrophoresis in 8 M urea and mercaptoethanol that they are made up of two types of polypeptide chains and have no chain in common. The carboxyl terminal sequence of the α-like and β-like chain of both components has been determined by carboxypeptidase (A or B) digestions. The α-like chain has the same C-terminal residues as human Hb (Arg-Tyr-); the β-like chain is: His-Tyr- for Hb trout IV and Phe-Tyr- for Hb trout I. Therefore in the latter component the C terminal His of the β-chains, which in mammalian hemoglobins is involved in the Bohr effect, is substituted by Phe.

Sedimentation analysis has shown that both components migrate as homogeneous species with a value of $S_{w,20}$ = 4.3 under physiological conditions (i.e. neutral pH and low ionic strength). The estimated molecular weight in solution is about 64,000, similar to mammalian hemoglobins (10). The value of the sedimentation constant,together with gel filtration experiments, indicate that both components are stable tetramers even at micromolar concentrations, so that the dissociation constants are much smaller than the corresponding ones for human hemoglobin.

The isoelectric points obtained by the electrofocussing technique are: 6.2 ÷ 6.4 for Hb trout IV and 8.2 for Hb trout I.

Analysis of the dichroic properties, carried out between 300 and 200 nm, gives a value of θ (reduced ellipticity) of 21800 for Hb trout I, and 22400 for Hb trout IV at 222 nm. No change in the CD spectrum of Hb trout I was observed going from the deoxygenated to the ligand bound (CO, O_2) derivative. On the other hand deoxygenation of Hb trout IV (at pH = 8.2) increases the negative ellipticity at 222 nm by about 5%. This finding recalls similar results obtained for human hemoglobin (4), and shows directly the existence of ligand linked structural changes in Hb trout IV.

Ligand binding properties. - Oxygen dissociation curves were measured by the method of Rossi Fanelli and Antonini (13), generally at a protein concentration of 3-5 mg/ml, in phosphate or borate buffer of ionic strength 0.1 -0.3 M. The O_2 binding curve of both Hb trout I and IV is given in Figure 3. For Hb trout I the oxygen affinity, as well as the shape of the ligand equilibrium curve, are completely pH insensitive (3). The value of the Hill coefficient (n) corresponds to about 2.5, indicating the presence of cooperative effects in

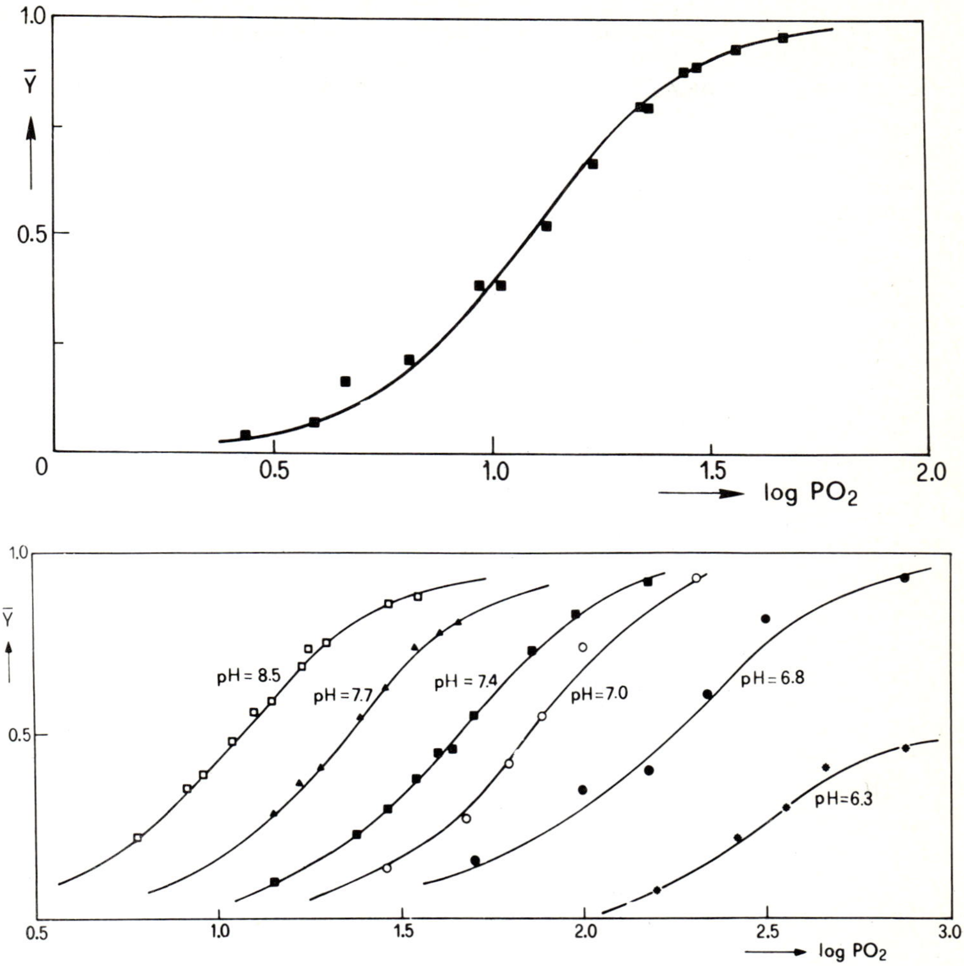

Figure 3 - a) Oxygen dissociation curve of Hb trout I at 20°C. Experiment carried out at pH 7.0. The smooth curve is a theoretical line corresponding to a value of n = 2.5.
b) Oxygen dissociation curves of Hb trout IV at different pH values in phosphate buffer 0.1M and T = 20°C.

the binding of O_2. On the other hand Hb trout IV shows drastic changes in oxygen affinity as the pH is lowered below ~7.5 (in phosphate buffer). At pH = 6.5 and 20°C, Hb trout IV is only 20% saturated at partial oxygen pressure of about 150 mmHg. Not only the affinity but also the shape of the oxygen binding curve is strongly affected by pH; the value of n, higher than 2 at pH 7.5, drops to 1 or less as the pH is decreased (Figure 3).

The dependence of O_2 affinity on temperature was also investigated at different pH values for both components (Figure 4). For Hb trout I the

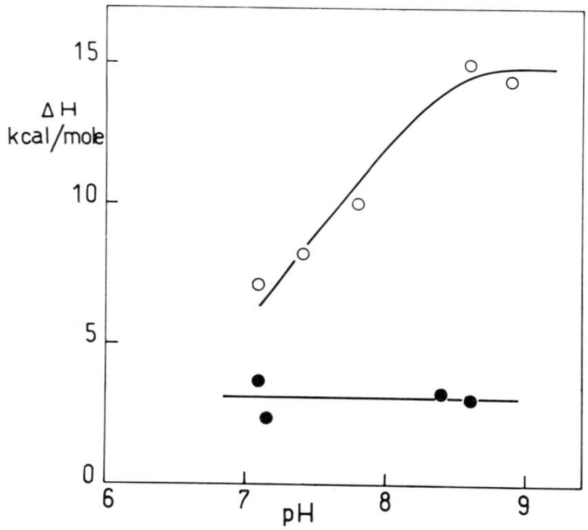

Figure 4 - Dependence on pH of the apparent heat of oxygenation for component IV (O) and component I (●). (From Binotti et al., ref.3).

enthalpy of O_2 binding is pH independent, and the intrinsic value is between zero and 1 Kcal/site. The behaviour of Hb trout IV appears to be very different. At pH $>$ 8.5 the value of ΔH is -14.5 Kcal/mole and therefore is very similar to that obtained for mammalian hemoglobins at similar pH (1); where the Bohr effect is operative, the apparent ΔH decreases strongly due to the ionization of the oxygen linked group (or groups) which have a positive ΔH.

These results show that the Root effect (14), a pH dependent phenomenon by which the oxygen saturation of blood of many fishes is drastically lowered at acid pH, is a characteristic property of Hb trout IV. Similar findings have been reported also for other teleost fish, like salmon (7).

In relation to a possible role of organic phosphates (1,2,5), it may be mentioned that the O_2 binding properties of Hb trout I are unaffected by solvent composition. On the other hand the properties of Hb trout IV, and in particular the Root effect, are very sensitive to the presence of ATP (unpublished experiments).

The kinetic features of oxygen and CO binding have been extensively investigated for both components which, as expected on the basis of the equilibrium experiments, at pH values above 7.5 display very similar behaviour. The main kinetic features show resemblances with those characteristic of human hemoglobin (1). They may be summarized as follows:

a) the CO combination process is clearly autocatalytic reflecting the presence of cooperative kinetic effects;

b) the oxygen relaxation kinetics, investigated by the temperature jump technique, is characterized by two relaxation times, both O_2 concentration dependent;

c) the oxygen dissociation follows satisfactory first order kinetics with an apparent rate constant which is higher than that of mammalian hemoglobins (K_{off} = 175 sec^{-1} for Hb trout I at 20°C and pH = 7.5).

The kinetic behaviour of Hb trout I is completely independent of pH, while that of Hb trout IV is markedly affected by changes in pH. For the latter component the slower relaxation time increases sharply as the pH is lowered, going from 1100 sec^{-1} at pH = 7.2 to 10,000 sec^{-1} at pH = 6.5. This effect follows the decrease in fractional saturation observed at constant oxygen pressure, i.e. the Root effect (Figure 5). As a general conclusion it appears that in Hb trout IV both the combination and dissociation velocity constants are affected by pH, but, at least for O_2, the increase in the dissociation rate constants occurring at low pH dominates over the decrease in the combination rate constant. These findings represent strong evidence to suggest that the kinetic basis of the Root effect resides, to a significant extent, in the "off" process,rather similar to that proposed for carp hemoglobin (11).

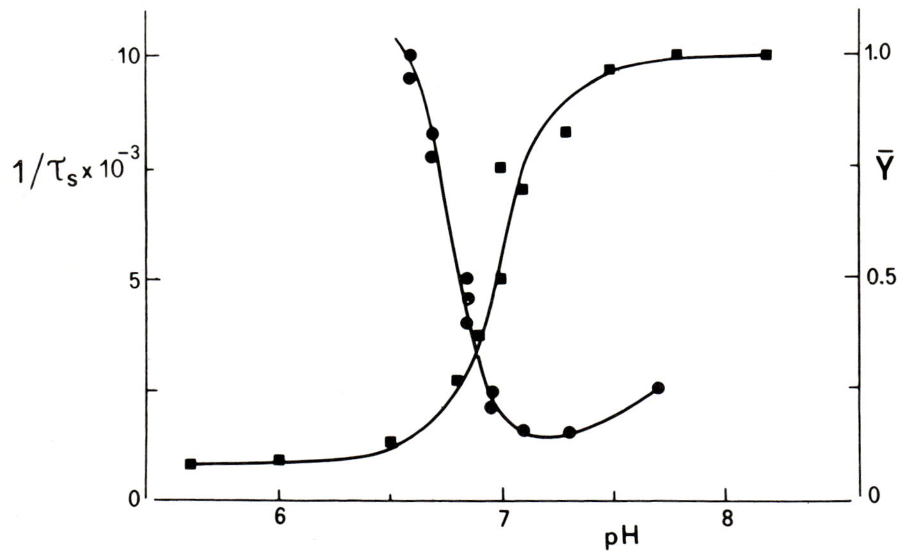

Figure 5 - Effect of pH on the slower relaxation time (●) and on the fractional saturation (■) of Hb trout IV at constant oxygen pressure (PO_2 = 152 mmHg). (From Giardina *et al.*, in preparation).

CONCLUDING REMARKS

The experiments reviewed in this paper clearly show that the various components of trout's blood are characterized by different functional properties. This finding is not unique for trout, but has been previously reported also for other teleost fish, like salmon (7).

An understanding of the molecular basis for such differences in functional properties may demand a complete characterization of the various components, and in particular a knowledge of the primary and tridimensional structure of the molecules. Such studies are in progress. However, already at this stage, a number of considerations are possible, some of which are relevant to the general problem of structure-function relationships in hemoglobin.

Comparison of the functional properties of Hb trout I and IV with their C-terminal sequences allows some considerations and stimulates some speculations. Both Hb components contain tyrosine in the penultimate position of both chains, a fact which appears to be in complete agreement with existing evidence which emphasizes the role of these residues in the molecular mechanism of homotropic interactions (12). Hb trout I, which lacks histidine at the terminal region of the β-chains, shows a complete independence of either affinity or cooperativity of pH, in spite of the invariance in the C-terminal sequence of the α-chains. On the basis of what is known for mammalian hemoglobins, it may be expected that the other groups involved in the Bohr effect, and in particular the α-NH_2 terminal of the α-chains, are either absent or blocked. The behaviour of Hb trout IV, which is identical to human hemoglobin in the C-terminal sequence of both chains, is unique in its pH dependence. It appears obvious that the Root effect, which to a first approximation may be considered a special case of the Bohr effect, must involve additional processes besides those operative in human hemoglobin to explain the marked dependence on pH of the shape of the ligand binding curve. In fact nothing is so far known of the molecular mechanism underlying the Root effect. It is however of great significance that removal of the C-terminal residues of the α or the β chains by digestion of Hb trout IV with carboxypeptidases (A or B) greatly reduces the Root effect. This finding points to the critical role of the C-terminal region of the molecule in the pH dependent change of O_2 binding properties characteristic of Hb trout IV.

The dependence of the shape of the ligand binding curve on pH may be explained on the basis of different models. Values of n below unity, which have been reported for fish hemoglobins at low pH, may be taken as an indication of the presence of negative heme-heme interaction effects operative in the molecule at low pH (9). However, at present, a pH-induced intrinsic heterogeneity as between the two types of chains, which may suffice to explain the observations, cannot he excluded.

It is indeed very suggestive to relate the differences in functional properties of the hemoglobin components from trout's blood to the role they may play in the respiratory economy of the animal. The O_2 binding properties of Hb trout I make this component suitable for providing oxygen to the tissues for respiratory demands. The independence of its functional properties of pH appear well suited for a hyperactive fish which lives in fast moving waters and needs a constant O_2 supply even under conditions of high lactic acid production. On the other hand the properties of Hb trout IV which is characterized by a Root effect,strongly imply its involvement in the secretion of oxygen into the swim bladder (5). Although the mechanism of action of the swim bladder is not completely understood (17), the production of lactic acid which(under nervous control) occurs in the gas gland brings about a pH drop which induces O_2 release in Hb trout IV. This may occur against very high O_2 back pressures thereby satisfying the requirement of pumping gases in the swim bladder under extreme conditions. Thus it appears that the pH induced allosteric effects operative in Hb trout IV are essential for the role that this hemoglobin component plays in the economy of the animal.

ACKNOWLEDGEMENTS

We wish to express our thanks to Mr.Mario Rossi (Troticoltura Rossi, Sefro, Macerata, Italy) for kindly providing trout from an inbred strain and for excellent facilities.

The skilful technical assistance of M.Coletta and G.Falcioni is gratefully acknowledged.

REFERENCES

1. Antonini,E., & Brunori,M. (1971) Hemoglobin and Myoglobin in their Reactions with Ligands. Frontiers of Biology, vol.21, North Holland Research Monographs.

2. Benesch,R. & Benesch,R.E. (1969) Intracellular

organic phosphates as regulators of oxygen release by haemoglobin. Nature 221, 618-622.

3. Binotti,I., Giovenco,S., Giardina,B.,Antonini, E., Brunori,M. & Wyman,J. (1971) Studies on the functional properties of fish hemoglobins, II. The oxygen equilibrium of the isolated hemoglobin components from trout blood. Arch.Biochem. Biophys. 142, 274-280.

4. Brunori,M., Engel,J. & Schuster,T.M. (1967) The effect of ligand binding on the optical rotatory dispersion of myoglobin, hemoglobin and isolated hemoglobin subunits. J.Biol.Chem. 242, 773-776.

5. Fänge,R. (1966) Physiology of the swim bladder. Physiol.Rev. 46, 299-322.

6. Giovenco,S., Binotti,I., Brunori,M. & Antonini, E. (1970) Studies on the functional properties of fish hemoglobins. I. The O_2 equilibrium of trout hemoglobin. Int.J.Biochem. 1,57-61.

7. Hashimoto,K., Yamaguchi,Y. & Matsuura,F. (1960) Comparative studies on two hemoglobins of salmon. IV. Oxygen dissociation curve. Bull. Japan.Soc.Sci.Fisheries 26, 827-834.

8. Haurowitz,F. (1949) In Haemoglobin pp.53-56, ed. Roughton,F.J.W. & Kendrew,J.C., London, Butterworths.

9. Levitzki,A. & Koshland,D.E.jr. (1969) Negative cooperativity in regulatory enzymes. Proc.Nat. Acad.Sci. 62, 1121-1128.

10. Manuscript in preparation.

11. Noble,R., Parkhurst,L. & Gibson,Q.H.(1970) The effect of pH on the reactions of oxygen and carbon monoxide with the hemoglobins of the carp, *Cyprinus carpio*. J.Biol.Chem. 245, 6628-6633.

12. Perutz,M.F. (1970) Stereochemistry of cooperative effects in Haemoglobin. Nature 228, 726-739.

13. Riggs,A. (1970) Properties of fish hemoglobins. In *Fish Physiology*, vol.IV, pp.209-252, ed.

Hoar,W.S. & Randall,D.J., London & New York Academic Press.

14. Root,R. (1931) The respiratory function of the blood of marine fishes. Biol.Bull. 61,427-456.

15. Rossi Fanelli,A., Antonini,E. & Caputo,A.(1964) Hemoglobin and Myoglobin. Advan. Protein Chem. 19, 73-222.

16. Roughton,F.J.W. (1964) Transport of oxygen and carbon dioxide. In Handbook of Physiology, vol.I, pp.767-825.

17. Scholander,P.F. & van Dam,L. (1954) Secretion of gases against high pressures in the swim bladder of deep sea fishes. I.Oxygen dissociation in blood. Biol.Bull. 107, 247-259.

18. Wyman,J. On allosteric models. Current topics in cellular regulation. (In press).

19. Wyman,J. (1964) Linked functions and reciprocal effects in hemoglobin. Advan. Protein Chem.19, 223-286.

Comparative Physiology, eds. L. Bolis, K. Schmidt-Nielsen and S. H. P. Maddrell

LIGAND-LINKED DISSOCIATION OF SOME INVERTEBRATE HEMOGLOBINS

J.BONAVENTURA[(O)] & G.B.KITTO
Department of Zoology and the Clayton Foundation Biochemical Institute,
Department of Chemistry, University of Texas at Austin, Austin, Texas, USA.

SUMMARY

Hemoglobin from the sea cucumber *Molpadia arenicola* shows cooperativity in oxygen binding, having a Hill constant "n" of 1.45 at pH 7.5. The oxy-form of this hemoglobin is predominantly dimeric ($S_{20,w}$ = 2.9) at protein concentrations above approximately 3 mg/ml. Below this concentration the oxy-hemoglobin dissociates to monomers. ($S_{20,w}$= 1.95). Upon deoxygenation the pigment aggregates markedly ($S_{20,w}$ = 6.5). This aggregation is reversed upon reoxygenation. Hemoglobin from other sea cucumber species also show similar, though not identical, aggregation upon deoxygenation and the phenomenon has also been demonstrated with the intracellular hemoglobin from the marine worm *Glycera dibranchiata*, the extracellular hemoglobin of the marine annelid *Thelepus crispus* and the larval hemoglobin of the horse bot-fly *Gastrophilus*.

INTRODUCTION

Although the amount of research carried out on the structure, properties and functions of hemoglobins undoubtedly exceeds that performed with any

---

(O) Present address: Duke University, Marine Laboratory, Beaufort, North Carolina, USA.

other single type of protein, the vast majority of these studies have been concerned with the vertebrate hemoglobins. By comparison relatively little is known of the properties and structure of invertebrate hemoglobins. The very scattered phylogenetic distribution, among the invertebrates, of organisms possessing hemoglobin supports the possibility of polyphyletic origins for these respiratory pigments (1-4). Recent studies on the sequence and three-dimensional structure of insect hemoglobins (5-8) and the x-ray diffraction studies on the hemoglobin of the marine annelid *Glycera dibranchiata* (9) are suggestive, however, of a common evolutionary origin of invertebrate and vertebrate hemoglobins. Beyond such indications,however,little is known of the relationships in structure and properties between these invertebrate and vertebrate hemoglobins and several proposed schemes for the evolution of hemoglobin have virtually ignored the vertebrate pigments (10-13). An important consideration in these schemes was the behaviour of agnathan hemoglobins. The monomeric oxyhemoglobin of lamprey and the dimeric hagfish hemoglobin were considered to be "primitive" hemoglobins. The aggregation of "primitive α chains" in the hagfish was considered to be an important evolutionary change in that it allowed for heme-heme interaction and consequent modulation of binding ligands. Since these models were published a variety of evidence has accumulated which indicates the need for considerable revision. First, the hagfish and lamprey hemoglobins are by no means as simple as at first believed, and have been shown to contain several types of globins (14-17). Further, lamprey hemoglobins have been shown to aggregate on deoxygenation forming a monomer-dimer-tetramer series, thus clouding the issue of the emergence of heme-heme interactions (18,19).

As part of a more general study on echinoderm proteins we have examined the hemoglobin of the sea cucumber *Molpadia arenicola* and find that ligand binding causes dissociation of this invertebrate respiratory pigment. This prompted further studies to determine whether such ligand-linked dissociation could be observed with other invertebrate hemoglobins.

MATERIALS AND METHODS

The following organisms were obtained from Pacific Biomarine Supply Company, Venice, California: Molpadia, Cucumaria, Thyone, Parastichopus, Glycera, Urechis, Dendrostomum, and Eudistylia. Golfingia and Glycera were obtained from Woods Hole Biological Laboratories, Woods Hole, Mass. Gastrophilus was obtained from Dr. T.J. Galvin, School of Veterinary Medicine, Texas A&M University, College Station, Texas. Paralithoides was collected at the University of Texas Marine Institute, Port Aransas, Texas. The Gastrophilus were kept in 0.9% NaCl at 40°C for approximately two weeks. The other organisms were either used upon arrival or kept in salt water aquariums at 15°C.

Coelomic fluid from the holothurians and annelids was filtered through cheesecloth into vessels containing cold 3.5% NaCl. The erythrocytes were washed by centrifugation three times with cold 3.5% NaCl. The buffy coat was removed and discarded. The erythrocytes were lysed in 0.1 M phosphate buffer, pH 7.5, with the aid of a tissue grinder. Stroma were removed by centrifugation. Hemolysates were then dialyzed vs. 0.1 M phosphate pH 7.5.

Tracheal material from Gastrophilus was washed in insect Ringer's (20). The washed tracheal material was then homogenized in 0.1 M phosphate, pH 7.5 in a tissue grinder. The homogenate was then centrifuged at 20,000 g for 30 minutes. The supernatant, which contained the hemoglobin, was dialyzed vs. 0.1 M phosphate, pH 7.5.

Hemerythrin from Dendrostomum and Golfingia was isolated as described by Ferrell and Kitto (21) and chlorocruorin from Eudistylia was isolated in the manner described for the holothurians and annelids.

Hemocyanin was obtained from Paralithoides by drawing blood from around the heart and allowing it to clot. The clot was then squeezed through cheesecloth three times. The blue hemocyanin solution was centrifuged and the supernatant dialyzed vs. 0.1 M phosphate pH 7.5.

Gel filtration on a calibrated column of Sephadex G-100 (2.5 cm x 50 cm) was carried out by the procedure of Whitaker (22) and Andrews (23).

Oxygen equilibria were determined by the procedure of Riggs and Wolback (24). The equilibria were done at 20°C in 0.1 M phosphate pH 7.5. Absorption spectra were obtained with a Cary 14 recording spectrophotometer.

Ultracentrifugation of the oxy- and deoxy-pigments was performed as described by Bonaventura and Riggs (25).

RESULTS

Holothurian Hemoglobins. - In the sea cucumber *Molpadia arenicola*, hemoglobin occurs in nucleated cells present in the coelomic fluid. The spectra of oxy- and deoxy- *M. arenicola* hemoglobins are very similar to those of vertebrate hemoglobins (Figure 1). The oxy- form, in addition to a Soret band at

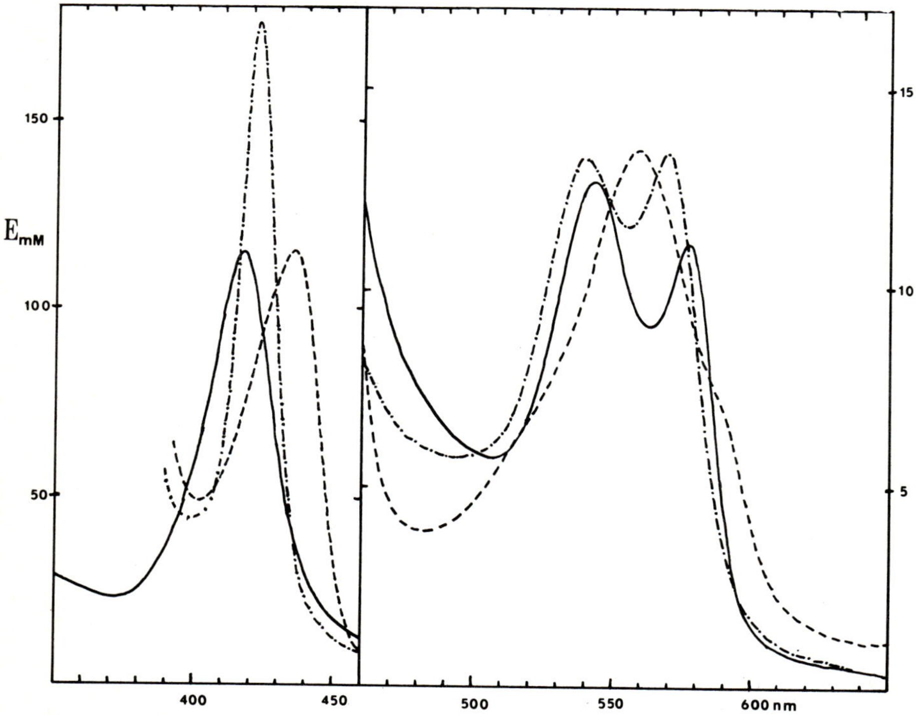

Figure 1 - Spectral properties of *M. arenicola* hemoglobin as oxy (—), deoxy (— —) and carbon monoxy (—·—) derivatives. pH 7, 20°C.

416 mμ, shows maxima at 544 and 577 mμ. In the deoxy-form the latter two peaks are replaced by a broad peak at 560 mμ. An estimation of the molecular size of the M. arenicola oxyhemoglobin was obtained by passage through a Sephadex G-100 gel filtration column calibrated with proteins of known molecular weight. We were aware of the possibility of dissociation of the hemoglobins at low concentrations and therefore carried out studies over a broad range of protein concentrations. At no concentration was a molecular weight greater than 38,000 observed for Molpadia oxy-hemoglobin (figure 2).

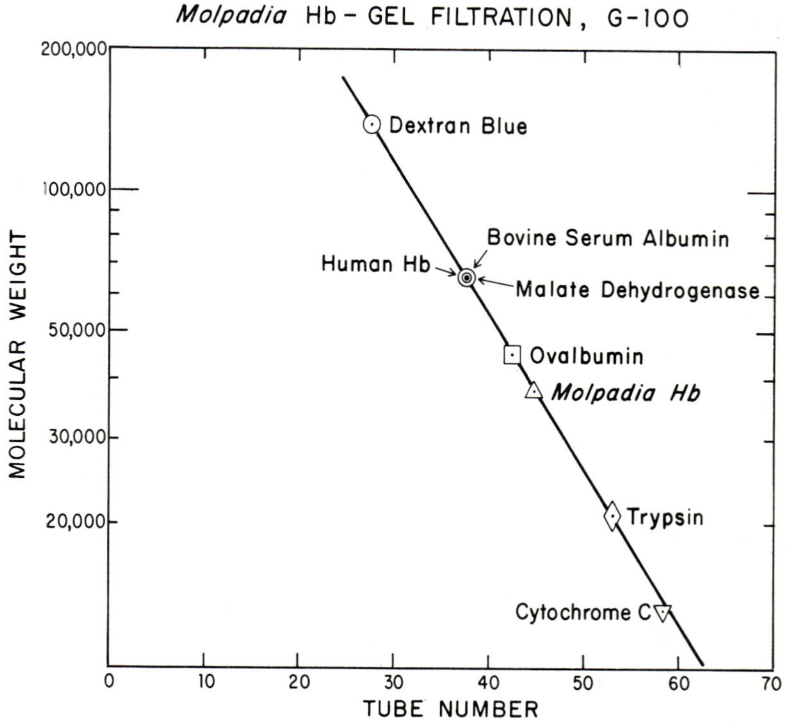

Figure 2 - Estimation of the molecular weight of M. arenicola hemoglobin by chromatography on a calibrated Sephadex G-100 column. The hemoglobin concentration was sufficiently high to avoid dissociation to monomers.

Dissociation to lower molecular weight forms was noted at protein concentrations below approximately 2 mg/ml. When examined in the ultracentrifuge at a concentration of 4 mg/ml the Molpadia oxyhemoglobin sedimented as a single symmetrical peak with an $S_{20,w}$ of 2.9. Preliminary sedimentation equilibrium experiments have given molecular weights of 35,000 for the oxy-form at this concentration and 17,800 for the globin chains.

Since these studies suggested that Molpadia oxy-hemoglobin existed predominantly in a dimeric form, it was of interest to determine whether the pigment showed cooperativity in ligand binding. A Hill plot of oxygen equilibrium data obtained with the Molpadia pigment in 0.1 M phosphate buffer at pH 7.5 gave a value for n = 1.45 (Figure 3). While

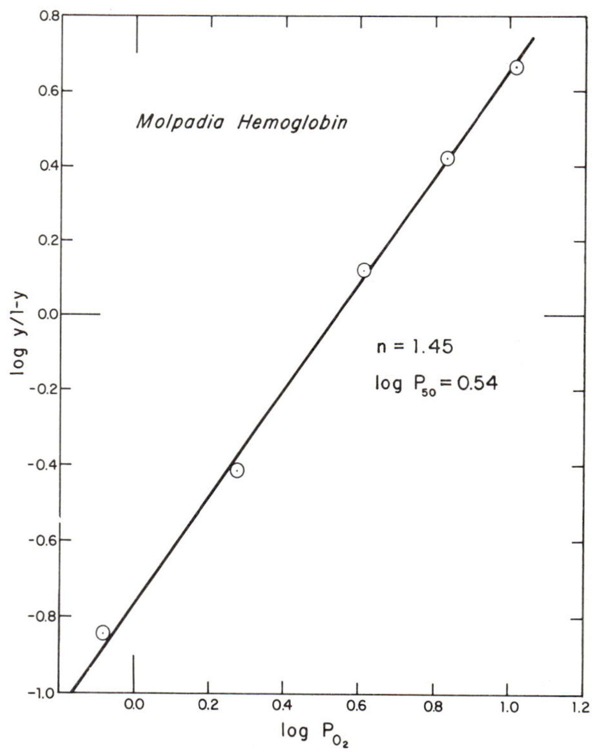

Figure 3 - A Hill plot of oxygen equilibrium data for *M. arenicola* hemoglobin.

this is considerably less than the value for vertebrate hemoglobins, it nevertheless represents significant cooperativity.

The previous studies by Briehl (18) on lamprey

hemoglobins led us to investigate the possibility that a reversible aggregation phenomenon might be associated with oxygenation-deoxygenation. The sedimentation behaviour of Molpadia oxy- and deoxyhemoglobins is illustrated in Figure 4. Binding of

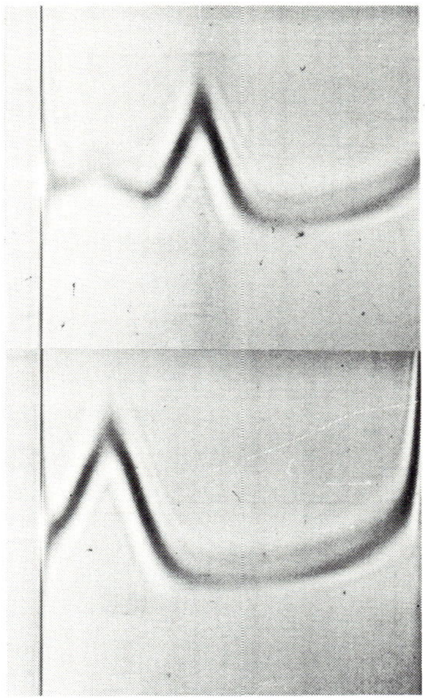

Figure 4 - Sedimentation velocity patterns of deoxy- (upper) and oxy- (lower) *M. arenicola* hemoglobins.

a ligand causes a marked change in the state of aggregation. With complete deoxygenation the hemoglobin sedimented as a single symmetrical peak with an $S_{20,w}$ of 6.5, compared with the $S_{20,w}$ of 2.9 found for the oxy-hemoglobin. In partially deoxygenated samples distinct components with $S_{20,w}$ values of 2.9 and 6.5 were observed. The association-dissociation is freely reversible. Identical results were obtained with carbon monoxide as the ligand. The concentration dependence of the sedimentation coefficients for the oxy- and deoxy-*Molpadia* hemoglobin is shown in Figure 5. The deoxygenated form is considerably more resistant to dissociation by dilution than is the oxy-hemoglobin. At concentrations below approximately 3 mg/ml the oxy-hemoglobin begins to show progressive dissociation to monomers with an $S_{20,w}$ of 1.95. No such dissociation was observed with Molpadia deoxyhemoglobin at concentrations as low as 0.5mg/ml.

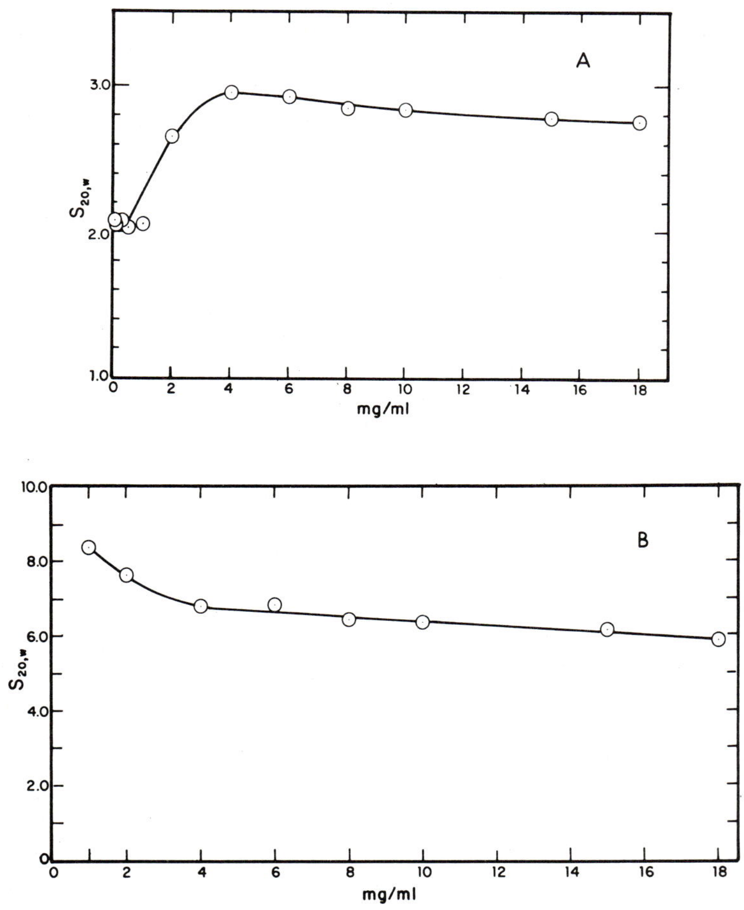

Figure 5 - Concentration dependence of the sedimentation coefficients for M. arenicola oxy-hemoglobin (a) and deoxy-hemoglobin (b).

In order to determine whether dissociation upon ligand binding was a characteristic feature of holothurian hemoglobins, the pigment from several other species was examined. In all cases so far studied a change in molecular size has been observed upon oxygenation of sea cucumber hemoglobins. Variation in the pattern of aggregation and disaggregation were noted, however, from species to species. For example, the hemoglobin of *Cucumaria miniata* has a sedimentation coefficient in the oxy- form of 2.9, like that of the *Molpadia* pigment. In the deoxygenated state a single component with an $S_{20,w}$ of 4.0 was observed. This may represent a dimer-tetramer relationship. Preliminary experiments with the hemoglobins of *Thyone briareus* and a *Parastichopus* species have given very similar results.

Other Invertebrate Hemoglobins. - The hemoglobins of several other invertebrates have been examined for ligand-induced dissociation. One species chosen for study was the marine worm *Glycera dibranchiata*, the same species with which Padlan and Love (9) have been carrying out hemoglobin X-ray crystallographic studies. Here, as in the sea cucumbers, hemoglobin is present within cells rather than free in solution. In crude lysates two forms of oxyhemoglobin, with sedimentation coefficients of 1.7 and 6.4, were noted on centrifugation. Upon deoxygenation, there appeared to be no change in the sedimentation of the light component but the 6.4 S species was replaced by a new component with an $S_{20,w}$ of 8.7. To obtain more definitive data about these hemoglobins a crude lysate was passed through a Sephadex G-100 column with the results shown in Figure 6. Starch gel electrophoresis showed the components to differ considerably in electrophoretic mobility (Figure 7). When the separate components were examined in the ultracentrifuge the light component showed no difference in sedimentation rates between the oxy- and deoxy- states. When the heavy component was examined separately in the deoxy- state we found that the situation was considerably more complex than was suspected from the preliminary studies with the crude lysate. Deoxygenation caused not only the apparent conversion of a 6.4 S component to an 8.7 S aggregate, but also the formation of several larger aggregates with sedimentation coefficients of 12 and higher. Identical results were obtained

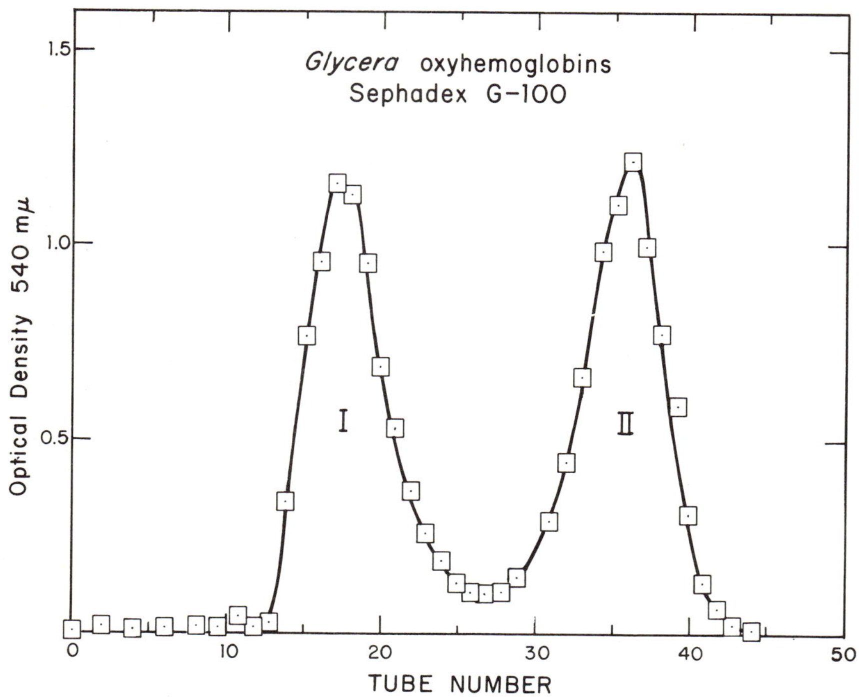

Figure 6 - Separation of *Glycera dibranchiata* oxyhemoglobins on a Sephadex G-100 column (2.5 x 50 cm).

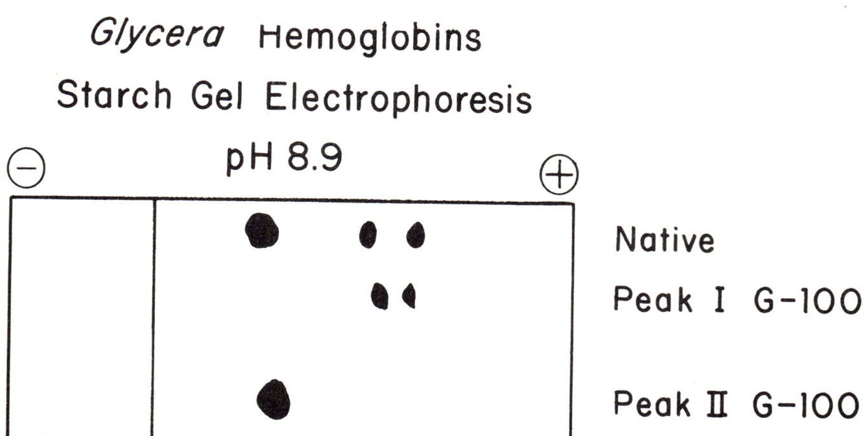

Figure 7 - A tracing of the starch gel electrophoresis pattern of the *Glycera dibranchiata* hemoglobins following separation on Sephadex G-100.

with specimens of *G. dibranchiata* from both Pacific and Atlantic coasts.

Dissociation upon oxygenation has also been observed with hemoglobins from the larvae of the horse bot-fly *Gastrophilus* and with an extra-cellular hemoglobin, that of the marine annelid *Thelepus crispus*. The phenomenon, however, is not a completely general property of invertebrate hemoglobins. For example, we found no change in the aggregation state, upon ligand binding, with the hemoglobin of the echiurid worm *Urechis caupo*. In both liganded and unliganded states an $S_{20,w}$ of 4.0 was observed and a tetrameric structure, like that of vertebrate hemoglobins, is inferred.

Three other invertebrate respiratory pigments, hemerythrin, hemocyanin and chlorocruorin have been examined to determine whether gross changes in the aggregation state occur on ligand binding. With samples of hemerythrin from the sipunculids *Dendrostomum pyroides* and *Golfingia gouldii* and with chlorocruorin from the sabellid *Eudistylia vancouveri*, no changes in molecular size were observed upon oxygenation, findings in accord with those of previous workers. With hemocyanin from the crab *Paralithoides* a very complex pattern of aggregation states was observed which was dependent upon pH and the degree of oxygenation. In general, the data are similar to those recently reported by Van Holde and coworkers (20,21) for *Loligo* and *Cancer* hemocyanins.

DISCUSSION

In recent years there has been considerable interest in associating protein systems, and a number of studies have been carried out on the association-dissociation properties of vertebrate hemoglobins. With most of the vertebrate hemoglobins the tetramer is in dynamic equilibrium with dimers and monomers. This equilibrium is markedly affected by the hemoglobin concentration and by the pH and ionic strength of the medium (22). Appreciable concentrations of monomers are obtained only under rather extreme conditions.

Intimately related to the association-dissociation phenomena are the changes in hemoglobin structure related to ligand binding and the cooperativity associated with this function. There has been considerable controversy over whether the dimeric or tetrameric structures of the higher vertebrate hemoglobins are the functional units (22). The liganded forms of most vertebrate hemoglobins show a greater tendency to dissociate than do deoxy- species. This tendency is much more marked with the lamprey hemoglobins. Here the liganded species is predominantly monomeric and association to dimers and tetramers occurs on deoxygenation (18,19). The present results indicate that such behaviour is not restricted to the lamprey hemoglobins, but is shared by a number of invertebrate hemoglobins, both intra- and extracellular. The pattern of aggregation states varies significantly from species to species and in some cases, for example *Molpadia* and *Glycera*, aggregates considerably larger than tetrameric structures are observed with the deoxygenated pigments. It seems reasonable to suggest that such ligand-linked association properties may be a common property of many invertebrate hemoglobins and that more detailed studies of these phenomena should provide useful information regarding the evolution of cooperativity in ligand binding.

The preliminary results reported here for the hemoglobins of *Molpadia arenicola* indicate that with the oxy- form there is a considerable concentration of monomers present under relatively physiological conditions of pH and protein concentration. Thus, this pigment offers an excellent medium for further studies on the monomer-dimer relationship and, together with structural analyses in progress, such work should prove helpful in furthering understanding of this important phase of hemoglobin association.

ACKNOWLEDGEMENTS

The authors wish to thank the following individuals for their assistance with various aspects of this work: Dr. Austen Riggs, Dr.Rimmon Fay, James Lemburg, Dale Henning, Vivian Sellers, Craig Whitfill and Robert Ferrell.

This work was supported in part by research grants from the National Institutes of Health (AM 09143 and HE 13882), the National Science Foundation (GB 27312) and the Clayton Foundation for Biochemical Research.

REFERENCES

1. Manwell,C. (1964) Comparative Physiology: Blood Pigments", Ann.Rev.Physiol. 22, 191.
2. Manwell,C. (1964) Chemistry, Genetics, and Function of Invertebrate Respiratory Pigments - Configurational Changes and Allosteric Effects In Oxygen in the Animal Organism, p.48, ed. Dickens,F. & Neil,E., New York, Pergamon Press.
3. Jones,J.D. (1963) The function of the respiratory pigments of invertebrates. In Problems in Biology, vol.1, p.11, Kerkut,G.A. ed.
4. Gratzer,W.B. & Allison,A.C. (1960) Multiple hemoglobins. Biol.Rev. 35, 459.
5. Braunitzer,G., Buse,G. & Braig,S. (1969) Intragene Duplikation der Peptid Ketten der Hämoglobine: Untersuchungen am Hämoglobinen (Erythrocruorinen) der Insekten (Chironomus th. th., Diptera). Hoppe Seyler's Z.Physiol.Chem. 350, 1477.
6. Buse,G., Braig,S. & Braunitzer,G. (1969) The constitution of a hemoglobin (erythrocruorin) of an insect (Chironomus thummi thumoni, Diptera) Hoppe-Seyler's Z.Physiol.Chem. 350, 1686.
7. Huber,R., Epp,O. & Formanek,H. (1969) Auf Klarung der Molekülstrukture des Insektenhämoglobins" Naturwissenschaften 56, 363.
8. Huber,R., Epp,O. & Formanek,H. (1969) The Environment of the haem group in erythrocruorin (Chironomus thummi). J.Mol.Biol. 42, 591.
9. Padlan,E.A. & Love,W.E. (1968) Structure of the hemoglobin of the marine annelid worm, Glycera dibranchiata, at 5.5. Å resolution. Nature, 220, 376.

10. Ingram,V.M. (1961) Gene evolution and the haemoglobins. Nature 189, 704.

11. Ingram,V.M. (1963) The Hemoglobins in Genetics and Evolution. New York, Columbia University Press.

12. Zuckerkandl,E. & Pauling,L. (1965) Evolutionary divergence and convergence in proteins. In Evolving genes and proteins, p.97, Bryson,V. & Vogel,H.J. ed., New York, Academic Press.

13. Zuckerkandl,E. (1968) Hemoglobins, Haeckel's 'Biogenetic Low, and molecular aspects of development. In Structural Chemistry and Molecular Biology, p.256, Rich,A. & Davidson,N. eds., San Francisco, W.H. Freeman.

14. Ohno,S. & Morrison,M. (1966) Multiple gene loci for the monomeric hemoglobin of the hagfish (Eptatretus stoutii). Science 154, 1034.

15. Quast,R. & Vesterberg,O. (1968) Isoelectric focusing and separation of hemoglobins of Myxine glutinosa L. in a natural pH gradient. Acta Chem.Scand. 22, 1499.

16. Ruman,N. & Love,W.E. (1963) The six hemoglobins of the sea lamprey (Petromyzon marinus). Arch. Biochem.Biophys. 103, 24.

17. Hendrickson,W.A., Love,W.E. & Murray,G.C.(1968) Crystal forms of lamprey hemoglobin and crystalline transitions between ligand states. J.Mol.Biol. 33, 829.

18. Briehl,R.W. (1963) The relation between the oxygen equilibrium and aggregation of subunits in lamprey hemoglobin. J.Biol.Chem. 238,2361.

19. Behlke,J. & Scheler,Q. (1970) Der Einfluss von Ligandenbufhen Assoziationsgrad des Desoxy-Hämoglobins der Flussineundugen (Lampetra fluviatilis L.). FEBS Letters 7, 177.

20. Stevenson,E. & Wyatt,G.R. (1962) The metabolism of silkmoth tissues. I.Incorporation of leucine into Protein. Arch.Biochem.Biophys. 99, 65.

21. Ferrell,R.E. & Kitto,G.B. (1970) Properties of Dendrostomum pyroides hemerythrin. Biochemistry 9, 3053.

22. Whitaker,J.R. (1963) Determination of molecular weights of proteins by gel filtration on Sephadex. Anal.Chem. 35, 1950.

23. Andrews,P. (1964) Estimation of the molecular weights of proteins by sephadex gel filtration. Biochem. J. 91, 222.

24. Riggs,A.F. & Wolback,R.A. (1956) Sulfhydryl groups and the structure of hemoglobin. J.Gen. Physiol. 39, 585.

25. Bonaventura,J. & Riggs,A.F. (1968) Hemoglobin Kansas, a human hemoglobin with a neutral amino acid substitution, and an abnormal oxygen equilibrium. J.Biol.Chem. 243, 980.

26. De Phillips,H.A., Nickerson,K.W., Johnson,M., & Van Holde,K.E. (1969) Physical studies of hemocyanins, IV. Oxygen-linked dissociation of *Loligo pealei* hemocyanin. Biochemistry 8,3665.

27. Ellerton, H.D., Carpenter,D.E. & van Holde,K.E. (1970) Physical studies of hemocyanins. V.Characterization and subunit structure of the hemocyanin of *Cancer Magister*. Biochemistry,9;2225.

28. Antonini,E. & Brunori,M. (1970) Hemoglobin. Ann.Rev. Biochem. 39, 977.

Comparative Physiology, eds. L. Bolis, K. Schmidt-Nielsen and S. H. P. Maddrell

THE CHEMICAL BASIS OF THE EVOLUTION OF HEMOCYANINS

F.GHIRETTI,A.GHIRETTI-MAGALDI & B.SALVATO
Centro per la Fisiologia e Biochimica delle Emocianine del Consiglio Nazionale delle Ricerche,
Istituto di Biologia Animale
Università di Padova, Italy

Among all copper proteins which occur in living organisms, animal and plants, only hemocyanins (Hcy) are able to combine reversibly with oxygen as do other respiratory proteins, i.e. the hemerythrins, the chlorocruorins and the hemoglobins (2,3,4,12,15).

Hcy are present in two phyla of Invertebrates: Mollusca and Arthropoda (Figure 1); within each phylum, however, as well as within each class, order and suborder, this protein occurs in few species only, closely related species having either blood or tissue hemoglobin or no respiratory pigment at all. In Mollusca it is common to find in the same animal Hcy and myoglobin as well, both active in respiration. No explanation can be given to the apparently accidental distribution of the respiratory proteins in invertebrates on the basis either of the morphological organization of the animals or of their evolutionary position in the zoological scale or of the environmental conditions in which they live. In Mollusca and Arthropoda Hcy is found in terrestrial and aquatic forms, marine and freshwater, having branchial or pulmonary respiration. The only constant exceptions are the higher tracheate forms which obviously do not need any oxygen carrying pigment (Tables 1 and 2).

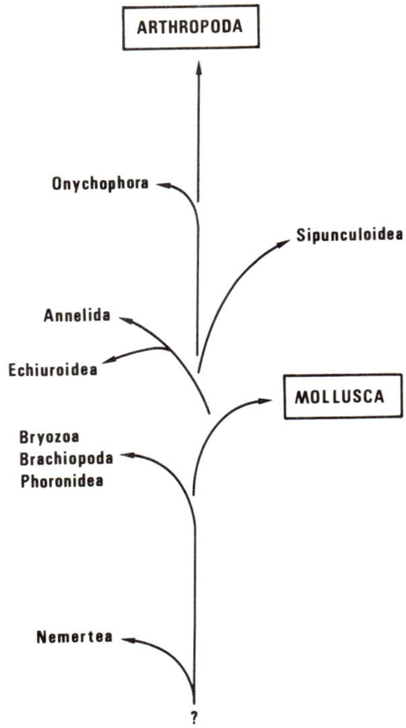

Figure 1 - Phylogenetic relationship between some phyla of invertebrates.

Table 1

THE PHYLUM MOLLUSCA

Class 1.	Amphineura	Hcy
Class 2.	Gasteropoda	Hcy
Class 3.	Scaphopoda	
Class 4.	Lamellibranchiata	
Class 5.	Cephalopoda	Hcy

Table 2

THE PHYLUM ARTHROPODA

Class	Order	Hcy
Class 1. Trilobita		
Class 2. Merostomida		
Class 3. Crustacea:	Branchiopoda	
	Ostracoda	
	Copepoda	
	Branchiura	
	Cirripedia	
	Malacostraca	Hcy
Class 4. Miriapoda		
Class 5. Insecta		
Class 6. Arachnomorpha:	Scorpionidea	Hcy
	Pseudoscorpionidea	
	Eurypterida	
	Xiphosura	Hcy
	Araneida	Hcy
	Palpigrada	
	Solifuga	
	Acarina	
	Phalangida	
	Pantopoda	

Molluscan and arthropod Hcy: dissimilarities

Molluscan and arthropod Hcy contain different amounts of copper: 0.25 and 0.17 per cent in weight respectively. The minimal molecular weight, calculated on the basis of one atom of copper,is 25,000 daltons for the Hcy of Mollusca and 37,000 daltons for the Hcy of Arthropoda. Since, during the oxygenation reaction at saturation, there are in every case two atoms of copper present per each molecule of oxygen bound (12), the minimal molecular weight of the'functional sububit'turns out to be equal to 50,000 and 74,000 daltons for molluscan and arthropod Hcy respectively. Moreover, according to Redfield (13) the equilibrium of the protein with oxygen indicates that a stable product is formed containing 4 molecules of oxygen, i.e. 4 binding sites or 8 atoms of copper. This would be evidence for a functional unit of 200,000 daltons for the Hcy of Mollusca and of about 300,000 for the Hcy of Arthropoda.

Hcy are remarkable in the variety of molecular aggregates that they form. These aggregates reach values up to 10 millions or greater and are among the largest functional structures known. In fact, according to the experimental conditions (concentration of the protein, pH, ionic strength, temperature, etc.) a given Hcy can exist in a number of discrete states of aggregation, as shown by sedimentation or light scattering measurements and, more recently, by electron microscopy. In the blood of the living animal only one or few components are present, while under certain conditions several aggregates can be found. However, for each phylum (Mollusca and Arthropoda) the number of the aggregation states in which Hcy can exist is limited and is typical of that phylum; moreover, the state of aggregation of the major Hcy component which is present in the blood seems to be distinctive for a given class, order or suborder of Mollusca and Arthropoda.

This was fully recognized by Svedberg in his pioneering work on Hcy in the thirties of this century; his classical paper published in 1936 (2) is the basis for all studies of Hcy as macromolecules.

In the hemolymph of Mollusca, Hcy exists primarily as structures of about 60 S and 100 S which are probably pentamers and decamers of the smallest subunit (11 S) that forms at alkaline pH values. Larger aggregates have also been observed in some species (130,155 and 175 S) which represent multiples of the 60 S component. Near the isoelectric point indefinite polymerization of the 60 S component can occur.

As shown in Table 3, the 60 S structure is the main component in Cephalopoda (Octopoda and Decapoda), whereas the 100 S is representative of the hemolymph of Gasteropoda. Only gastropod Hcy is able to give higher structures as further states of aggregation.

Table 3

Molecular weights of molluscan Hcy

Species		S	M.W.
CEPHALOPODA			
Octopoda:	*Octopus vulgaris*	60	
Decapoda:	*Rossia owenii*	56.2	3.316.000
	Loligo pealei	58.7	3.750.000
	Sepia officinalis	55.9	-
GASTEROPODA			
Prosobranchiata:	*Murex trunculus*	102.7	8.885.000
	Archatina marginata	102,3	9.080.000
Opistobranchiata:	*Dolabella auricularia*	105	-
Pulmonata:	*Helix pomatia* (Hcy)	104.3	8.910.000
	Helix pomatia (Hcy)	105.8	8.950.000
	Pila leopoldvillensis	101.3	8.660.000
AMPHINEURA ?			

Arthropod Hcy can exist as structures with sedimentation coefficients of about 5,16,24,34 and 60 S. The 24 S and 34 S components are the result of two successive dimerizations of the 16 S component, whereas the 60 S particle is believed to be an octamer of the 16 S unit (Table 4). This is the main component of Hcy in the hemolymph of *Limulus polyphemus*; the estimated molecular weight (3.300,000) is in fact about eight times that

Table 4
Molecular weights of arthropod Hcy

Species		S	M.W.
MALACOSTRACA			
Caridea:	*Pandalus borealis*	17.4	397,000
	Palaemon fabrici	16	-
Palinura:	*Palinurus vulgaris*	16.4	446,000
	Jasus lalandii	17.1	450,000
Anomura:	*Pagurus striatus*	16	-
	Eupagurus bernhardus	17	-
Astacura:	*Nephrops norvegicus*	24.5	812,000
	Homarus vulgaris	22.6	752,000
	Homarus americanus	24.5	825,000
	Astacus fluviatilis	23.3	-
Brachiura:	*Hyas araneus*	23	-
	Maja squinado	27	-
	Cancer pagurus	23.6	-
	Carcinus moenas	23.3	-
	Cancer magister	25.6	940,000
	Eriphia spinifrons	24	950,000
SCORPIONIDEA:	*Hadagonese bicolor*	34	-
ARANEIDA:	*Dugesiella hentzi*	34	1.300,000
	Avicularia metallica	34	-
XIPHOSURA:	*Limulus polyphemus*	60	3.300,000

found for the 16 S particle (400,000). The 5 S subunit is a dissociation product which forms at high pH values or in the presence of denaturing agents and is equal to roughly one sixth of the

16 S unit. The corresponding molecular weight (75,000) is that of the physiological subunit containing two atoms of copper, i.e. a single oxygen binding site.

Also in Arthropoda, species belonging to the same zoological group appear to have all the same Hcy aggregate as the main component in the blood. In Crustacea the principal constitutents are 16 S and 24 S; the first is distinctive of Caridea, Palinura and Anomura; the second of Astacura and Brachiura (lobsters and crabs). The 34 S structure is found as the main component in scorpions and spiders, whereas the 60 S plays the major role in the hemolymph of Limulus.

Unfortunately, studies on the aggregation state of Hcy are limited to few species only. (Figure 2).

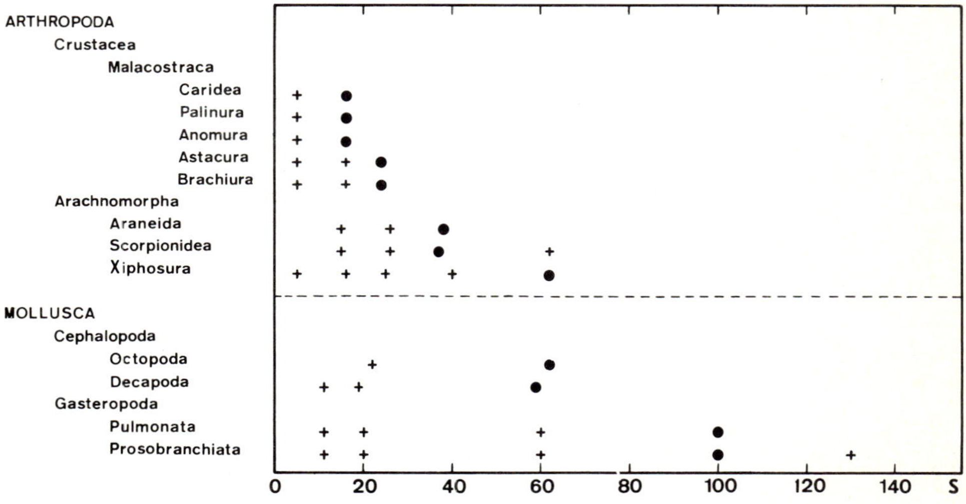

Figure 2 - Molecular aggregates of molluscan and arthropod Hcy. Full circles: the main component present in the blood. Crosses: minor and major components.

Nothing is known about the Hcy of Amphineura among Mollusca and of Amphipoda and Isopoda among Crustacea. Moreover, very little attention has been given to the class of Arachnomorpha where Hcy is largely distributed. Even so, the few data available lead to the general conclusion that the Hcy of Mollusca and Arthropoda are built on an entirely different plan of organization, from the minimal subunit containing a single oxygen binding site (50,000 and 74,000) up to the highest states of aggregation.

This conclusion seems to be confirmed by the electron micrographs of the protein (3,11,14). The Hcy molecule of Mollusca exhibits in every case circular and rectangular projections which clearly correspond to a cylindrical structure. The main component in Cephalopoda (*Octopus* and *Loligo*) is a cylinder 350 Å (diameter) x 170 Å (height) with a fivefold rotation axis, built from three layers of subunits perpendicular to the axis. The cylinder has a thickness of about 65 Å. The 100 S structure which occurs as the main component in Gasteropoda (*Helix* and *Busycon*) ia a cylinder 350 Å x 380 Å with a tenfold rotation axis, made up by six parallel rows of subunits. Higher structures are stacked cylindrical aggregates of the same diameter and increasing number of rows: 9 (130 S), 12 (155 S) and 15 (175 S).

The 16 S structure which is the building block of the arthropod Hcy appears as a cube or hexagonal prism. Electron micrographs of the 24 S structure, a dimer of the 16 S component, show a combination of a hexagon with a square or rectangle. The 60 S structure is probably a two layered octameric aggregate, each layer consisting of four 16 S structures.

These observations clearly underline the absence of any structural relationship between the Hcy of Mollusca and Arthropoda at the quaternary level of organization. They, together with the differences in the copper content, in the sedimentation coefficients and in the distribution of the protein in nature, have led us to believe that molluscan and arthropod Hcy are in fact unrelated proteins which have evolved independently.

Molluscan and arthropod Hcy: similarities

In contrast to these findings, the amino acid composition of Hcy from different species shows a striking similarity not only within the classes and orders of Mollusca and Arthropoda, but also between the two phyla. The analyses of Hcy from many molluscan and arthropod species (8,9,10,16) show that the protein is characterized by the presence of the most common amino acids (cysteine and tryptophan included) and by relatively large amounts of dicarboxylic acids. The similarity of the amino acid composition of molluscan and arthropod Hcy is shown in Figure 3, drawn from the analytical data.

These results raise some doubts for the belief that molluscan and arthropod Hcy are made by polypeptide chains of different molecular weight, as calculated from the different copper content. Consequently, the large difference of the functional subunit (M.W. 50,000 and 74,000) derived from the oxygen to copper ratio does not imply that they are made by two polypeptide chains of different molecular weight, but could well be explained by assuming the presence of a third polypeptide chain in the functional subunit of the Hcy from Arthropoda.

As mentioned before, in molluscan Hcy the smallest subunit which can be observed by ultracentrifugation is of about 11 S, i.e. one tenth of the original molecular weight; the Hcy of Arthropoda goes down to about 5 S which corresponds approximately to 75,000 daltons. However, all attempts for dissociating Hcy down to these values and for obtaining homogeneous solutions of these sizes, failed.

With the aim of dissociating the protein into minimal subunits which are still able to bind oxygen, it was found (13) that only in the presence of 2 M urea and 0.75 M guanidine-HCl, the Hcy of <u>Octopus vulgaris</u> and <u>Carcinus moenas</u> give functional aggregates of molecular weights roughly corresponding to a subunit containing four oxygen binding sites or 8 atoms of copper. By increasing the concentration of guanidine, further dissociation of these aggregates occurs but the ability to bind oxygen is completely lost.

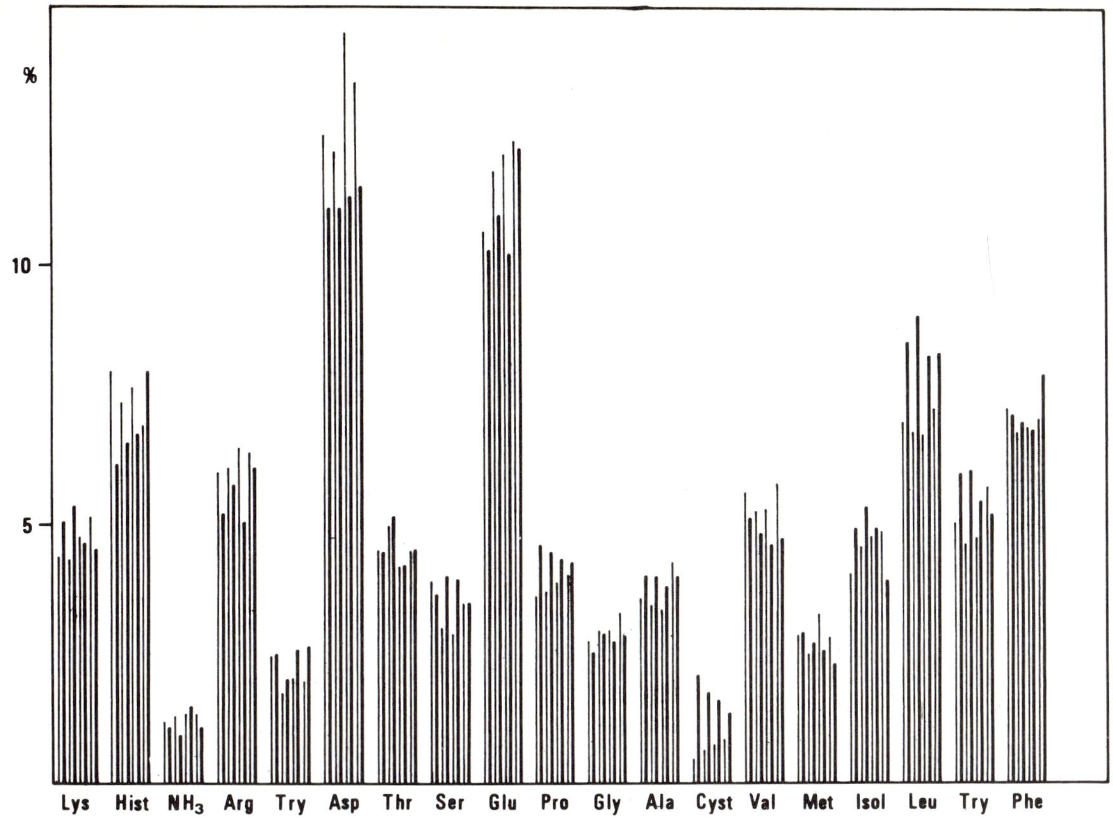

Figure 3 - The amino acid composition of Hcy from molluscan and arthropod species. Thick lines: Mollusca. Thin lines: Arthropoda.

A higher degree of dissociation into non-functional subunits can also be obtained by treating with 0.1% SDS the protein previously acetylated with acetic anhydride at pH 8.6. The separation by gel filtration on Sephadex G-200 gives fractions of the following size: 200,000 and 50,000 for Octopus Hcy and 75,000, 50,000 and 25;000 for Carcinus Hcy. After reduction of the acetylated protein with dithioerythrol or mercaptoethanol, followed by cyanoethylation with acrylonitrile,two new components of 100,000 and 25,000 are found from Octopus Hcy and only two subunits of 75,000 and 25,000 from the Hcy of Carcinus (Figure 4). It must be recalled that in crustacean Hcy one S-S bridge and one SH group are present, whereas the Hcy of Octopus contains at least five disulfide bridges per functional unit (7).

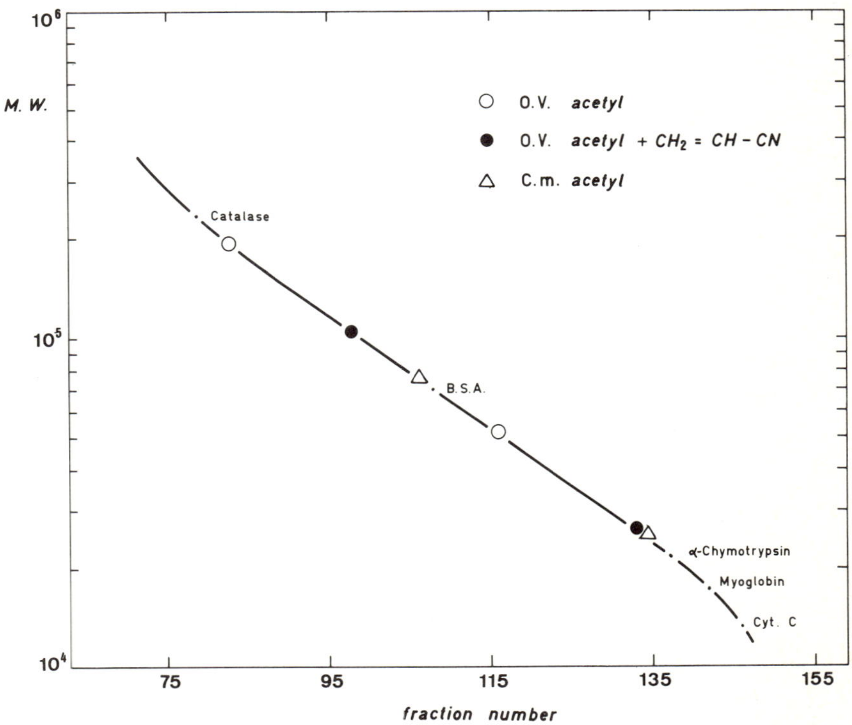

Figure 4 - Molecular weight determination by gel filtration of the subunits obtained from the Hcy of Octopus vulgaris (O.v.) and Carcinus moenas (C.m.).

These results indicate that the 50,000 aggregate of the Hcy of *Octopus vulgaris*, containing one oxygen binding site, i.e. two atoms of copper, is formed by two polypeptide chains of 25,000 daltons each linked together by S-S bridges. In *Carcinus moenas* Hcy the subunit of 50,000 is also formed by two polypeptide chains of 25,000 linked together by a S-S bridge; the 75,000 aggregate, however, results by the addition of a third polypeptide chain of the same size carrying probably the free SH group and which, since it is free of copper, does not participate directly in the reaction with oxygen.

Conclusions

While decisive conclusions must await more experimental data (until now the only results have been obtained with the Hcy from *Limulus polyphemus*) it seems likely that Hcy are made by polypeptide of 25,000 daltons; in Mollusca the functional subunit is formed by two chains each containing one atom of copper; in the Hcy of Arthropoda the oxygen binding unit is made by three chains of equal molecular weight, two of which are "functional" and one is "structural", i.e. without copper.

Recently van Holde & colleagues (1) suggested that the cubic structure derived from electron micrographs of *Cancer magister* Hcy could be better explained by assuming the 16 S as formed by 18 subunits,26,000 daltons each, six pairs placed on the 12 edges of the cube which are oxygen binding and six "structural", centered behind the faces on an octahedral core. The biochemical findings here reported strongly support this hypothesis. Studies are in progress to ascertain whether the polypeptide chains are the same or not in molluscan and arthropod Hcy.

References

1. Ellerton,D.H., Carpenter,D.E. & van Holde,K.E. (1970) Physical studies of Hemocyanins. V.Characterization and subunit structure of the hemocyanin of *Cancer magister*. Biochemistry 9,2225-2232.

2. Eriksson-Quensel,I.B. & Svedberg,T.(1936) The molecular weights and pH stability regions of the hemocyanins. Biol.Bull. 71, 498-574.

3. Fernandez-Moran,H., van Bruggen,E.F.J. & Ohtsuki,M. (1966) Macromolecular organization of Hemocyanins and Apohemocyanins as revealed by electron microscopy. J.Mol.Biol. 16, 191-207.

4. Ghiretti,F. (1962) In Oxygenases, pp.517-553, ed. Hayaishy,O., New York & London, Academic Press.

5. Ghiretti,F. (1966) In The Physiology of Mollusca, pp,233-248, vol.II, ed. Wilbur,K.M. & Yonge,C.M., New York & London, Academic Press.

6. Ghiretti,F. & Ghiretti-Magaldi,A. (1972) Respiratory Proteins in Mollusks, In Chemical Zoology, pp.201-217, vol.VII, ed.Florkin,M. & Scheer,B.T., New York & London, Academic Press.

7. Ghiretti-Magaldi,A. & Nuzzolo,C. (1965) Thiol groups in hemocyanins. Comp.Biochem.Physiol. 16, 249-252.

8. Ghiretti-Magaldi,A., Nuzzolo,C. & Ghiretti,F. (1966) Chemical studies on hemocyanins. I.Amino acid composition. Biochemistry 5, 1943-1951.

9. Gruber,M.(1968) Structure and function of Helix pomatia hemocyanin. In Physiology and Biochemistry of Hemocyanins, pp.49-59, ed. Ghiretti,F., New York & London Academic Press.

10. Makino,N. (1971) Hemocyanin from Dolabella auricularia. J.Biochem. (Japan) 70, 149-155.

11. Mellema,J.E. & Klug,A. (1972) Quaternary structure of gastropod haemocyanin. Nature 239, 146-150.

12. Redfield,A.C.(1950) Hemocyanin. In Copper Metabolism, pp.174-190, ed.McElroy & Glass,B. Baltimore.

13. Salvato,B.,Sartore,S., Rizzotti,M. & Ghiretti-Magaldi,A. (1972) Molecular weight determination of polypeptide chains of molluscan and arthropod hemocyanins. FEBS Letters 22, 5-8.

14. van Bruggen,E.F.J. (1968) Electron Microscopy of hemocyanins. In Physiology and Biochemistry of Hemocyanins, pp.37-48, ed. Ghiretti,F., New York, & London, Academic Press.

15. van Holde,K.E. & van Bruggen,E.F.J. (1971) The hemocyanins. In Subunit in Biological Systems, pp. 1-53, ed.Timasheff,N.S. & Fashman,G.D.

16. Witters,R. & Lontie,R. (1968) Stability regions and amino acid composition of gastropod hemocyanins. In Physiology and Biochemistry of Hemocyanins, pp.61-73, ed. Ghiretti,F., New York & London, Academic Press.

Comparative Physiology, eds. L. Bolis, K. Schmidt-Nielsen and S. H. P. Maddrell

INVERTEBRATE HEMOGLOBINS

E.CHIANCONE, B.GIARDINA, M.BRUNORI, and E.ANTONINI
C.N.R.Centre of Molecular Biology, Institutes of Biochemistry and Chemistry, University of Rome (Italy), and Laboratory of Molecular Biology, Institute of Biochemistry, University of Camerino (Italy).

It is known that, while hemoglobin occur as respiratory pigments in all vertebrates, these proteins are present only in some of the invertebrate phyla, since in others (such as molluscs or arthropods) the O_2 carrying pigments are hemocyanins and hemerythrins (see paper by Ghiretti et al. in this volume). In the phyla in which hemoglobins occur, proteins with widely different molecular properties are encountered (12). A study of the structural and functional properties of invertebrate hemoglobins may offer the opportunity to investigate the molecular basis of physiological adaptation to the environment. This type of study, besides its physiological implications, is of interest for the information it may yield on the general problem of structure - function relationships in hemoproteins (2).

For practical purposes, the hemoglobins from invertebrates may be grouped into two main categories, according to their intra- or extra-cellular location:

a) hemoglobins contained in cells (both in tissues and in circulating fluids), are normally characterized by low molecular weights (less than 50,000) and, very often, by a simple functional behaviour.

b) Extracellular pigments, freely dissolved in body fluids, are generally gigantic proteins, with molecular weights of the order of 10^6 or higher, and a complex functional behaviour.

Although this correlation between molecular size and location in the body is not absolute, hemoglobins with low molecular weight are generally not found in the circulating fluids. The existence of extracellular hemoglobins in the form of very large complexes is related to a number of factors, such as the reduction of osmotic pressure brought about by the aggregation of smaller subunits and the prevention of loss of pigments through the excretory filters.

In what follows, we shall review some of the structural and functional properties of representative hemoglobins from these two categories.

Intracellular hemoglobins

As already mentioned above, these include pigments from widely different sources, which may also be quite variable in their molecular weights. All of them, however, contain only protoheme, which is the prosthetic group of vertebrate hemoglobins as well (2).

Some of the more simple proteins have been extensively investigated both from a structural and a functional view point, and for a few of them detailed information on the amino acid sequence and on the tridimensional structure is available. In particular the hemoglobin (erythrocruorin) from *Chironomus thummi thummi* (11) and that from *Glycera dibranchiata* (13,14) are known at the atomic level.

The results hitherto available show that the overall arrangement of the molecule is indeed very similar for the various monomeric hemoglobins; in particular a common feature is the number of helical and non-helical segments in the molecule, as well as their approximate length. In turn the tridimensional structure is for these invertebrate hemoglobins very similar to that of mammalian myoglobin or to human hemoglobin   chains.

Comparison of primary structures has also shown that in several regions of the polypeptide chain there are clear homologies in that some of the residues occupy identical positions in the various cases. It may be mentioned that while the residue "proximal" to the heme iron is in all cases the imidazole ring of a histidine, the so called "distal" residue which is in position E 7 Sperm Whale myoglobin, is not necessarily a histidine, as shown by Chironomus (the corresponding residue being Val) and Aplysia (the corresponding residue being Glu). This implies that the formation of a reversible O_2 complex, which is the basis for the physiological function of hemoglobins, is not exclusively dependent on the presence of a histidyl residue in the distal position (11,21).

The minimum molecular weight, and the weight in solution are given in Table 1 for a number of invertebrate hemoglobins.

TABLE 1 - Some molecular properties of hemoglobins from various invertebrates (at pH ~7 and 20°C).

Source	M.M.W.	$s_{20,w}$	M.W.
Chironomus	16,000	-	16,000
Aplysia	16,500	2.0	16,500
Glycera	18,200	-	18,200
Gastrophilus	17,600	2.9	35,000
Molpadia	17,000	2.9	35,000
Daphnia	17,500	-	4.20×10^5
Lumbricus	23,000	59.6	3.45×10^6
Arenicola	17,000	57.4	3.0×10^6
Spirographis	35,000	57.5	2.8×10^6

The O_2 equilibrium curve of the monomeric hemoglobins conforms, in the absence of molecular heterogeneity, to a simple formulation of the mass low, with a single equilibrium constant:

$$\mathrm{Fe} + \mathrm{X} \rightleftharpoons \mathrm{FeX}; \qquad K(\mathrm{X}) = \frac{(\mathrm{Fe\ X})}{(\mathrm{Fe})} = \frac{\bar{Y}}{1 - \bar{Y}} \qquad (1$$

The absolute O_2 affinity shows changes from one hemeprotein to another (see Table 2), while the overall enthalpy of O_2 binding is in most cases very much the same ($\sim$ -13 Kcal/mole for O_2 in the gas phase).

Monomeric hemeproteins are devoid of a Bohr effect, and more generally of ligand-linked binding of third components. However, in the case of hemoglobin from Chironomus thummi thummi some of the purified components display a significant Bohr effect, although the O_2 binding curve remains at all pH values hyperbolic (1). The residue responsible for such an effect is probably one of the histidines in the molecule, and its pK in the oxy and deoxy derivatives has been directly determined by NMR (19).

The kinetics of O_2 binding conforms to a simple one-step process for all monomeric hemeproteins. The reciprocal relaxation time follows the behaviour characteristic of a bimolecular reaction, as shown in Figure 1 for Chironomus Hb and sperm whale Mb.

The absolute values of the equilibrium and rate constants change very markedly from one protein to another, presumably as a reflection of the accessibility of the ligand binding site. Table 2 gives a summary of equilibrium velocity constants for the reaction with O_2 of different hemoglobins.

A special situation may be present in polymeric hemoglobins, such as those from Molpadia and Gastrophilus larvae. In both cases it is known that at concentration above $\sim 10^{-4}$M (heme) the monomer is largely associated into higher molecular weight complexes. As reported elsewhere in this volume, (Bonaventura and Kitto),in the case of Molpadia Hb the ligand linked change in polymerization is associated with a significant cooperativity in O_2 binding. However extensive investigation of the kinetics of ligand (O_2, CO, ElC) binding by Molpadia Hb has failed to show any significant deviation of the behaviour of the macromolecule from that of a simple non-interacting system (unpublished work from this laboratory).Similarly in Gastrophilus Hb the kinetics of ligand binding also conforms to simple expectations, although the protein is dimeric at concentrations above 1 mg/ml (15).

Two points of major importance for discussion

<u>TABLE 2</u> - Functional properties of simple hemoglobins from invertebrates (at pH ~ 7 and 20°C); from refs.: (1,2,15).

Source	$P_{1/2}$ (mm/Hg)	K (M^{-1}x10^{-6})	k_{on} ($M^{-1}sec^{-1}$)	k_{off} (sec^{-1})	l_{on} ($M^{-1}sec^{-1}$)	l_{off} (sec^{-1})	L (M^{-1}x10^{-6})
<u>Chironomus</u>	0.7	0.29	30 x 10^7	218	2.7x10^7	0.095	0.28
<u>Aplysia</u>	2.5	0.22	1.5 x 10^7	70	4.9x10^5	0.02	25
<u>Gastrophilus</u>	< 0.02 (°	10	0.97x10^7	0.89	6.5x10^5	0.14	4.5

(° T = 39°C

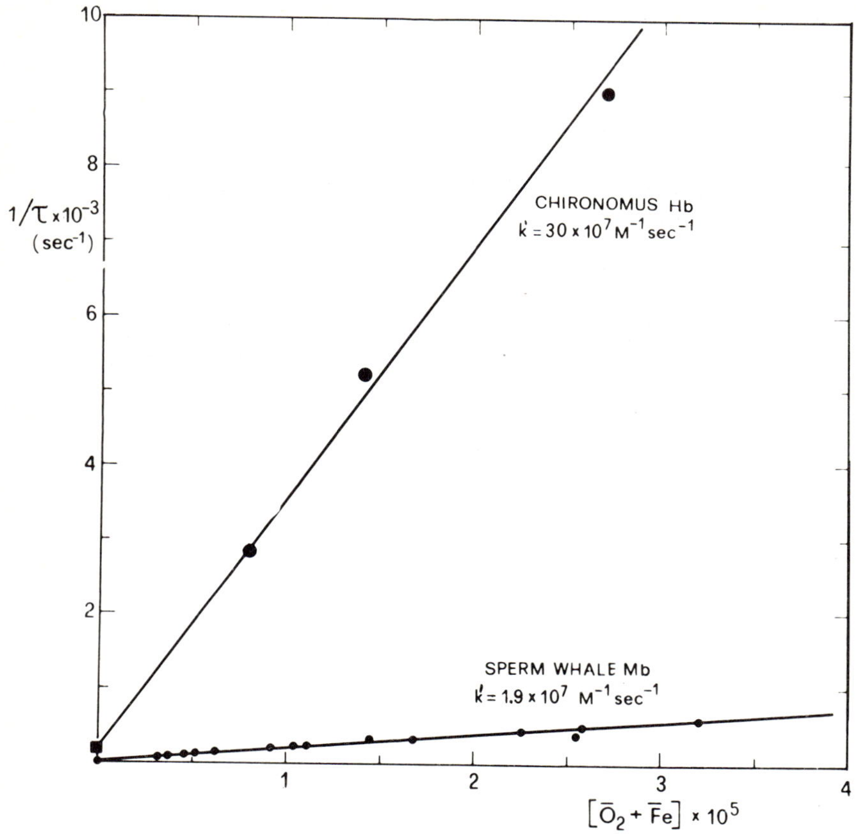

Figure 1 - Concentration dependence of $1/\tau$ for the reaction of oxygen with native Chironomus hemoglobin and native sperm whale myoglobin at pH 7,phosphate buffer 0.1M and 20°C. Intercepts on the ordinate axis have been obtained independently by rapid mixing measurements. (From Amiconi et al. (1)).

emerge from consideration of these results.

In the first instance, correlation of structural features with functional properties in various simple hemoproteins may be used for understanding in detail the role of the protein in controlling the reactivity of the heme (2).Physiologically the essential property of an O_2 carrier is the ability to form a reversible complex with this ligand. Thus, while free heme is very rapidly

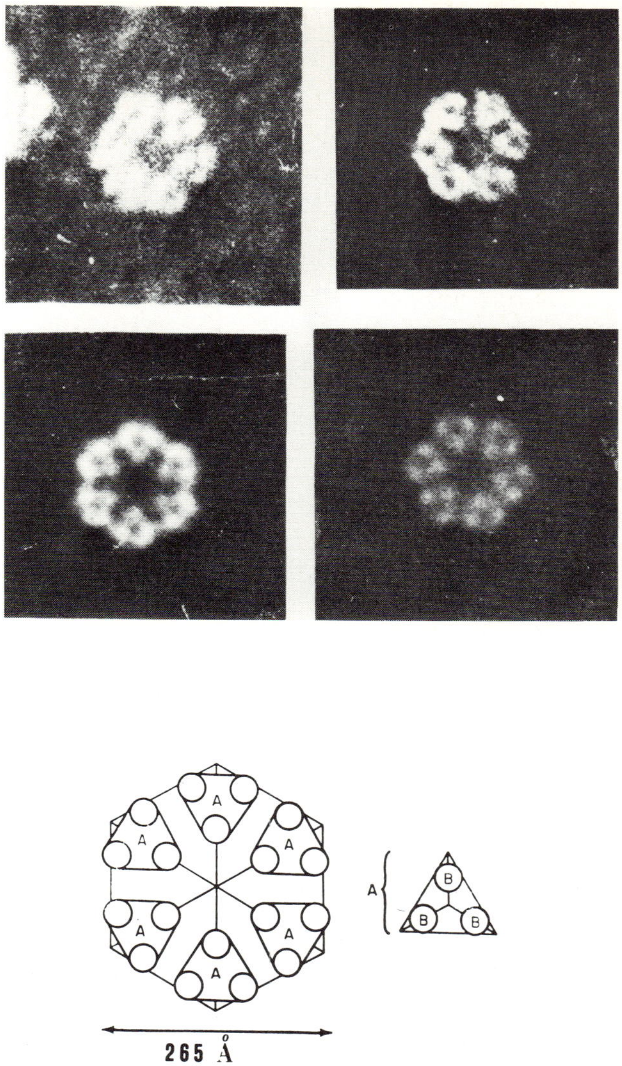

Figure 2 - a. Electron microscope pictures of oxy (left) and deoxy (right) *Spirographis* chlorocruorin at pH 7.2. x 450,000. Lower pair of pictures were printed by rotating the negative through an angle of 60° about the presumed axis of symmetry (i.e. the centre of a hexagon), between 6 successive exposures of the paper. b. Schematic representation of a face-view of the hexagons (from Guerritore et al., ref.9).

oxidized in the presence of O_2, hemes embedded in a protein are capable of forming a wholly reversible O_2 complex. The stability of such a complex is largely, although not exclusively, determined by the non-polar environment provided by the polypeptide chain (2,22). This prerequisite is fulfilled, as far as we know, by all known respiratory hemeproteins. Studies of a comparative nature provide additional information. Thus it is now proved that the distal imidazole is not necessary for reversible O_2 binding, since at least two of the simple cases considered above do not have a distal histidine. This is in good agreement with findings on abnormal human hemoglobin, such as Hb Zürich (24).

A second point of interest is the physiological role played by some of the invertebrate hemoglobins. In particular, it may be asked to what extent the functions of these proteins are transport of O_2 from the outer environment to the tissues, or O_2 storage. A complementary function of myoglobins, probably important for other tissue proteins, may be that of facilitating O_2 diffusion through the cell (18, 25). For this purpose some of the simple hemeproteins appear beautifully suited in view of their very high rate constants, and in particular of the velocity at which O_2 may be loaded or unloaded by the molecule.

Extracellular hemoglobins

The extracellular respiratory pigments exhibit very high molecular weights (0.5 - 3 x 10^6 daltons). They may show similar molecular architectures in spite of differences in the source and in the nature of the prosthetic groups (protoheme in the case of *Lumbricus terrestris*, *Arenicola marina*, and chloroheme in the case of *Spirographis spallanzanii*). For these proteins the general molecular architecture is that of two superimposed hexagonal discs, each consisting of six similar particles (A) located at the corners of the hexagon; the six particles are in turn made up of three subunits (B) which appear to form a triangle (Figure 2) (9,16). An estimate of the number of hemes per molecule, which is of the order of 100-200, has been obtained on the basis of the structure from electron microscopy together with knowledge of the molecular weight in solution and of the heme content. From this type of studies major differences in the detailed structure of the smaller subunits (B) have become apparent; they arise mainly from variation in the minimum molecular weight which ranges from 17,000 to 23,000 for the erythrocruorins of *Arenicola cristata* and *Lumbricus* respectively, to 35,000 for *Spirographis* chlorocruorin (23,17,4). Of course in view of the large variability in the minimum molecular weight among these various proteins, the total number of hemes carried by a molecule is very different. Thus in *Lumbricus* erythrocruorin there are about 144 hemes/molecule of MW 3.4×10^6, while in *Spirographis* chlorocruorin there are about 72 hemes/molecule of MW 2.8×10^6. For the *Arenicola* and *Spirographis* respiratory pigments there is chemical evidence for the presence of 2 types of polypeptide chains (23,10). In *Arenicola* there are 2 polypeptide chains per heme, a feature which is in contrast with that of the low molecular weight invertebrate hemoglobins and of the mammalian ones, where the prosthetic group is always associated with one polypeptide chain.

As already shown by Svedberg and Eriksson (20), the whole molecule tends to dissociate at alkaline pH values into various types of components of smaller size. In fact, in sedimentation velocity experiments, on increasing the pH above 8, two new components, with sedimentation coefficient of $\sim$10S and $\sim$3.5S, progressively substitute for the undissociated molecules. The 10S component which is the first product of dissociation, corresponds to 1/12 of the whole molecule (subunit A); this dissociation step is irreversible and is not affected by the presence of oxygen. The dissociation behaviour of the intact molecules provides clear indications for molecular micro-heterogeneity in a protein system. Dissociation of subunit A gives in turn rise to subunit B and to fractions of it (3.5 S component); these dissociation phenomena, contrary to the first step, represent a reversible equilibrium situation and are oxygen linked. Qualitatively the effect of oxygenation is to promote dissociation as in the case of mamma lian hemoglobins and in hemocyanins. This means that the smaller subunits have a higher oxygen affinity than the larger ones (8).

The functional behaviour has been extensively studied in the case of *Lumbricus* erythrocruorin and *Spirographis* chlorocruorin. It exhibits the same general characteristics, as exemplified by heme-heme interaction and Bohr effect, which are typical of (polymeric) mammalian hemoglobins (3, Giardina *et al.*, in preparation). Thus the curve describing the reversible binding of oxygen are not hyperbolas, but are sigmoid in shape and may be represented by the Hill equation

$$Fe + n\, X \rightleftharpoons FeX_n; \quad K\,(X)^n = \frac{(FeX_n)}{(Fe)} = \frac{\bar{Y}}{1 - \bar{Y}} \qquad (2$$

The high cooperativity in O_2 binding at physiological pH values is associated with a characteristic sigmoid shape of the $\bar{Y}$ versus log pO_2 plots; the central part of the curves have a very steep slope, corresponding to values of $n > 4$, while at low and high saturation n approaches a value of 1 (Figure 3). As shown by Wyman (26) from a Hill

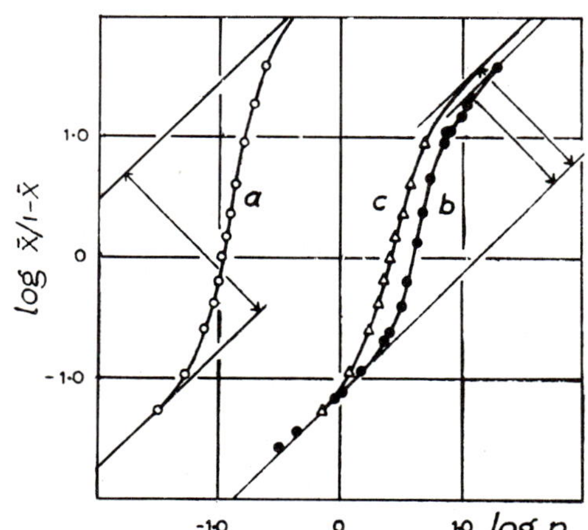

Figure 3 - Hill plots of the oxygen equilibrium of three invertebrate pigments. Curve a: *Arenicola cristata* erythrocruorin; curve b: *Spirographis* chlorocruorin; curve c: *Homarus americanus* hemocyanin. Temperature 20°C; pH 7.6-7.8 in phosphate buffer (curves a and b), in tris buffer (curve c). (From Wyman, ref.27).

plot it is possible to obtain an estimate of the minimum free energy of interaction in the binding of O_2. For these giant molecules the overall free energy of interaction ($\Delta F_i \sim 2.5$ Kcal/mole),is somewhat smaller than that found for mammalian Hbs ($\Delta F_i \sim 3$ Kcal/mole), although the value of n in the middle region of saturation is considerably higher (27).

The effect of pH on the O_2 equilibrium has been studied in detail with *Lumbricus* erythrocruorin. The oxygen linked binding of protons is reflected not only in changes of the O_2 affinity, as in the case of mammalian hemoglobins, but also in changes of the shape of the O_2 binding curve.Thus, outside the pH range 7-8.4, the curves become much less sigmoid and the value of n drops to$\sim$2 (Figure 4, 5a). However when the pH is brought back to neutrality from either the alkaline or the

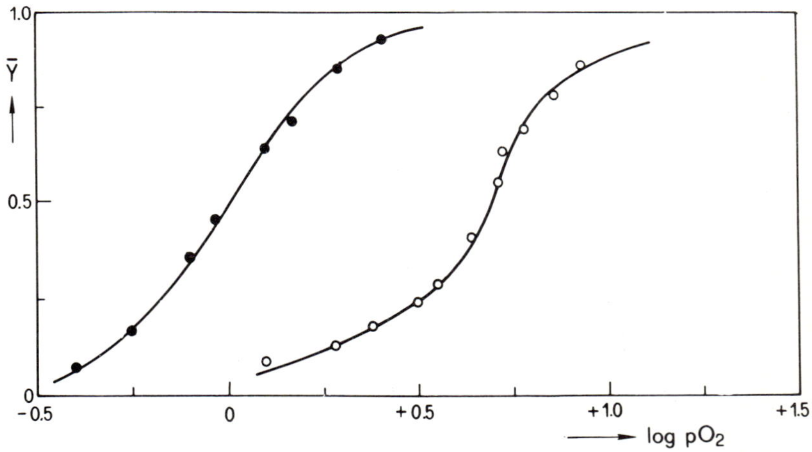

Figure 4 - Oxygen equilibrium of *Lumbricus* erythrocruorin.Protein concentration was 3-5 mg/ml at 20°C. Results obtained at pH 9.2 in 2% Borate (●); at pH 7.8 in 0.1 M phosphate buffer (o).

acid range the shape of the curves remains unchanged (n = $\sim$2), while the oxygen affinity returns nearly to the values of erythrocruorin freshly dissolved at neutral pH. A behaviour somewhat similar to that of *Lumbricus* erythrocruorin was reported for *Spirographis* chlorocruorin, although, for this protein, experiments were limited to a more restricted pH span in view of its relati-

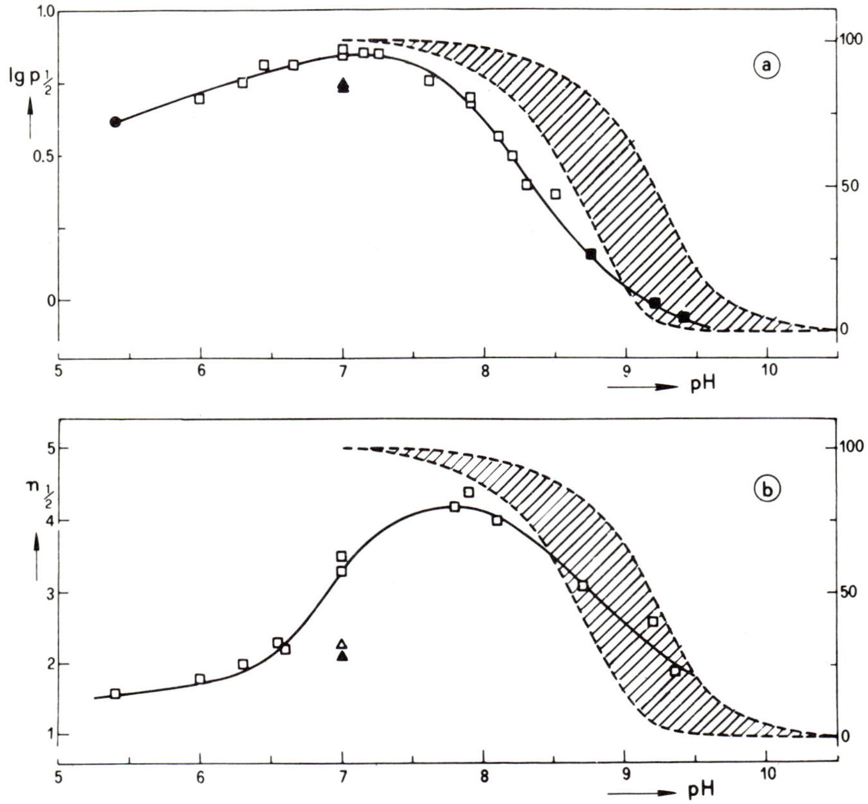

Figure 5 - a. Oxygen Bohr effect of *Lumbricus* erythrocruorin. Protein concentration was 3-5 mg/ml. Results in 0.1 M acetate buffer (●), results in 0.1 M phosphate buffer (□); results in 2% borate-NaOH buffer (■); sample previously brought to pH 9 for 1 hr (△); sample previously brought to pH 6 for 1 hr (▲). Temperature 20°C.

b. Effect of pH on n at 50% saturation in *Lumbricus* erythrocruorin. Protein concentration was 3-5 mg/ml at 20°C. Sample previously brought to pH 9 for 1 hr (△); sample previously brought to pH 6 for 1 hr (▲).

The shaded areas in a and b represent the percentage of undissociated molecules for several erythrocruorin preparations as obtained by light scattering. (From Chiancone *et al.*, ref.8).

ve instability at alkaline pH values (3).

At least for Lumbricus erythrocruorin, the effect of pH on the O_2 equilibrium reflects only in part the state of aggregation of the protein. This is shown by a comparison of the pH dependence of n and of the state of aggregation. In fact, at acid pH values where the protein is still undissociated, cooperativity drops, the values of n being 1.8-2.0 (Figure 5b). As reported above, in the alkaline pH region (pH $>$ 8.0) the molecule (MW 3.4×10^6 daltons) dissociates into lower molecular weight species. Figure 5 shows a direct comparison of the equilibrium parameters (log p 1/2 and n 1/2) and of the pH induced dissociation of the whole molecule (MW 3.4×10^6). It appears that there is a certain phenomenological correlation between the drop in n and the dissociation in the alkaline range, although cooperativity is still present in the subunits of MW lower than 270,000.

The kinetics of reaction with O_2 and CO displays overall similarities but also differences with the behaviour characteristic of mammalian hemoglobins. For Lumbricus erythrocruorin, the kinetic behaviour with both ligands is, under all conditions, complex, the time course of reaction being biphasic or even multiphasic. Temperature jump experiments on the reaction of this protein with O_2 show two relaxation times, similar in a broad sense to what is known for cooperative hemoglobins from mammals. The faster relaxation time, which may be more easily interpreted, depends linearly on O_2 concentration. The second order rate constant calculated from the data shown in figure 6 is $3 \times 10^7 M^{-1} sec^{-1}$ at 25°C and pH 7. This value is indeed similar to that of the isolated α and β chains of human Hb or to that of the quickly reacting species of Hb ($k_{on} = 4.7 \times 10^7$ $M^{-1} sec^{-1}$) (7). For Lumbricus erythrocruorin, Boelts and Parkhurst (5) have also shown that the ratio of the dissociation velocity constant by dithionite to that obtained by replacement is larger than that characteristic of human Hb. This finding has been interpreted as an indication of the more marked cooperativity in comparison to human Hb ($n \sim 4$ compared to $n \sim 2.8$). The values obtained by us for the same protein at neutral pH are in reasonable agreement

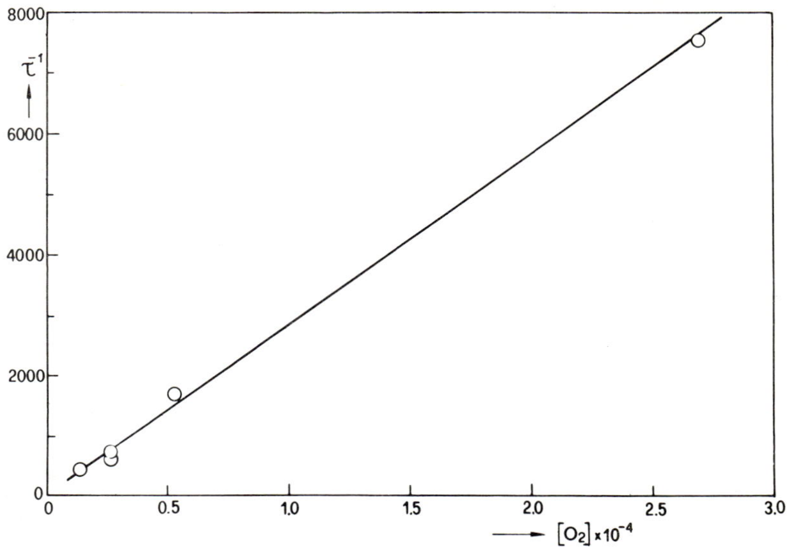

Figure 6 - Concentration dependence of τ^{-1} for the reaction of oxygen with *Lumbricus* erythrocruorin at pH 7, phosphate buffer 0.1 M and 20°C. Protein concentration 1-2x10^{-5}M heme.

with those reported by the latter authors. It may be remarked that the dissociation rate constant by dithionite is at some wavelenghts (e.g. 555 nm) a homogeneous process (k_{off} = 33 sec^{-1}).

Kinetic experiments with CO by rapid mixing and flash photolysis give further indications of the complexity of the system. At neutral pH, where the molecule is wholly polymeric, there is direct evidence for the presence of two reacting species with an identical time course in both rapid mixing and flash photolysis experiments. The CO concentration dependence of the slow process yields a second order rate constant of 1.5 x $10^5M^{-1}sec^{-1}$ at 20°C and pH 7 (Figure 7). The rate constant for the faster process is 2.7 x $10^6M^{-1}sec^{-1}$.

The CO combination of *Spirographis* chlorocruorin corresponds, on the other hand, to a simple process with very similar values of the second order rate constants at pH 7 (0.6x$10^5M^{-1}sec^{-1}$) and 9.2 (1x$10^5M^{-1}sec^{-1}$) (6).

For *Lumbricus* erythrocruorin the time course is independent of protein concentration, at least

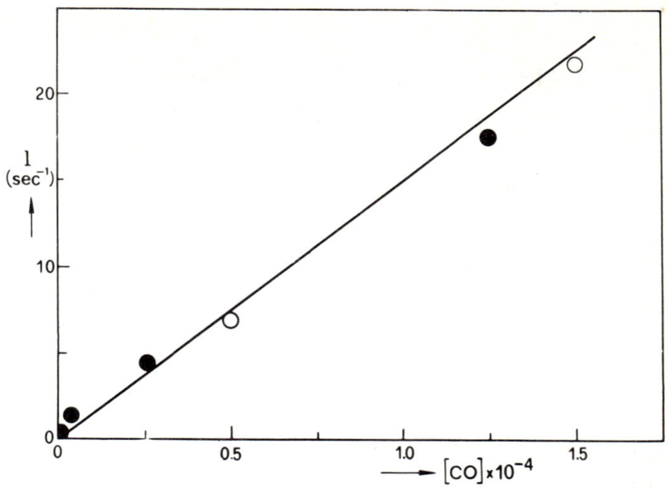

Figure 7 - Dependence on CO concentration of the pseudo first order rate constant (1) for the combination of the ligand with deoxyerythrocruorin from Lumbricus. Experiments by rapid mixing (●), by flash (○). Temperature 20°C. 0.1 M phosphate buffer at pH 7.

at neutral pH. Moreover the whole blood and the purified protein display identical behaviour. The kinetic behaviour by flash has been studied over a pH range from 5.6 to 10.7. In the region around pH 8.5 to 9.0 there is a sharp change of the rate constants for both kinetic components, the absolute values increasing from 3 to 6 fold over this range. However the ratio of the fast to slow component is approximately constant over the whole pH span. The spectral features of the various species change in a parallel fashion. Thus, at wavelengths close to the isosbestic point, it is obvious that the two components have on CO combination opposite contributions at pH ~ 7. Such spectral feature tends to change with pH, and at pH > 8.5 (e.g. 9.65) sharp isosbestic points are obtained. It remains to be seen how far these findings are correlated with pH dependent changes in aggregation, ligand affinity and cooperativity.

Consideration of the functional properties of the giant respiratory proteins raises a number of points. Functionally the presence of many hemes per molecule brings about a regulatory sensitivity which is absent in the monomeric hemoglobins. Just as for mammalian hemoglobins the questions one may ask are how much cooperativity is related to the number of hemes and what is the minimum functional unit of these large complexes. The experiments with *Lumbricus* erythrocruorin clearly show that there is no unique correlation between cooperativity and the number of hemes per molecule since it is possible to obtain (e.g. at acid pH) an undissociated polymer with a low value of n ($n \sim 2$). Similar considerations on the state of aggregation and the value of n in the alkaline pH region allow conclusion that cooperativity is still operative in the A subunits (1/12 of the whole molecule), although this result cannot be taken as evidence that such subunits represent the minimum functional unit of the protein.

Another aspect of the behaviour of the giant respiratory pigments is the presence of ligand linked changes in the state of aggregation of the molecule in the alkaline pH region. The enhanced dissociation produced by the presence of ligands, which is qualitatively similar to that observed for mammalian hemoglobins (2), must reflect ligand linked conformational changes, which as yet have not been observed directly.

A last point to be considered is whether there is evidence of intrinsic heterogeneity between different hemes in a molecule. No general conclusions can be drawn from the data available to date since *Spirographis* chlorocruorin and *Arenicola* erythrocruorin behave as homogeneous systems, while *Lumbricus* erythrocruorin always exhibits heterogeneous kinetic behaviour.

REFERENCES

1. Amiconi,G., Antonini,E, Brunori,M.,Formaneck,H. & Huber,R. (1972) Functional properties of native and reconstituted hemoglobins from *Chironomus thummi thummi*. European J.Biochem. (in (press).

2. Antonini,E. & Brunori,M. (1971) Hemoglobin and

myoglobin in their reactions with ligands, ch. 13, p.348, Amsterdam, North Holland Publishing. Company.

3. Antonini,E., Ross Fanelli,A. & Caputo,A. (1962) Studies on Chlorocruorin I. The oxygen equilibrium of *Spirographis* chlorocruorin. Arch. Biochem.Biophys. 97, 336-342.

4. Antonini,E., Rossi Fanelli,A. & Caputo,A.(1962) Studies on Chlorocruorin II. Some physicochemical properties of *Spirographis* chlorocruorin. Arch.Biochem.Biophys. 97, 343-350.

5. Boelts,K.J. & Parkhurst,L.J. (1971) Ligand binding kinetics of hemoglobin from the earth worm. Biochem.Biophys.Res.Comm. 43, 637-643.

6. Brunori,M., Guerritore,D., Antonini,E.,Wyman,J. & Rossi Fanelli,A. (1965) Studi sulla clorocruorina IV. Osservazioni sulla cinetica delle reazioni della clorocruorina di *Spirographis spallanzanii* con i gas. Pubbl.Staz.Zool.Napoli 34, 521-527.

7. Brunori,M. & Antonini,E. (1972) Kinetics of the reaction with oxygen of mixtures of oxy-and carbon monoxide hemoglobin. J.Biol.Chem. 247, 4305-4308.

8. Chiancone,E., Vecchini,P., Rossi Fanelli,M.R., & Antonini,E. (1972) Studies on erythrocruorin II. Dissociation of earth worm erythrocruorin. J.Mol.Biol. 70, 73-84.

9. Guerritore,D., Bonacci,M.L., Brunori,M.,Antonini,E., Wyman,J. & Rossi Fanelli,A. (1965) Studies on Chlorocruorin III. Electron microscope observations on *Spirographis* chlorocruorin. J.Mol.Biol. 13, 234-237.

10. Guerritore,D. & Zito,R. (1971) The terminal groups of chlorocruorin. Biochim.Biophys.Acta 229, 720-723.

11. Huber,R., Epp,O. & Formaneck,J. (1970) Structures of deoxy and carbomonoxy erythrocruorin. J.Mol.Biol. 52, 349-353.

12. Jones,J.D. (1972) *Comparative Physiology of Respiration*, London, E.Arnold Publishers.

13. Love,W.E. (1970) Haemoglobins from the annelid worm *Glycera dibranchiata* and from the sea lamprey *Petromyzon marinus*. 8th International Congress of Biochemistry, Switzerland.

14. Padlan,E.A. & Love,W.E. (1968) Structure of the haemoglobin of the marine annelid worm *Glycera* dibranchiata at 5.5. Å resolution. Nature 220, 376.

15. Phelps,C.F., Antonini,E., Brunori,M. & Kellett, G.L. (1972) The kinetics of binding of oxygen and carbonmonoxide to *Gastrophilus* hemoglobin. Biochem.J. 129, 891-895.

16. Roche,J.B. (1965) Electron-Microscope Studies on high molecular weight erythrocruorins (Invertebrate haemoglobins and chlorocruorins of Annelids) In *Studies in Comparative Biochemistry*, pp. 62-80, Munday,E.A. ed., Oxford Pergamon Press.

17. Rossi Fanelli,M.R., Chiancone,E., Vecchini,P. and Antonini,E. (1970) Studies on erythrocruorin I.Physicochemical properties of earthworm erythrocruorin. Arch.Biochem.Biophys.141, 278-283.

18. Scholander,P.F. (1960) Oxygen transport through hemoglobin solutions. Science 131, 585-590.

19. Sick,M., Gersonde,K., Thompson,J.C.,Maurer,W., Maar,W., & Riiterjans,M. (1972) The Bohr proton-binding site in a monomeric haemoglobin. Eur.J.Biochem. 29, 217-223.

20. Svedberg,T. & Eriksson,J.B. (1933) The molecular weight of erythrocruorin. J.Am.Chem.Soc. 55, 2834-2841.

21. Tentori,L., Vivaldi,G., Carta,S., Marinucci,M. Massa,A., Antonini,E., & Brunori,M. (1971) Primary structure of *Aplysia* myoglobin: sequence of A 63 residue fragment. FEBS Letters 12, 181-185.

22. Wang,J.H. (1958) Hemoglobin studies. II A synthetic material with hemoglobin like property. J.Am.Chem.Soc. 80, 3168-3169.

23. Waxman,L. (1971) The hemoglobin of *Arenicola cristata*. J.Biol.Chem. 246, 7318-7327.

24. Winterhalter,K.H., Anderson,N.M., Amiconi,G., Antonini,E., & Brunori,M. (1969) Functional properties of hemoglobin Zürich. European J. Biochem. 11, 435-440.

25. Wittenberg,J.B. (1966) The molecular mechanism of hemoglobin. Facilitated oxygen diffusion. J.Biol.Chem. 241, 104-114.

26. Wyman,J. (1964) Linked functions and reciprocal effects in hemoglobin. Adv.Protein Chem. 19, 223-286.

27. Wyman,J. (1967) Allosteric linkage. J.Am.Chem. Soc. 89, 2202-2218.

Comparative Physiology, eds. L.,Bolis, K. Schmidt-Nielsen and S. H. P. Maddrell

THE EFFECTS OF CALCIUM AND LEAD ON THE POTASSIUM PERMEABILITY OF HUMAN ERYTHROCYTES AND ERYTHROCYTE GHOSTS.

J.R. RIORDAN & H. PASSOW
Max-Planck-Institut für Biophysik,
Frankfurt/Main, Germany.

In the fifties, Shaw (42) and Glynn (12) discovered that the relationship between potassium concentration in the medium and potassium transfer across the red blood cell membrane can be interpreted as a superposition of two flux components : a saturable component which depends on the metabolic state of the cell, and a linear component which is independent of metabolism. These findings formed the basis of the hypothesis that alkali ions cross the red blood cell membrane by pump and leak. The assumption of the two independent and parallel pathways proved its worth in many general considerations on permeability, as for example in the work of Tosteson and Hoffman (43) on volume regulation of the red blood cell. However, since the pump and leak concept has been developed, the question has arisen of whether or not it is sufficient to postulate only two pathways for cation transfer. For example, in addition to the coupled potassium/sodium movements which lead to active transport of the two cation species, an ATP mediated Na/Na exchange system has been described (11,13,14). This system is incapable of accomplishing net movements and its relationship to the potassium/sodium pump is still not clear. Similarly, there are ATP independent, net cation movements (e.g. in solution of low ionic strength; 3a, 21a, 46) whose relationship to the leak fluxes as defined by Shaw and Glynn is not yet established. Hence it remains to be decided whether or not pump and leak each represent more than one cation pathway in the red blood cell membrane.

The present article deals with an experimentally induced specific increase of potassium permeability which cannot yet be definitely related to a modification of either the pump or the leak. It is not even known whether or not the observed effects are due to an opening up of channels which are normally impermeable to potassium. The conditions which lead to the observed increase of potassium permeability are highly artificial. They require the presence of calcium or lead and, in all instances studied so far, can be inhibited by oligomycin. Calcium in contrast to lead does not act on normally metabolizing intact red blood cells. The observed increase of K^+ permeability only takes place if Ca^{++} is added after substrate depletion or in conjunction with metabolic poisons or special drugs. The results obtained depend on a large number of variables and thus are difficult to relate and even more difficult to interpret. Nevertheless, the observed phenomena have received a great deal of attention in many laboratories around the world. The earliest work was done by Wilbrandt (44,45) in Heidelberg and Bern; the essential role of calcium was first demonstrated by Gardos in Budapest (6-10). Continuing contributions have come from Hoffman at Yale (1,2,17,19,21) and Ekman *et al.* (5) and Manninen (30) in Helsinki. There are recent reports from Glynn and Warner (15) and Romero and Whittam (40) in England and Lew (25,26) in Argentina. Further investigations were done in many other laboratories. In the following pages we shall give a brief review of certain aspects of the problem which were of particular interest to our own work in this field.

I. The involvement of cell metabolism and alkaline earth ions.

The first example of a selective increase of potassium permeability was discovered by Wilbrandt (44) in 1937. After addition of fluoride to red cell suspensions he found a considerable increase of net K^+ efflux which was accompanied by a shrinkage of the cells, and a concomitant increase of osmotic resistance. Since the shrinkage indicated a diminution of the osmotic content of the cells he concluded that the potassium exit was not accompanied by sodium uptake but by a release of equivalent amounts of chloride. This inference was later confirmed by Dunker and Passow (4) by direct determination of the intracellular sodium content (figure 1c). In ingenious experiments Wilbrandt proved the existence of a causal relationship between potassium loss and the inhibition of glycolysis by fluoride (45).

Many years later it was shown that, in addition to the metabolic inhibitor, the presence of alkaline earth ions in the medium is required to induce the permeability change. Very low concentrations of Ca^{++} (0.1 mmoles/l) or Mg^{++} (1.0 mmoles/l) were sufficient to evoke the response (7-9, 22,23).

Since the rate of potassium exit induced by the combined actions of calcium and fluoride is more than one order of magnitude faster than the rate of potassium efflux in the intact red blood cells, the action of fluoride cannot be interpreted on the assumption of a simple cessation of the activities of the Mg-activated potassium-sodium pump. At a time when the existence of the potassium/sodium pump was still unknown, Wilbrandt (45) stipulated that the inhibition of enolase by fluoride leads to a diversion of glycolysis to abnormal pathways, resulting in the formation of abnormal side products which in turn alter the permeability of the membrane to potassium. This view was also adopted by Gardos (6). However, later findings required a revision of this view: prior to adding the fluoride, erythrocytes were depleted of metabolic intermediates by prolonged incubation in the absence of substrates. Subsequently, fluoride and calcium were added. Although under these conditions inhibition of enolase could not lead to the formation of abnormal side products, the cells still responded to the combined actions of fluoride and calcium. In fact, the sensitivity of the cells was considerably increased. From these experiments it was concluded that fluoride exerts a dual action. It inhibits glycolysis and thereby renders the membrane susceptible to the combined action of the

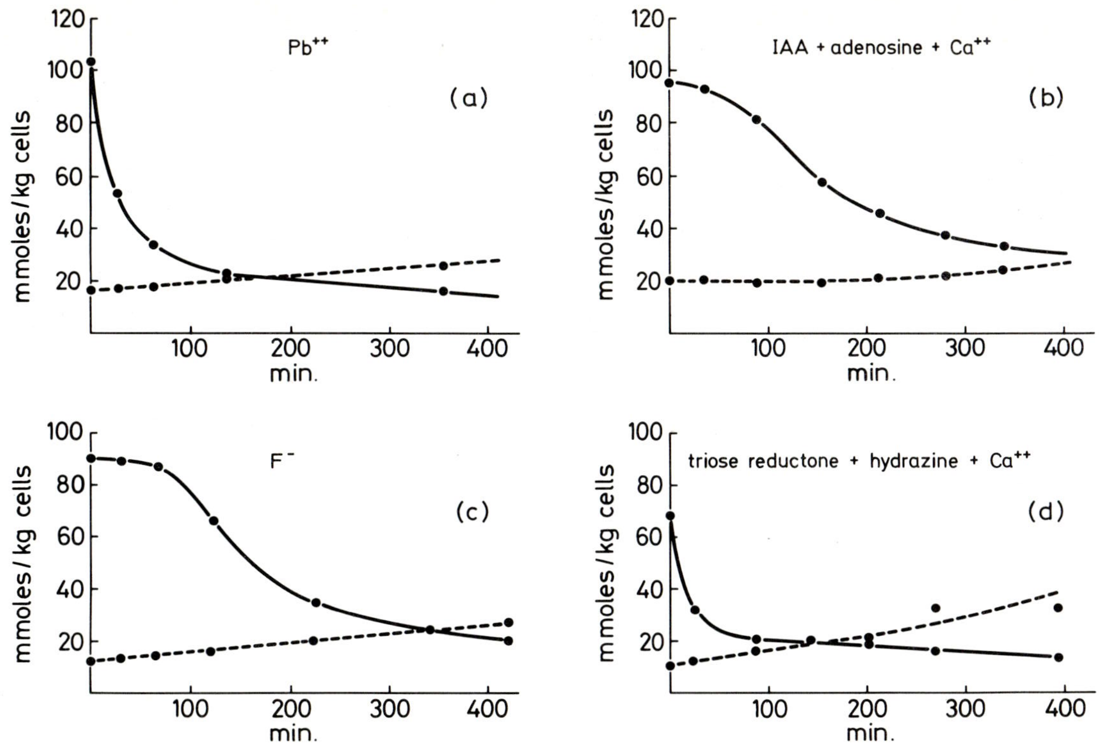

Figure 1 - K^+ loss (●——●) and Na^+ uptake (●- - -●) by cells treated with a.) 0.06 mmoles/l Pb^{++} at 21°C, b.) 2.5 mmoles/l iodoacetate + 10.0 mmoles/l adenosine + 0.5 mmoles/l $CaCl_2$ at 37°, c.) 40 mmoles/l NaF at 37°C and d.) 2.5 mmoles/l triose reductone, + 10 mmoles/l hydrazine + 0.5 mmoles/l $CaCl_2$ at 37°C (according to Passow (33)).

two agents on the membrane itself (28). Subsequent work concentrated on the role of ATP in fluoride poisoning. The ATP content of fluoride poisoned cells was found to decrease immediately after the addition of the agent, while potassium loss started only after a lag period (Figure 2). This finding

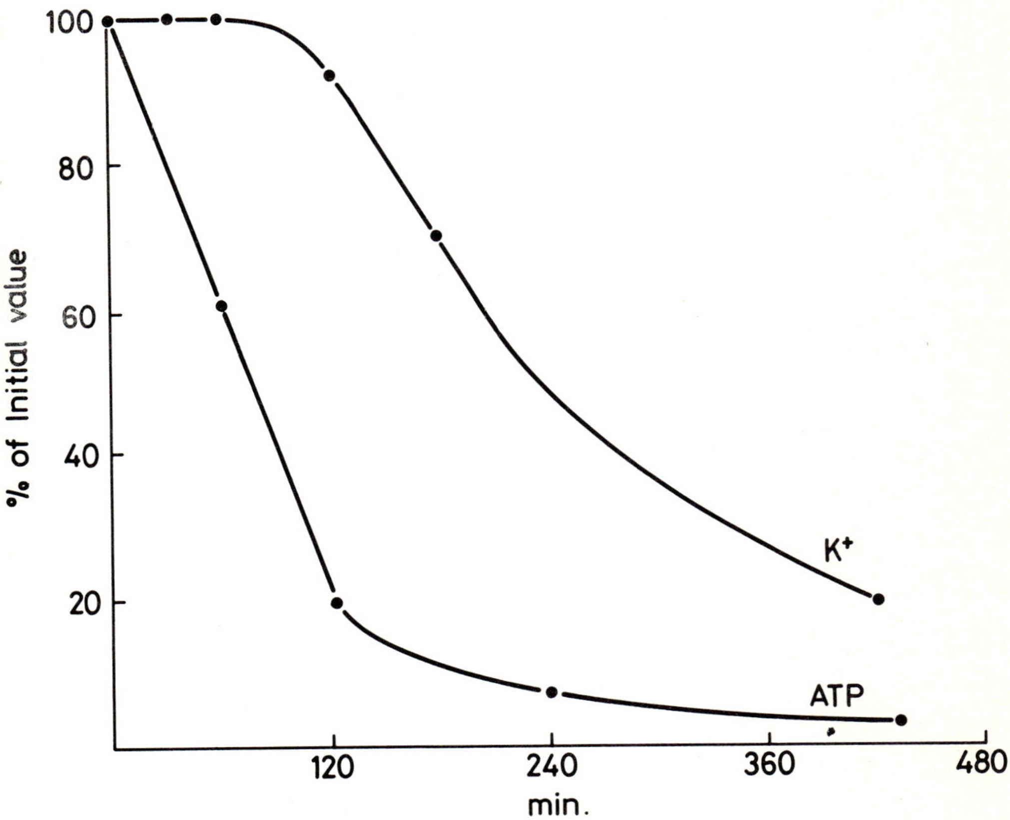

Figure 2 - The effect of 40 mmoles/l NaF + 0.5 mmoles/l $MgCl_2$ on the K^+ (upper curve) and ATP (lower curve) content of human erythrocytes suspended in 126 mmoles/l NaCl (according to Lepke and Passow (22)).

naturally raised the question of whether or not the decrease of ATP was a prerequisite for the potassium loss. Such an interrelationship was suggested by the discovery of a correlation between initial ATP content and the duration of the lag period preceding the onset of potassium loss in cells which were incubated prior to the addition of fluoride in the absence of substrate (Figure 3). However, since

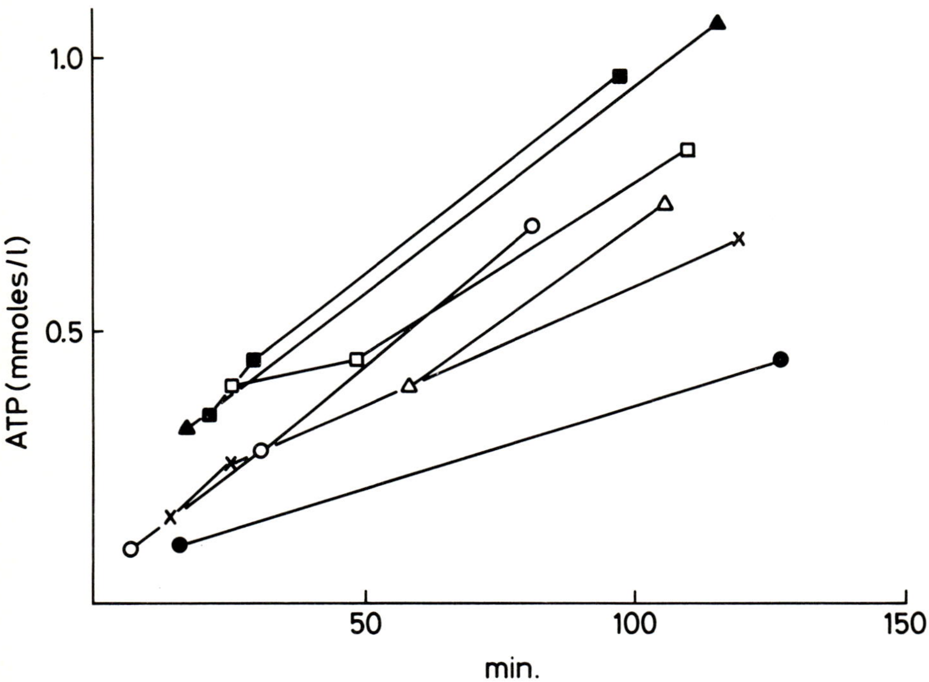

Figure 3 - Relationship between duration of lag period preceding K^+ loss after F^- treatment (40 mmoles/l NaF) and initial ATP content of cells starved for various lengths of time. The incubation media (126 mmoles/l NaCl) also contained 0.01 mmoles/l $CaCl_2$ (according to Lindemann and Passow (28)).

it was found that the addition of (non penetrating) EDTA to the medium inhibited the permeability change by removal of calcium but did not prevent the breakdown of cellular ATP (figure 4) it was conclu-

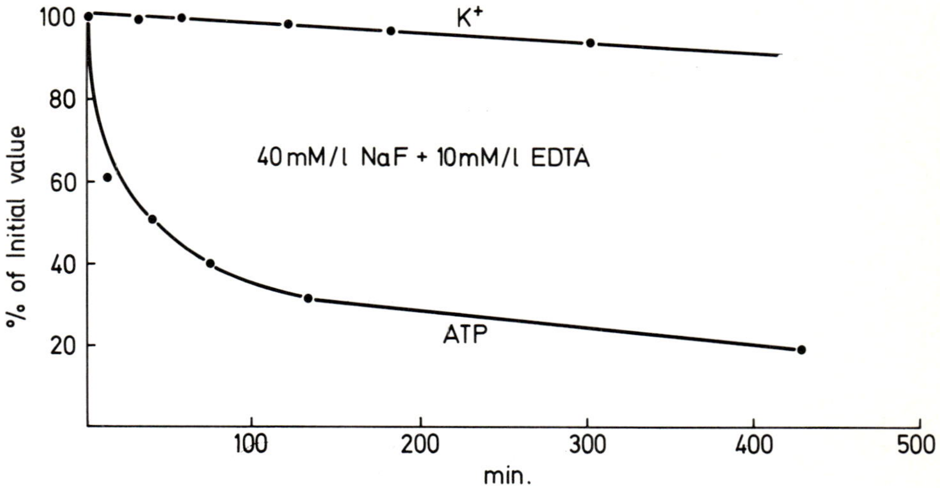

Figure 4 - The effect of external EDTA (10 mmoles/l) on K^+ (upper curve) and ATP (lower curve) content of cells incubated in 126 mmoles/l NaCl + 40 mmoles /l NaF at 37°C (according to Lepke and Passow (24))

ded that if such a relationship exists, it is an indirect one. This suggested that a decrease of intracellular ATP is necessary for the successful action of fluoride and calcium on the membrane, but is not itself the cause of the permeability change. This was confirmed by experiments with erythrocyte ghosts. In the absence of ATP, red cell ghosts respond to the action of calcium and fluoride like substrate depleted intact red cells. However, if the ghosts are loaded with ATP, calcium and fluoride are ineffective. There is no potassium loss (Figure 5). Similar effects as with ATP were observed with incorporated EDTA (24).

The effects of fluoride were described in some detail since they demonstrate most features which are typical for the subject of the present review. The permeability change is fairly specific for potassium and the rate of potassium loss is much faster than one would expect if the effect were

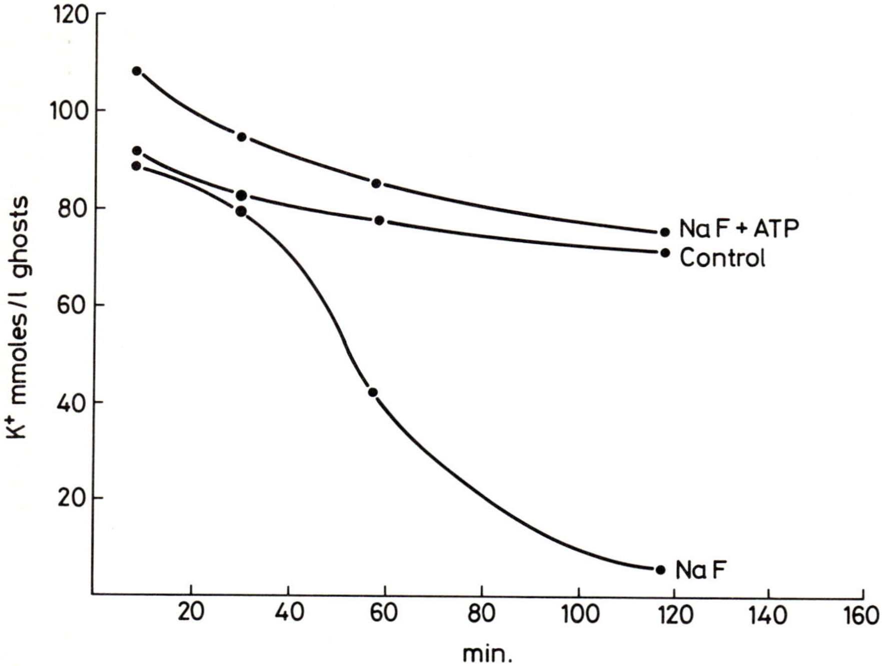

Figure 5 - Prevention of F^--induced K^+ loss from ghosts by incorporation of ATP (4 mmoles/l). Ghosts also contained 4 mmoles/l $MgCl_2$. Media contained 0.05 mmoles/l $CaCl_2$ and either 166 mmoles/l NaCl ("control") or 126 mmoles/l NaCl plus 40 mmoles/l NaF ("NaF" and "NaF + ATP") (according to Lepke and Passow (24)).

simply due to a blockage of the pump. The site of action is the membrane and calcium is required to induce the permeability change. Complexing agents are capable of preventing the permeability change regardless of whether they are present on the inside or the outside of the membrane.

Permeability changes which are similar to those described above were also observed if in place of fluoride, iodoacetate plus adenosine were added to the red blood cells together with calcium (7,8). In contrast to fluoride poisoning, Mg^{++} could not replace Ca^{++}. Again it could be shown that the permeability changes are specific for potassium and associated with a decrease of intracellular ATP (20a, 32). If the cells are starved prior to the initiation of the permeability change, their sensitivity to the poison increases just as in fluoride poisoning. Blum and Hoffman (1) observed that in starved cells no addition of substrate is required to induce the permeability change. Iodoacetate plus calcium or even calcium alone are sufficient to evoke the response. Similarly, ATP-free red cell ghosts made from previously starved (21) or intact cells (37) respond to iodoacetate plus calcium or calcium alone much as the starved red cells do. Again, incorporation of complexing agents like ATP, ITP, EDTA or EGTA prevents the action of the alkaline earth ion (Figure 6).

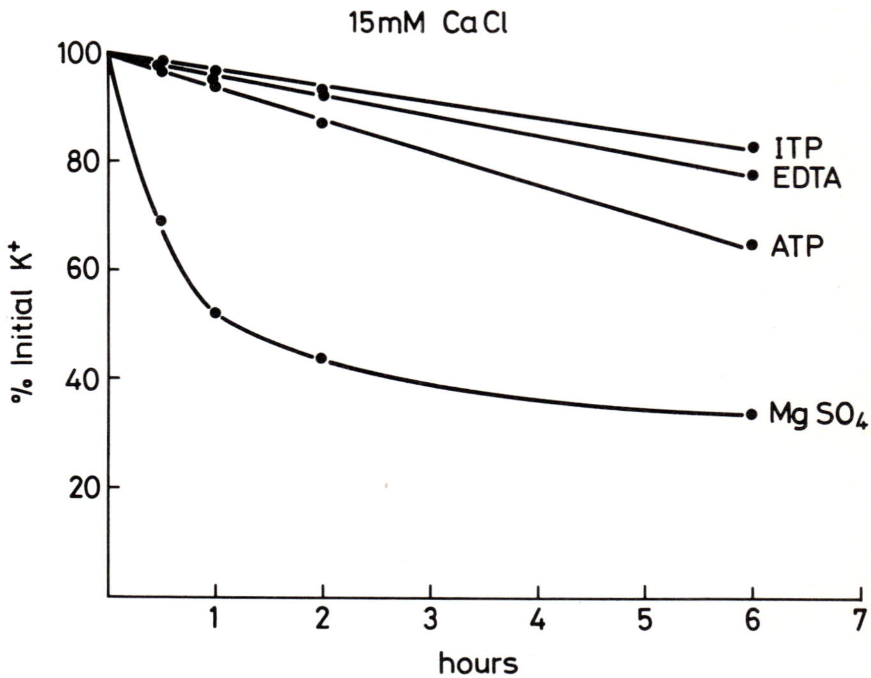

Figure 6 - Prevention of Ca^{++}-induced K^+ loss from ghosts by incorporation of complexing agents (2 mmoles/l). Ghosts also contained 4 mmoles/l $MgSO_4$. Incubation media consisted of 120 mmoles/l NaCl plus 15 mmoles/l $CaCl_2$ (according to Riordan and Passow (37)).

In view of these experiments one may be tempted to conclude that the function of the metabolic poisons is simply to deplete the cells of intracellular ATP and thus to prevent its protective action against calcium. This does not seem to be the case, since the calcium concentrations required to induce the permeability change in ATP-free ghosts are about 2 orders of magnitude lower if applied in combination with fluoride than if applied alone. Iodoacetate has a similar potentiating effect but is considerably less powerful than fluoride. Moreover, the effects of fluoride and iodoacetate on fresh red blood cells can be modified considerably by the addition of the nucleosides adenosine and inosine. In fluoride poisoning, K^+-loss is inhibited by adenosine (8,27) and enhanced by inosine (40) while in iodoacetate poisoning the situation is reversed; inosine inhibits and adenosine enhances (32,33). Such and similar findings lead Gardos (8) and Passow (33) to suggest that under the special condition where the action of the metabolic poison is modified by the presence of a substrate, such as a nucleoside, the formation of some abnormal side products of intermediary metabolism is involved in bringing about the permeability change. This suggestion was analoguous to the original one of Wilbrandt for explaining the effects of fluoride. On the other hand, on the basis of their experiments with starved cells described above, Lindemann and Passow (28) rejected the idea that a diversion of metabolic reactions to abnormal pathways constitutes the <u>major</u> cause of K^+-loss in fluoride poisoning. Nevertheless, the evidence at hand did not rule out the possibility that poisoning of intact cells with F^- or iodoacetate leads to effects which are not strictly identical with those seen after addition of Ca^{++} to ATP-depleted cells or ghosts. This view was further supported by the striking augmentation of the iodoacetate + inosine induced increase of K^+-leakage by ammonium salts, hydrazine (figure 1c) and other N-containing compounds. With these remarks we intend to emphasize that although there are important similarities between the action of calcium and of the combinations of calcium with various metabolic inhibitors, there are also distinct differences in detail which require separate investigation.

Basic similarities and even more significant differences in detail become apparent if we turn to a number of other phenomena: the effects of propranolol (5,30) and triose reductone (33,33a). Both agents produce potassium loss with little concomitant sodium uptake and both require calcium in the medium to exert this action. However, calcium acts even if the intracellular ATP level is not lowered. In fact, propranolol may even increase the intracellular ATP content while inducing the permeability change. Finally, lead poisoning of red blood cells undoubtedly belongs to the class of phenomena discussed in this paper (32,34). The effect is specific for potassium (figure 1a) and resembles in many other respects that of calcium (see below). The heavy metal induces a degradation of intracellular ATP by some unknown mechanism, but this degradation is much slower than the response of the permeability barrier. Hence, the effect on potassium permeability is fully developed while the ATP content is still near the normal level (29). Lead poisoning leads to potassium loss in the absence of any calcium in the medium. Apparently, lead takes the place of the alkaline earth ion.

By touching upon the differences between the effects of lead, calcium or the various combinations of calcium with other agents, we wish to caution against unqualified generalisations. On the other hand, we discussed a whole spectrum of related phenomena in order to be able to derive from each one some information which may be relevant for an understanding of the nature of the selective increase of potassium permeability which constitutes the subject of the present article.

II. The site of action of calcium and lead

The calcium concentration in the intact red blood cell is so low that it is difficult to determine accurately. This low level of internal calcium is maintained by the combined action of binding to intracellular phosphate esters, notably ATP and of the calcium pump. The pump is capable of extruding intracellular calcium quite effectively into the medium thereby establishing and maintaining large concentration gradients across the membrane (41). Inhibition of red blood cell metabolism and the consequent degradation of ATP prevents the pump from extruding calcium which enters the cell. Hence the question must be asked whether or not the described sensitivity of the red blood cell membrane to the action of calcium is related to calcium entry and an action of the alkaline earth metal on the inside of the membrane. A number of authors are in favour of such an idea (2,21,25,40). Metabolically depleted red blood cells which respond to calcium with the described permeability change to potassium took up calcium from the medium. The rate of calcium uptake was characteristically dependent on pH. At pH values below 7 there was little calcium uptake and little potassium efflux. Above pH 7, the rate of calcium uptake increased sharply and so did the rate of potassium movements. If ATP synthesis was induced in the substrate depleted cells by the addition of suitable nucleosides, calcium was pumped out and potassium efflux slowed down considerably (40). In the presence of fluoride or iodoacetate, calcium uptake was increased. However, this increase did not account completely for the increased sensitivity of the membrane to the combined actions of calcium and either one of the two poisons. This can also be inferred from the work of Blum and Hoffman (2) who found that in the presence of iodoacetate in the medium as little as 9 µmoles/l of calcium inside the ghosts caused potassium loss. In the absence of iodoacetate, calcium concentrations up to about ten times this value do not induce potassium loss but facilitate the resealing of the membrane for alkali ions (3).

In view of these findings it is reasonable to assume that all conditions which are known to be associated with calcium uptake lead to an increase of potassium efflux. Even though the site of action may not be the internal surface of the red blood cell membrane, it is likely to be a place within the membrane which may become accessible only under the conditions of increased calcium permeability. A location of calcium sensitive sites at a place which is not in direct contact with the internal membrane surface is suggested by the finding that after addition of propranolol or triose reductone calcium evokes the permeability change for potassium in the presence of high concentrations of ATP inside the cell. Similarly, intracellular ATP and other phosphate esters should reduce the concentration of lead to a negligible concentration, yet lead is a very powerful promoter of potassium efflux.

III. The relationship between concentration and effect and the problem of an "all-or-none" versus a "graded" response.

If the response of a cell population to a poison is studied by analysis of samples which are composed of large numbers of cells, one is left with an ambiguity in the interpretation of the result. The result of the analysis may either represent the average composition of modified and unmodified cells or the composition of the average cells all of which are modified to nearly the same extent. In the former case, an increase of the response with increasing concentration would primarily reflect an increase of the number of damaged cells, in the latter case an increase of the damage to the average cell.

With these considerations in mind we turn to an inspection of figure 7 which represents the relationship between lead concentration in the medium and the time course of potassium loss. The most prominent feature of these curves consists of the finding that at intermediate lead concentrations the rate of potassium exit slows down considerably before the diffusion equilibrium is reached. The time required to reach the point at which this abrupt decrease of the rate of potassium exit takes place is nearly independent of the lead concentration. Hence the best measure for the effect of lead is the potassium concentration at the inflection point of the curves (or better, the concentration which is obtained if one extrapolates the nearly linear portion of the curves at long times back to

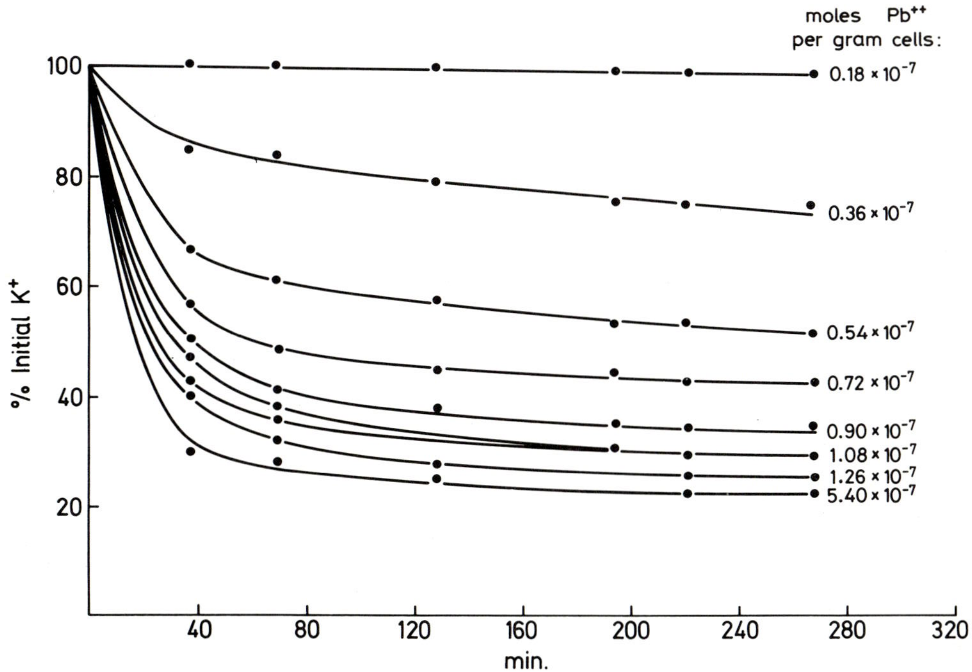

Figure 7 - The effect of varying doses of Pb^{++} on the time course of K^+ loss from cells suspended in 166 mmoles/l NaCl at 25°C (according to Passow and Tillman (31)). The cell density was 10 vol.%

zero time of the experiment) rather than the rate of loss.

The peculiar time course of the potassium loss could be explained by assuming that all of the cells of the population are capable of recovering from the effect of lead and that this recovery takes place at a rate which is nearly independent of the lead concentration in the medium. Alternatively one could assume that the cells respond in an "all-or-none" fashion: increasing the lead concentration has little effect on an individual cell until a threshold value is exceeded. A slight further increase of concentration of the poison leads to the maximal rate of potassium exit. Figure 8 shows that the cells respond in an "all-or none"

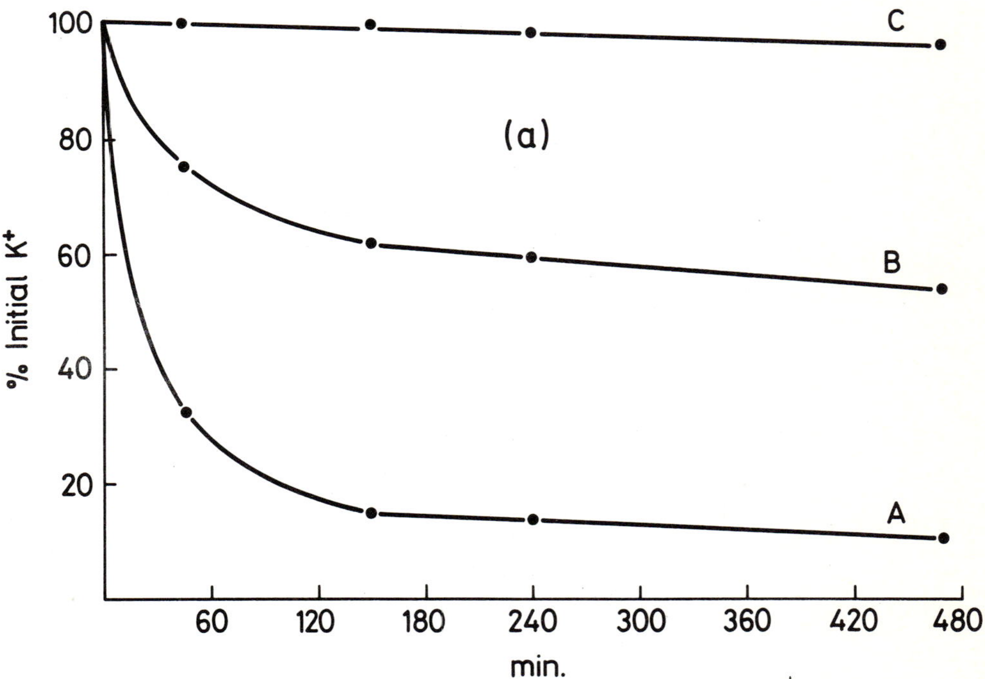

Figure 8 - The osmotic resistance of and K^+ loss from cells treated with A.) a maximally effective Pb^{++} concentration (0.06 mmoles/l). B.) a Pb^{++} concentration (0.008 mmoles/l) causing an intermediate degree of K^+ loss and C.) no Pb^{++}. Part a.) shows the loss of K^+ from cells for each experimental condition.

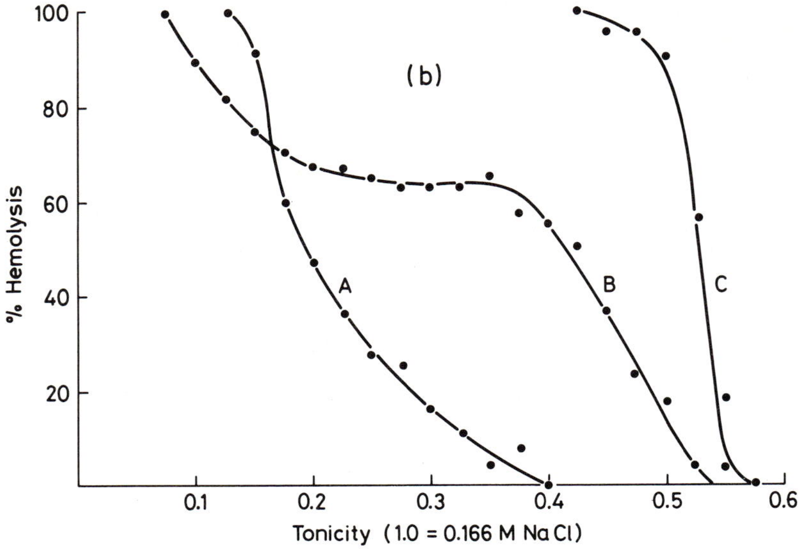

Figure 8 - Part b.) shows the relationships between the tonicity of the hemolysing solution in fractions of isotonic NaCl and degree of hemolysis for the three experimental conditions.

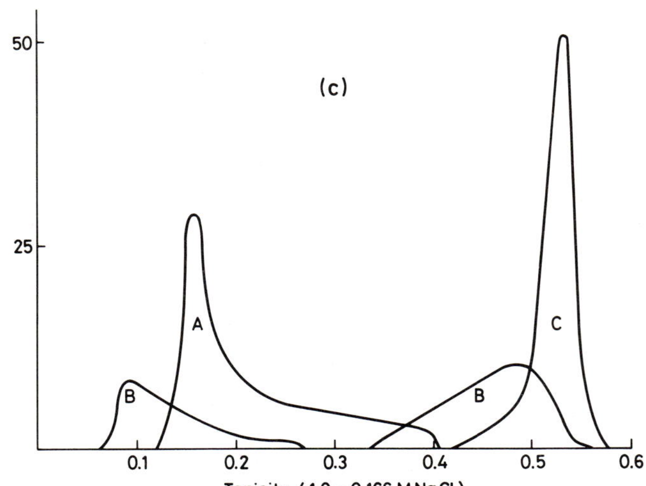

Figure 8 - Part c.) represents the first derivatives of the curves in part b.). Cells were incubated at 21°C. Samples for determination of osmotic resistance were taken after 240 min. of exposure to lead (according to Passow and Tillman (31)).

fashion. In the experiment represented in this figure, we measured the osmotic resistance after transition from the fast to the slow phase of potassium loss was completed. In the control without lead the cells exhibited normal fragility with a frequency distribution around a single peak. If exposed to maximally effective doses, the osmotic resistance is increased. Although the frequency distribution is rather skewed (presumably due to some sodium uptake) there is only a single peak. However, at an intermediate lead concentration which caused about 50% potassium loss before the curve reached the plateau, the frequency distribution shows two peaks of about equal size. This indicates the coexistance of about equal numbers of shrunken and nearly normal cells. Direct potassium determination showed beyond doubt that the shrunken cells contain much less potassium than the unaltered cells. We conclude therefore that the response to lead is primarily of an "all-or-none" character or, in more modern language, the effect of lead seems to be a highly cooperative phenomenon with different threshold values in different cells.

Interestingly enough, relationships between concentration and effect essentially similar to those described for lead poisoning are found when red cell ghosts are exposed to increasing concentrations of calcium (Figure 9a). Although no detailed investigation of the behaviour of the individual cells has yet been made it seems reasonable to suspect that the calcium induced potassium efflux is also of the "all-or-none" type. Moreover, many of the curves on the effects of propranolol and calcium represented in Manninen's paper (30) are also strongly reminiscent of those on lead or calcium poisoning described above. Hence it seems possible that all the phenomena which we are dealing with in this paper represent an "all-or-none" rather than a graded response.

Although the time course of potassium loss observed after poisoning with lead or calcium can be satisfactorily explained on the assumption of an "all-or-none" effect, a spontaneous recovery from the initial insult cannot be excluded. In the case of lead poisoning such recovery has actually been observed. However, the recovery process is slow compared to the rate of potassium exit after the first contact with lead. The evidence for a recovery process is based on experiments like those represented in Figure 10. The cells were suspended in a medium whose potassium concentration was equal to the intracellular concentration. Under these conditions the addition of lead evokes the permeability change but since there is no concentration gradient, no potassium exit can take place. If after various lenghts of time, concentration gradients for potassium are reinstituted by dilution of the cell suspension with isotonic saline solution, the potassium efflux can be followed. Within the first hour, the rate of loss was independent of the time of incubation in the potassium chloride

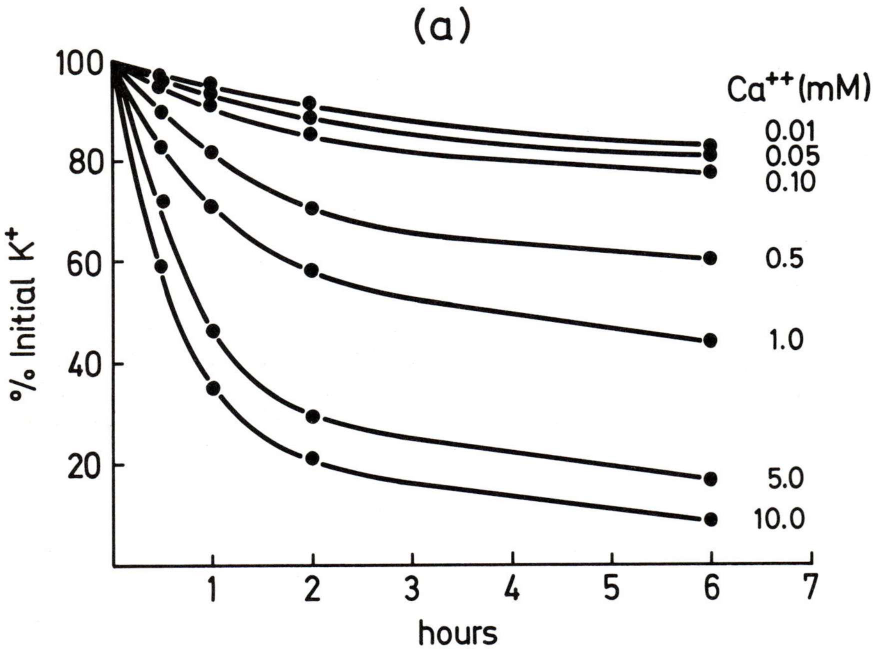

<u>Figure 9</u> - The effect of increasing media Ca^{++} concentrations of K^+ loss from ghosts suspended in isotonic NaCl. <u>Part a.</u> shows the time course of K^+ efflux for each Ca^{++} concentration.

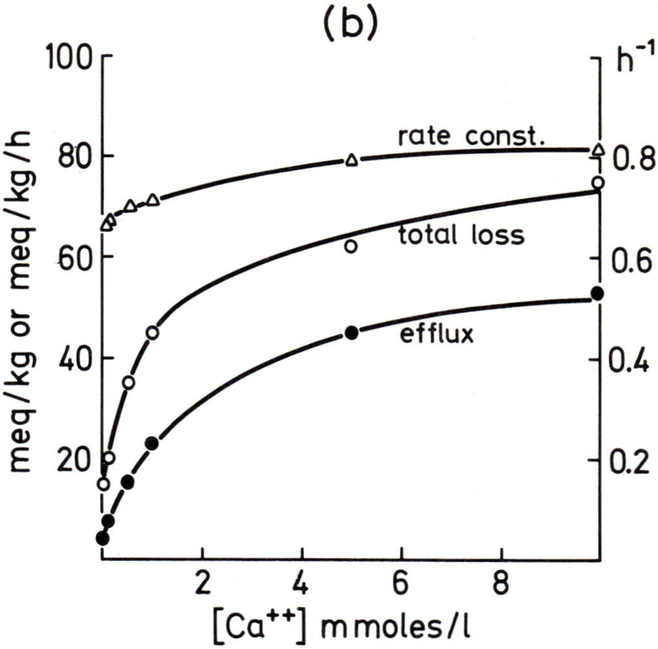

Figure 9 - Part b. shows the relationships between the Ca^{++} concentration and the initial rates of K^+ loss ("efflux"), the amount of K^+ lost during the rapid phase ("total loss") and the rate constants characterizing the rate of K^+ efflux during the initial rapid phase of K^+ exit. The abscissa indicates the media Ca^{++} concentration. The ordinate to the left indicates the initial rates in meq/kg cells/hour or total K^+ loss in meq/kg cells. The ordinate to the right indicates the rate constants in reciprocal hours (according to Riordan and Passow (37)).

medium. However, as time went on, the rate of potassium loss into the saline media decreased. After about 5 to 6 hours, the poisoned cells were about as impermeable to potassium as intact ones. Moreover, they did not respond any more to the action of a second dose of lead. In contrast to former speculations, we suspect today that the "recovery" of the membrane is related to some medium uptake from the incubation medium which contained, in

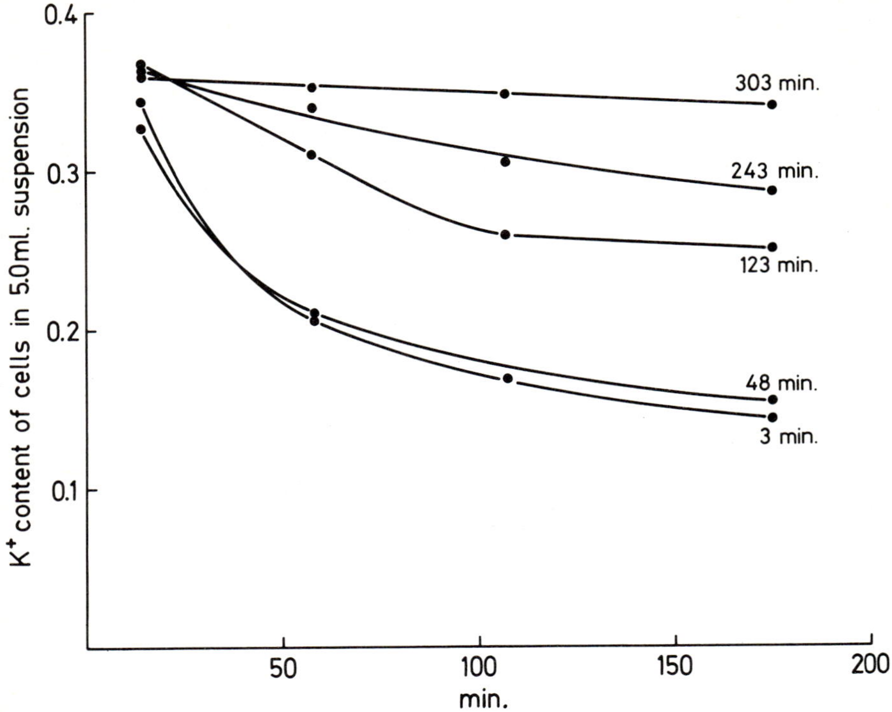

Figure 10 - K^+ efflux from cells exposed to Pb^{++} (0.1 mmoles/l) for various lengths of time. During exposure to Pb^{++} the cells were suspended in 90 mmoles/l KCl plus 76 mmoles/l NaCl. At the end of these times of exposure, aliquots were diluted 1:5 in 166 mmoles/l NaCl and K^+ efflux was monitored at 22.7°C (according to Passow and Tillman (31)).

addition to 90 mmoles/l KCl, about 76 mmoles/l NaCl. As has been shown previously, intracellular sodium inhibits net potassium exit (35). These potassium-sodium interactions will be dealt with extensively in a later section.

In conclusion of the present section we should like to mention a practical consequence of our considerations of "all-or-none" effects for the evaluation of the relationship between the concentration of the poison and the response. If one assumes that all of the ghosts of the population are similarly affected, the results of experiments such as those in figures 7 and 9a can be expressed by plotting the initial rates of potassium efflux versus the concentration of the poison in the medium. However, if one assumes that the curves primarily reflect an "all-or-none" effect, it would be preferable to plot the rate constant, which describes the time course of the transition from the rapid to the slow rate of potassium efflux, against the concentration of the poison (figure 9b). This rate constant primarily reflects the response of the ghosts whose threshold was exceeded. This quantity varies only little with calcium or lead or propranolol concentration. By extrapolation of the relationship between potassium concentration and time observed after the completion of the transition from rapid to slow leakage, one obtains the number of cells whose threshold value has been exceeded.

IV. Uphill movements in an electrical potential gradient or carrier mediated counterflow?

It had been shown by Joyce, Moore and Wheatherall (20) and Grigarzik and Passow (16) that ^{42}K is accumulated against its concentration gradient in lead poisoned erythrocytes. Figure 11 illustrates this phenomena in an experiment in which, for convenience, the movements of the long lived ^{86}Rb instead of the short lived ^{42}K were followed. After a few minutes the ^{86}Rb concentration in the cells exceeds that in the medium several fold. This accumulation is energized by the net efflux of potassium. It has also been observed after poisoning of fresh red blood cells with calcium and propranolol (30) and after exposure of substrate depleted cells to calcium plus iodoacetate (1). Hence it seems to represent a phenomenon which is pertinent to all types of specific increase of net potassium efflux as described in this paper.

In our system the energy which could lead to a concentration of ^{42}potassium or ^{86}Rb ions can only be derived from the downhill movements of potassium and chloride. Two possibilities for the transformation of this energy into uphill movements of ion species other than potassium and chloride were considered. Ekman <u>et al.</u> (5), Manninen (30), Hoffman (1) and Passow (36) discussed a coupling of potassium efflux with ^{42}potassium influx by a common carrier system; Glynn and Warner (15) suggested a coupling through a common diffusion potential which is primarily generated by the leakage of potassium chloride from the cells into the medium.

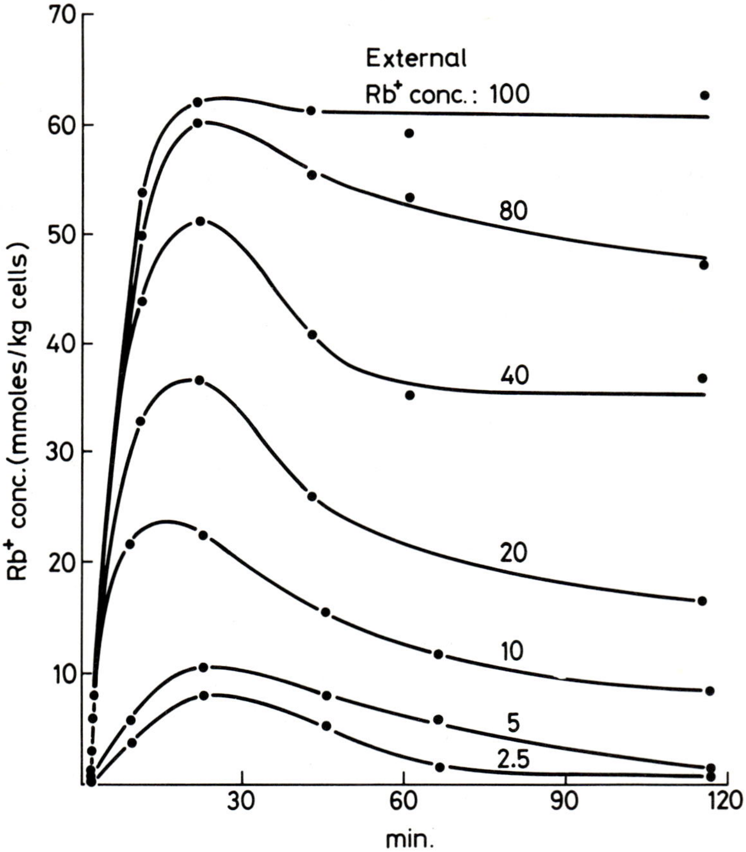

Figure 11 - Rb^+ uptake and K^+ loss in the presence of 0.08 mM/l Pb^{++} at 25°C. The media consisted of RbCl in the concentrations indicated on the curves plus sufficient NaCl to maintain isotonicity (according to Passow (36)).

We shall consider the latter suggestion first. Recent work on anion permeability has shown that net movements of anions across the red blood cell membrane are very much slower than anion exchange. They are, in fact, so slow that under conditions of high potassium permeability the anions may become rate limiting (15,19). This leads to the establishment of electrical fields and hence of electrical driving forces. If chloride is rate limiting, the potassium exit is associated with an increase of the potential, inside negative. This would accelerate the influx of cations from the medium and provide the energy for an accumulation within the cell. Glynn and Warner (15) could interpret the data on uphill movements of ^{42}potassium, induced by propranolol and calcium, by making more or less arbitrary assumptions about the relative permeabilities for potassium, sodium and chloride in the poisoned cells.

Additional evidence for the participation of electrical driving forces in these effects was obtained by Hoffman and Knauf (19). They showed that a typical inhibitor of anion permeability, dipyridamole, inhibits the calcium plus iodoacetate induced net potassium efflux into saline solutions. The interpretation is obvious. The inhibition of anion efflux prevents the only penetrating cation species, potassium, from leaving the cells. When sufficient extracellular potassium was added to the medium to allow a potassium/potassium exchange across the membrane, then there was much less inhibition by dipyridamole. This suggests strongly that the effects of the agent was not due to a direct action on the potassium channels but to an inhibition of anion permeability. However, a more quantitative evaluation of such and other experiments lead Hoffman and Knauf (19) to believe that the potential is only one component which participates in inducing the observed counterflow.

We turn now to a discussion of a carrier mediated uphill transport of potassium or Rb. At a time when the low rate of net anion transfer was not yet recognized and hence the participation of an electrical driving force seemed rather far fetched,the participation of an uphill transport was inferred from experiments like the following (1): substrate depleted cells were loaded with ^{42}potassium and suspended in media containing, besides iodoacetate and calcium, increasing concentrations of potassium. At low concentrations the presence of external ^{39}K facilitated ^{42}K efflux. The relationship between the rate constant for ^{42}K exit and external ^{39}K concentration was found to pass through a flat maximum. This was interpreted to show that the non-radioactive potassium in the external medium replaced radioactive potassium on the carrier molecules which appeared at the external surface of the cell and thus enhanced the rate of potassium exit. In other words, it was assumed that external potassium induced an acceleration by coupling via a common transport system with a limited number of transfer sites.

Since the increase of the external potassium concentration in these experiments should be associated with a decrease of the membrane potential, we repeated Blum and Hoffman's work with a slightly different design of the experiments: using ghosts we established identical concentrations of all ion species present in the system on both sides of the membrane. Hence, there were no driving forces; the system was at equilibrium. Variations of potassium concentration were performed in cells and medium alike such that over the whole concentration range there was never a significant deviation from equilibrium. If the ^{42}potassium efflux was measured under these conditions, we still found a considerable increase of the rate constants for ^{42}K movements with increasing potassium concentration (Figure 12). At about 5-10 mmoles/l, a maximum was reached. Towards higher potassium concentrations the rate constant decreased again gradually. The flux (equal to the rate constant multiplied by the concentration) tended to saturate. This experiment strongly suggests that we are dealing with a potassium transport system which has only a limited transport capacity. In addition it confirmed the existence of a maximum under conditions where the rate constants were measured at equilibrium.

In summing up the evidence presented in this

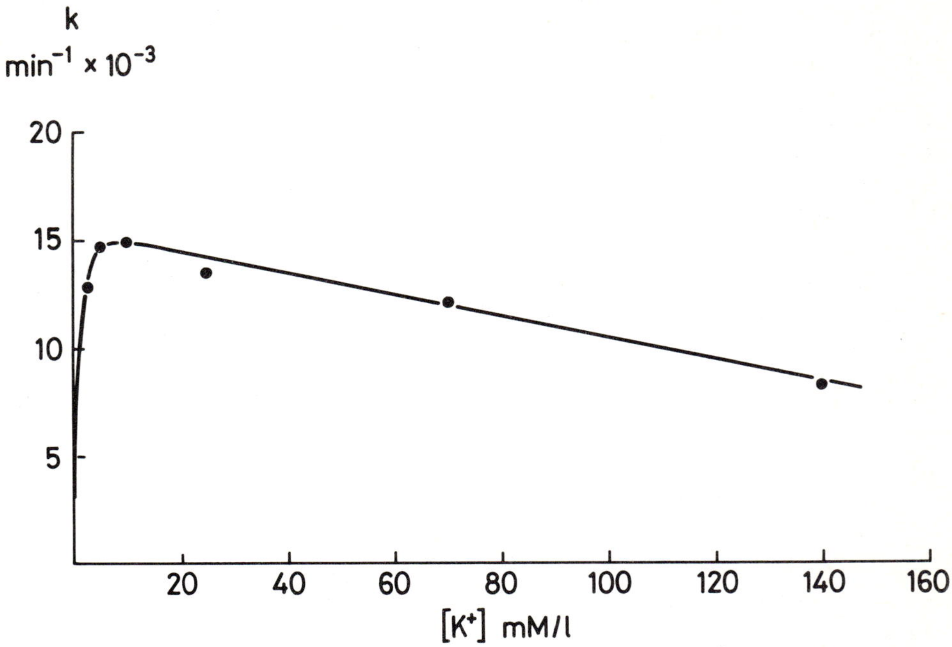

Figure 12 - The effect of increasing K^+ concentrations on the rate of ^{42}K efflux from ghosts containing equal concentrations of K^+ inside and in their incubation media. The latter contain 8 mmoles/l $CaCl_2$. The abscissa indicates the K^+ concentrations. Isotonicity was maintained by the addition of sufficient choline chloride to give a total concentration of 166 mmoles/l inside the ghosts and in the medium. The ordinate indicates the rate constant of ^{42}K movements (according to Riordan and Passow (38)).

section one feels compelled to agree with Hoffman and Knauf (19) and to conclude that both carrier mediated as well as potential driven countertransport, are involved in the uphill movements described above.

V. The involvement of the potassium/sodium pump

Recently, Blum and Hoffman (1) have demonstrated a partial inhibition by ouabain and furosemide and an almost complete inhibition by oligomycin of the effects of calcium and iodoacetate on substrated depleted cells. This would suggest that under their conditions the increased potassium efflux is associated with an alteration of a system which serves as the potassium-sodium-pump in the normal cell. We could confirm and extend the findings with oligomycin. This agent very effectively inhibited potassium efflux from ghosts treated with calcium alone or in the presence of fluoride or iodoacetate plus adenosine. It was less effective but still inhibitory in lead poisoning. Hence, the inhibition by oligomycin defines to a certain extent whether or not an increase in potassium efflux belongs to the class of phenomena discussed in the present paper (37). However, we were unable to discover an inhibitory action of ouabain in ghosts which were prepared from ouabain treated cells, even though we could demonstrate that the potassium/sodium sensitive magnesium ATPase of these ghosts was considerably (40-65%) inhibited. Moreover, in intact cells, pretreated with high doses of ouabain under optimal binding conditions, lead was as fully active as in cells which were never exposed to the glycoside. Finally Manninen (30) also failed to observe an effect of ouabain on a propranolol plus calcium induced potassium efflux. Hence we feel that the evidence at hand does not yet permit one to assume that the potassium/sodium pump mechanism in whole or in part is involved in the described permeability change.

Although the previously described evidence does not allow one to decide conclusively whether or not the potassium movements we are dealing with in the present paper proceed via a partial reaction of the sodium/potassium pump or through some other pathway, they clearly show the involvement of a specific saturable transfer system. So far we can only speculate why this transfer system is stimulated by lead or by the various combinations of calcium with metabolic depletion, metabolic inhibitors, or drugs like propranolol.

If one assumes that a carrier system is involved two possibilities come to mind for explaining the enormous increase of the rate of potassium movements: 1) An equilibrium between bound and free carrier molecules in the membrane is shifted towards higher concentrations of the free form (36) or 2) the rate of turnover of each carrier molecule is increased by an alteration of the environment through which the carrier moves or by a change of the conformation of the carrier molecule itself. Whatever may eventually turn out to be the molecular basis of the effect, the transfer system has properties which one would not expect to find in simple systems. The interactions between sodium and potassium to be considered below clearly demonstrate a high degree of complexity.

VI. Interactions between sodium and potassium

As has already been stated in the introduction, one of the reasons why we originally suggested that the permeability change induced by a variety of seemingly unrelated measures is due to a modification of the same transfer mechanism was the specificity of the effect for potassium: the enormous increase in potassium efflux was accompanied by comparatively little sodium uptake (32-34). However, later work showed that even though there is no spectacular effect on sodium movements, sodium ions exert striking effects on the potassium movements induced by lead (35), calcium and fluoride (24) or calcium on ATP-free ghosts (37) or substrate depleted intact cells (19). The interactions between sodium and potassium are very complex and the interrelations between the many variables involved are far from clear. Nevertheless, the rather incomplete evidence at hand suggests that these interactions take place primarily, although probably not exclusively, on the inside of the red blood cell membrane and depend on the direction of a potassium gradient across the membrane. Below we shall briefly summarize those of our observations which, at the present time, seem to us more relevant than others for an understanding of the mechanism of the interactions between the two alkali ion species. For a more complete presentation of our recent experimental work on this complicated subject, the reader is referred to the original communication (39).

In view of the complexities of the system it is necessary to state exactly the details of the experimental conditions. We worked with ATP-free ghosts which, in addition to 4 mmoles/l magnesium sulfate contained 70 mmoles/l KCl, 70 mmoles/l choline chloride and 20 mmoles/l TRIS, pH 7.6. These ghosts were exposed to a nearly maximally effective concentration of calcium, 8 mmoles/l. The calcium was dissolved in TRIS-buffered (20 mmoles /l) media of the same ideal osmolarity and pH as the cell interior. The effects of Na were studied by replacing equimolar concentrations of intracellular or extracellular choline chloride by sodium chloride.

In the described system, net potassium efflux into potassium free media was inhibited by intracellular sodium (table 1). The inhibition was seen

Table 1 - Inhibition of Ca^{++}-induced net K^+ efflux by intracellular sodium. Ca^{++} concentration in the medium: 8 mmoles/l

Inside (mM/l)			Outside (mM/l)		$k \cdot 10^{-3}$ (min.$^{-1}$)
K^+	Na^+	Chol.	Na^+	Chol.	
70	0	70	0	140	14.4 (12)
70	70	70	0	140	4.6 (12)
70	0	70	140	0	20.1 (6)
70	70	0	140	0	5.8 (6)

The numbers in brackets indicate the number of experiments from which the averages of the rate constants, k, were calculated.

In the experiments summarized in this table and in the following ones, all cations were added in the form of their chlorides. The pH was adjusted to 7.5 by 20 mmoles/l Tris-Cl. The buffer was present at equal concentrations on both sides of the membrane.

regardless of the presence or absence of Na in the external medium. However, under the specified conditions, net K efflux was accelerated if the external medium contained sodium chloride in place of choline chloride. The acceleration by external sodium depended on the absolute value of the intracellular K concentration. At 140 mmoles/l KCl inside, a replacement of extracellular choline by sodium did not affect the rate of K exit (table 2).

Table 2 - Stimulation of Ca^{++}-induced net K^+ efflux by extracellular sodium. Ca^{++} concentration in the medium: 8 mmoles/l.

Inside (mM/l)		Outside (mM/l)		$k \cdot 10^{-3}$ (min.$^{-1}$)	
K^+	Chol.	Na^+	Chol.		
70	70	0	140	11.6	(8)
70	70	140	0	18.2	(8)
140	0	0	140	25.4	(4)
140	0	140	0	25.5	(4)

The numbers in brackets indicate the number of experiments from which the averages of the rate constants, k, were calculated.

Inhibition by internal sodium was only seen when the intracellular K^+ concentration exceeded the extracellular concentration (table 3). When the concentrations of all ion species present, including K^+, were made equal inside the cells and outside, instead of inhibition by sodium we observed the same rate of K^+ efflux as in sodium free cells. In a number of experiments we even found some acceleration by internal Na^+. This did not only apply to our standard conditions (70 mmoles/l KCl inside) but also if the KCl concentration on both sides of the membrane was varied between 5 and 140 mmoles/l. When the K gradient across the membrane was reversed by suspending cells loaded with 70 mmoles/l KCl in media which contained 140 mmoles/l KCl, there was neither an acceleration nor an inhibition by sodium: the replacement of intracellular choline by an equivalent amount of NaCl at constant internal KCl concentration did not influence the rate of

^{42}K efflux.

Table 3 - Effect of extracellular potassium on sodium inhibition of Ca^{++}-induced potassium efflux. Ca^{++} concentration in the medium: 8 mmoles/l.

Inside (mM/l)			Outside (mM/l)			$k \cdot 10^{-3}$ (min.$^{-1}$)
K^+	Na^+	Chol.	K^+	Na^+	Chol.	
70	0	70	0	0	140	13.2 (3)
70	70	0	0	0	140	5.3 (4)
70	0	70	70	0	70	15.8 (5)
70	70	0	70	0	70	15.9 (2)
70	0	70	140	0	0	12.6 (2)
70	70	0	140	0	0	10.3 (2)

The numbers in brackets indicate the number of experiments from which the averages of the rate constants, k, were calculated.

On the basis of these findings it would seem possible that sodium inhibition can only take place if there exists a gradient for potassium from the inside towards the outside. If this gradient is reduced to zero or reversed, then the sodium inhibition of potassium movements disappears. This is in accord with the observed absence of sodium inhibition when the ^{42}K movements are measured at equilibrium or when the external potassium concentration is made higher than that inside the ghosts (table 3). On the basis of such a hypothesis one would predict that, in contrast to a net potassium efflux, a net potassium influx should not be inhibited by intracellular sodium. Measurements of potassium influx are subject to greater experimental error than those of efflux. Nevertheless, our experiments show that there is little if any effect of internal sodium on potassium entry (table 4). Moreover, external sodium is also ineffective. This shows that the situation is not symmetric.

This latter finding is of interest with respect to our previous discussion of the mechanism of sodium inhibition of potassium flux in lead poisoned red blood cells (35). A theoretical investi-

Table 4 - Lack of effect of intracellular or extracellular sodium on Ca^{++}-induced net potassium efflux. Ca^{++} concentration: 8 mmoles/l.

Inside (mM/l)			Outside (mM/l)			Influx (mM/l/min.)
K^+	Na^+	Chol.	K^+	Na^+	Chol.	
0	0	140	70	0	70	0.76 (2)
0	0	140	70	70	0	0.73 (2)
0	140	0	70	0	70	0.79 (2)

The numbers in brackets indicate the number of experiments from which the averages of the fluxes were calculated.

gation of competition of potassium and sodium for a common carrier site had shown that maximal discrimination should take place if the two ion species are present on the same side of the membrane whereas in single file diffusion the maximal discrimination should become apparent if the two species migrate in opposite directions. Our experimental investigations on the sidedness of the potassium/sodium interactions show them to be asymmetric. Hence they cannot be explained by single file diffusion or carrier transport across a homogeneous membrane. We can rule out, therefore, the previously discussed possibility of a flow coupling by a common single file system for the two ion species.

The finding that net K efflux is inhibited by intracellular Na, while net K influx is not, provides stronger evidence for the existence of a causal relationship between sodium inhibition and the direction of the potassium gradient than the observed reversal of sodium inhibition of K efflux by increasing the extracellular potassium concentration to the same level as inside the ghosts or above. Increasing the external potassium concentration initiates a potassium/potassium exchange flux whose rate is no longer dependent on the relatively slow rate of anion permeation. This exchange flux may not necessarily be mediated by the same transfer system as the net potassium flux into potassium free media. However, as long as there is no evidence in support of the assumption of separate pathways for net movements and exchange flows, we shall stipulate that the disappearance of Na inhibition upon reversal of the direction of K^+ net movements is under both conditions due to the same mechanism.

What could be the molecular basis of such a mechanism?

Since calcium induced net movements of potassium across the erythrocyte membrane are likely to proceed at a rate which is comparable to that of net anion movements (15,18,19), we can assume that the net potassium exit is associated with the establishment of an electrical field. It would seem possible that inhibitory sodium binding sites are buried in the absence of such a field or when the outside is negative i.e. when the external potassium concentration is equal to or greater than the internal concentration. If, however, the internal K^+ concentration exceeds the external concentration, then the direction of the field would be reversed. This could lead to an exposure to the inside of sodium binding sites which control potassium movements across the membrane. In such a system internal sodium could inhibit potassium exit only if there exists an outwardly directed concentration gradient for potassium (see figure 13).

The above discussion primarily intends to show the complexity of the potassium-sodium interactions rather than to offer a definite hypothesis about the mechanism of these interactions. Since the rate of potassium movements depends on the internal and external concentrations of potassium, sodium, calcium and hydrogen ions, on the existence of gradients across the membrane (at least for potassium), and on the presence of more than one binding site for potassium on the transport system, the number of variables is extraordinarily great. A considerable amount of additional work is still required before the response of potassium permeability to all relevant combinations of these variables will have been tested. However, only when this has been accomplished will it be feasible to develop a more

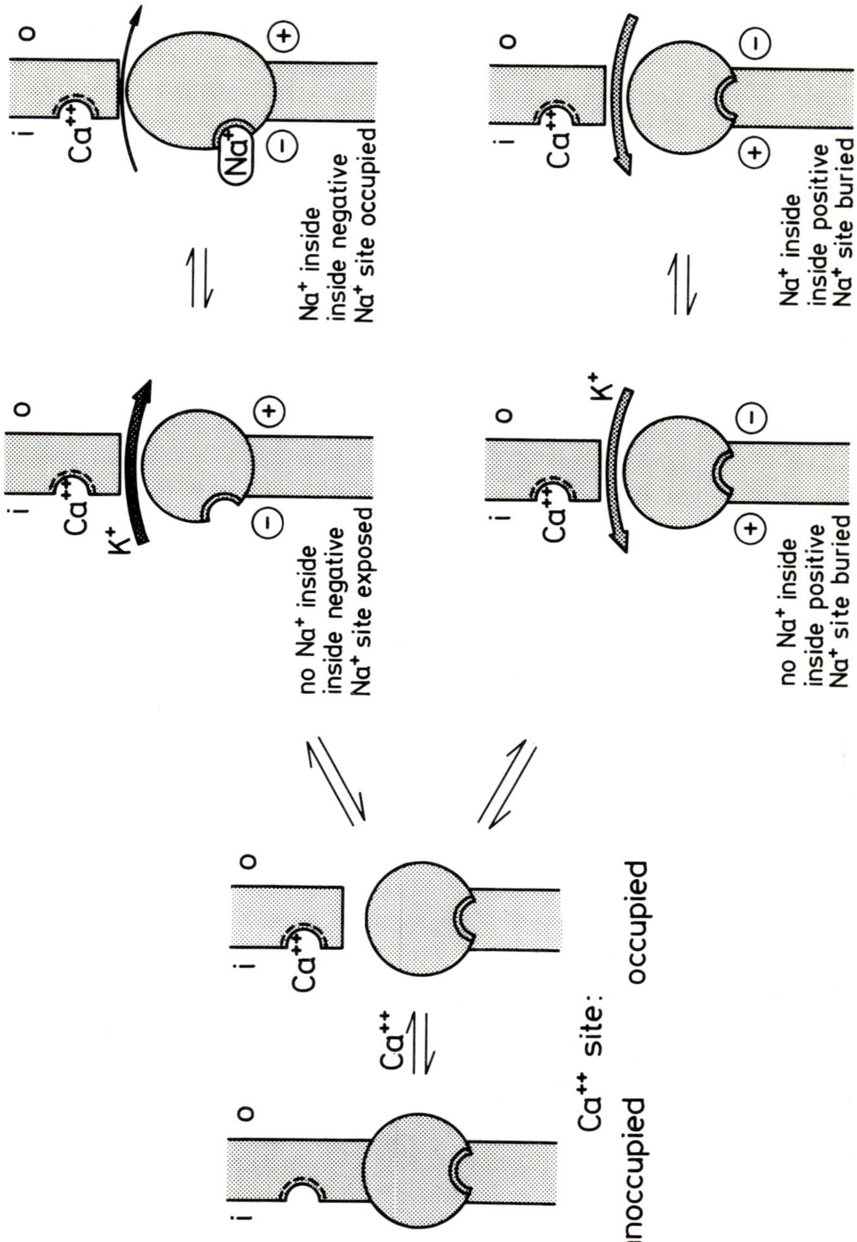

Na+ inside
inside negative
Na+ site occupied
no Na+ inside
inside negative
Na+ site exposed
Na+ inside
inside positive
Na+ site buried
no Na+ inside
inside positive
Na+ site buried
Ca++ site:
unoccupied
occupied

Figure 13

definitive hypothesis of the mechanism.

In conclusion of this review one may ask why we and others devote so much time to a study of a complicated system whose functional significance is still completely unknown. We ourselves were attracted by the specificity of the system for potassium and by the variety of intriguing features,such as the all-or-none response to lead, the possible dependence of potassium-sodium interactions on the direction of the potassium gradient, etc. Above all, we suspect this system to play a significant functional role. In a system with a coupled sodium potassium pump, where the coupling ratio is fairly constant and independent of the metabolic state of the cell, for each pumping rate only one potassium sodium ratio can be established in the cell. Only if the ratio of the leak fluxes for potassium and sodium can be varied independently, could the cell adjust K and Na concentrations separately. Perhaps, the mechanism under study has something to do with such a control function (33).

ACKNOWLEDEGEMENTS

We thank Miss S.Lepke and Miss B.Schuhmann for their collaboration. The work was supported by the Deutsche Forschungsgemeinschaft and by a fellowship of the Canadian Cystic Fibrosis Foundation to J.R.R.

REFERENCES

1. Blum,R.M. & Hoffman,J.F. (1972) The membrane locus of Ca-stimulated K transport in energy depleted human red blood cells. J.Membrane Biol. 6, 315-328.

2. Blum,R.M. & Hoffman,J.F. (1972) Ca-induced K transport in human red cells: location of the Ca-sensitive site to the inside of the membrane. Biochem.Biophys.Res.Commun. 46, 1146-1151.

3. Bodemann,H. & Passow,H. (1972) Factors controlling the resealing of the membrane of human

erythrocyte ghost after hypotonic hemolysis. J.Membrane Biol. 8, 1-26.

3a. Donlon,J.A. & Rothstein,A. (1969) The cation permeability of erythrocytes in low ionic strength media of various tonicities. J.Membrane Biol. 1, 37-52.

4. Dunker,E. & Passow,H. (1950) Verteilung von anionen und Kationen bei Fluoridvergiftung menschlicher Erythrocyten. Pflügers Archiv. 252, 542-550.

5. Ekman,A., Manninen,V. & Salminen,S. (1969) Ion movements in red cells treated with propranolol Acta Physiol.Scand. 75, 333-344.

6. Gardos,G. (1956) The permeability of human erythrocytes to potassium. Acta Physiol. 10, 2-4 (Budapest).

7. Gardos,G. (1958) Effect of the ethylendiamine-tetraacetate on the permeability of human erythrocytes. Acta Physiol. 14, 1-5 (Budapest).

8. Gardos,G. (1958) The function of calcium in the potassium permeability of human erythrocytes. Biochim.Biophys.Acta 30, 653-654.

9. Gardos,G. (1959) The role of calcium in the potassium permeability of human erythrocytes. Acta Physiol. 15, 121-125 (Budapest).

10. Gardos,G. (1961) The function of calcium in the regulation of ion transport. In Membrane Transport and Metabolism, ed. A.Kleinzeller & A.Kotyk, New York, Academic Press, pp.553-558.

11. Garrahan,P.J. & Glynn,I.M. (1966) Exchange ratios for the cation pump in red cells during Na:Na and Na:K exchange. J.Physiol. 185, 31-32.

12. Glynn,I.M. (1957) The ionic permeability of the red blood cell membrane. Prog.Biophys.Chem. 8, 241-253.

13. Glynn,I.M. & Hoffman,J.F. (1971) Nucleotide requirements for Na-Na exchange catalysed by the sodium pump in human red cells. J.Physiol. 218, 239-256.

14. Glynn,I.M., Hoffman,J.F. & Lew,V.L. (1971) Some partial reactions of the sodium pump. Phil. Trans.Roy Soc.Lond.B. 262, 91-102.

15. Glynn,I.M. & Warner,A.E. (1972) Nature of the calcium dependent potassium leak induced by (+)-propranolol, and its possible relevance to the drug's antiarhythmic effect. Br.J. Pharmacol. 44, 271-278.

16. Grigarzik,H. & Passow,H. (1958) Versuche zum Mechanismus der Bleiwirkung auf die Kalium-permeabilität roter Blutkörperchen. Pflügers Archiv. 267, 73-92.

17. Hoffman,J.F. (1966) The red cell membrane and the transport of sodium and potassium. Am.J. Med. 41, 666-680.

18. Hoffman,J.F. & Lassen,U.V. (1971) Plasma membrane potentials in amphiuma red cells. Proc. XXV Internat.Union Physiol.Sci. 9, 253.

19. Hoffman,J.F. & Knauf,P.A. (1972) The mechanism of the increased K transport induced by Ca in human red blood cells. 2nd Int.Symp. Metabolism and Membrane Permeability of Erythrocytes, Thrombocytes and Leucocytes, Wien (in press).

20. Joyce,C.R.B., Moore,H. & Weatherhall,M.(1954) The effect of lead, mercury and gold on the potassium turnover of rabbit red cells. Br.J. Pharmacol. 92, 463-470.

20a. Kregenow,F.M. & Hoffman,J.F. (1962). Metabolic control of passive transport and exchange diffusion of Na and K in human red cells. (Abstract) Biophys. Society, Washington D.C.

21. Kregenow,F.M. & Hoffman,J.F. (1972) Some kinetic and metabolic characteristics of calcium-induced potassium transport in human red cells. J.Gen.Physiol. 60, 406-429.

21a. LaCelle,P. & Rothstein,A. (1966) The passive permeability of the red blood cell to cations. J.Gen.Physiol. 50, 171-188.

22. Lepke,S. & Passow,H. (1960) Hemmung des Kaliumverlustes fluoridvergifteter Erythrocyten durch Komplexbildner. Pflügers Archiv. 271,389-396.

23. Lepke,S. & Passow,H. (1960) Die Wirkung von Erdalkalimetallionen auf die Kationenpermeabilität fluoridvergifteter Erythrocyten. Pflügers Archiv. 271, 473-487.

24. Lepke,S. & Passow,H. (1968) Effects of fluoride on potassium and sodium permeability of the erythrocyte membranes. J.Gen.Physiol. 51,365s-372s.

25. Lew,V.L. (1971) On the ATP dependence of the Ca^{2+}-induced increase in K^+ permeability observed in human red cells. Biochim.Biophys. Acta 233, 827-830.

26. Lew,V.L. (1971) Effect of ouabain on the Ca^{2+}-dependent increase in K^+ permeability in depleted guinea pig red cells. Biochim.Biophys. Acta 249, 236-239.

27. Lindemann,B. & Passow,H. (1960) Versuche zur Aufklärung der Beziehung zwischen Glycolysehemmung und Kaliumverlust bei der Fluoridvergiftung von Menschenerythrocyten. Pflügers Archiv. 271, 497-510.

28. Lindemann,B. & Passow,H. (1960) Die Wirkung von Erdalkalimetallionen auf die Hemmung des Kaliumverlustes fluoridvergifteter Erythrocyten durch Substrate und Stoffwechselgifte. Pflügers Archiv. 271, 488-496.

29. Lindemann,B. & Passow,H. (1960) Kaliumverlust und ATP-Zerfall in bleivergifteten Menschenerythrocyten. Pflügers Archiv. 271, 369-373.

30. Manninen,V. (1970) Movements of sodium and potassium ions and their tracers in propranolol-treated red cells and diaphragm muscle. Acta Physiol.Scand.Suppl. 355, 1-76.

31. Passow,H. & Tillmann,K. (1955) Untersuchungen über den Kaliumverlust bleivergifteter Menschenerythrocyten. Pflügers Archiv. 262, 23-36.

32. Passow,H. (1961) Zusammenwirken von Membranstruktur und Zellstoffwechsel bei der Regulierung der Ionenpermeabilität roter Blutkörperchen. In Biochemie des aktiven Transportes (Colloquium der Gesellschaft für Physiologische

Chemie in Mosbach/Baden), pp.54-99, Berlin, Springer Verlag.

33. Passow,H. (1963) Metabolic control of passive cation permeability in human red cells. In Cell Interface Reactions, ed.Brown,H.D.,pp.57-90, New York, Scholar's Library.

33a. Passow,H. & Vielhauer,E. (1966) Die Wirkung von Trioseredukton auf die K^+- und Na-Permeabilität roter Blutkörperchen. Pflügers Archiv. 288, 1-14.

34. Passow,H. (1964) Ion and water permeability of the red blood cell. In The Red Blood Cell, ed. Bishop,C. & Surgenor,D.M., pp.71-145, New York, Academic Press.

35. Passow,H. (1969) The molecular basis of ion discrimination in the erythrocyte membrane. In The Molecular Basis of Membrane Function, ed. Tosteson,D.C., pp.319-351, Englewood Cliffs, New Jersey, Prentice-Hall,Inc.

36. Passow,H. (1970) The red blood cell: penetration, distribution and toxic action of heavy metals. In Effects of Metals on Cells, Subcellular Elements and Macromolecules, ed. Maniloff,J., Coleman,J.R. & Miller,M., pp.291-340, Springfield,Ill. Charles C.Thomas Publ.

37. Riordan,J.R. & Passow,H. (1971) Effects of calcium and lead on potassium permeability of human erythrocyte ghosts. Biochim.Biophys.Acta 249, 601-605.

38. Riordan,J.R. & Passow,H. (1972) Asymmetry of inhibition by Na of Ca-induced net potassium fluxes in human red cell ghosts. Proc. IV Int. Biophysics Congress IX-XV, 134 (Moscow).

39. Riordan,J.R. & Passow,H. (1972) Calcium-induced permeability of the human erythrocyte membrane to potassium: Sodium-Potassium interactions. Submitted for publication.

40. Romero,P.J. & Whittam,R.R. (1971) The control by internal calcium of membrane permeability to sodium and potassium. J.Physiol. 214,481-507.

41. Schatzmann,H.J. & Vincenzi,F.F. (1969) Calcium movements across the membrane of human red cells. J.Physiol. 201, 369-395.

42. Shaw,T.I. (1955) Potassium movements in washed erythrocytes. J.Physiol. 129, 464-475.

43. Tosteson,D.C. & Hoffman,J.F. (1960) Regulation of cell volume by active cation transport in high and low potassium sheep red cells. J.Gen.Physiol. 44, 169-194.

44. Wilbrandt,W. (1937) A relation between the permeability of the red cell and its metabolism. Trans.Faraday Soc. 33, 956-959.

45. Wilbrandt,W. (1940) Die Abhängigkeit der Ionenpermeabilität der Erythrocyten vom glycolytischen Stoffwechsel. Pflügers Archiv. 243, 519-536.

46. Wilbrandt,W. & Schatzmann,H.J. (1960) Changes in the passive cation permeability of erythrocytes in low electrolyte media. In Ciba Foundation Study Group, Symposium n.5 Regulation of the Inorganic Ion Content of Cells, pp.34-52, London, Churchill.

Comparative Physiology, eds. L. Bolis, K. Schmidt-Nielsen and S. H. P. Maddrell

COMPARATIVE TRANSPORT OF SUGARS ACROSS RED BLOOD CELLS

L. BOLIS,
Istituto di Fisiologia Generale,
Università di Roma,
Rome, Italy.

The transport of sugars across red blood cell membranes has been investigated since the beginning of the century (14, 22). Red blood cells have a highly specialized nature: in mammals a lack of subcellular structure and very reduced metabolic activity (mainly glycolysis) concentrate interest on membrane phenomena. The cells originate from fully developed and highly organized cell types but in spite of their highly specialized function, to transport metabolites among the organs, they maintain important physiological properties found in other cells. Experimentally they are convenient because there is no fixed intercellular space, the extracellular milieu can be very carefully controlled. Also, no particular allowance needs to be made for the metabolism of monosaccharides, since this is such a slow process: in human erythrocytes glucose may be transported 250 times faster than would be the requirement for metabolism (33).

Glucose penetrates the human erythrocyte much faster than one would anticipate on the basis of its molecular size and solubility in the membrane. Polyhydric alcohols of similar size to glucose, such as mannitol, usually penetrate many orders of magnitude more slowly.

The transport of monosaccharides in erythrocytes is usually a facilitated process, (the term "facilitated diffusion" is used synonymous to carrier diffusion). The kinetics of glucose movement in the human erythrocytes show the essential features of carrier transport: 1) saturation kinetics; 2) competition; 3) counter-transport and competitive acceleration; 4) stereospecificity. It is important to note, however, that the transport is not dependent upon energy-yielding reactions. In red cells glucose moves down its own chemical gradient: no spontaneous "uphill movement" occurs.

The nature of the carrier molecule is not known but some data suggest it is a protein or lipoprotein. The lipid in the membrane represents a barrier to penetration of non-electrolytes of low lipid-solubility, like glucose (6,7). It has been proposed that the carrier functions by masking the hydrophilic group and thus renders it soluble in the membrane. A recent hypothesis involves allosteric reactions of a specific membrane protein elicited by the binding of a sugar. The proposed carrier system operates by a sequence of intermolecular reactions which allow the sugar to diffuse through the membrane (21,29). Observations of conformational changes attributed to the carrier involved in sugar transport have been made (15). At any rate the transfer is highly specific and shows several characteristics of enzyme reactions.

A large number of inhibitors have been found to block sugar transport. They often act upon SH and NH_2 groups. Non-competitive inhibitors are known, like N-ethylmaleimide (NEM), parachloromercuribenzoate, other mercuric compounds and 1-fluoro-2,4-dinitrofluorobenzene (DNFB). The latter reacts irreversibly so that penetration cannot be restored. Kinetically competitive inhibitors include phlorizin, phloretin, polyphloretin-phosphate and urethane:the latter can also penetrate the cell (5,8,35,39).

A thorough comparison of the transport of different sugars,and analysis of the kinetics have been made in human erythrocytes in the past 25 years (4,12,13,18,19,26,27,29,32,33,37,38,39,40). There is a higher affinity for most hexoses, such as D-glucose, D-mannose and D-galactose, than for pentoses, such as D-ribose, D-xylose, L-arabinose and D-sorbose.

The degree of specificity of the system was demonstrated with the erythrocytes of *Macacus rhesus*, which have a similar monosaccharide transport to human red blood cells. It was shown that isomeric sugars penetrate at different rates (14).

Among different animals there is often a pronounced variation in sugar transport parameters. The sugar and polyhydric alcohol permeability has been studied in different animals and a different rate of penetration as measured by hemolysis was found (30), being faster in mouse than in man and dog.

A slow degree of penetration of cat and dog erythrocyte membranes by monosaccharides has been found (36). In the dog L-arabinose, and D-xylose penetrate much faster than fructose, sorbose and galactose. The penetration of hexose and pentose monosaccharides into erythrocytes of the ground-hog (*Marmota monax*) has been attributed to a free-diffusion process (10). More recently an unexpectedly high affinity, comparable to other mammalian cells, for D-glucose and D-3-O-methylglucose has been shown in beef erythrocytes (11): being the maximal rate of transport about 1/2 with respect to rabbit erythrocytes and 1/2000 with respect to human erythrocytes. It has also been shown that beef erythrocytes can be divided into two populations (17), between which there is a 5 to 10 fold difference in the rate of penetration by fructose, but no difference in glucose penetration. These two classes respond differently, however, to the inhibitory effect of glucose upon the fructose penetration. The "slow fructose" population is sensitive to glucose inhibition, the other not. The inhibitory effect of glucose is present in human (37), but not in rabbit erythrocytes (25).

Glucose penetration of rabbit erythrocytes shows all the characteristics of a facilitated process (24,25). The transport, however, is not inhibited by parachloromercuribenzoate, N-ethylmaleimide, Hg^{2+}, or Cu^{2+}. Dinitrofluorobenzene causes a slight increase in the rate of glucose entry, whereas in human erythrocytes DNFB is an inhibitor. The rate of entry of glucose into rabbit erythrocytes is slower compared to fructose, but glucose and fructose do not compete. It is therefore assumed that there must be a striking difference in the operation and nature of the carrier system compared with that in other erythrocytes, such as those of humans.

The sugar penetration of foetal erythrocytes

has also been studied in different species (31,34). Foetal cells of deer, sheep, rabbit, pig, guinea pig and man, show an evident penetration by glucose. Foetal cat red blood cells are on the contrary impermeable to glucose. Erythrocytes from the guinea pig foetus showed a higher penetration than those from the adult: similar findings have been made in man, where the rates differ by a factor of two. The characteristics of facilitated transport are shown by the kinetics of hexose penetration of foetal erythrocytes.

The red blood cells from a lower vertebrate, the fresh-water teleost *Salmo trutta* L. have been investigated by Bolis *et al.* (1) and shown to be impermeable to glucose, D-mannose and D-galactose. The situation is different for pentoses: D-ribose, L-arabinose and D-xylose penetrate at different rates, D-ribose penetrating fastest. Isocountertransport and isotope exchange experiments upon ribose showed a difference between human and trout erythrocytes: for the latter simple diffusion kinetics may apply (2).

In view of the striking difference between the sugar permeability of mammalian and trout erythrocytes, the lipid composition of the membrane was investigated (3). It was shown that, unlike in other animals, most of the lipids of the trout erythrocyte membrane are highly unsaturated (around 50% of the total lipid).

In the light of these findings, it may appear at present that the sugar permeability of trout erythrocytes can be interpreted better in terms of a porous membrane structure than of a carrier system.

REFERENCES

1. Bolis,L., Luly,P., Baroncelli,V. (1971) D(+)glucose permeability in brown trout *Salmo trutta* L. erythrocytes. J.Fish Biol. 3, 273-275.

2. Bolis,L., Luly,P. (1972) Monosaccharide permeability in brown trout *Salmo trutta* L. erythrocytes. In *Role of Membranes in Secretory Pro-*

cesses, ed. L.Bolis, R.D.Keynes & W.Wilbrandt, pp. 215-221,Amsterdam,North Holland Publishing Company.

3. Bolis,L., Luly,P. (1972) Membrane lipid pattern and non-electrolytes permeability in Salmo trutta L. In Passive Permeability of Cell Membranes ed. Kreuzer,F. and Slegers,J.F.G., pp. 357-362, Plenum Press,New York & London.

4. Bowyer,F. (1957) The kinetics of the penetration of non-electrolytes into the mammalian erythrocytes. Int.Rev.Cytol. 6, 469-511.

5. Bowyer,F., & Widdas, W.F. (1958) The action of inhibitors on the facilitated hexose transfer system in erythrocytes. J.Physiol. 141,219-232.

6. Danielli,J.F. (1941) Cell permeability and diffusion across the oil-water interface. Trans.Faraday Soc. 37, 121-124.

7. Davson,H. & Danielli,J.F. (1943) Permeability of natural membranes., Cambridge, Cambridge University Press.

8. Dawson,A.C. & Widdas,W.F. (1963) Inhibition of the glucose permeability of human erythrocytes by N-methylmaleimide. J.Physiol. 168, 644-659.

9. Dawson,A.C. & Widdas,W.F. (1964) Variations with temperature and pH of the parameters of glucose transfer across the erythrocyte membrane in the foetal guinea pig. J. Physiol. 172, 107-122.

10. Faust,R.G. & Parpart,A.K. (1961) Permeability studies on the red blood cell of the ground-hog (Marmota monax). J.Cell Comp.Physiol. 57,1-3.

11. Hoos,R.T., Tarpley,H.L., Regen,D.M. (1972) Sugar transport in beef erythrocytes. Biochim. Biophys.Acta 266, 174-181.

12. Jacobs,M.H., Glassman,H.N. & Parpart,A.K.(1938) Osmotic properties of the erythrocyte. X. Differences in the permeability of the erythrocytes of two closely related species. J. Cell Comp.Physiol. 11, 479-494.

13. Jacobs,M.H., Parpart,A.K., Corson,S.A. (1937) Osmotic properties of the erythrocytes. IX. The effect of low concentrations of electrolytes on hemolysis by penetrating non-electrolytes and on cell volume. J.Cell Comp.Physiol. 9,177-190.

14. Kozawa,S. (1914) Beiträge zum arteigenen Verhalten der roten Blut-Körperchen. Biochem. Z. 60, 231-256.

15. Krupka,R.M. (1971) Evidence for a carrier conformational change associated with sugar transport in erythrocytes. Biochemistry 10, 1143-1148.

16. Lacko,L. & Burger,M. (1962) Interaction of some disaccharides with the carrier system for aldoses in erythrocytes. Biochem. J. 83, 622-625.

17. Laris,Ph.C. (1967) Variation in fructose transport among beef erythrocytes. J.Cell Physiol. 70, 1-6.

18. LeFevre,P.G. (1961) Sugar transport in red blood cells: structure activity relationship in substrate antagonists. Pharmacol. Rev. 13, 39-70.

19. LeFevre,P.G. (1962) Rate and affinity in human red blood cell sugar transport. Am.J.Physiol. 203, 286-290.

20. Levine,M., & Stein,W.D. (1966) The kinetic parameters of the monosaccharide transfer system in the human erythrocyte. Biochim.Biophys. Acta 127, 179-193.

21. Lieb,W.R. & Stein,W.D. (1970) Quantitative prediction of a non carrier model for glucose transport across the human red cell membrane. Biophys.J. 10, 585-609.

22. Masing,E. (1914) Über die Verteilung von Traubenzucker in Menschenblut und ihre Abhängigkeit von der Temperatur. Pflügers Arch.ges.Physiol. 156, 401-425.

23. Miller,D.M. (1969) Monosaccharide transport in erythrocytes. In Red Cell Membrane Structure and Function. ed. Jamieson,G.A. & Greenwalt,J.

pp. 240-290, Philadelphia, Lippincott Co.

24. Morgan,H.E., Kalman,C.F., Post,R.L. & Park,C.R. (1955) Kinetics of glucose transport across cell membrane. Fed.Proc. 14, 103-104.

25. Regen,D.M., Morgan,H.E. (1964) Studies of the glucose-transport system in the rabbit erythrocytes. Biochim.Biophys.Acta 79, 151-166.

26. Rosenberg,T., Wilbrandt,W. (1963) Carrier transport uphill. I.General. J.Theoret.Biol. 5, 288-305.

27. Rosenberg,T. (1961) Membrane transport of sugars, a survey of kinetical and chemical approaches. Biol.Path. 9, 795-802.

28. Sen,A.K., Widdas,W.F. (1962) Variation of the parameters of glucose transfer across the human erythrocyte in the presence of inhibitors of transfer. J.Physiol. 160, 404-416.

29. Stein,W.D. (1967) The Movement of Molecules Across Cell Membranes, New York & London, Academic Press.

30, Ulrich,H. (1934) Untersuchungen über die Permeation lipoidunlöslicher Nichtleiter von relativ grossem Molekularvolumen in die Erythrocyten von Säugetieren. Pflügers Arch.ges.Physiol. 234, 42.

31. Widdas,W.F. (1953) Hexose permeability in mammalian foetal erythrocytes. Abstr. XIX Int. Physiol.Congr., pp. 885-886.

32. Widdas,W.F. (1953) Kinetics of glucose transfer across the human erythrocyte membrane. J.Physiol. 120, 23-24P.

33. Widdas,W.F. (1954) Facilitated transfer of hexoses across the human erythrocyte membrane. J.Physiol. 125, 163-180.

34. Widdas,W.F. (1955) Hexose permeability of foetal erythrocytes. J.Physiol. 127, 318-327.

35. Widdas,W.F. (1972) Aspects of competitive inhibition. In Passive Permeability of Cell Membranes, ed. Kreuzer,F. & Slegers,J.F.G., pp. 101-105, New York & London, Plenum Press.

36. Wilbrandt,W. (1938) Die Permeabilität der roten Blutkörperchen für einfache Zucker. Pflügers Archiv. 241, 302-309.

37. Wilbrandt,W. (1954) Secretion and transport of non-electrolytes, Symp.Soc.Exp.Biol. 8, 136-162.

38. Wilbrandt,W., Frei,S. & Rosenberg,T. (1956) The kinetics of glucose transport through the human red cell membrane. Exp.Cell.Res. 11, 59-66.

39. Wilbrandt,W. & Rosenberg,T. (1961) The concept of carrier transport and its corollaries in pharmacology. Pharmacol.Rev. 13, 109-183.

40. Wilbrandt,W. (1972) Carrier Diffusion. In Passive Permeability of Cell Membranes, ed.Kreuzer, F. & Slegers,J.F.G., pp. 79-99, New York & London, Plenum Press.

Comparative Physiology, eds. L. Bolis, K. Schmidt-Nielsen and S. H. P. Maddrell

A NEW MODEL FOR RECORDING SMOOTH MUSCLE ACTIVITY OF PERFUSED PERIPHERAL BLOOD VESSELS.

J.G. PERISTIANY & H.J. HUGGEL
Laboratoire d'Anatomie et Physiologie Comparées,
Université de Genève,
Genève, Switzerland.

ABSTRACT

The rhythmically contractile veins of the Bat's wing possess suitable properties for recording *in situ* the activity of perfused peripheral vascular smooth muscle by means of a photometric method.

A physiological solution C.S.3., determined from the serum analysis, is used to replace the blood by perfusion. It is thus possible to investigate directly the influence of several haematological, pharmacological and physical parameters, on peripheral smooth muscle function and autoregulatory flow mechanisms, in the absence of any influence from the surrounding circulation.

Extra-systoles and extra-diastoles may participate in aiding flow. The latter exert a suction pressure by reversing the normal sequence of contraction. Both these responses may be reproduced experimentally by simulated intravascular pressure changes.

Visco-elastic properties of the vessel wall play an important role in determining the form of contractile response to different stimuli *in vivo.*

Denervation modifies the activity of a venous segment, isolated from the arterial pole by means of an anastomosis.

Some doubt is cast as to the presence of Na-based action potentials. Contractions cease immediately in calcium-free (+ EGTA) C.S.3. solution. In the absence of EGTA, this reaction is complete only after approximately one hour of perfusion.

INTRODUCTION

Though considerable progress has been made in determining the nature of vascular tone (11-17,21, 26,34,36-38,50,57,60,75,76,80) and the dynamic behaviour of isolated vascular smooth muscle (18-20, 24,31-35, 41,54,64-67,69), relatively little research has been devoted to the effects of nervous, humoral, electrical and mechanical stimulation of blood vessels in situ. This has led us to examine a new model and a novel method for recording the activity of perfused vascular smooth muscle *in vivo*.

The tendency of certain vessels to exert spontaneous rhythmical contractions, and the usefulness of the latter in studying vascular mechanisms, has been extensively described in different territories. Such vasomotion, as it is termed, may be found in the umbilical vessels of man, the anterior mesenteric veins of cats, rats, rabbits, guinea-pigs and babboons, the subcutaneous arteries of the dog, the aorta and portal veins of rats and rabbits, the turtle vena cava and the peripheral veins of the bat's wing (3,9,19,27,34,39,49,51,52,54,64,70).

The pulsatile metacarpal veins of the bat's wing are subdivided into independent functional units (59) possessing intricate mechanisms for controlling blood flow (59,60). These venous segments are endowed with certain properties of singular interest to the study of peripheral smooth muscle and circulatory flow mechanisms. Among these properties may be included: the superficial organisation within a highly translucent and accessible membrane; the well defined extrinsic and intrinsic innervation; the high frequency and regularity of spontaneous vasomotion and the very specialised manner of adapting to local conditions.

More general points of interest include similarities of behaviour with cardiac-cell function, such as the occurrence of extra-systoles (59-60), and the presence of a particularly well defined pacemaker activity and depolarisation-contraction coupling process (28).

The relative proportion of circular to longitudinal muscle within all vessels may not only vary from one area to another (68), but also may have entirely different qualities within the same artery or vein (70). The metacarpal vein of the bat is entirely devoid of longitudinal muscle fibres (46). This limits all its contractile activity to a helicoidally arranged muscle layer, whose physiology is still largely unexplored.

A detailed analysis of the blood and serum properties of two species of Megachiroptera (62, 63) has led to the reconstruction of a physiological solution (C.S.3) capable, by itself, of restoring normal activity to the perfused digital veins of these animals.

Venous vasomotion is recorded by a photometric method modified after an original technique described elsewhere (62).

MATERIAL AND METHODS

A. The physiological solution

IONIC VALUES m.eq/l.	Na^+	Cl^-	Ca^{++}	K^+	Mg^{++}	$PO_4^{\equiv}$
WITH NAFAG	139 ± 4	110 ± 2	4,6 ± 0,7	5,9 ± 0,9	2,7 ± 0,3	3,5 ± 1,6
WITHOUT NAFAG	133 ± 10	106 ± 5	4,5 ± 0,4	6,7 ± 0,8	2,8 ± 0,4	3,9 ± 1,4
GLUCOSE	1,5 g/l		ACIDITY	7,38		
PROTEIN	68,8 g/l		OSMOLARITY	315 m.Osm		
PHYSIOLOGICAL SOLUTION in g/l	NaCl: 5,647	KCl: 0,438	$CaCl_2 \cdot 2H_2O$: 0,333	$NaHCO_3$: 3,448	$MgCl_2 \cdot 6H_2O$: 0,267	Na_2HPO_4: 0,073; Glucose: 2,00

Figure 1 - A serum analysis of *Pteropus giganteus*.

Ionic Values m.eq/l.	Na^+	Cl^-	Ca^{++}	K^+	Mg^{++}	$PO_4^{\equiv}$
Pteropus giganteus with complement	139	110	4,6	5,9	2,7	3,5
Pteropus giganteus without complement	133	106	4,5	6,7	2,8	3,9
Rousettus aegyptiacus	138	106	4,7	7,9	2,9	4,6
$\bar{X}$ of 13 mammals	149	112	5,6	6,5	1,8	2,9

Figure 2 - A comparison of the mean ionic values of 13 Mammals with those of two Megachiroptera.

The results of this analysis (tables 1 and 2) were compared with ionic values gathered from a wide range of Mammals (17). It was found that both groups of bats possessed unusually high concentrations of Na, Cl and Ca, and low concentrations of K, Mg and PO_4, and this could explain the difficulty with previous "Ringer solutions" in adequately maintaining vasomotion. Table 2 compares the ionic composition (in mEq/l) of Pteropus and Rousettus with average values (X) calculated for each ion from 13 random species of mammal(including man, pig, rat, and hedgehog). The physiological solution C.S.3. is adjusted to pH 7.4 by the addition of 0.16 ml conc. sulphuric acid corresponding to approximately 1.13 mEq/l SO_4^-.

Animals caged for periods exceeding two months displayed symptoms of nutrient deficiency. A second group was therefore analysed after the addition of a nutrient adjuvant, "Nafag" (Futterzusatz für Fleischfresser No 168, Gossau, S.G.), rich in vitamins, fats and proteins, to their diet. This latter group improved in health, and was found after a further period of two months, to possess significantly higher levels of serum proteins, haemoglobin and haematocrit. This represented an increase of

18%, 21 and 22% respectively.

The C.S.3. solution prepared from this final analysis was tested in vivo and found to preserve intact the mechanical and excitable properties of the vessel over several hours (63).

B. The photometric measurement of venomotion. (Figure 3).

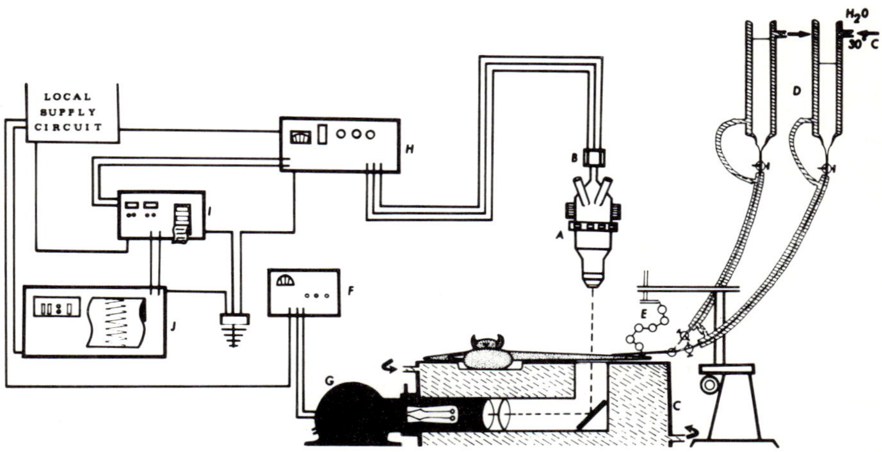

Figure 3 - Schematic diagram of photometric circuit used for recording venomotion:
A. Stereoscopic binocular M5 Wild; B. Double photoresistor (Clairex CL730L/2); C. operating chamber; D. Temperature-controlled graduated cylinders; E. Prior micromanipulator; F.Transformer; G.Mercury lamp (projectina 5561 C); H.D-C bridge amplifier; I. Frequency counter (Sodeco);J.Sefram servotrace.

The bat is narcotised with Nembutal (50 mg/kg I.P.) and heated 5-10 minutes under a 60 watts lamp during the brief period of pre-narcotic muscular hypersensitivity. After 1 1/2-2 hours, 0.2 cc (50 mg/ml) solution is injected at approximately one hour intervals. The animal is placed on its back, and a wing is spread and immobilised over an operating chamber maintained at 35°C. A controlled and calorifically filtered mercury light beam (G) is projected through the wing. A section of vein, approximately 1 mm in length, is optically isolated by means of a Wild M-5 binocular incorporating a double photoelectric cell (Clairex). The current from the bridge circuit is amplified and d.c.-recorded in parallel onto a Sefram servo-trace and a frequency counter (Sodeco) modified to record total number of contractions per minute.

Any change of diameter of the vessel acts as a diaphragm in the path of light and modifies the resistance of the photosensitive unit. A contraction increases (upward deflection) and a dilatation decreases (downward deflection) the total light energy reaching the cells.

C. Canulation and Perfusion.

The skin is liberated (on one side of the bone) over an area of approximately 10-15 mm in length at the distal and proximal ends of the metacarpal bone. The neurovascular bundle is covered with the physiological solution to avoid drying of the tissues, and the vein, artery and nerve are carefully separated by means of a surgical thread. Special care is taken not to damage the nerve bundle lying between the two vessels.

The artery is ligatured at the proximal, and the vein at both the proximal and distal regions. The vein is then canulated at both ends, leaving a section 8-10 cm long in between. The distal cannula is used to introduce, and the proximal one to collect the perfusate (test solution).

All measurements are made at the center of a segment between two valves. Secondary venules irrigating the anastomosed section are cauterised to avoid mixing of the blood with the test solution. An experiment is always preceded by the recording of the activity of the intact vessel, followed by that with the normal physiological solution (C.S. 3.).

THE MECHANOGRAMME

D'Agrosa (10) working with Microchiroptera, separated vasomotion into two groups. These, he appropriately termed "continuous" and "discontinuous", according to the form and duration of the relaxing phase of the cycle. This definition, however, made no distinction between the "diastolic filling" and "resting period" components of the relaxation phase. In order to clarify this point, and to reduce confusion between the terms "filling pressure, diastole, relaxation, resting period, return to basal-tone level" etc., we proposed to review this classification in more detail.

Though basically similar in pattern to that of the Microchiroptera, the normal venogramme of *Pteropus giganteus* (Megachiroptera) is subject to complex variations. Some of these may be reproduced in the laboratory,

a. following artificially induced intramural pressure changes (60);
b. in "fatigued preparations" perfused *in vivo* or suspended *in vitro*, and
c. after certain changes in the ionic composition of the physiological solution.

Normal venous vasomotion is subdivided into *regular*(A) and *irregular* rhytms (B).

The following observations have been made on approximately 100 live fruit-eating Megachiroptera (*Pteropus giganteus*) originating from Bombay, and kept in two cages of six cubic meters each, at high relative humidity and at 22-24°C.

A. *Regular rhythms*.

Under "regular rhythms"are included all venogrammes in which no variation in amplitude is exhibited by either the systolic or diastolic phases. The contraction-dilatation cycle may ("discontinuous") or may not ("continuous") be separated by a "perceptible" resting period.

Continuous and discontinuous venous vasomotion may alternate periodically within the same segment with no apparent motive. The change from one pattern to another takes place by a gradual shortening or lengthening of the iso-volumetric resting period.

(a) *"Discontinuous"venomotion* is characterised by the presence of a definite relaxation period which includes two separate phases: one anisotonic and one isovolumetric. In all, four consecutive phases may be distinguished (Figure 4).

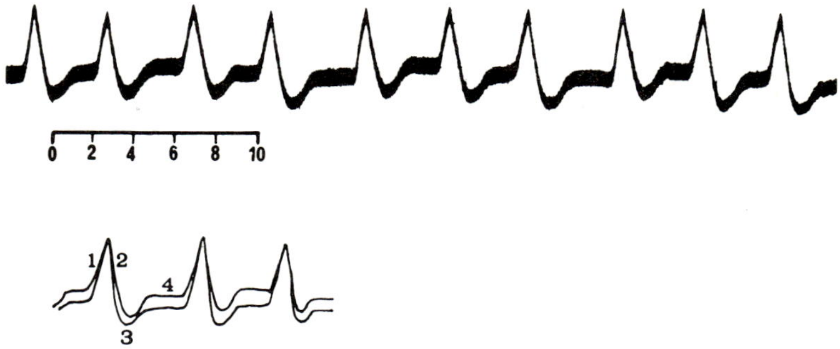

Figure 4 - Discontinuous venomotion in the intact metacarpal vein of *Pteropus giganteus.*

1. Contraction (systole)	- rapid in origin, duration and passage to phase 2.
2. Dilatation (diastole)	- Surpassing basal-tone level. Phase rapid at first, then tapering off to resting period
3. Anisotonic resting period	- which returns dilated vessel to a basal-tone resting level of regularly oscillating amplitude (phase 4).
4. Isovolumetric resting period	- of variable, but generally longer duration than phase 3.

(b) "Continuous" venomotion. The continuous form is composed of regularly alternating contractions (systoles) and dilatations (diastoles).The duration of the systolic phase may be equal (Figure 5A), superior (Figure 5B) or inferior (Figure 5C) to that of the diastolic component. The passage from one phase to the other is smooth and without visible interruption.

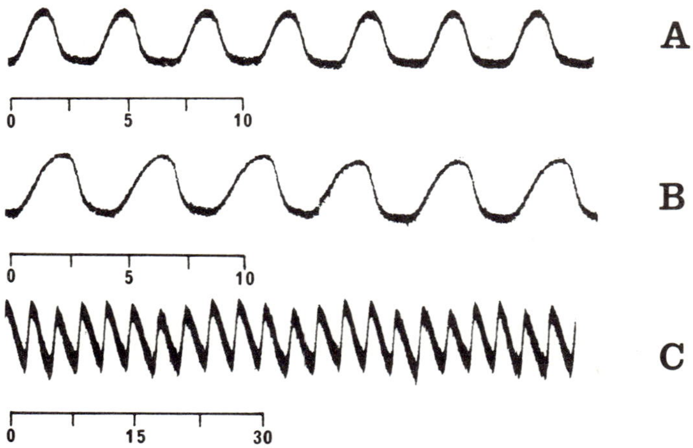

Figure 5 - Continuous venomotion of the metacarpal vein of *Pteropus giganteus*. The systolic phase is of equal (A), superior (B), or inferior (C) duration to that of the diastolic.The figures 0,5,10, etc., represent the passage of time in seconds.

Contractions of the "continuous" type may be grouped together in numbers varying from one to five (more rarely up to eight), each group being separated from the other by a long "compensatory" isotonic relaxation period (Figure 6). This effect is known as the "Wenkebach period" and has formerly been described in the heart of the dogfish (72).

Similar effects have been observed in our laboratory (Figure 7) following longitudinal tension changes of the isolated frog's auricle (*Rana temporaria*). In the digital veins of the bat, several conditions including simulated intravascular pressure changes, denervation and different changes in the ionic composition of the perfusate, have been seen to bring about this "Wenkebach-phenomenon".

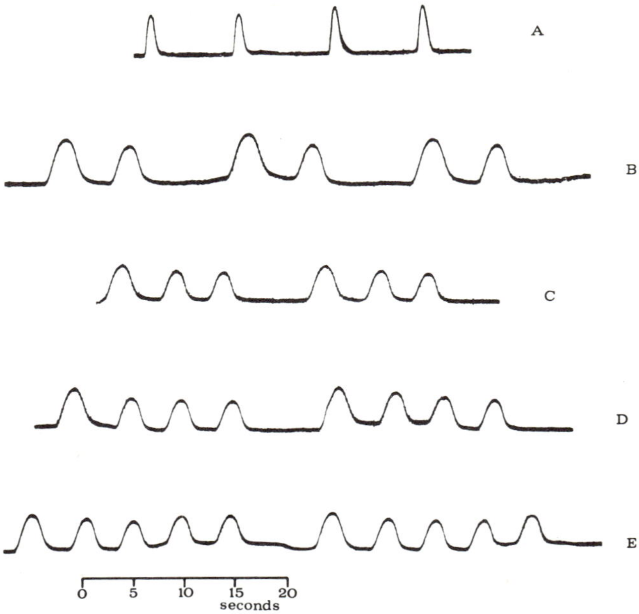

Figure 6 - Grouping of contractions of the "continuous" type (Pteropus giganteus) recorded at various pressures. A: 75 cm H_2O; B: 80 cm H_2O; C: 90 cm H_2O; D: 90 cm H_2O; E: 100 cm H_2O.

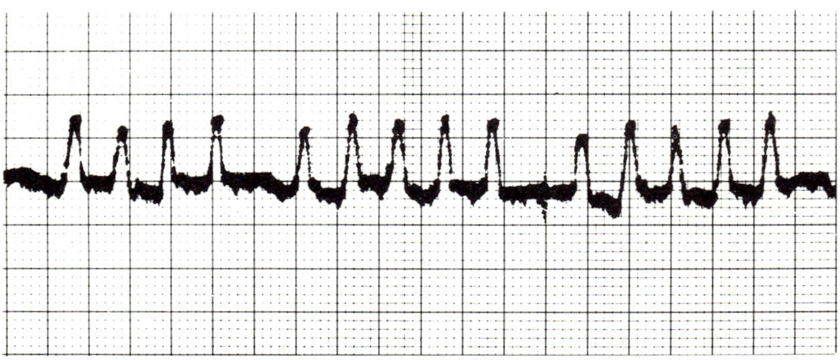

Figure 7 - Grouping of contractions in the isolated auricle of Rana temporaria. Each square represents 5 seconds.

B. Irregular rhythms.

Irregular activity(arhythmia) may be one of two different types: "transitional" or "complete". Both these forms are characterised by the interruption of a regular pattern of contraction through a change in amplitude, frequency and/or basal tone.

Irregular rhythms of the "transitional" type include extra-systoles and extra-diastoles. These phenomena arise spontaneously following sudden changes in systolic or diastolic pressures, and may be instrumental in restoring normal activity to the vein.

"Complete" arhythmia is associated with the total disruption from any basic rhythm. For unknown reasons, a vessel may periodically be unable to adapt to any definite rhythm. This may be due to a desynchronisation of contraction.

Extra-systoles. - An extra-systole (E-S) appears in the majority of cases after a premature contraction of low amplitude (Figure 8. E-S 1). The E-S is never preceded by a resting period, and, in contrast to cardiac extrasystoles, is not followed by a "compensatory" resting period.

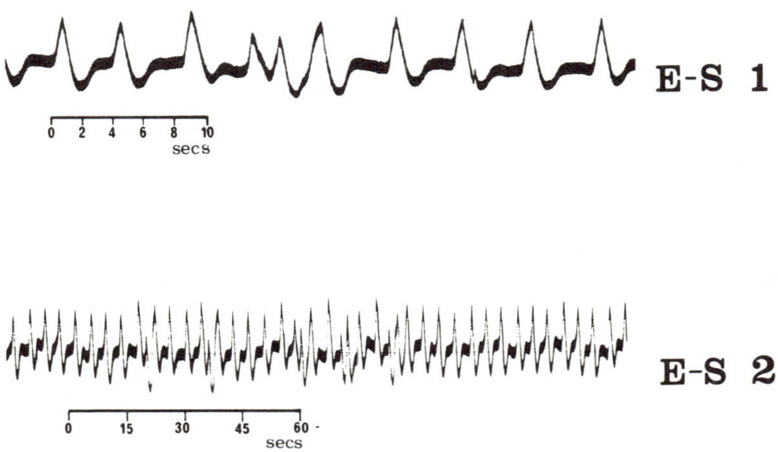

Figure 8 - Extra-systoles of the digital vein of Pteropus giganteus.

The systolic contraction my be of variable height, whilst the diastolic amplitude is always superior to that of the preceding one, and tends to increase the diastolic filling volume. The following contraction re-establishes the normal systolic level of activity.

Extra-diastoles. - It has also been observed, in the intact preparation, that a rapidly rising basal tone, accompanied by a diminishing diastolic filling phase, may result in the appearance of a varying number of extra-diastoles (E-D).

Extra-diastoles (Figure 9) are distinguished

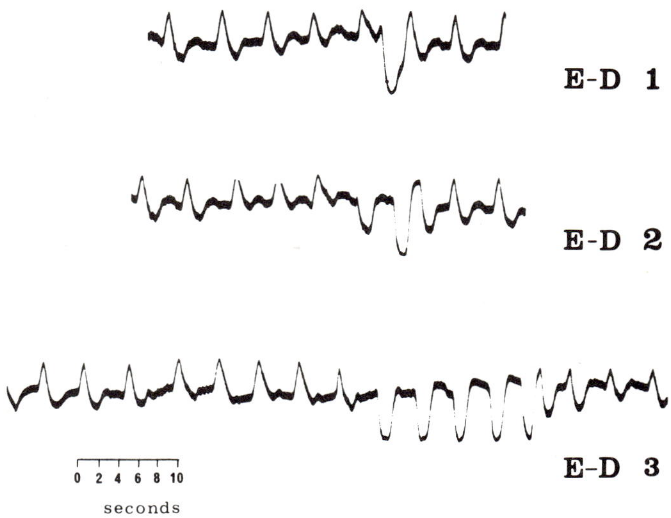

Figure 9 - Extra-diastoles of the metacarpal vein of Pteropus giganteus. Note one (E-D 1), two (E-D 2) and five (E-D 5) extra-distoles compensating previous insufficient filling phases.

by the complete reversal of the normal contraction-dilatation cycle of the mechanogramme (diastole-systole-isovolumetric period). Total diastolic volume displacement may increase by 100% with no noticeable change in frequency of contraction.

Given that the normal phases of the mechanogramme are reversed, that the diastole replaces the systole and that the segment is closed at the proximal end by a valve, it is believed that this "diastolic contraction" may act as a suction-pump and participate in aiding flow.

DISCUSSION AND REVIEW

The study of the venous system has always been something of an enigma! There are few parameters of sufficiently constant value on which to base dynamic experimental studies, and the number of "natural" influences simultaneously affecting any given research, greatly outnumber those isolated phenomena one endeavours to explore.

The physiological and hydrodynamic attributions of "pulsatile" flow are also considerably complicated by the visco-elastic properties of the vessel walls. These properties perform a crucial role in determining the dynamic geometry and energy dissipation of the "contracting" vein, and care must be taken in interpreting variations in chronotropic and inotropic responses of such vessels to different stimuli.

The venous system may accomodate large changes in blood volume without significant change in distending pressure. Such variations in total blood volume, or "shifts" in part of this volume from one area of the body to another, are usually accomplished through capacity changes or dilatations of the venal walls (77,79,2,50). Thus, while venous pressure is important in determining venous return, its relation to actively regulated changes in diameter and tension (such as compensation for blood loss), or to changes in flow distribution (during various forms of stress), is much more complex. Variations in the viscous state of the walls therefore make both "pressure-volume" and "flow velocity" data very difficult to analyse in the venous circulation.

It has been found that the visco-elastic moduli of vessel walls are markedly "time" (2) and "frequency-dependent" (22), as manifested by <u>stress-relaxation</u> (the decrease in stress associated with constant strain), by *creep* (the change in strain associated with constant stress), and lastly by *hysteresis*, when the vessel is alternately filled and then emptied (59).

The effects of hysteresis on the flow rate, frequency and amplitude of contraction of a venous segment are particularly evident (Figure 10) when the perfusion pressure is first raised and then reduced from 60-105 cm water column height. It has been demonstrated *in vitro*, by means of a mechanical transducer, that for water column pressures not surpassing 60 cm H_2O, the active tension or "increase in basal tone" (i.e. reduction in diameter) arising in response to "stretch", is still maintained when the pressure of perfusion is reduced soon after. This may explain the "reversal" of the hysteresis loop of the frequency (and to a lesser degree, of the amplitude) to be found below these values. *In vitro*, the hysteresis is reversed for passive tensions of 200-640 dynes/cm^2.

Nicoll (55) stated, that the "sharp contraction" produced by an increase in intraluminal pressure..." is not, in his words, "a result of increased volume within the segment since equivalent filling at a very slow rate will not excite a contraction".

The following responses of the perfused metacarpal vein of the bat to artificial changes in distending pressure demonstrate, however, how very different reactions may be evoked depending on the rate and intensity at which these stimuli are applied:

- the *rapid* distension of the vein, whether by the occlusion of close-lying segments (80) or by the artificial increment in perfusion pressure (59), may result in a contraction of such magnitude that the vessel is momentarily closed.
- a *sudden* but *less intense* increase in pressure, however, provokes the occurrence of extra-systoles, similar in form to those observed infrequently in the intact preparation (60). This demonstrates, incidentally, how a more powerful stimulus arising during the relative refractory period of the cells allows these veins as it does the heart, to adapt rapidly to sudden increments in distending pressure.

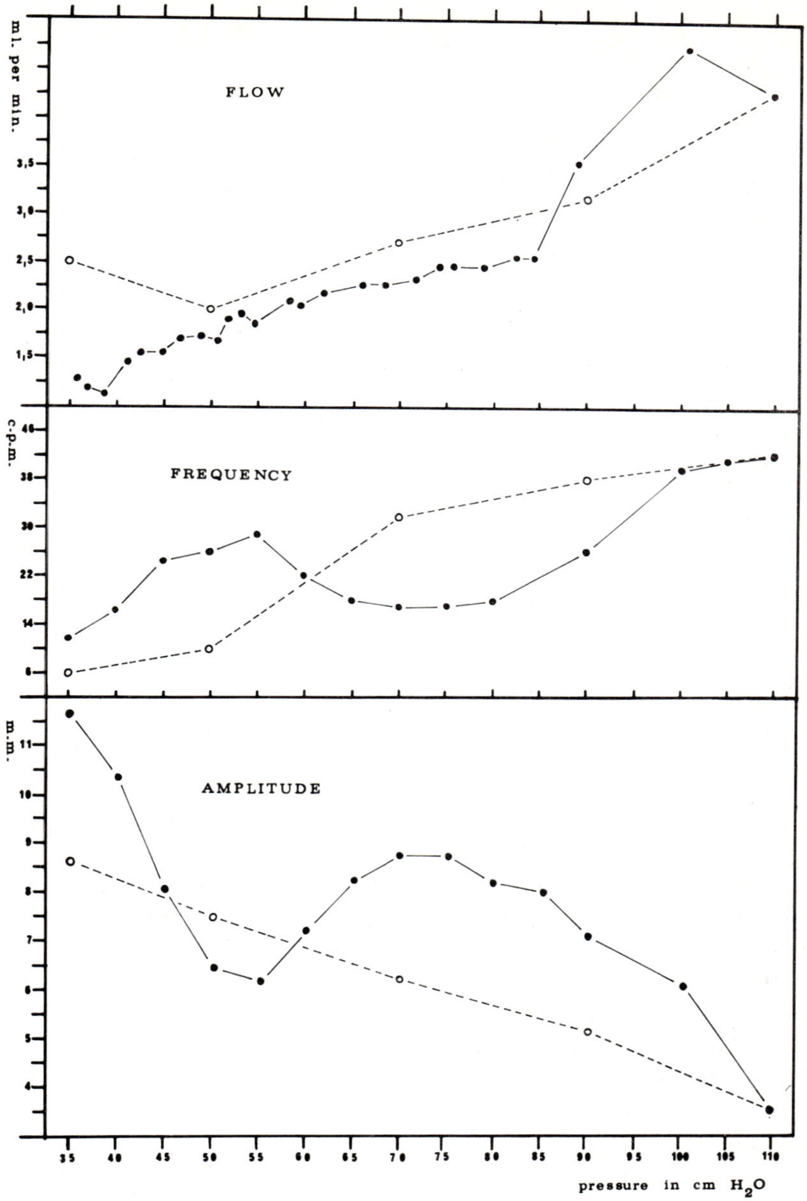

Figure 10 - The effects of both "hysteresis" and "active tension" development on flow (debit), frequency and amplitude, during the gradual elevation and reduction of inflow pressures in the perfused digital vein of Pteropus g. The elevation of pressure is represented by a continuous, and the reduction in pressure by a dotted line.

- finally, if the pressure change is applied *gradually*, the rhythm of vasomotion accomodates progressively to each increment, and the flow rate is modified accordingly (see effects of rising pressure in figure 10).

The results of the above experiments strongly suggest that such active vasomotor responses are influenced by the visco-elastic content of the vessel wall. A biophysical explanation of such phenomena, following observations of suspended vessel strips *in vitro* has been put forward by Sparks (23). This author postulated that if elastic elements lying in series with smooth muscle cells were more viscous than the latter, then "a fast pull would stretch the cells more than a slow pull" and "the slow pull would not capture as much of the "time-dependency" portion of the series elasticity and, therefore, would not stretch the cells as much as a rapid pull".

It is believed by the present authors that the "time-dependency" of the visco-elastic-muscular response is at least in part responsible for the *rapid accomodation* (closure of the vessel, supramaximal contractions,extra-systoles or extra-diastoles) or *gradual accomodation* (progressive variations in chronotropic and inotropic responses) which takes place following stimuli of different types.

Various authors have examined the peripheral circulation of the bat, firstly in relation to local mechanisms of controlling blood flow (27,40,53, 56,57,76) and secondly, in response to various stimuli such as pressure (55,57,59,60,78) temperature (4,7,42,44), pharmacological agents (10,55,73,74) and denervation (10,55,59,62,71,73,74).

The venous system has always been credited with a secondary role in the peripheral circulation, either by accomodating blood-volume reserves, or by myogenic autoregulatory responses to variations in distending pressure. The persistence of such responses in the peripheral circulation of the bat's wing after denervation, has led to the widely held belief, of a total absence of nervous control of the venous side.

Wiedeman postulated that a greater resting diameter of the larger arterial vessels, following denervation, was responsible for an increase of flow reaching the precapillary arterioles and enhancing their contractile activity (73,78). The same author later reported (75) that some veins (approximately 40%) increased, whilst others (approximately 60%) decreased their spontaneous activity subsequent to brachial denervation. It was again suggested that the positive chronotropism observed was solely brought about by a distension of the walls following the increase in flow. It has since been demonstrated in the Megachiroptera, however, (62,63) that local denervation (at the proximal end of the neurovascular bundle) results in a reduction of tone, frequency and amplitude, not only in the intact, but also in the perfused metacarpal vein, whose flow pressure is entirely independent from that of the arterial pole (figure 11).

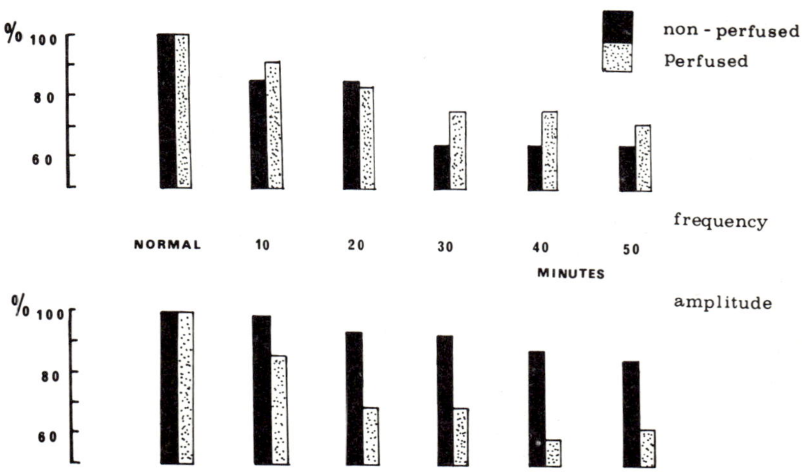

Figure 11 - The chronotropic and inotropic effects of local denervation on the spontaneous contractile activity of the metacarpal vein (Megachiroptera).

Kluger and Heath were able to initiate vasodilatation in the bat (*Eptesicus fuscus*) by means of internally heated rectal thermodes (43) and heating of the Preoptic Anterior Hypothalamus or PCAH (44). Lesions in the PCAH region, however, (45) led to only slight thermoregulatory deficits, and the bats continued to respond to internal "heat-stress" by peripheral vasodilatation. Such results as these do not define whether peripheral responses are due to direct vasomotor stimulation, or to local accomodation through heating of the circulating blood. Denervation, and isolation of vasoactive vessels from the remaining circulation *in situ*, could help to resolve such problems.

Until 1966, the metacarpal vein of the bat was considered to be a non-innervated structure, and its responses to various stimuli such as temperature and pressure, therefore, only "myogenic" in origin.

Mislin (53) demonstrated that nerve fibres did in fact cover the veins in digital regions of the bat's wing, although neither he, or Nicoll (55) were able to elicit a response to metacarpal nerve stimulation. Since the development of more precise histochemical methods for the determination of nerve structures, Lane and Huggel (47) established the presence (in *Rousettus*) of cholinergic nerve fibres arising from the central nerve bundles and ending as varicosities at the medio-adventitial border. An adrenergic nerve plexus, was also found to end as varicosities at the same level (48).

Denervation of the brachial nerve bundles has been shown in *Pteropus* (29) to be associated with a partial loss in catecholamine and cholinesterase activity within dense nervous plexi situated in proximity to the muscle layer of the venal and arterial walls.

Luchsinger tried as early as 1881 to maintain vasomotion in the isolated wing of the bat by perfusing it with beef blood. Various attempts to replace the blood in an isolated vessel by a physiological solution were handicapped by the non-existence of an ionic analysis of the serum. It was, therefore, impossible to sustain contraction for more than a few moments in the absence of a tissue extract (53) or of diluted beef serum (27).

The elaboration of a physiological solution (C.S.3.) capable of replacing the blood by perfusion (62,63), has played a critical role in the gathering of experimental data concerning the reactivity of these veins *in vivo* and *in vitro*. Several parameters including frequency, amplitude, flow rate,pressure,and temperature as well as variations in the composition of the perfusate, may be controlled and measured at all moments,independently of the surrounding conditions *in situ.*

Investigation with varying concentrations of Ca, K and Na, as well as with Adrenaline, Noradrenaline, Acetylcholine, etc., has shown that the metacarpal vein of the bat is a particularly responsive preparation, well suited to the dynamic study of vascular smooth muscle.

Preliminary results obtained *in vivo* (59,61) and *in vitro* (28,61) have cast some doubt as to the presence of Na-based action potentials in the metacarpal vein of the bat. This is demonstrated by the fact that the cells remain electrically active two hours after the perfusion of Na-free (plus Tris) C.S.3. solution (both in normal, fatigued vessels, and veins that have previously been exposed to long periods of Ca-free medium). On the other hand, the tissue is immediately inactivated in Ca-free + EGTA C.S.3. bathing solution (as opposed to an EGTA-free solution, where the time of inactivation is dramatically increased).

Such responses are similar to those evoked by intestinal smooth muscle (6) and the portal vein of the rat (5), and in contrast to such vascular tissues as the carotid artery of sheep (20). Here, on the contrary, smooth muscle cells become electrically active in simple Ca-free solution, yet remain inexcitable when Na in such solution is replaced by Tris or choline. This may reflect individualities among different vascular smooth muscles, either in their ability to store Ca, or in the effectiveness of Ca-free environment in releasing bound Ca (65).

Following the observation by Paton and Rothschild (58) that the calcium content of smooth muscle decreases, if the calcium concentration of the bathing fluid is reduced; and by Hurwitz *et al.* (30) that a reduction in the extracellular calcium ion concentration results in an increased migration of calcium from "cellular stores"; Sitrin and Bohr (65) showed Ca efflux to be depressed within a Na-free solution, and postulated that Ca released from storage depots might remain in the cell and activate contraction.

These experiments indicate some of the complexities involved in studying the role of electrolytes in vascular smooth muscle and the differences of behaviour from one tissue to another. It is hoped that dynamic studies of the bat digital vein, by means of the model described in this paper, will help to clarify the role played by such ions in governing myogenic and "other" responses *in vivo*. The precise and regular activities of both the pacemaker and contraction coupling processes of this tissue (28) also make it a superb model for the investigation of such parameters *in vitro*.

ACKNOWLEDGEMENTS

This study was supported by grants from the Swiss National Scientific Research Fund No 3490.70, and the Georges et Antoine Claraz-Schenkung Fund.

We wish to thank Professors B.Johansson, C.Owman (Lund) and Bülbring (Oxford), as well as Drs. H.Lane and N.Schönenberger (Geneva) for useful discussions concerning this study.

We are also grateful to the Directors of Air India and the Indian Tourist Office (Geneva) for their kindness in supplying the animals.

REFERENCES

1. Altman,Ph. & Dittmer,D. (1961) Blood and other body fluids. Fed. of Am.Soc.f.Exp.Biol.

2. Attinger,E.O. (1969) Wall properties of veins. IEEE Transactions on Biomedical Engineering, vol. BME-16, n.4, 253-261.

3. Axelsson,J., Wahlström,B., Johansson,B. & Jonsson,O. (1967) Influence of the ionic environment on spontaneous electrical and mechanical activity of the rat portal vein. Circulation Research, 21, 609-618.

4. Bartholomew,G.A., Leitner,P. & Nelson,J.E.(1964) Body temperature, oxygen consumption and heart rate in three species of Australian Flying Foxes. Physiol.Zool.37, 179-198.

5. Biamino,G. & Johansson,B. (1970) Effects of calcium and sodium on contracture tension in the smooth muscle of the rat portal vein. Pflügers Arch. 321, 143-158.

6. Bulbring,E. & Kuriyama,H. (1963) Effects of change in the external Na and Ca concentrations on spontaneous electrical activity in smooth muscle of guinea-pig taenea coli. J.Physiol. (London) 166, 21-58.

7. Cowles,R.B. (1947) Vascular changes in the wings of bats. Science 105, 362-363.

8. Cuthbert,A.W. & Sutter,M.C. (1965) The effect of drugs on the relation between the action potential discharge and tension in a mammalian vein. Brit.J.Pharmacol.Chemother.25,590-601.

9. Cuthbert,A.W., Matthews,E.K. & Sutter,M.C. (1965) Spontaneous electrical activity in a mammalian vein. J.Physiol. (London) 176,22-23.

10. D'Agrosa,L. (1970) Patterns of venous vasomotion in the bat wing. Am.J.Physiol. 218, 530-535.

11. Davis,D.L. & Hammond,M.C. (1969) Effects of constant inflow and constant pressure of perfusion on vascular responses. Am.J.Physiol.216, 1292-1299.

12. Folklow,B. (1964) Description of the myogenic hypothesis. Circulation research. suppl.I., 279-285.

13. Folklow,B. (1952) A study of the factors influencing the tone of denervated blood vessels perfused at various pressures. Acta Physiol. Scandinav. 27, 99-117.

14. Folklow,B. (1955) Intravascular pressure as a factor regulating the tone of small vessels. Acta Physiol. Scandinav. 17, 289-310.

15. Folklow,B. (1955) Nervous control of blood vessels. Physiol.Rev. 35, 629-663.

16. Folklow,B. (1960) Role of the nervous system in the control of vascular tone. Circulation 21, 760-768.

17. Folklow,B. & Mellander,S. (1964) Veins and venous tone. Amer.Heart J. 68, 397-408.

18. Funaki,S. (1958) Studies on membrane potentials of vascular smooth muscle with intracellular microelectrodes. Prof.Jap.Acad. 34, 534-536.

19. Funaki,S. & Bohr,D.F. (1964) Electrical and mechanical activity of isolated vascular smooth muscle of the rat. Nature (London) 203, 192-194.

20. Graham,J.M.& Keatinge,W.R. (1970) Sucrose-gap recording of prolonged electrical activity from arteries in Ca-free solution containing EDTA at low temperature. J.Physiol.(London) 208, 2-3P.

21. Green,H.D. & Kepchar,J.H. (1959) Control of peripheral resistance in major systemic vascular beds. Physiol.Rev. 39, 617-686.

22. Hardung,V. (1962) Propagation of pulse waves in visco elastic tubings. In Handbook of Physiology, sect.2: Circulation, vol.I, p. 107.

23. Harvey,S.Jr. (1964) Effects of quick stretch on isolated vascular smooth muscle. Circulation Res.14-15, suppl.1, 1-250 to 1-260.

24. Harvey,S.jr & Bohr,D.F. (1962) Effect of stretch on passive tension and contractility of isolated vascular smooth muscle. Am.J.Physiol. 202, 835-840.

25. Hill,L. (1921) The pressure in the small arteries, veins and capillaries of the bat's wing. Proc.Physiol.Soc. feb.12, p.cx/iv - J.Gen. Physiol. 54.

26. Hilton,S.M. (1962) Local mechanisms of regulating peripheral blood flow. Physiol.Rev. 42, 265-275.

27. Huggel,H. (1959) La pression sanguine du système veineux autonome de l'aile de la Roussette Eidolon helvum Kerr (Macrochiroptera). Rev. Suisse de Zool. 66, 315-321.

28. Huggel,H., Johansson,B. & Peristiany,J.G.

(1972) Unpublished observations.

29. Huggel,H., Owman,C. & Peristiany,J.G. (1972) Unpublished observations.

30. Hurwitz,L., Joiner,P.D. & von Hagen,S.(1967) Mechanical responses of intestinal smooth muscle in a calcium-free medium. Proc.Soc. Exptl.Biol.Med. 125, 518-522.

31. Johansson,B. & Bohr,D.F. (1966) Rhythmic activity in smooth muscle from small subcutaneous arteries. Am.J.Physiol.210, 801-806.

32. Johansson,B. & Jonsson,O. (1963) Cell volume as a factor influencing electrical and mechanical activity of vascular smooth muscle. Acta Physiol. Scandinav. 72, 456-468.

33. Johansson,B. & Jonsson,O. (1968) Similarities between the vascular smooth muscle responses to sudden change in external potassium, sodium and chloride ion concentrations. Acta Physiol. Scandinav. 73, 365-378.

34. Johansson,B. & Ljung,B. (1967) Sympathetic control of rhythmically active vascular smooth muscle as studied by a nerve-muscle preparation of portal-vein. Acta Physiol.Scaninav. 70, 299-311.

35. Johansson,B. & Ljung,B. (1967) Spread of excitation in the smooth muscle of the rat portal-vein. Acta Physiol.Scandinav. 70, 312-322.

36. Johnson,P.C. (1959) Myogenic nature of increase in intestinal vascular resistance with venous pressure elevation. Circulation Res. 7, 992-999.

37. Johnson,P.C. (1964) Symposium on autoregulation of blood flow. Circulation Res. 15, 1-291, (suppl. 1).

38. Johnson,P.C. (1965) Effect of venous pressure on mean capillary pressure and vascular resistance in the intestine. Circulation Res. 16, 294-300.

39. Jones,W.(1852) Discovery that the veins of the bat's wing (which are furnished with valves)

are endowed with rhythmical contractibility, and that onward flow of blood is accelerated by each contraction. Phil.Trans.Roy.Soc.London, part 1, 131-136.

40. Karfunkel,X (1905) Untersuchungen über die sogenannten Venenherzen der Fledermaus. Arch. Physiol. 538-546.

41. Keantinge,W.R. (1968) Ionic requirements for arterial action potential. J.Physiol. (London) 194, 168-182.

42. Kluger,M.J. (1969) Fluctuations in the wing temperature of the bat in response to internal heating. Comp.Biochem.Physiol. 32, 219-226.

43. Kluger,M.J. & Heath,J.E. (1970) A thermoregulatory response to internal heat loading. Am. Zool. 9, 591.

44. Kluger,M.J. & Heath,J.E. (1971) Thermoregulatory response to preoptic-anterior hypothalamic heating and cooling in the bat *Eptesicus fuscus* 2. Vergl.Physiol. 74 340-352.

45. Kluger,M.J. & Heath,J.E. (1971) The effect of preoptic-anterior hypothalamic lesions on thermoregulation in the bat. Am.J.Physiol. 221,144-149.

46. Lane,H.C. (1972) La structure histologique et l'organisation nerveuse de la veine autonome et de l'artère métacarpiennes de *Rousettus aegyptiacus* (Mégachiroptera). Thèse n.1589, Faculté des Sciences, Université de Genève.

47. Lane,H.C. & Huggel,H. (1969) L'innervation cholinergique de la veine et de l'artère digitales de la Chauve-souris (*Rousettus aegyptiacus*). J. de Physiol. 61, 330-331.

48. Lane,H.C., Schönenberger,N. & Huggel,H. (1971) Quelques données sur l'innervation de l'artère et de la veine digitales des Megachiroptères (*Rousettus aegyptiacus* et *Pteropus giganteus*). Revue suisse de Zool. 78, 655-660.

49. Maloff,G.A. (1934) Zur Pharmakologie der Venen. Über selbstandige Venekontraktionen. Arch. Intern.Pharmacodyn.Ther. 48, 333-353.

50. Mellander,S. & Johansson,B. (1968) Control of resistance, exchange and capacitance functions in the peripheral circulation. Pharmacol.Rev. 20, 117-196.

51. Mislin, H. (1941) Über die Veneperistaltik der Chiroptera. Revue Suisse de Zool. 48, 563-568.

52. Mislin,H. (1966) Zum Problem der Innervation und Erregungsausbreitung bei den activ pulsierenden Flughautvenen der Chiroptera *Pteropus vampyrus* und *Myotis myotis*. Revue suisse de Zool. 73, 534-547.

53. Mislin,H. & Helfer,H. (1958) Vergleichende quantitativ anatomische Untersuchungen an glatten Muskelzellen der Fluhautgefässe. Revue suisse de Zool. 65, 384-389.

54. Nakajima,A. & Horn,L. (1967) Electrical activity of single vascular smooth muscle fibres. Am. J.Physiol. 213, 25-30.

55. Nicoll,P.A. (1964) Structure and function of minute vessels in autoregulation. Circulation Res. 24-25, suppl. 1, 245-253.

56. Nicoll,P.A. & Webb,R.L. (1946) Blood circulation in the subcutaneous tissue of the living bat's wing. Ann.N.Y. Acad.Sci. 46, 697-711.

57. Nicoll,P.A. & Webb,R.L. (1955) Vascular patterns and active vasomotion as determiners of flow through minute vessels. Angiology 6, 291-308.

58. Paton,W.D.M. & Rothschild,A.M. (1965) The changes in response and in ionic content of smooth muscle produced by acetylcholine action and by calcium deficiency. Br.J.Pharmac.Chemother. 24, 437-448.

59. Peristiany,J.G. & Huggel,H. (1971) Le mécanogramme et quelques effets de la pression intérieure et du calcium sur la motricité des veines métacarpiennes du *Pteropus giganteus* (Mégachiroptères). Revue suisse de Zool. 78, 1219-1229.

60. Peristiany,J.G. & Huggel,H. (1972) Quelques

effets spéciaux intravasculaires sur l'activité dynamique de la musculature lisse des veines pé riphériques de la Chauve-souris. J.Physiol. (in press).

61. Peristiany,J.G. & Huggel,H. (1972) Unpublished observations.

62. Peristiany,J.G., Lane,H.C. & Huggel,H. (1969) Analyse ionique du serum de la Chauve-souris (Rosettus aegyptiacus) et l'éelaboration d'une solution physiologique (C.S.1.) pour la perfusion de la veine alaire digitale. J.de Physiol. 61, suppl. 2, 370-371.

63. Peristiany,J.G., Lane,H.C. & Huggel,H. (1971) Elaboration d'une solution de perfusion pour les veines alaires de Pteropus giganteus (Mégachiroptères) par rapport au régime alimentaire et l'analyse hématologique. Revue suisse de Zool. 78: 1202-1208.

64. Roddie,I.C. (1962) The transmembrane potential changes associated with smooth muscle activity in turtle arteries and veins. J.Physiol.(London) 163, 138-150.

65. Sitrin,M.D. & Bohr,D.F. (1971) Ca and Na interaction in vascular smooth muscle contraction. Am.J.Physiol. 220, 1124-1128.

66. Somylo,A.V. & Somylo,A.P. (1968) Electrochemical and pharmacological coupling in vascular smooth muscle. Am.J.Physiol. 202, 385-840.

67. Sparks,H.V. & Bohr,D.F. (1962) Effect of stretch on passive tension and contractility of isolated vascular smooth muscle. Am.J. Physiol. 202, 835-840.

68. Speden,R.N. (1970) Excitation of vascular smooth muscle. In Smooth Muscle, pp. 558-588, Bülbring,E., Brading,A., Jones,A. & Tomita,T. ed., London, Edward Arnold Publishers Ltd.

69. Steedman,W.M. (1966) Microelectrode studies on mammalian smooth muscle. J.Physiol.(London) 186, 382-400.

70. Sutter,M.C. (1965) The pharmacology of isolated veins. Br.J.Pharm.Chemother. 24, 742-751.

71. Webb,R.L. & Nicoll,P.A. (1952) Persistence of active vasomotion after denervation. Federation Proc. 11, 169.

72. Wenkebach,K.F. (1903) Über die Dauer der kompensatorischen Pause nach Reizung der Vorkammer des Säugetierherzens. Arch. f.Physiol. 57.

73. Wiedeman,M.P. (1954) Reactivity of arterioles following denervations of subcutaneous areas of the bat wing. Am.J.Physiol. 177, 308-314.

74. Wiedeman,M.P. (1955) Effect of denervation on diameter and reactivity of arteries in the bat wing. Circulation Res. 3, 618-622.

75. Wiedeman,M.P. (1957) Effect of venous flow on frequency of venous vasomotion in the bat wing. Circulation Res. 5, 641-644.

76. Wiedeman,M.P. (1959) Response of subcutaneous vessels to venous distension. Circulation Res. 7, 238-242.

77. Wiedeman,M.P. (1963) Dimensions of blood vessels from distributing artery to collecting vein. Circulation Res. 12, 375-378.

78. Wiedeman,M.P. (1966) Contractile activity of arterioles in the bat wing during intraluminal pressure charges. Circulation Res. 19, 559-563.

79. Wiederhielm,C.A. (1965) Distensibility characteristics of small blood vessels. Federation Proc. 24, 1075-1084.

80. Wiederhielm,C.A. (1967) Effects of temperature and transmural pressure on contractile activity of vascular smooth muscle. 4th Europ.Conf.Microcirculation, Cambridge 1966. Bibl.Anat. 9, 321-327.

Comparative Physiology, eds. L. Bolis, K. Schmidt-Nielsen and S. H. P. Maddrell

CLINIC-CHEMICAL METHODS APPLIED TO FISH BLOOD WITH REFERENCE TO EFFECTS OF CHLORINATED HYDROCARBONS.

A.LARSSON
Department of Zoophysiology,
University of Göteborg,
Göteborg, Sweden.

Many aquatic animals, including fishes, have the ability to remove from the environment and store mainly in their fat tissues substances present in the surrounding water at levels having no visible or detectable short term effect. Such a slow accumulation process may continue in the animal until the body content of the pesticide reaches such levels that several different measurable effects can be detected. During this accumulation period there may be a lot of insidious physiological changes, which after some time can result in chronic effects. Examples of chronic effects of chlorinated hydrocarbons on fish are damage to liver tissue, kidneys, pancreas and adrenal cortex, reduced reproduction, slowed reaction to external stimuli, loss of appetite and impaired growth (2,5)

At our laboratory, studies have been conducted to show effects of chlorinated hydrocarbons on fish. The aim of our studies is to obtain information about internal metabolic changes which occur before the fish shows apparent external signs of poisoning. As the chemical picture of the blood reflects to some extent the over-all metabolism of the animal, the blood chemistry can be used as a tool for investigations of metabolic disturbances caused by environmental poisons. In order to obtain information about metabolic disorders or disturbances, a wide routine analytical program has been developed. This program includes mainly different blood metabolites, but also tissue metabolites and other clinical parameters (7). Figure 1 shows

FIG. 1. ANALYTICAL ROUTINE PROGRAM FOR FISH BLOOD.

I.	Carbohydrate metabolism:	Blcod glucose
		Blood lactate
II.	Lipid metabolism:	triglycerides (plasma)
		free fatty acids (plasma)
		esterified cholesterol (plasma)
		free cholesterol (plasma)
III.	Electrolyte balance:	Na^+
		K^+
		Ca^{2+}
		Cl^-
		inorg. phosphate
IV.	Other clinical parameters:	hematocrit
		hemoglobin
		total plasma protein

the different metabolic and physiological parameters which are investigated in the blood from each fish individual. By use of these analytical methods changes in carbohydrate, lipid or electrolyte metabolism can be detected. Furthermore, the blood analyses are combined with estimations of the content of tryglycerides and glycogen in liver and muscle tissue, and calculation of liver-somatic index (= liver weight x 100/body weight). Most of the analytical methods are based upon well-known clinical procedures used within Clinical Medicine, but often special adaptations have been made before application to fish blood samples. Among other things the analytical procedures have been scaled down to micro or ultra-micro level to make it possible to conduct all the analyses on each individual fish. The whole analytical program for blood (15 parameters) requires about 1.3-1.4 ml whole blood from each fish. A complete description of the different analytical methods, which all are simple and rapid procedures, has been given recently (7).

The present clinic-chemical analysis program has been used during the last 4-5 years in studies on the over-all metabolism of eels under different physiological conditions. These studies have given us some knowledge of the fluctuations in the different parameters under "normal" conditions in fish. So far, the following investigations have been conducted at our laboratory: metabolic differences between yellow and silver phase of the eel (1,6), metabolic effects of starvation in eels (9), metabolic effects of different hormones in eels (8,10). Thus, the methods have been used in studies on the "normal" metabolism of fishes before they have been applied in the toxicological studies.

Two different chlorinated hydrocarbons have so far been tested Pentachlorophenol (PCP) (3) and Polychlorinated biphenyls (PCB) (4). In the pentachlorophenol investigation, eels (Anguilla anguilla L.) were exposed to sea water or fresh water containing 0.1 ppm PCP (technical pentachlorophenol, sodium salt). The accumulation of the pesticide in muscle and liver tissue, as well as the effects on different blood metabolites were studied during 8 days in the sea water and 4 days in the fresh water experiment. After the exposure time the content of

PCP (on fresh tissue basis) in muscle tissue was 9.4 ppm in the sea water study and 0.81 ppm in the fresh water study. In the liver corresponding levels were 33.4 ppm and 8.8 ppm respectively. At these PCP levels the visible symptoms of poisoning were increased respiratory movements and hypersensitivity to external stimuli. During the pesticide accumulation pronounced alterations in the level of several blood metabolites occurred. On the whole the metabolic effects of PCP were the same in the fresh water and sea water experiments.The changes in different blood parameters, which occurred gradually even before the outward symptoms of poisoning could be detected, are summarized in Figure 2. After the exposure time the eels were allowed to recover in clean water for 8 days in the sea water and 55 days in the fresh water experiment. During the recovery periods, the level of PCP decreased both in liver and muscle tissue. In the fresh water experiment the PCP content decreased from 8.8 to 1.3 ppm in the liver and from 0.81 to 0.08 ppm in the muscle. In spite of the recovery in clean water and the accompanying decrease in PCP content, the effects obtained (summarized in figure 2) seemed to persist. In summary the results show that PCP *in vivo* causes an increased utilization of tissue energy reserves which is reflected in the blood by elevation of glucose, lactate and free fatty acids (FFA). This hypermetabolic state is probably a result of an uncoupling action of PCP. The increase in respiratory movements and the polycythemic response observed during the PCP treatment are also in accordance with this hypermetabolic state. The biochemical mechanism of PCP's action on mitochondria has been studied by Weinbach and Garbus (11,12).They found in *in vitro* studies that PCP is very tightly bound to mitochondrial proteins and exerts a very strong uncoupling effect on the oxidative phosphorylation. Figure 3 shows the observed clinical picture in the PCP treated eels, as well as the probable action of PCP *in vivo*, on eels.

The effect of PCP on plasma cholesterol might partly be explained by the uncoupling effect and the subsequent hypermetabolic state. However, since the metabolism and excretion of cholesterol are

FIG.2. EFFECTS OF PENTACHLOROPHENOL ON BLOOD PARAMETERS.

Hematocrit	rapid and persisting increase
Hemoglobin	rapid and persisting increase
Plasma electrolytes	no effects
Inorg. phosphate	initial increase
Plasma protein	no effects
Blood glucose	rapid and persisting increase
Blood lactate	rapid increase
Plasma triglycerides	increase
Plasma FFA	slow increase
Plasma esterif. cholesterol	slight increase
Plasma free cholesterol	slight increase

FIG.3. ACTION OF PCP ON EELS IN VIVO.

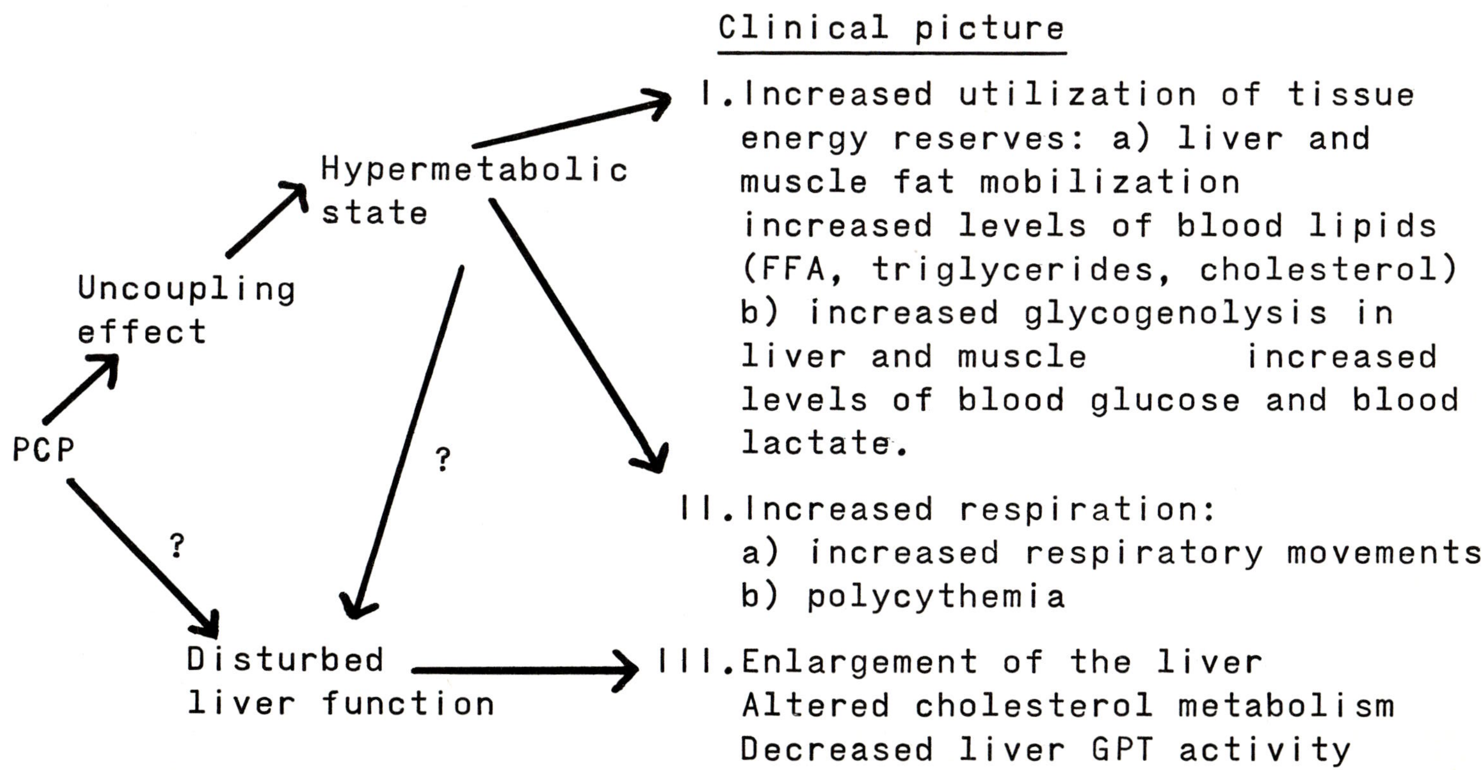

primarily functions of the liver, it cannot be excluded that the altered content of esterified and free cholesterol in PCP treated eels also might be a result of an impaired liver function. Such an assumption is supported by other effects obtained (liver enlargement, decreased activity of liver GPT) in the same experiment (3).

In the PCB investigation, the experimental animals were brown trouts (*Salmo trutta*). The treated fishes (average weight 50 g) were given two doses (each of 5 µg/g body weight) of PCB (Chlophen A 50 dissolved in ethanol) in gelatine capsules directly into the stomach. The control fishes were treated in the same way with pure ethanol. After 43 days, blood and tissue samples were taken out from both groups of fishes. No apparent visible symptom of poisoning could be observed at this moment. However, as shown in Figure 4 ,the PCB treatment caused "internal effects" such as blood anemia, hyperglycemia and altered cholesterol metabolism. Although the PCB treated fishes after 43 days seemed to be in good condition, metabolic changes had obviously taken place. After the sampling on day 43 the PCB treated trouts were starved during the winter (116 days) followed by a new feeding period until day 203. At this time there were still metabolic differences between the treated fishes and the control fishes (4). The PCB treated trouts showed an increase in body weight, a more pronounced decrease in liver triglycerides, lower muscle glycogen level, and contrary to the starved and refed control trouts the PCB treated fishes obtained no hyperglycemic response. The different results between the two sampling occasions, might be due to the fact that on day 203 the fishes had been subjected to a 4-month starvation, during which a redistribution of accumulated PCB cannot be excluded.

It may be concluded from these two experiments that simple blood samples and blood chemical analyses can give valuable information about the metabolic changes which occur in the animal before apparent external signs of poisoning can be detected. In combination with estimation of tissue enzymes and metabolites, the blood chemical studies give relatively detailed information about metabolic

FIG. 4. EFFECTS OF PCB TREATMENT ON BLOOD PARAMETERS.

Hematocrit	decrease
Hemoglobin	decrease
Plasma electrolytes	no effects
Plasma protein	no effects
Blood glucose	increase
Blood lactate	no effect
Plasma FFA	slight increase
Plasma esterified cholesterol	increase
Plasma free cholesterol	slight decrease

disturbances caused by environmental poisons. So far the technique has only been used on fishes, but certainly it can also be applied to other vertebrates.

REFERENCES

1. Dave,G., Johansson,M., Larsson,A, Lewander,K. & Lindman, U. (1972) Metabolic differences between yellow and silver eel. (Manuscript in preparation).

2. Holden,A.V. (1965) Contamination of freshwater by persistent insecticides and their effects on fish. Ann.Appl.Biol. 55, 332-335.

3. Holmberg,B., Jensen,S., Larsson,A.,Lewander,K. & Olsson,M. (1972) Metabolic effects of technical pentachlorophenol (PCP) on the eel *Anguilla anguilla* L. Comp.Biochem.Physiol. 43B, 171-183.

4. Johansson,N., Larsson,A. & Lewander,K. (1972) Metabolic effects of PCB (polychlorinated biphenyls) on the brown trout (*Salmo trutta*). Comp. gen.Pharmac. 3, 310-314.

5. Johnson,D.W. (1968) Pesticides and fishes - a review of selected literature. Trans.Am.Fish Soc.97, 398-424.

6. Larsson,A. & Fänge,R. (1969) Chemical differences in the blood of yellow and silver phase of the European eel (*Anguilla anguilla* L.). Arch. intern.Physiol.Biochem. 77, 701-709.

7. Larsson,A. (1972) Metabolic studies on eels. A presentation of routine procedures for metabolic studies on fish, and of their use in investigations of alterations in the over-all metabolism of the eel after different treatments. Ph.dr. thesis from the Department of Zoophysiology, Göteborg, Sweden.

8. Larsson,A. (1972) Metabolic effects of epinephrine and norepinephrine in the eel, *Anguilla anguilla* L. Gen.Comp.Endocrinol. (in press).

9. Larsson,A. & Lewander,K. (1972) Metabolic effects of starvation in the eel, *Anguilla*

anguilla L. Comp.Biochem.Physiol. (in press).

10. Larsson,A. & Lewander,K. (1972) Effects of glucagon administration to eels, *Anguilla anguilla* L. Comp.Biochem.Physiol. (in press).

11. Weinbach,E.C. (1956) The influence of pentachlorophenol on oxidative and glycolytic phosphorylation in snail tissue. Arch.Biochem.Biophys. 64, 129-143.

12. Weinbach,E.C. & Garbus,J. (1965) The interaction of uncoupling phenols with mitochondria and mitochondrial protein. J.Biol.Chem. 240, 1811-1819.

SUBJECT INDEX